A Systems Biology Approach to Study Metabolic Syndrome

Matej Orešič • Antonio Vidal-Puig
Editors

A Systems Biology Approach to Study Metabolic Syndrome

Springer

Editors
Matej Orešič
VTT Technical Research Centre of Finland
Espoo
Finland

Antonio Vidal-Puig
Institute of Metabolic Science
University of Cambridge
Cambridge
UK

ISBN 978-3-319-01007-6 ISBN 978-3-319-01008-3 (eBook)
DOI 10.1007/978-3-319-01008-3
Springer Dordrecht Heidelberg London NewYork

Library of Congress Control Number: 2013947066

Printed on acid-free paper

Springer is part of Springer Science+Business Media (www.springer.com)

Preface

The prevalence of the Metabolic Syndrome (MetS), an entity defined as the coincidence in the same individual of obesity, insulin resistance, dyslipidemia, hypertension and increased cardiovascular morbidity is increasing with a profile that could be considered of epidemic proportions. It is clear that on top of its undeniable genetic contribution other environmental factors such as lack of physical activity, diet, gut microflora and/or ageing are likely contributors to its progressive acceleration. Despite the obvious public health and economic implications of MetS as well as global research efforts to treat the disorder, the molecular and pathophysiological mechanisms linking the manifestations of MetS are still elusive. This may be in part because of the difficulty in elucidating the primary pathogenic events given the dynamic involvement of multiple interacting organ systems along the disease processes.

Systems biology views and studies the biological systems in the context of complex interactions between their building blocks and processes, as well as with the environment. Despite many traits which define the Metabolic Syndrome being highly heritable, it is evident that the genetic contribution to these traits is mediated via gene—gene and gene—environment interactions across several spatial and temporal scales, and that some of these traits may even be a product of long-term adaptation to the environmental factors including changes in energy balance. This makes the Metabolic Syndrome a strong case for the adoption of systems approach.

The aim of this book is to provide the readers an overview about how the Metabolic Syndrome can be tackled using a systems biology approach, identifying its challenges and opportunities. The emphasis is on pathophysiology of MetS, not on the role of genetic factors behind it. The specific aims correspond to the four main sections of the book: (1) to give an introduction to pathophysiology of the Metabolic Syndrome and medical systems biology, (2) to introduce the key biological processes involved in the pathophysiology of MetS, (3) to introduce the emerging technologies utilized using systems approached to study MetS, and (4) to introduce the novel mathematical modeling approaches to study metabolic syndrome.

The present volume has two main purposes. It brings together current hypotheses about the mechanisms of MetS and its co-morbidities as well as introduces emerging systems biology approaches to tackle the underlying complexity of MetS. This will be of interest to researchers in the fields of medicine, biochemistry, biophysics and computational biology. The book also provides a convenient reference work summarizing published work in the area of medical systems biology. We hope that this will assist the research of a new generation of investigators drawn to this rapidly growing field.

While the work leading to this book was supported from several sources, ten of the chapters are from participants involved in the EU-funded project 'Characterization and modelling of dietary effects mediated by gut microbiota on lipid metabolism' — ETHERPATHS (FP7-KBBE-222639, http://www.etherpaths.org/). ETHERPATHS is a project aiming to develop a platform for nutritional systems biology. Special thanks therefore go to the European Commission for the financial support of the research which had a major impact on this volume.

March 2013 Matej Orešič and Antonio Vidal-Puig

Contents

Part III Key Technologies to Study Metabolic Syndrome

Part IV Modeling Metabolic Syndrome

Contributors

Martin Adiels Department of Molecular and Clinical Medicine, University of Gothenburg, Wallenberg Laboratory, 413 45 Gothenburg, Sweden

Michael Affolter BioAnalytical Science Department, Nestlé Research Centre, Vers-chez-les-Blanc, 1000, Lausanne, Switzerland

Julie Amyot Department of Biochemistry, University of Montreal, Montreal, QC, Canada

Pedro de Atauri Department of Biochemistry and Molecular Biology, Faculty of Biology, University of Barcelona, Av Diagonal 643, Edifici Nou, planta -2, 08028 Barcelona, Spain

Anna-Marja Aura VTT Technical Research Centre of Finland, Tietotie 2, Espoo, P.O. Box 1000, 02044 VTT, Finland

Fredrik Bäckhed Sahlgrenska Center for Cardiovascular and Metabolic Research/Wallenberg Laboratory, Department of Molecular and Clinical Medicine, Sahlgrenska University Hospital, University of Gothenburg, Bruna Stråket 16, 413 45 Gothenburg, Sweden

Adrián Benito Department of Biochemistry and Molecular Biology, Faculty of Biology, University of Barcelona, Av Diagonal 643, Edifici Nou, planta -2, 08028 Barcelona, Spain

Jan Borén Department of Molecular and Clinical Medicine, University of Gothenburg, Wallenberg Laboratory, 413 45 Gothenburg, Sweden

Marta Cascante Department of Biochemistry and Molecular Biology, Faculty of Biology, University of Barcelona, Av Diagonal 643, Edifici Nou, planta -2, 08028 Barcelona, Spain

Josep J. Centelles Department of Biochemistry and Molecular Biology, Faculty of Biology, University of Barcelona, Av Diagonal 643, Edifici Nou, planta -2, 08028 Barcelona, Spain

Simona D'Amore Laboratory of Lipid Metabolism and Cancer, Consorzio Mario Negri Sud, Santa Maria Imbaro (CH), Italy

Loïc Dayon Molecular Biomarkers Core, Nestlé Institute of Health Sciences SA, EPFL Campus Quartier de l'innovation, bâtiment H, 1015 Lausanne, Switzerland

Oleg Demin Institute for Systems Biology SPb, Leninskie Gory 1/75, Moscow, Russia, 119992

Carlos Diéguez Department of Physiology, CIMUS, University of Santiago de Compostela- Instituto de Investigación Sanitaria, 15782 Santiago de Compostela, Spain

Ghislaine Fontés Department of Medicine, University of Montreal, Montreal, QC, Canada

Nail Gizzatkulov Institute for Systems Biology SPb, Leninskie Gory 1/75, Moscow, Russia, 119992

Ilkka Huopaniemi The Charles Bronfman Institute for Personalized Medicine, The Icahn School of Medicine at Mount Sinai, New York, NY, USA

Tuulia Hyötyläinen VTT Technical Research Centre of Finland, Tietotie 2, P. O. Box 1000, 02044 VTT, Espoo, Finland

Livnat Jerby The Blavatnik School of Computer Science, Tel Aviv University, Tel Aviv, Israel

Samuel Kaski Helsinki Institute for Information Technology (HIIT), Department of Information and Computer Science, Aalto University, Espoo, Finland

Artturi Koivuniemi VTT Technical Research Centre of Finland, PO Box 1000, 02044, Espoo, Finland

Martin Kussmann Molecular Biomarkers Core, Nestlé Institute of Health Sciences SA, EPFL Campus Quartier de línnovation, bâtiment H, 1015 Lausanne, Switzerland

Miguel López Department of Physiology, CIMUS, University of Santiago de Compostela- Instituto de Investigación Sanitaria, 15782 Santiago de Compostela, Spain

Igor Marín de Mas Department of Biochemistry and Molecular Biology, Faculty of Biology, University of Barcelona, Av Diagonal 643, Edifici Nou, planta -2, 08028 Barcelona, Spain

Luís Martins Department of Physiology, CIMUS, University of Santiago de Compostela- Instituto de Investigación Sanitaria, 15782 Santiago de Compostela, Spain

Eugeny Metelkin Institute for Systems Biology SPb, Leninskie Gory 1/75, Moscow, Russia, 119992

Anibal Miranda Department of Biochemistry and Molecular Biology, Faculty of Biology, University of Barcelona, Av Diagonal 643, Edifici Nou, planta -2, 08028 Barcelona, Spain

Antonio Moschetta Clinica Medica "A. Murri", Aldo Moro University of Bari, Consorzio Mario Negri Sud, Via Nazionale 8A, 66030 Santa Maria Imbaro (Chieti), Italy

Rubén Nogueiras Department of Physiology, CIMUS, University of Santiago de Compostela- Instituto de Investigación Sanitaria, 15782 Santiago de Compostela, Spain

Matej Orešič VTT Technical Research Centre of Finland, Tietotie 2, PO Box 1000, 02044 Espoo, Finland

Giuseppe Palasciano Clinica Medica "A. Murri", Aldo Moro University of Bari, Consorzio Mario Negri Sud, Via Nazionale 8A, 66030 Santa Maria Imbaro (Chieti), Italy

Alexandre Panchaud BioAnalytical Science Department, Nestlé Research Centre, Vers-chez-les-Blanc, 1000, Lausanne, Switzerland

Vincent Poitout Montreal Diabetes Research Center, CRCHUM—Technopôle Angus, 2901 Rachel Est, Montréal, H1W 4A4, QC, Canada

Montserrat Romero Institute for Research in Biomedicine (IRB Barcelona), C/Baldiri Reixac 10, 08028 Barcelona, Spain

Isabel Rubio-Aliaga BioAnalytical Science Department, Nestlé Research Centre, Vers-chez-les-Blanc, 1000, Lausanne, Switzerland

Eytan Ruppin The Sackler School of Computer Sciences & School of Medicine, Tel Aviv University, 69978 Tel Aviv, Israel

David Sanchez-Infantes Pennington Biomedical Research Center, Louisiana State University, Baton Rouge 70808, LA, USA

David Sebastián Institute for Research in Biomedicine (IRB Barcelona), C/Baldiri Reixac 10, 08028 Barcelona, Spain

Meriem Semache Department of Biochemistry, University of Montreal, Montreal, QC, Canada

Irma Silva-Zolezzi BioAnalytical Science Department, Nestlé Research Centre, Vers-chez-les-Blanc, 1000, Lausanne, Switzerland

Sergey Smirnov Institute for Systems Biology SPb, Leninskie Gory 1/75, Moscow, Russia, 119992

Oleg Stepanov Institute for Systems Biology SPb, Leninskie Gory 1/75, Moscow, Russia, 119992

Jacqueline M. Stephens Pennington Biomedical Research Center, Louisiana State University, Baton Rouge 70808, LA, USA

Ilpo Vattulainen Department of Physics, Tampere University of Technology, PO Box 692, 33101 Tampere, Finland

Antonio Vidal-Puig Metabolic Research Laboratories Level 4, NIHR Cambridge Biomedical Research Centre, Addenbrooke's Hospital, Institute of Metabolic Science, University of Cambridge, Box 289, Cambridge CB2 0QQ, United Kingdom

Andrew J. Whittle Metabolic Research Laboratories Level 4, NIHR Cambridge Biomedical Research Centre, Addenbrooke's Hospital, Institute of Metabolic Science, University of Cambridge, Box 289, Cambridge CB2 0QQ, United Kingdom

Bader Zarrouki Department of Medicine, University of Montreal, Montreal, QC, Canada

Antonio Zorzano Institute for Research in Biomedicine (IRB Barcelona), C/Baldiri Reixac 10, 08028 Barcelona, Spain

List of Abbreviations

2D	two-dimensional
3-OHPPr	3-(3×-Hydroxyphenyl)propionic acid
3-OHPAc	2-(3×-Hydroxyphenyl)acetic acid
3,4-OHBA	3,4-Dihydroxybenzoic acid (Protocatechuic acid)
3,4-OHPPR	3-(3×,4×-Dihydroxyphenyl) propionic acid
3,4-OHPAc	2-(3×,4×-Dihydroxyphenyl)acetic acid
4-OHPPr	3-(4×-Hydroxyphenyl)propionic acid
ABC	aTP-binding cassette
ACC	acetyl-CoA carboxylase
ADPN	adiponectin
AgRP	agouti-related protein
AIC	akaike information criterion
Akt	protein kinase B
ALT	alanine aminotransferase
AMPK	AMP-activated protein kinase
AMT	accurate mass tag
ANOVA	analysis of Variance
ANCOVA	analysis of Covariance
ApoA-I	apolipoprotein A-I
ApoA-II	apolipoprotein A-II
ApoAV	apolipoprotein A-V
ApoB-100	apolipoprotein B-100
ApoC-II	apolipoprotein C-II
ApoC-III	apolipoprotein C-III
ARC	arcuate nucleus of the hypothalamus
ASCA	ANOVA-Simultaneous Component Analysis
AST	Aspartate aminotransferase
ATF6	activating transcription factor-6
ATP	adenosine triphosphate
ATP III	ault Treatment Panel III
BA	bile acid
BAT	brown adipose tissue

BBB	blood brain barrier
BIC	bayesian Information Criterion
BMI	body mass index
BMP8b	bone morphogenetic protein 8b
CA	cholic acid
CAR	catch-and-release
CART	cocaine and amphetamine regulated transcript
CDCA	chenodeoxycholic acid
CD36	cluster differentiation protein-36
CETP	cholesteryl ester transfer protein
CJD	creutzfeldt-Jakob disease
CHOP	C/EBP homologous protein
ChREBP	carbohydrate response element-binding protein
CCA	canonical Correlation Analysis
CNS	central nervous system
CPT	carnitine palmitoyltransferase
CPT1	carnitine palmitoyltransferase 1
CSC	cell surface-capture
CSF	cerebrospinal fluid
CT	computed tomography
CVD	cardiovascular disease
CYP7A1	cholesterol 7α-hydroxylase
DDA	data-dependent acquisition
DF	dietary fibre
DIA	data-independent acquisition
DIGE	difference gel electrophoresis
DMH	dorsomedial nucleus of the hypothalamus
DNL	de novo lipogenesis
DPP4	dipeptidyl-peptidase 4
ECD	electron-capture dissociation
END	enterodiol
ENL	enterolactone
ER	endoplasmic reticulum
eIF2α	eukaryotic translation initiation factor 2 α
ESI	electrospray ionisation
ETD	electron-transfer dissociation
FA	Factor Analysis
FA	fatty acid
FABPs	fatty acid binding proteins
FAO	fatty acid oxidation
FAS	fatty acid synthase
FAT	fatty acid translocase
FATP	fatty acid transport protein
FDR	False Discovery Rate
FFA	free fatty acid

FOXO1	forkhead box protein O1
FP	false positive
FT-ICR	fourier-transform ion cyclotron resonance
FXR	farsenoid X receptor
GE	gel electrophoresis
GC	gas chromatography
GGT	gamma-glutamyltransferase
GIP	gastric inhibitory polypeptide
GLM	general Linear Model
GLP-1	glucagon-like peptide-1
GLUT1	glucose transporter type 1
GLUT4	glucose transporter type 4
G-6-Pase	glucose-6-phosphatase
GPF	gas-phase fractionation
GSEA	gene Set Enrichment Analysis
HDL	high-density lipoprotein
HFD	high fat diet
HL	hepatic lipase
HMG-CoA	hydroxymethyl-glutaryl coenzyme A
HPP	human proteome project
HSL	hormone-sensitive lipase
ICAT	isotope-coded affinity tagging
ICPL	isotope-coded protein labelling
IDL	intermediate-density lipoprotein
IR	insulin resistance
IRF	isoelectric focusing
IRS	insulin receptor substrate
IRE1	inositol-requiring protein-1
IT	ion trap
iTRAQ	isobaric tag for relative and absolute quantification
JNK	c-Jun N-terminal kinase
KNN	K-Nearest Neighbors
KSR2	kinase suppressor of RAS 2
LC	liquid chromatrography
LCFA	long chain fatty acid
LCFAs-CoA	long chain fatty acids-CoA
LDA	linear Discriminant Analysis
LDL	low-density lipoprotein
LDLr	low-density lipoprotein receptor
LHA	lateral hypothalamic area
LMM	Linear Mixed Models (LMM)
LPK	liver pyruvate kinase
LPL	lipoprotein lipase
LOD	limit of detection
LRH-1	liver receptor homolog-1

LSPAD	localized statistics of protein abundance distribution
LXR	liver X receptor
MetS	Metabolic Syndrome
MS	mass spectrometry
MALDI	matrix-assisted laser desorption/ionisation
MANOVA	multivariate ANOVA
MANCOVA	multivariate Analysis of Covariance
MAPK	mitogen activate protein kinase
MetS	metabolic syndrome
MCD	malonyl-CoA decarboxylase
MetS	metabolic Syndrome
MRI	magnetic resonance imaging
MS	mass spectrometry
MS/MS	tandem mass spectometry
MS^E	precursor-independent acquisition of intact and fragment masses
MudPIT	multidimensional protein identification technology
mTOR	mammalian target of rapamycin
MTP	microsomal transfer protein
MUFA	monounsaturated fatty acid
NAFLD	nonalcoholic fatty liver
NASH	nonalcoholic steatohepatitis
NEFA	non-esterified fatty acid
NMR	nuclear magnetic resonance
NPC1L1	niemann-Pick C1-like 1
NPY	neuropeptide Y
NR	nuclear receptor
O-DMA	O-Desmethylangolensin.
ODE	ordinary differential equations
OA	oleic acid
OT	orbitrap
OxLDL	oxidized low-density lipoprotein
Ox-Phos	oxidative phosphorylation
PAcIFIC	precursor acquisition independent from ion count
PAI	protein abundance index
PCA	principal Component Analysis
PCR	polymerase chain reaction
PDK4	pyruvate dehydrogenase kinase 4
PEPCK	phosphoenolpyruvate carboxykinase
PGC-1α	peroxisome proliferator-activated receptor-gamma coactivators 1alpha
PI3K	phosphatidylinositol 3-kinases
PKA	protein kinase A
PLS-DA	partial Least Squares Discriminant Analysis
PMF	peptide mass fingerprinting
PMS	peptide mass sequencing

PMSS	peptide match score summarization
PPAR	peroxisome proliferator-activated receptor
PP2A	protein Phosphatase 2A
POMC	proopiomelanocortin
PTP1B	protein tyrosine phosphatase 1B
PUFA	polyunsaturated fatty acid
PVH	paraventricular nucleus of the hypothalamus
Q	quadrupole
ROS	reactive oxygen species
RSTN	resistin
RT-PCR	reverse-transcriptase polymerase chain reaction
SCD1	stearoyil-CoA desaturase 1
SCFA	short-chain fatty acid
SFA	saturated fatty acid
SHP	short heterodimer partner
SID	stable-isotope dilution
SILAC	stable-isotope labelling with amino acids in cell culture
sLDL	small low-density lipoprotein
SOCS3	suppressor of cytokine signaling
SpS	spectrum sampling
SREBPs	sterol regulatory element binding proteins
SVM	support Vector Machines
T2DM	type 2 Diabetes Mellitus
T2D	type-2 diabetes mellitus
T3	triiodothyronine
TG	triglyceride
TGs	triacylglycerols
TIM-1	TNO Intestinal Model 1 simulating the upper intestine
TIM-2	TNO Intestinal Model 2 simulating the colon
TNF	tumor necrosis factor
TOF	time-of-flight
TOFA	5-(tetradecyloxy)-2-furoic acid
TSC	tuberous sclerosis complex
TUDCA	tauroursodeoxycholic acid
TZD	thiazolidinedione
UCP1	uncoupling protein 1
UCP2	uncoupling protein 2
UCP3	uncoupling protein 3
UGT1A3	UDP-glucuronosyltransferase 1–3
UPR	unfolded protein response
US	ultrasound
VLDL	very low-density lipoprotein
WAT	white adipose tissue
Xbp1	X-box binding protein 1

Part I
Introduction

Chapter 1
The Metabolic Syndrome and its Complex Pathophysiology

Antonio Vidal-Puig

Abstract The Metabolic Syndrome (MetS) represents the association of common pathologies such as obesity, diabetes, insulin resistance, dyslipidaemia, fatty liver and associated cardiovascular complications. The MetS is important because of its prevalence, potential severity and costs. The MetS is a challenge due to its undefined pathogenesis, clinical heterogeneity and complexity, lack of good risk predictors and therapeutic options. Here we propose an adipocentric pathogenic model supported by the concepts of adipose tissue expandability and peripheral lipotoxicity. We also put forward the need for an allostatic perspective and a systems biology approach to understand its pathogenesis, natural history, subphenotypes, potential diagnostic and prognostic biomarkers and new therapeutic targets.

Keywords Obesity · Insulin resistance · Lipotoxicity · Inflammation · Adipose tissue expandability · Allostasis

1.1 The Metabolic Syndrome (MetS) and Systems Biology

It is not uncommon to diagnose obesity, insulin resistance, diabetes, dyslipidaemia, and fatty liver clustered in the same individual. The high prevalence of this association indicates that it is not a random event that occurs just by chance. Thus, it is likely that these apparently diverse pathologies may share some common etiopathogenic mechanisms. For this reason, and despite their pathogenic link not having been yet defined, it has been considered practical from the point of view of diagnostic, risk assessment, educational and global treatment purposes, to integrate these manifestations under the concept of the Metabolic Syndrome (MetS) (Eckel et al. 2010; Alberti et al. 2009; Eckel et al. 2005).

Globally considered, the pathological entities included within the concept of MetS are the most important risk factors for cardiovascular disease, the major cause of morbimortality in obese and Type 2 Diabetes mellitus patients. Another useful aspect

A. Vidal-Puig (✉)
Metabolic Research Laboratories Level 4, Institute of Metabolic Science,
University of Cambridge, Addenbrooke's Hospital Cambridge, Box 289,
CB2 0QQ Cambridge, United Kingdom
e-mail: ajv22@cam.ac.uk

M. Orešič, A. Vidal-Puig (eds.), *A Systems Biology Approach to Study Metabolic Syndrome*,
DOI 10.1007/978-3-319-01008-3_1, © Springer International Publishing Switzerland 2014

related to the MetS concept is that despite its apparent simplicity, it is considered complex and heterogeneous. This provides a degree of flexibility that will allow, particularly as the information technology becomes more sophisticated and patient epidemiological data are better integrated, to expand the pathological spectrum of the MetS to include new pathologies likely to span from psychiatric disorders to some specific forms of cancer. For these same reasons, its heterogeneity and complexity, there is some disagreement about the suitability of the MetS as a risk prediction model given that each of its component factors has different predictive strengths, different importance, and that the traits used for T2DM and CVD risk prediction are distinct(Eckel et al. 2010; Alberti et al. 2009, Eckel et al. 2005). However, it can be argued that despite these limitations, the MetS provides diagnostic and therapeutic awareness and this represents an advantage that outweighs the disadvantages resulting from its relatively obscure pathogenesis and unclear link between its components. In any case, it is hoped that further research into the MetS will simultaneously identify the root causes for these metabolic diseases identifying specific nosological entities within the Syndrome. In this respect the two major goals of the research around the MetS are: (1) to be able to accurately predict the risk of metabolic diseases and thus optimally institute preventative intervention. (2) to identify the underlying pathophysiology of these related diseases with the hope of uncovering new therapeutic and preventative measures.

By definition, a syndrome represents a specific pathological state originated by different aetiologies. The MetS is not an exception, and it is likely to have different aetiologies or entrances representing specific pathogenic subgroups with distinctive early disease histories. The specific phenotype of the individual will probably depend on the balance between organ specific pathogenic effector mechanisms and adaptive allostatic responses (see below) ultimately leading to a global homeostatic failure. Accumulating evidence suggests that defects in beta cells leading to primary hyperinsulinaemia, hepatic defects leading to fatty liver and insulin resistance, defects in skeletal muscle leading to insulin resistance or primary defects in hypothalamus causing defects in peripheral energy homeostasis, may all trigger the development of the MetS. There is substantial experimental data supporting all these organs as primary entrances that can initiate or contribute to the development of the MetS. However, from the point of view of the diagnostic or therapeutic implications, the key question is not only what these entrances to the MetS are, but how relevant these entrances are. Ideally we want to know which ones represent marginally relevant factors affecting a relatively small proportion of patients and which ones represent major pathogenic "highways" affecting a high percentage of patients. On the other hand, to define the entrance to the MetS may become totally irrelevant once the individual has developed a fully established severe MetS. It is unclear if at that stage it is possible to identify the primary vulnerable organs or a specific clinical feature, which is specifically related to the triggering factor. From this point of view, investigating the pathogenesis of the MetS in individuals that have already developed the full spectrum of the MetS, in whom it is difficult to identify the early pathogenic mechanisms, their primary effectors or the secondary allostatic adaptations, may simply be impossible and irrelevant. At this stage a global diagnostic of MetS may be practical with respect to setting the spectrum of potential pathologies to be investigated in a specific individual.

These considerations raise important points about the natural history of and the best timing to study the MetS. In some way the best timing will depend on the main question investigated. Ideally, to understand the natural history of the disease it is interesting to combine both, (a) studies of the early stages to identify the potential entrances to the syndrome, the specific pathogenic effectors linked to a specific entrance as well as the allostatic adaptations to maintain the homeostasis of the system and (b) studies of advanced stage studies which may reveal late pathogenic effectors common to different entrances, which may provide the basis for global blockbuster therapeutic approaches. The attractiveness of focussing on the early stages of the disease is the potential identification of the early mechanisms stressing the homeostatic system, the early allostatic adaptations and the signs suggestive of the individual's organ vulnerability. These early pathogenic mechanisms may actually provide the basis for the personalised diagnosis and the rational for individualised treatments. At the present moment we do not know how the syndrome evolves from its early manifestations to its ultimate global metabolic collapse. We do not know what the specific interactions, the relative relevance, and hierarchical order of organ failure are. This information may have an impact on identifying the best, evolutive stage specific, therapeutic approaches and in helping to determine priori non obvious preventive measures that could reinforce specific functions that may prevent or at least delay the development of the MetS. These are good reasons to try to characterise the natural history of the MetS, delineating the evolution from its initial specific manifestations towards its progressive dedifferentiation and global homeostatic metabolic failure.

1.2 Allostasis as an Essential Concept to Understand the Pathogenesis and Natural History of the

Allostasis refers to the concept of maintenance of normality at the expense of adaptive changes. However, these changes are not for free, they may have a cost in terms of energy requirements and in terms of collateral damage or side effects. In this respect we think the concept of allostasis can be applied to the study of the MetS (Virtue and Vidal-Puig 2010; Virtue and Vidal-Puig 2008). When studying the MetS we need to consider not only the effector pathogenic mechanisms that drive the progression of the pathology but also the attempts of the organism to maintain the homeostasis of their metabolic systems. It may appear that homeostasis and allostasis are similar concepts. However allostasis is a more useful concept in terms of including not only the concept of a steady state that represents normality or health status (homeostasis) but also the mechanisms required to maintain it, the metabolic stress induced by these adaptive mechanisms and the concept of failure induced by these adaptive mechanisms. An illustrative example is the struggle to maintain normal glycemia levels in the context of positive energy balance at the expense of increasing insulin secretion to overcome the progressive impairment of its insulin resistance. In this respect the increase in the secretion of insulin will be able to maintain glucose levels under control for a while until the beta cells will fail under the metabolic pressure to produce and secrete huge amounts of insulin causing ER stress. We consider the MetS as a toxic syndrome induced by a relative excess of nutrients and whose specific individual manifestations are determined by organ specific genetic vulnerability to nutritional insults. From this definition we can introduce concepts such as allostatic load defined as the level of stress induced by the adaptation to overcome the toxic effects derived from overnutrition, and the concept of allostatic overload that refers to a state of excessive allostatic load (overnutrition maintained for a long period of time) that will inevitably lead to metabolic failure. Globally considered, a systems approach to the metabolic syndrome should include and quatitate the level of allostatic load and refer it to the organ vulnerability to determine the maximal threshold of tolerance determining an allostatic overload state.

1.3 Taking Advantage of the Heterogeneity and Paradoxes of the MetS to Understand its Pathogenesis

The pathologies of the MetS have a strong ***genetic component***, with an estimated 40–70 % of the variation in obesity risk attributed to genetic factors. However, despite this stable genetic influence, which represents a stable vulnerability of the system, the prevalence of obesity and MetS has steadily increased over the last 40–50 years, outlining the pathogenic relevance of an ***obesogenic environment*** contributing to the allostatic load. Although the MetS manifestations tend to coincide in many individuals, it is also true that not all patients develop the whole spectrum of MetS manifestations, nor is there strong evidence that develops the encompassed pathologies in any specific or sequential order. Moreover, the degree of obesity is not consistently linked with metabolic complications. Although it is true that the higher the fat mass the higher the prevalence of the MetS, large numbers of apparently healthy obese individuals can be found in addition to significant numbers of normal weight individuals who develop metabolic complications. On top of this, the combination of changing globalised environments on specific ethnic groups may contribute to this complexity, since metabolic complications occur at a lower body fat level in Asians and the major diabetes epidemic currently occurs in Asia.

1.4 An Adipocentric Model of the Pathogenesis of the MetS

To approach these intricate questions and to develop a working model of the MetS, it is necessary to start somewhere and to enable this, some informed prioritisation is required. From a simplistic point of view and departing from the evidence that

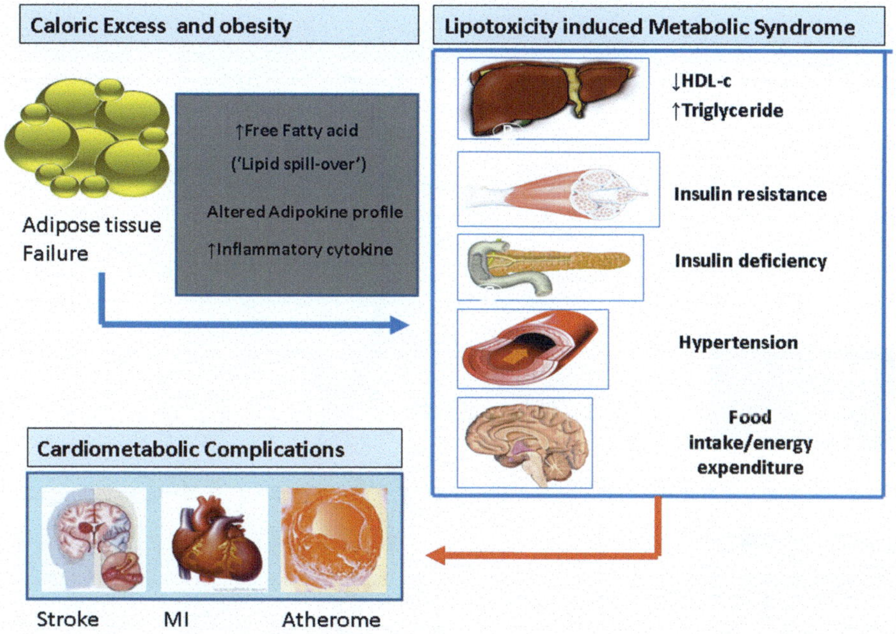

Fig. 1.1 Adipocentric perspective of the metabolic syndrome. Excessive accumulation of lipids in adipose tissue may overcome the storage capacity of this organ resulting in increased lipid spill over and lipid induced inflammatory effects associated with qualitative and quantitative alterations in the adipokine profile. Lipids and adipokines may contribute to induce a overnutrition related toxicity in important metabolic organs configuring manifestations typically associated with the MetS ultimately leading to cardiometabolic complications

most patients diagnosed of MetS are obese, we have proposed an adipocentric perspective of the MetS. We consider that defects in the expansion and functionality of the adipose tissue, associated with obesity, are fundamental for the development of the most common forms of the MetS. Most of the metabolic effects of obesity are due to its deleterious effect on insulin sensitivity and we propose three potential mechanisms contributing to obesity related metabolic complications. During states of positive energy balance, adipose tissues will expand, occasionally until its maximal expandability is reached and eventually fail to store excess calories adequately as fat. This results in: (i) the deposition of lipids in other metabolically relevant tissues (e.g. muscle, liver, beta cells) not normally suited to fat storage, resulting in tissue dysfunction via a phenomenon known as lipotoxicity (Vidal-Puig and Unger 2010) (Fig. 1.1); (ii) adipocytes become dysfunctional and produce an altered quantitatively and qualitative adipocytokine profile [decrease in adiponectin, increase in resistin, increase in tumour necrosis factor α (TNFα), and interleukin 1 (IL-1) production] which promotes IR; (iii) adipose tissue inflammation with macrophage infiltration (ATMs) results in the production of pro-inflammatory cytokines, which further prevents adipose tissue expansion and aggravates IR (Fig. 1.2).

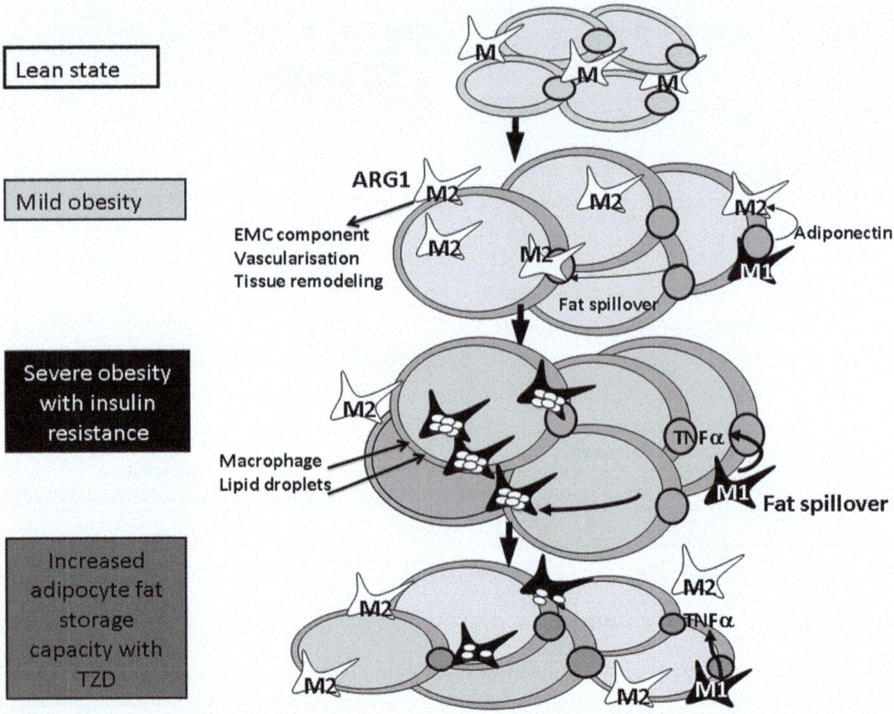

Fig. 1.2 Adipose tissue expandability and inflammatory changes in adipose tissue. The MetS is typically associated with macrophage infiltration of the adipose tissue associated to a proinflammatory phenotype. This model suggests the failure of the adipocyte capacity to accumulate fat as a trigger promoting macrophage infiltration and subsequent fat accumulation in macrophages leading to the formation of foam cells. This model proposes a functional role of M2 antinflammatory macrophages mediating adipose tissue remodelling in the context of positive energy balance and a pathophysiological effect secondary to M1 proinflammatory polarisation as a result of lipid induced toxicity in macrophages. Finally our model proposes that regaining control of the adipocyte function prevents or reverse the M1 proinflammatory effects

In our adipocentric model, when the adipose tissue develops insulin resistance and functionally fails, this is associated with inappropriately increased rates of lipolysis and overall non-esterified (‘free’) fatty acid flux. Increases in non-esterified fatty acid delivery in combination with progressive development of hepatic IR is followed by increases in hepatic glucose production (impaired fasting glycaemia), hepatic lipid accumulation (hepatosteatosis) and hepatic production of large triacylglycerol-rich lipoprotein particles (hypertriglyceridaemia). In turn, the combination of raised plasma triglycerides and alteration in high-density lipoprotein (HDL) lipidation contributes to the reduction in HDL particles (low HDL). This overall increase in lipid delivery to skeletal muscles promotes IR and reduces insulin-mediated glucose disposal (IMGD) rates (impaired glucose tolerance) therein. Finally, systemic IR and the associated hyperinsulinaemia contribute to endothelial dysfunction and hypertension. Exacerbating the shift towards a pathological state, the interdependence of these tissues promotes a cycle of progressive deterioration in metabolism, which in turn, worsens insulin resistance and fuels a metabolic vicious cycle.

Although we think this adipocentric model probably represents an important model to the MetS, it is recognised that the concept of MetS represents a heterogeneous spectrum of clinical presentations. In some ways it would be expected that a given individual's spectrum is determined by their organ-specific vulnerability modulated by genetic and environmental factors that interact with potentially confounding factors such as age, ethnicity or gender. This level of complexity poses the challenge of creating a disease model that captures the complexity of the syndrome and simultaneously is able to explain its apparent paradoxes. In this respect, focussing on phenotypes where expansion of fat mass is uncoupled from metabolic complications, may provide new understanding of the link between obesity and these metabolic complications. By 'uncoupled', we specifically refer to states characterized by severe obesity in otherwise metabolically healthy individuals or conversely to those lean individuals with severe metabolic syndromes.

These obese healthy individuals exist and represent around 10–15 % of the morbidly obese population. They have excessive body fatness predominantly accumulated in their subcutaneous adipose tissue and associated with less visceral and ectopic fat and lower lean mass. These patients are remarkably insulin sensitive, exhibit a healthy lipid profile characterised by high HDL cholesterol and low triglycerides, and their adipose tissue is remarkably healthy despite it being maximally overstretched to store the excess fat. Typically their adipose tissue does not present macrophage infiltration, nor proinflammatory markers and exhibits a favourable adipokine profile with high adiponectin levels. Moreover, their healthy adipose tissue is formed of relatively small adipocyte cells, which suggests facilitated adipocyte differentiation and an excellent capacity to expand and meet the demands for fat storage. The converse phenotype is typically observed in the lean metabolically compromised patient. In these patients the adipose tissue is characterised by large adipocytes, extensive macrophage infiltration and elevated inflammatory markers. All such factors are compatible with impaired adipose tissue expansion and functionality, ultimately leading to adipose tissue failure and the accumulation of lipids in peripheral metabolically relevant organs.

Altogether, the realisation of the heterogeneity and apparent paradoxes associated to the MetS, allows hypothesising that (a) there must be specific factors linking obesity and its metabolic complications which are not directly related to the amount of fat stored adipose and (b) the most unknown mechanisms that protect some obese people from developing metabolic complications, that if identified may provide a new conceptual direction in the way we approach the therapy of the MetS. The relevance of these questions stems from the possibility to identify biomarkers with diagnostic and prognostic value able to target obese populations mechanistically at high risk of developing metabolic complications and the possibility of identifying biomarkers suitable for the development of novel therapeutic interventions.

1.5 Reconciling the MetS with the Obese Healthy Phenotype: The Adipose Tissue Expandability Hypothesis

There appears to be a difference between the amount of fat accumulated in adipose tissue and its overall biological function. Alteration in the ***distribution of fat*** between subcutaneous and visceral adipose depots without changes in the total amount of fat can contribute to differences in metabolic complications. Similarly, interventions in humans and murine models which facilitate fat deposition and increase expansion of fat mass can paradoxically reduce metabolic complications. We consider that it is not the amount of fat accumulated in the adipose tissue, but rather how well our adipose tissue functions that counts. To explain these apparent paradoxes we have proposed the ***adipose expandability hypothesis*** that states that the link between obesity and metabolic complications does not depend on the total fat mass accumulated, but depends on the remaining capacity for expansion and functioning of the adipose tissue. When the adipose tissue cannot expand further and becomes dysfunctional, excess nutrients cannot be stored appropriately within adipose tissue and are stored ectopically, in other metabolically relevant organs. Ectopic lipid deposition, particularly of lipid reactive species such as DAGs and ceramides, per se can cause insulin resistance, fatty liver, dyslipidaemia, beta cell failure and cardiovascular complications as typically seen associated with obesity.

From our model it can be hypothesised that: (a) therapeutic strategies that optimise the expansion and fat deposition in adipose tissue may uncouple obesity from its associated metabolic complications and may provide a palliative therapeutic strategy that may at least prevent/delay the deleterious metabolic effects of obesity. (b) that strategies that prevent inappropriate ectopic fat deposition in metabolic relevant organs and/or promote lipid oxidation may prevent or delay the development of the MetS, and (c) we could envisage that strategies focused on modifying specific metabolic routes diverting lipids from more toxic to less toxic pathways or from more vulnerable organs to less vulnerable, may provide an additional buffer capacity preventing the deleterious effects of lipid induced toxicity.

1.6 What Factors Influence the Capacity of WAT Expandability and Function? (Fig. 1.3)

Factors that may determine the expandability and functionality of the adipose tissue include:

a) **Insufficient capacity to generate new adipocytes from preadipocytes** (Tang et al. 2008) to cope with the nutrient storage demands. There is evidence from genetic experiments that a limit on adipose tissue expansion under conditions of positive energy balance leads to insulin resistance. The fact that there is a relatively fixed turnover of adipocytes in humans (Arner et al. 2011; Spalding et al. 2008) which is defective in the subcutaneous adipose tissue of obese individuals as they

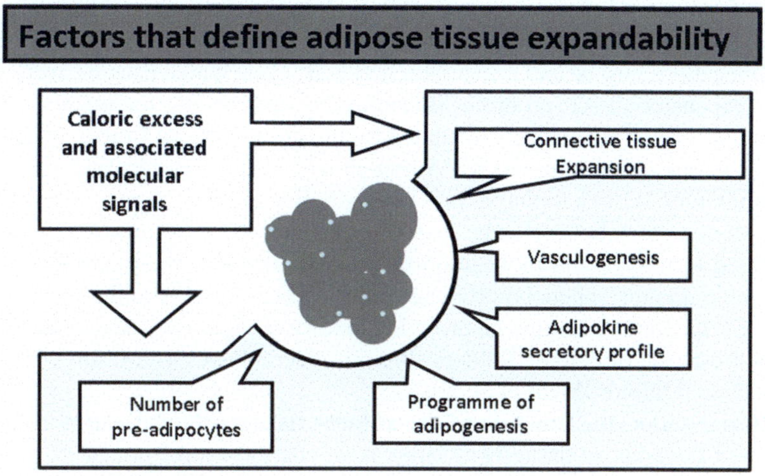

Fig. 1.3 Factors controlling adipose tissue expandability. Spectrum of factors that can determine the adipose tissue expandability limit and may be manipulated for therapeutic purposes

age (25 % decrease between the age of 45 and 65 years) suggests that the number of adipocytes is genetically determined and that this process may be modulated by toxic environmental factors associated to positive energy balance and age. It could be argued that this genetic insufficiency may be related to increased storage demands accompanying the increased prevalence of obesity and that the same adipose tissue genetic make up that was functionally competent 50 years ago may now, under mounting nutritional pressures and in ageing population, fail to cope with increasing storage demands.

b) **Defective adipocyte membrane remodelling and functionality in response to obesity.** We have identified two phases in the progression towards the functional failure of the adipose tissue: The first phase represents a stage of metabolic stress still compensated by allostatic responses that optimise the function of the adipose tissue through adipocyte membrane lipid remodelling (Pietilainen et al. 2011). We have recently shown that adipose tissue in obesity is characterized by enrichment of specific ether phospholipids containing arachidonic acid. Using computational simulation of lipid plasma membranes based on lipidomics data, we have shown that initial AT lipid remodelling is part of an adaptive process, maintaining the normal membrane function but at the cost of higher vulnerability to adipose tissue to inflammation. The second phase represents a more advanced stage characterised by the breakdown of allostatic adaptations subjected to the allostatic overload induced by adipocyte overloading. This leads to the failure of adipose tissue to store fat appropriately and subsequently the ectopic accumulation of lipid in metabolically relevant organs such as liver, muscle, beta cells or brain that contributes to the MetS.

c) **Role of inflammation** (Neels and Olefsky 2006). However, besides the genetic component, there is evidence that as the MetS evolves, there is a progressive development of inflammation in WAT and liver. In this respect inflammation represents a second hit that can directly compromise insulin sensitivity, the expansion of the adipose tissue and accelerates a vicious cycle which progressively impairs the expansion of AT, promotes inflammation and extends the insulin resistance systemically. This effect is mediated by the attraction and functional activation of macrophages within the AT. Although it is clear that accumulation of proinflammatory M1 macrophages in AT can amplify a proinflammatory cascade with deleterious metabolic effects, it is still unclear why these macrophages are attracted to the AT and more importantly why they become polarised towards a proinflammatory phenotype(Olefsky and Glass 2010). We and others have proposed that these macrophage proinflammatory changes may be another lipotoxic manifestation of ectopic fat deposition in macrophages as a result of adipocyte functional failure(Prieur et al. 2011).

d) **Altered endocrine function of the adipose tissue.** Although the essential function of the adipose tissue is to serve as an energy storage organ, its function is certainly more complex as it is required to maintain close communication with the centres that control energy homeostasis, namely the CNS and key peripheral metabolic organs such as skeletal muscle, liver or BAT, to efficiently take or release lipids on demand. To fulfil this and other missions, the WAT produces and secretes proteins such as leptin and adiponectin. However, in the context of the MetS there is a qualitative and quantitative alteration in the repertoire of these proteins, which compromises the function of the adipose tissue and its peripheral target organs The reasons for these alterations are unclear but it can be argued that as they tend to happen in the context of obesity, adipocyte hypertrophy, changes in membrane lipid composition, inflammation and that they are reminiscent of immunological responses to infective agents, they may represent artefacts or inappropriate regulation of physiological pathways not necessarily related to their primary role. Whatever the mechanism, the dysregulation of these adipokines can secondarily compromise the expansion and functionality of the adipose tissue ultimately leading to adipose tissue failure.

e) **Inappropriate angiogenesis and vascularisation** (Rupnick et al. 2002; Cao 2007; Brakenhielm et al. 2004) may compromise the expansion and functionality of the adipose tissue. It is clear that the expansion of the adipose tissue requires good blood supply to support the tissue, to facilitate the accessibility of oxygen and the nutrients as well as their release when required. The relevance of angiogenesis in the control of adipose tissue mass is becoming the focus of attention, in part from the paradoxical results of pharmacological interventions with TNP-470, angiostatina y endostatina, angiogenic inhibitors that prevent the expansion of the adipose tissue and but also result in improved insulin sensitivity (Cao 2007). These unexpected findings raise the possibility that these drugs may have off target effects that may result in negative energy balance. Irrespectively of this controversial evidence, genetic factors affecting angiogenesis/vasculogenesis as well as epigenetic changes determined by the excess of nutrients and metabolic stress may contribute to limit the expansion and functionality of the AT.

f) **Defective extracellular matrix (ECM).** Accumulating evidence indicates that the AT of the obese patient requires remodelation to allow its expansion. This requires the concerted action of specific metalloproteases (MMPs) and appropriate secretion of collagen fibres. The relevance of the effect of the ECM for the control of adipose tissue expansion and associated metabolic disturbances has been elegantly demonstrated through genetic experiments where the inhibition in the biosynthesis of collages VI resulted in hypertrophic adipocyte cells and paradoxically improved insulin sensitivity (Khan et al. 2009). Conversely disruption of MMP14 levels results in small adipocytes constrained within excessive accumulation of collagen deposits. This process may also be relevant in common forms of obesity where signs of fibrosis in adipose tissue have been identified. It is unclear whether this represents a primary dysregulation of the remodelling of EMC or more likely it is secondary to the excessive expansion of AT and the process of tissue repair in the context of adipocyte cell death, inflammation or hypoxia (Chun et al. 2006). It is possible that under these adverse metabolic conditions the EMC remodelling is affected.

1.7 What are the Effectors of Lipotoxicity in Peripheral Organs?

An essential question is whether the toxic effects derived from the excessive accumulation of lipids/nutrients in peripheral organs and/or CNS occurs through similar mechanisms. Alternatively, each organ may have its own peculiar lipid metabolic network and genetic vulnerability derived from activation of different organ specific pathogenic mechanisms. It is conceivable that studies performed at specific time points may identify organ-specific lipid species (e.g DAGs, ceramides, LysPCs) or lipid networks accordingly with the peculiarities of the organ and/or specific metabolism of nutritional interventions. However, these time windows may simply capture snapshots of a heterogeneous organ-specific response, which is temporally evolving through a progression that reflects a hierarchy of organ susceptibility. Both possibilities are not necessarily exclusive and their elucidation has important therapeutic and diagnostic implications (see below). Our point of view is that lipid-induced toxicity may activate stress mechanisms affecting different organelle responses, such as ER stress, autophagy, ROS production and mitochondria toxicity or lipid-induced signalling defects. Although it is clear that lipids have essential biological functions including energy storage, signalling, or structural purposes, when in excess, in inappropriate proportions, or mislocalised, these lipid alterations have been associated with insulin resistance and apoptosis. For instance, whereas ceramides, DAGs or specific phospholipids are normal cellular constituents, when in excess they may trigger the activation of stress related proinflammatory kinases (e.g Jun N-terminal quinasa (JNK), IKK/NF-κB, mTOR or proteín kinase C delta (PKCθ) ultimately leading to insulin resistance and cellular stress(Summers 2006). However the relative relevance of these mechanisms may differ amongst the target organs and the evolutive state of the MetS depending on the specific characteristics of the target organ and original metabolic insult. The characterisation of the effectors linking accumulation of reactive lipid species with specific mechanisms of toxicity may offer global/organ targeted therapeutic approaches to the pathologies of the Metabolic Syndrome.

1.8 The Output: Diagnostic and Prognostics Biomarkers

From the biomedical point of view, a key challenge is to identify whom amongst the obese/lean population will develop metabolic complications. This information helps to avoid human suffering and has enormous financial implications. Although IR appears to play a major role in the development of obesity induced metabolic complications, it is not yet useful as a tool to diagnose the MetS, neither is it helpful as a point of reference for comparisons between studies. Early identification of those healthy obese patients may allow non-urgent, relatively conservative and less expensive therapeutic approaches. Conversely, early identification of those individuals likely to develop severe metabolic complications, despite not being severely obese will identify opportunities for more aggressive interventions. Also if confirmed that different organs have specificities in their lipid network, it may be feasible to identify specific lipid signatures in plasma correlating with lipid induced toxic insults in specific organs. This may identify within an individual those organs most susceptible to more severe pathologies and de facto address the problem of the heterogeneity of the MetS. Useful biomarkers could predict the evolution from steatosis to steatohepatitis, exacerbated insulin resistance or accelerated beta cell failure leading to diabetes.

1.9 Implications of the Adipose Tissue Expandability Model for Biomarker and Therapeutic Target Selection

Based on our current knowledge it can be hypothesised that the following candidate processes may provide markers with diagnostic and prognostic value and are likely specific points for therapeutic intervention:

a) **Mechanisms facilitating efficient fat deposition** in adipose tissue and away from liver, muscle and beta cells should prevent nutrition-induced toxic events in the latter organs and protect from the development of metabolic obesity complications. This attractive model removes the principal driving factor behind multiple metabolic complications including insulin resistance, NAFLD, adipose tissue inflammation and demand on beta cell expansion.
b) **Mechanisms involved in removing excess circulating lipid and increased oxidation.**Increased nutrient oxidation should protect from obesity-associated complications. The strategy of removing excess circulating lipids is particularly attractive. This includes strategies mimicking exercise and strategies facilitating the development and activation of brown fat to oxidize the excess energy.

c) **Mechanisms that facilitate storage of ectopic lipids in less harmful forms.** The type of ectopic lipid residing in tissues is probably more important than the amount of lipid accumulated. It is therefore conceivable to treat diabetes in obesity by driving lipids deposited in ectopic depots into safe storage forms (such as TAG) and away from more harmful forms, such as ceramides, DAGs and LPCs.

1.10 A Systems Biology Approach to the MetS

For a given degree of obesity, IR or adipose tissue dysfunction, different individuals seem to have different susceptibility to the development of metabolic complications. Moreover, unlike other systemic disorders, major tissues involved in metabolism (liver, fat, muscle, pancreas and brain) are highly interdependent. Metabolically relevant tissues are closely interlinked, and each tissue's response to adipose dysfunction cannot be considered truly independent. Studies involving one organ should be interpreted in the context of other tissues, whereas reactive and causative findings should also be clearly distinguished. Thus if we are going to understand the pathogenesis of the MetS a systems-based approach, that integrates findings from all organs and analyses the evolution of the MetS over time to capture the relative contribution of each organ to the overall phenotype is required. The challenge posed by this task is immense, since it requires an integrated multidisciplinary approach that brings together biomedical expertise in each one of the components of the MetS, technological analytical platforms to accurately measure things and sophisticated informatics tools to integrate extremely heterogeneous data to deliver useful information to answer relevant biomedical questions and propose new ones amenable to statistical validation. Even more challenging is to get the scientists behind all these expertises to be able to communicate efficiently. This book is an attempt to do exactly this, to bring together the different perspectives and tools required to address probably the most common, most important and most complex human health problem.

References

Alberti KG, Eckel RH, Grundy SM, Zimmet PZ, Cleeman JI, Donato KA, Fruchart JC, James WP, Loria CM, Smith SC Jr (2009) Harmonizing the metabolic syndrome: a joint interim statement of the International Diabetes Federation Task Force on Epidemiology and Prevention; National Heart, Lung, and Blood Institute; American Heart Association; World Heart Federation; International Atherosclerosis Society; and International Association for the Study of Obesity. Circulation 120:1640–1645

Arner P, Bernard S, Salehpour M, Possnert G, Liebl J, Steier P, Buchholz BA, Eriksson M, Arner E, Hauner H, Skurk T, Ryden M, Frayn KN, Spalding KL (2011) Dynamics of human adipose lipid turnover in health and metabolic disease. Nature 478:110–113

Brakenhielm E, Cao R, Gao B, Angelin B, Cannon B, Parini P, Cao Y (2004) Angiogenesis inhibitor, TNP-470, prevents diet-induced and genetic obesity in mice. Circ Res 94:1579–1588

Cao Y (2007) Angiogenesis modulates adipogenesis and obesity. J Clin Invest 117:2362–2368
Chun TH, Hotary KB, Sabeh F, Saltiel AR, Allen ED, Weiss SJ (2006) A pericellular collagenase directs the 3-dimensional development of white adipose tissue. Cell 125:577–591
Eckel RH, Grundy SM, Zimmet PZ (2005) The metabolic syndrome. Lancet 365:1415–1428
Eckel RH, Alberti KG, Grundy SM, Zimmet PZ (2010) The metabolic syndrome. Lancet 375:181–183
Khan T, Muise ES, Iyengar P, Wang ZV, Chandalia M, Abate N, Zhang BB, Bonaldo P, Chua S, Scherer PE (2009) Metabolic dysregulation and adipose tissue fibrosis: role of collagen VI. Mol Cell Biol 29:1575–1591
Neels JG, Olefsky JM (2006) Inflamed fat: what starts the fire? J Clin Invest 116:33–35
Olefsky JM, Glass CK (2010) Macrophages, inflammation, and insulin resistance. Annu Rev Physiol 72:219–246
Pietilainen KH, Rog T, Seppanen-Laakso T, Virtue S, Gopalacharyulu P, Tang J, Rodriguez-Cuenca S, Maciejewski A, Naukkarinen J, Ruskeepaa AL, Niemela PS, Yetukuri L, Tan CY, Velagapudi V, Castillo S, Nygren H, Hyotylainen T, Rissanen A, Kaprio J, Yki-Jarvinen H, Vattulainen I, Vidal-Puig A, Oresic M (2011) Association of lipidome remodeling in the adipocyte membrane with acquired obesity in humans. PLoS Biol 9:e1000623
Prieur X, Mok CY, Velagapudi VR, Nunez V, Fuentes L, Montaner D, Ishikawa K, Camacho A, Barbarroja N, O'rahilly S, Sethi JK, Dopazo J, Oresic M, Ricote M, Vidal-Puig A (2011) Differential lipid partitioning between adipocytes and tissue macrophages modulates macrophage lipotoxicity and M2/M1 polarization in obese mice. Diabetes 60:797–809
Rupnick MA, Panigrahy D, Zhang CY, Dallabrida SM, Lowell BB, Langer R, Folkman MJ (2002) Adipose tissue mass can be regulated through the vasculature. Proc Natl Acad Sci U S A 99:10730–10735
Spalding KL, Arner E, Westermark PO, Bernard S, Buchholz BA, Bergmann O, Blomqvist L, Hoffstedt J, Naslund E, Britton T, Concha H, Hassan M, Ryden M, Frisen J, Arner P (2008) Dynamics of fat cell turnover in humans. Nature 453:783–787
Summers SA (2006) Ceramides in insulin resistance and lipotoxicity. Prog Lipid Res 45:42–72
Tang W, Zeve D, Suh JM, Bosnakovski D, Kyba M, Hammer RE, Tallquist MD, Graff JM (2008) White fat progenitor cells reside in the adipose vasculature. Science 322:583–586
Vidal-Puig A, Unger RH (2010) Special issue on lipotoxicity. Biochim Biophys Acta 1801:207–208
Virtue S, Vidal-Puig A (2008) It's not how fat you are, it's what you do with it that counts. PLoS Biol 6:e237
Virtue S, Vidal-Puig A (2010) Adipose tissue expandability, lipotoxicity and the metabolic syndrome—an allostatic perspective. Biochim Biophys Acta 1801:338–349

Chapter 2
Systems Biology in Human Health and Disease

Matej Orešič

Abstract If we are to study complex multi-factorial disorders such as Metabolic Syndrome (MetS) by applying the state-of-the-art 'omics' technologies, the reductionist approach commonly applied in life sciences is no longer suitable. In order to understand the adaptive changes in molecular networks in different stages of the disease pathogenesis, a comprehensive view of the system is needed because activation of different pathways may still lead to the same functionality, but with different metabolic costs. Systems biology emerged as an inter-disciplinary field of study that focuses on complex interactions within and between biological systems, using a more holistic perspective approach to biological and biomedical research. While the importance of the systems approach has already been recognized decades ago, the experimental and modeling techniques have matured to the level where comprehensive characterization of biological systems at the molecular level is now feasible.

Keywords Allostasis · Allostatic load · Bioinformatics · Genomics · Gut microbiota · Lipid metabolism · Metabolic networks · Metabolomics · Proteomics · Systems biology

2.1 Need for Systems Approach to Study Health and Disease

Molecular biology contributed many essential experimental tools used in today's life science research. However, its early days also introduced the still pervasive reductionist approach to study the biological systems. While the concepts such as 'metabolic control analysis' (Kacser and Burns 1973) and 'systems theory' (von Bertalanffy 1969) to describe the biological systems had been introduced already in 1970s, their practical utilization was limited due to the lack of quantitative experimental data needed to parameterize the mathematical models. Instead, molecular biologists resorted to a simpler experimental paradigm, focusing on the elucidation of function of single molecular components such as genes and their products by studying them in isolation. In such setting, dependencies of specific biological functions on specific

M. Orešič (✉)
VTT Technical Research Centre of Finland, Tietotie 2, PO Box 1000, 02044 Espoo, Finland
e-mail: matej.oresic@vtt.fi

M. Orešič, A. Vidal-Puig (eds.), *A Systems Biology Approach to Study Metabolic Syndrome*,
DOI 10.1007/978-3-319-01008-3_2,

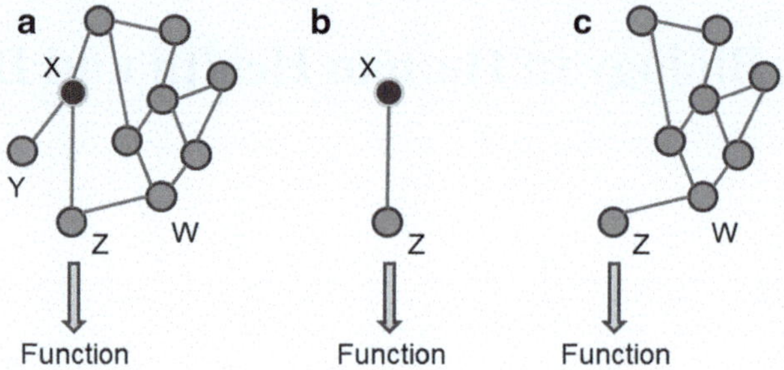

Fig. 2.1 Limitations of the reductionist approach when studying complex interconnected systems. **a** Real biological network, showing nodes as molecular entities and edges as their interdependencies. **b** Focus on single component only, X, will by experimentally modulating X lead to conclusion that X controls functionality of Z. **c** Potential other networks that may regulate Z as well as modulate function of X are disregarded by the reductionist approach

molecular components are usually sought, e.g., as established by singe-component interventions such as by gene knock-down experiments. Following a series of well thought-through experiments of that kind, a 'mechanistic insight' can be gained; which in the field of molecular biology means that in a specific context a specific component such as a gene controls a specific biological function.

The fundamental limitation of the reductionist approach as applied to molecular biology has been highlighted in an entertaining essay 'Can a biologist fix a radio? Or, what I learned while studying apoptosis' (Lazebnik 2002), where the author also remarked that "*an approach that is inefficient in analyzing a simple system is unlikely to be more useful if the system is more complex*". Particularly, the experimental paradigm used in molecular biology does not account for global interconnectivity of the system and is thus strongly context dependent. As a simplistic but illustrative example, Fig. 2.1a shows a molecular network, where the nodes denote the interacting molecular components (genes, proteins, metabolites), while edges show their interdependencies (e.g., *via* molecular interactions or biochemical reactions). The investigation is focusing on the elucidation of the regulation of the molecule Z, which is associated with a specific biological function. The hypothesis being investigated is that Z is controlled by the molecule X. In order to test the hypothesis, function of X is modulated (e.g., by knock-down), which following the experiment leads to the conclusion that indeed function of Z depends on X (Fig. 2.1b). However, such an approach disregards potential other networks that may regulate Z as well as modulate the function of X (Fig. 2.1c). The so-obtained 'mechanistic models' therefore primarily describe the binary dependencies of molecular components and their functions. Given these binary relationships are usually acquired in different experimental contexts, i.e., the 'other networks' cannot be controlled for, it is thus not surprising that reproducibility of conclusions from molecular biology experiments is strikingly poor (Begley and Ellis 2012). Perhaps most troublingly, putting these 'mechanistic models' into practical use is very challenging because the global systemic context is lacking, i.e., it is difficult to know under what circumstances a particular binary dependency really holds.

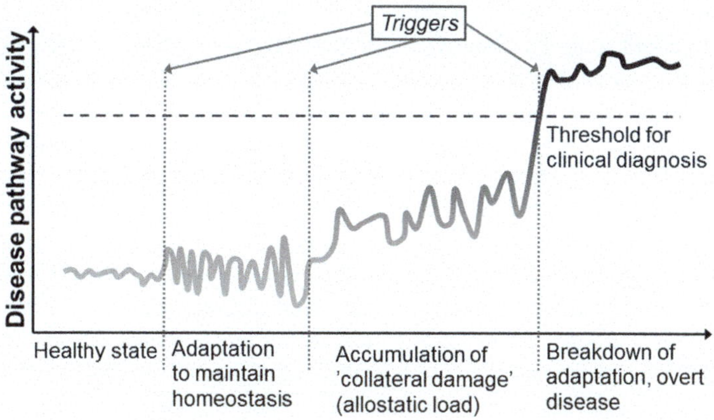

Fig. 2.2 Progression to complex disease, conceptionalized by a single imaginary variable denoted as 'disease pathway activity'. Adaptation and its metabolic cost play a key role in this process

While the discussion above mainly referred to the studies at the levels of cells, the challenges are even bigger when attempting to apply the reductionist approach at the whole-organism physiological level (Joyner and Pedersen 2011). In such setting, the system's components and their interactions occur at many levels and timescales, from individual molecules to tissues and organs. The organisms have built-in robust mechanisms which help to maintain the essential physiological functions such as metabolism under the varying environmental challenges. When considering the experimental paradigm to address specific hypothesis at the whole-organism level, the familiarity with the physiological concepts such as homeostasis, allostasis, adaptation, robustness and resilience is thus essential. For example, maintenance of lipid composition of the cell membranes is essential for cell functionality and survival. Cellular lipid homeostasis is regulated by a family of membrane-bound transcription factors designated sterol regulatory element–binding proteins (SREBPs) (Horton et al. 2002). While SREBP1c regulates the genes of membrane phospholipid metabolism, SREBP2 preferentially activates the genes of cholesterol metabolism. Surprisingly, knock-down of SREBP1c *in vivo* does not lead to disruption of phospholipid metabolism, which is because the loss of SREBP1c function is compensated by overexpression of SREBP2. However, as a cost of such adaption, mice lacking SREBP1c tend to accumulate more cholesterol (Horton et al. 2002). This is a good example of allostatic adaptation (the concept introduced in Chap. 1) aimed at induction of short-term corrective changes to regulatory systems. However, when such an adaptive response remains activated for long periods of time, the maintenance of metabolic homeostasis might actually come at a metabolic cost, or 'collateral damage', defined by McEwen as allostatic load (Korte et al. 2005). In the case of SREBP1c knock-down, the allostatic load is for example the accumulation of cholesterol due to the adaptive activation of SREBP2.

Development of a complex disorder, from early prodromal phases to clinically manifest disease, is usually a complex process which proceeds in several phases in which allostatic adaptations play an important role (Fig. 2.2, see also Chap. 1).

Environmental triggers such as change in lifestyle may impose a pressure on the organism to adapt (e.g., by changes in the underlying molecular networks) in order to maintain the system homeostasis. In the case of Metabolic Syndrome (MetS), this early phase corresponds to metabolically compensated obesity (or 'healthy obesity'). However, extended duration of the activated allostatic response eventually leads to the accumulation of allostatic load, e.g., progressively losing the ability to store lipids in obesity. In this phase also the disease vulnerability increases, i.e., the organism is more sensitive to any triggers which may cause the disease, because it is reaching the limit of adaptability. At a certain point, this limit is reached and the organism is no longer able to adapt, leading to the overt disease (Fig. 2.2). The timelines of this progression vary between the individuals and also depend on the genetic make-up and the environment, including individual's gut microbiota (Chap. 9). For example, some individuals can become very obese but still remain metabolically compensated while others soon develop metabolic co-morbidities of obesity such as type 2 diabetes (Virtue and Vidal-Puig 2008). Since the allostatic load accumulates over time, the earlier the stage of the progression to the disease, more likely the trend can be reversed. For this reason, it is important to detect the disease early in the process, prior to the appearance of clinical symptoms.

If we are to study diseases using the dynamic and physiological framework as described above, the reductionist approach is no longer viable. In order to understand the adaptive changes in molecular networks in different stages of the disease pathogenesis, a comprehensive view of the system is needed because different pathways may still lead to the same functionality, but with different allostatic load. Systems biology emerged as an inter-disciplinary field of study that focuses on complex interactions within and between biological systems, using a more holistic perspective approach to biological and biomedical research. While the importance of systems approach has already been recognized decades ago (Kacser and Burns 1973; von Bertalanffy 1969), the experimental techniques have also matured to the level where comprehensive characterization of biological systems at the molecular level is possible.

2.2 Key Enabling Technologies and Modeling Approaches of Systems Biology to Study Health and Disease

The 'omics' revolution empowered us with the tools for comprehensive characterization of biological systems. For example, genomics, proteomics and metabolomics each cover a specific layer of biological organization (Fig. 2.3). At the genome level, in addition to host genome one must also consider the microbial genomes which together carry about 150-times as many genes as the host genome, primarily in the gut (Chap. 9). Gut microbiota is sensitive to environmental factors including the diet and can be considered as a 'mediator' between the environment and host biology. Gene products such as proteins regulate many biological processes in the cells including the biochemical networks involving metabolites. Small changes in enzyme concentrations and fluxes through their pathways may produce large changes in the

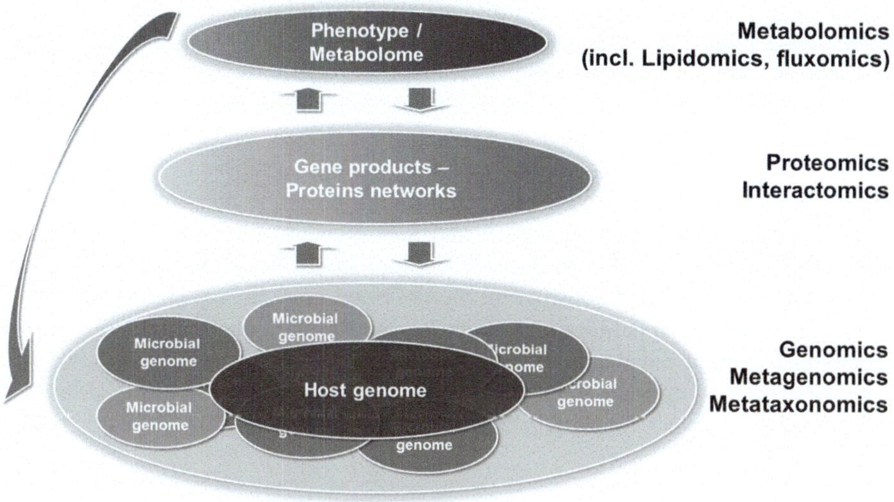

Fig. 2.3 Factors influencing the metabolome and proteome and the key analytical platforms for systems biology to study health and disease. The metabolome is sensitive to genetic and environmental factors which may together contribute to the disease. Metabolomics is thus a powerful phenotyping platform in biomedical studies

concentrations of metabolites which are the end products of these pathways (Kell 2006). Gut microbes have distinct metabolomes and proteomes, which interact with the host as well as contribute to the regulation of the host metabolism (Tremaroli and Backhed 2012). In this context, the *in vitro* colon model described in Chap. 13 is a particularly valuable tool to study how the food metabolome is transformed by the gut microbes, thus providing physiologically relevant information about the food-derived metabolites entering the systemic host metabolism. In Part III, this book will introduce the key emerging technologies which support the studies of MetS using the systems biology approach, including in the context of nutrition. These include proteomics, metabolomics, fluxomics as well as the *in vitro* colon model.

In order to interpret the 'omics' data in the physiological context, models are needed which capture the relevant topology and dynamics of biological networks and processes under investigation. Global reconstruction of human metabolic network (Duarte et al. 2007; Thiele et al. 2013) has for example allowed for tissue-specific modeling of metabolic networks, as dependent e.g., on genomic, proteomic and metabolomics data (Chap. 14). Metabolic modeling is rapidly emerging as a powerful tool which can also help in the identification of targets for interventions as well as in the prediction of specific biomarkers (by predicting outgoing metabolic fluxes).

Not all metabolic functions can be conceptualized at the network level, however. For example, lipids are key building blocks of cellular membranes and lipoprotein particles. Changes of lipid levels in these structures lead to changes in their biophysical properties and thus also potentially affecting their function. While network-based modeling involves statistical inference as dependent on the network structure, biophysical modeling requires *in silico* assembly of relevant molecular structures such as membranes by using, e.g., molecular dynamics simulations. As an example of such approach, recent study has shown that adipose tissue in obesity is characterized by enrichment of specific ether phospholipids containing arachidonic acid, despite the lower dietary intake of polyunsaturated fatty acids (Pietiläinen et al. 2011). Using a novel computational approach to simulate lipid plasma membranes based on lipidomics data, the study found that this lipid remodeling is part of an adaptive process, maintaining the normal membrane function but at the cost of higher vulnerability of adipose tissue to inflammation. Such an insight could not have been gained without considering the observed lipid changes in obesity at the level of their effect on cellular membrane properties. In general, modeling of lipid metabolism at biophysical and physiological levels is very challenging and an emerging area of systems biology. In this book, both topics are covered in Chap. 15 and 17, respectively.

2.3 Conclusions

In order to apply a systems approach to study specific disorders such as MetS, one needs three essential components: (a) a system, (b) experimental techniques, and (c) modeling techniques. These components are inter-connected by the so-called 'systems biology cycle' where measurements on a system are fed to a mathematical model, which is then further refined, leading to novel hypotheses and experiments etc. This book includes all three components, with the specific topics selected based on relevance to the study of MetS. Part II reviews 'the system' as relevant to MetS, which includes liver (Chap. 3), adipose tissue (Chap. 4), beta cell (Chap. 5), skeletal muscle (Chap. 6), central nervous system (Chap. 7), lipid metabolism (Chap. 8) and gut microbiota (Chap. 9). Selected emerging experimental techniques are introduced in Part III, including proteomics (Chap. 10), metabolomics (Chap. 11), fluxomics (Chap. 12) and *in vitro* colon model (Chap. 13). Part IV introduces specific modeling approaches which are particularly relevant to study MetS, including genome-scale metabolic modeling (Chap. 14), biophysical modeling of lipid membranes and lipoproteins (Chap. 15), methods of computational statistics (Chap. 16), and modeling of tissue cross-talk at the level of lipid metabolism (Chap. 17). Needless to say, the experimental and modeling techniques covered are not exclusive and some widely adopted approaches such as genomics are not explicitly included in this book. However, the methods covered in the book are particularly important if one is to adopt the physiological framework described in this chapter. How these techniques are connected into the 'systems biology cycle' ultimately depends on the questions asked and the specific system studied. Some practical examples will be provided in this book.

Acknowledgments This work was supported by the EU-funded project ETHERPATHS (FP7-KBBE-222639, http://www. etherpaths. org/).

References

Begley CG, Ellis LM (2012) Drug development: raise standards for preclinical cancer research. Nature 483:531–533

Duarte NC, Becker SA, Jamshidi N, Thiele I, Mo ML, Vo TD, Srivas R, Palsson BO (2007) Global reconstruction of the human metabolic network based on genomic and bibliomic data. Proc Natl Acad Sci USA 104:1777–1782

Horton J, Goldstein J, Brown M (2002) SREBPs: activators of the complete program of cholesterol and fatty acid synthesis in the liver. J Clin Invest 109:1125–1131

Joyner MJ, Pedersen BK (2011) Ten questions about systems biology. J Physiol 589:1017–1030

Kacser H, Burns JA (1973) The control of flux. Symp Soc Exp Biol 27:65–104

Kell DB (2006) Metabolomics, modelling and machine learning in systems biology—towards an understanding of the languages of cells. Delivered on 3 July 2005 at the 30th FEBS Congress and 9th IUBMB conference in Budapest. FEBS J 273:873–894

Korte SM, Koolhaas JM, Wingfield JC, Mcewen BS (2005) The Darwinian concept of stress: benefits of allostasis and costs of allostatic load and the trade-offs in health and disease. Neurosci Biobehav Rev 29:3–38

Lazebnik Y (2002) Can a biologist fix a radio?–Or, what I learned while studying apoptosis. Cancer Cell 2:179–182

Pietiläinen KH, Róg T, Seppänen-Laakso T, Virtue S, Gopalacharyulu P, Tang J, Rodriguez-Cuenca S, Maciejewski A, Naukkarinen J, Ruskeepää A-L, Niemelä PS, Yetukuri L, Tan CY, Velagapudi V, Castillo S, Nygren H, Hyötyläinen T, Rissanen A, Kaprio J, Yki-Järvinen H, Vattulainen I, Vidal-Puig A, Orešič M (2011) Association of lipidome remodeling in the adipocyte membrane with acquired obesity in humans. PLoS Biol 9:e1000623

Thiele I, Swainston N, Fleming RM, Hoppe A, Sahoo S, Aurich MK, Haraldsdottir H, Mo ML, Rolfsson O, Stobbe MD, Thorleifsson SG, Agren R, Bolling C, Bordel S, Chavali AK, Dobson P, Dunn WB, Endler L, Hala D, Hucka M, Hull D, Jameson D, Jamshidi N, Jonsson JJ, Juty N, Keating S, Nookaew I, Le Novere N, Malys N, Mazein A, Papin JA, Price ND, Selkov E Sr, Sigurdsson MI, Simeonidis E, Sonnenschein N, Smallbone K, Sorokin A, Van Beek JH, Weichart D, Goryanin I, Nielsen J, Westerhoff HV, Kell DB, Mendes P, Palsson BO (2013) A community-driven global reconstruction of human metabolism. Nat Biotechnol 31:419–427

Tremaroli V, Backhed F (2012) Functional interactions between the gut microbiota and host metabolism. Nature 489:242–249

Virtue S, Vidal-Puig A (2008) It's not how fat you are, it's what you do with it that counts. PLoS Biol 6:e237

Von Bertalanffy L (1969) General systems theory. George Braziller, New York

Part II
Pathophysiology of Metabolic Syndrome

This section provides background to key components of metabolic syndrome. The contributions show examples how systems approached were applied to study these specific aspects of MetS.

Chapter 3
The Liver in Metabolic Syndrome

Simona D'Amore, Giuseppe Palasciano and Antonio Moschetta

Abstract The liver plays a major role in the regulation of pathways that govern systemic metabolic homeostasis. Hepatic regulation of glucose and lipid homeostasis is under the control of a complex system of hormones, transcription factors, and signaling pathways. Hepatic steatosis is a condition characterized by an increased amount of fat in the liver, mainly in the form of triglycerides, in a quantity greater than 5 % of liver weight, or histologically defined when 5 % or more of hepatocytes contains visible intracellular triglycerides, that cannot be attributed to alcohol consumption. Fatty liver is closely related to impairment in glucose, fatty acid, and lipoprotein metabolism, and it is now widely recognized as the hepatic manifestation of the Metabolic Syndrome. In the present chapter we will critically present the potential mechanisms that link obesity, type 2 diabetes, hepatic lipid accumulation, and insulin resistance. Indeed, although the correlation between fatty liver and insulin resistance is well established, we will specifically discuss the combination of elevated serum levels of glucose, fatty acids, and peripheral insulin resistance for the pathogenesis of liver dysfunction in the Metabolic Syndrome.

Keywords Insulin resistance · Lipid metabolism · Metabolic Syndrome · Nonalcoholic fatty liver disease · Nonalcoholic steatohepatitis

3.1 Can Metabolic Syndrome Exist Without Fatty Liver?

Metabolic Syndrome (MetS) and nonalcoholic fatty liver disease are the clinical expressions of the same metabolic impairment, both sharing the common pathogenic factor and leading to an increased risk for diabetes and cardiovascular disease (CVD).

A. Moschetta (✉) · S. D'Amore · G. Palasciano
Clinica Medica "A. Murri", Aldo Moro University of Bari,
Consorzio Mario Negri Sud, Via Nazionale 8A,
66030 Santa Maria Imbaro (Chieti), Italy
e-mail: moschetta@negrisud.it

A. Moschetta · S. D'Amore
Laboratory of Lipid Metabolism and Cancer,
Consorzio Mario Negri Sud, Santa Maria Imbaro (CH), Italy

M. Orešič, A. Vidal-Puig (eds.), *A Systems Biology Approach to Study Metabolic Syndrome*,
DOI 10.1007/978-3-319-01008-3_3,

Table 3.1 Criteria for clinical diagnosis of Metabolic Syndrome (ATP III, 2001). (Grundy et al. 2005)

Clinical measure	Categorical cut-points (any 3 of the following 5 features constitute diagnosis of Metabolic Syndrome)
Abdominal obesity	Waist circumference (WC) 102 cm in men or 88 cm in women
Elevated triglycerides	Triglycerides ≥ 150 mg/dL or on drug treatment for elevated triglycerides
Reduced HDL-C	HDL-C < 50 mg/dL in men or < 40 mg/dL in women or on drug treatment for reduced HDL-C
Blood pressure	≥ 130/85 mmHg or on antihypertensive drug treatment for hypertension
Fasting glucose	≥ 100 mg/dL[a] or diabetes or on drug treatment for elevated glucose

[a] The 2001 definition identified fasting plasma glucose of ≥ 110 mg/dL (6.1 mmol/L) as elevated. This was modified in 2004 to be ≥ 100 mg/dL (5.6 mmol/L), in accordance with the *American Diabetes Association*'s updated definition of impaired fasting glucose

MetS is a complex syndrome which involves different metabolic conditions, and according to the *Third Report of the National Cholesterol Education Program Expert Panel on Detection, Evaluation, and Treatment of High Blood Cholesterol in Adults* (*Adult Treatment Panel III, ATP III*), is defined by the presence of three or more of the following features; visceral adiposity, hypertriglyceridemia, hypertension, glucose intolerance or type 2 diabetes, and low high-density lipoprotein (HDL) levels (Table 3.1) (Anon 2002). Insulin resistance (IR) represents the key component and the putative pathogenetic hit that links all features of the MetS (Balkau and Charles 1999; Olufadi and Byrne 2008). Although fatty liver is still not included in the diagnostic criteria of MetS, it is the primary hepatic complication of obesity and MetS.

The term "nonalcoholic steatohepatitis" was coined in 1980 to describe "the pathological and clinical features of nonalcoholic disease of the liver associated with the pathological features most commonly seen in alcoholic liver disease itself" (Ludwig et al. 1980). Nonalcoholic fatty liver disease represents a heterogeneous cluster of diseases, ranging from hepatic steatosis alone (nonalcoholic fatty liver, NAFLD), to nonalcoholic steatohepatitis (NASH), fibrosis, and cirrhosis (Matteoni et al. 1999; Powell et al. 1990; Teli et al. 1995; Wanless and Lentz 1990a). The common denominator is the presence of macrovescicular hepatic steatosis, resulting from accumulation of triglycerides (TGs) within hepatocytes, in the absence of significant alcohol intake.

If steatosis is a normal feature in the liver of migratory birds, that need to store energy in order to face long periods of fasting, this condition is not a physiological finding in humans (Capeau 2008). As a consequence, NAFLD is not a benign condition with a non-progressive course, as initially believed, but a progressive condition leading to NASH, and finally to fibrosis, cryptogenic cirrhosis, and increased risk for hepatocellular carcinoma (Bugianesi 2005; Caldwell et al. 1999; Marrero et al. 2002). When compared to general population, NAFLD is associated to a five times higher mortality and morbidity, with an increased risk for CVD (Hamaguchi et al. 2007; Matteoni et al. 1999; Musunuru 2010). In a simplistic way of thinking, NAFLD can be considered the hepatic manifestation of the MetS, since both MetS and NAFLD originate from the combination of modern diet, lifestyle, and socio-economic factors. Over-nutrition, cheap high-energy food, and sedentary habits represent the starting point of the pathogenetic mechanisms involved in hepatic lipids accumulation, and in both hepatic and whole body IR (Bugianesi et al. 2005a; Korenblat et al. 2008; Marchesini et al. 2001; Musso et al. 2008). However, temporal and causal relationships between hepatic steatosis and IR remain uncertain, since both these states can potentiate each other, also it remains still unclear wheter IR or steatosis arises first. If IR and the elevated plasma levels of glucose and fatty acids (FAs) are responsible for increased hepatic lipogenesis, impaired β-oxidation, and hepatic lipid accumulation (Marchesini et al. 2001; Sanyal et al. 2001), on the other hand hepatic IR can develop also independently from peripheral IR. As a confirmation, hepatic IR and abnormal liver lipid storage can occur also in the absence of peripheral IR (Kraegen et al. 1991; Larter et al. 2010; Rinella and Green 2004; Rizki et al. 2006; Schattenberg et al. 2005). Rats fed with a short term high fat diet develop hepatic fat accumulation and hepatic IR, in absence of changes in the whole body IR (Kraegen et al. 1991; Samuel et al. 2004). On the other hand, hyperglycemia, hyperinsulinemia, and decreased peripheral insulin sensitivity are also associated to the development of both MetS and NAFLD, since they drive hepatic lipogenesis, and promote the release of TGs in serum (hypertriglyceridemia), and the intrahepatic (NAFLD) and peripheral (obesity and atherosclerosis) lipid accumulation, further worsening IR (Unger 2003).

The following paragraphs cover an overview of the complex physiological mechanisms controlling glucose and lipid metabolism, together with a critical view on the pathological role of the liver in the onset of the clinical manifestation of MetS and NAFLD.

3.2 Role of the Liver in the Pathophysiology of Metabolic Syndrome

The liver plays a major role in the regulation of functions necessary for the maintenance of systemic metabolic homeostasis. Thus, a metabolic imbalance of glucose and lipid metabolism may results in hepatic steatosis. On the other hand, insulin is responsible for the maintenance of glucose and lipid homeostasis in the liver, which is its main site of action in addition to adipose tissue and skeletal muscle. A condition of IR determines different pathogenetic effects depending on the metabolic pathway involved, thus a refractory insulin-response leads to increased glucose production in liver, and to activation of the lipid synthesis and dyslipidemia, which can lead to hepatic steatosis. Fatty liver and IR present various reciprocal influences and can potentiate each other.

3.2.1 Glucose Metabolism and Insulin Resistance

During fast and fed conditions the role of the liver is to control different metabolic processes to maintain serum glucose concentrations within physiological limits. Glycogenolysis and gluconeogenesis increase plasma glucose; fatty acid oxidation (FAO) generates available energy for gluconeogenesis, while lipogenesis provides FAs (Fig. 3.1). Insulin controls serum glucose concentrations and induces the transcription of most of the metabolic enzymes involved in hepatic glucose production and in glycogen synthesis (Foufelle and Ferre 2002). Glycogenolysis is activated to produces glucose during short period of fast, whereas is suppressed by insulin after food intake (Gastaldelli et al. 2001). During longer periods of fasting, glycogen stores are depleted in the liver and, thus, gluconeogenesis is stimulated to maintain normal serum glucose concentration (Boden 2003). Insulin is the primary hormone promoting the storage of energy in the fed state and is secreted in response to nutrients into the portal circulation. High levels of insulin promote the fast-fed shift in hepatic carbohydrate metabolism, stimulating glycogen synthesis and suppressing both gluconeogenesis and glycogenolysis (Fig. 3.1).

In pathological states of absolute insulin deficiency, such as type 1 diabetes, or of relative insulin deficiency, such as obesity and type 2 diabetes, gluconeogenesis results activated via an impaired inhibition of its key enzymes, phosphoenolpyruvate carboxykinase (PEPCK), responsible of the conversation of oxaloacetic acid to phosphoenolpyruvate, and glucose-6-phosphatase (G-6-Pase), responsible for dephosphorylation of glucose 6-phosphate to free glucose (Bischof et al. 2002; Roden and Bernroider 2003). However, when the liver becomes fatty, insulin also fails to inhibit gluconeogenesis (Bugianesi et al. 2005a), causing an increase of serum glucose and, then, of hyperinsulinemia. Therefore, in a state of IR, the major alterations in glucose metabolism are represented by decreased ability of insulin to stimulate glucose consumption and to suppress hepatic glucose production.

3.2.2 Lipid Metabolism and Hepatic Insulin Resistance

The FA and TG content of hepatocytes is the result of the balance between their uptake and synthesis (dietary intake, increased lipolysis of adipose tissue, *de novo* synthesis with esterification of TG) and their catabolism and export (impaired mitochondrial oxidation and export of TGs); in general, lipids accumulation in the liver can occur when the rate of FA input is greater than the rate of FA output. There are four primary mechanisms that can be responsible for lipids accumulation in the liver : (1) increased hepatic lipid uptake (2) increased intrahepatic lipid synthesis (3) impaired hepatic lipid export, and (4) reduced hepatic FAO.

All these mechanisms are under the control of insulin signaling; insulin modulates the degradation of lipids by FAO (Randle 1998), the free fatty acid (FFA) flux into the liver, the lipoprotein synthesis and lipogenesis of FFA, and the very low-density lipoprotein (VLDL) secretion (Horton et al. 2003). IR in skeletal muscle and adipose

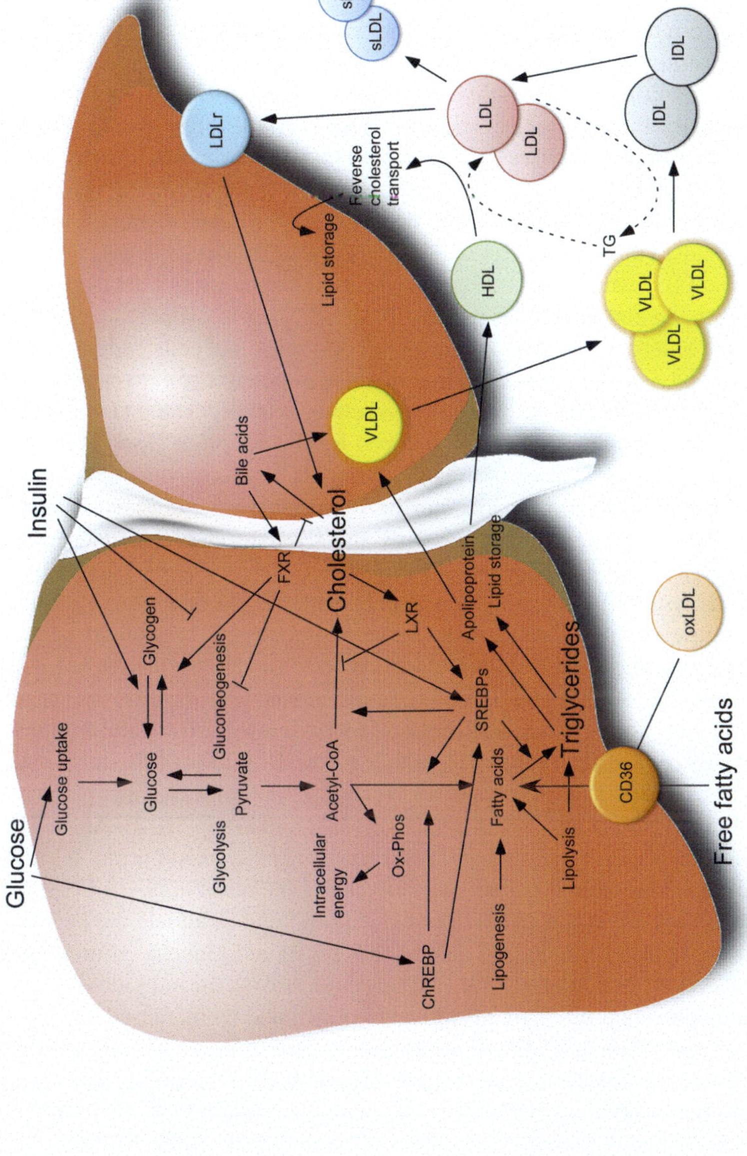

Fig. 3.1 Role of the liver in the control of glucose and lipid metabolism. *CD36* cluster differentiation protein-36, *ChREBP* carbohydrate response element-binding protein, *FXR* farsenoid X receptor, *HDL* high-density lipoprotein, *IDL* intermediate-density lipoprotein, *LDL* low-density lipoprotein, *LDLr* low-density lipoprotein receptor, *LXR* liver X receptor, *OxLDL* oxidized low-density lipoprotein, *Ox-Phos* oxidative phosphorylation, *sLDL* Small low-density lipoprotein, *SREBPs* sterol regulatory element binding proteins, *TG* triglyceride, *VLDL* very low-density lipoprotein

tissue is responsible for the impairment of the insulin signal transduction in skeletal muscle cells, adipocytes and also in hepatocytes. A consequence of the peripheral IR is hepatic steatosis. Conversely, abnormalities in lipid and lipoprotein metabolism, abnormal lipid accumulation and hepatic IR may determine and worsen total body IR (Larter et al. 2010; Rinella and Green 2004; Rizki et al. 2006; Schattenberg et al. 2005).

Hepatic Lipid Uptake In obese subjects with NAFLD approximately 15 % of hepatic lipids comes from dietary FAs; 25 % are from *de novo* synthesis and 60 % from FFAs, mainly from adipose tissue (Donnelly et al. 2005). Therefore, one of the major sources of TGs are the dietary FAs delivered to the liver, a pathway activated in a condition of IR, and the dietary glucose content, which is converted in glycogen and, when in excess, enter in the glycolysis pathway and then "turbo boost" the FA synthesis. The FA synthesis pathway includes several enzymes induced by ingestion of a high carbohydrate diet, such as ATP citrate lyase, acetyl-CoA carboxylase (ACC), and fatty acid synthase (FAS) (Towle et al. 1997). FAs thus produced can be further used to synthetize TGs, which can be stored within hepatocytes or packages into VLDL released into the circulation (Fig. 3.1).

Dietary fat and cholesterol are directly absorbed into intestinal cells and incorporated in chylomicrons, which are secreted into bloodstream and then into capillaries of several tissues, such as adipose tissue and skeletal muscle. The TGs content of chylomicrons and circulating VLDL is reduced by the action of lipoprotein lipase (LPL, the rate-limiting step of tissue circulating TG uptake), thus leading to the formation of chylomicrons remnants and LDL enriched in cholesterol (Westerbacka et al. 2007); LDL can be captured by the liver through the binding to LDL receptor (LDLr) (Fig. 3.1) (Brown and Goldstein 1986; Tulenko and Sumner 2002). Patients with NAFLD present an increased VLDL production, elevated plasma LDL (Zoltowska et al. 2001), and a reduced lipoprotein hydrolysis of the TGs into chylomicrons due to inhibition of exportation of LPL to the vessel surface, which causes an uptake of TGs-rich chylomicrons remnants by the liver (Capeau 2008). In addition LPL activity may be decreased by high levels of its inhibitor, the apolipoprotein C-III (ApoC-III) (Jong et al. 1999); ApoC-III inhibits LPL-mediated hydrolysis and clearance of TGs-rich lipoproteins, thus modulating plasma lipid profiles (Cohn et al. 2004; Ginsberg and Brown 2011; Yao and Wang 2012). ApoC-III also promotes hepatic VLDL assembly and secretion; this mechanism promotes the elimination of excess TGs from the liver, thus preventing the onset of NAFLD (Sundaram et al. 2010; Yao and Wang 2012). Multiple stimuli can modulate the hepatic gene expression of ApoC-III, including insulin (inhibitor) and glucose (stimulator) (Altomonte et al. 2004; Caron et al. 2011; Chen et al. 1994). In individuals with IR, ApoC-III cannot be suppressed by insulin and is promoted by increased glucose, leading to an overproduction of VLDL. Furthermore, recent studies have suggested a genotype–phenotype relationship between ApoC-III and NAFLD, showing that some single-nucleotide polymorphisms (SNPs) in the gene encoding APOC-III, may be associated with hypertriglyceridemia and MetS (Guettier et al. 2005; Miller et al. 2007; Petersen et al. 2010). In fact, the ApoC-III polymorphisms T-455C and C-482T were found associated with increased plasma ApoC-III and TG concentrations, and with increased prevalence of NAFLD (Miller et al. 2007; Petersen et al. 2010); these findings have not been confirmed in each ethnic backgrounds (Kozlitina et al. 2011). On the other hand, people with hereditary deficiencies in ApoC-III generally have lower plasma TG levels (Pollin et al. 2008).

In the post absorptive and fasting states the major rate of FFA delivered to the liver derived from lipolysis of peripheral tissue, mainly from adipose tissue. Lipolysis of adipose tissue is under the control of the hormone-sensitive lipase (HSL), which is an enzyme activated when the body needs to mobilize energy stores, whereas is strongly inhibited by insulin in the post prandial feeding state. Adipose tissue IR of NAFLD patients, which is characterized by the lack of suppression of lipolysis by insulin, determines increased HSL levels with a consequent deregulation of adipose-derived FFA flux to the liver in the fasting state (Mittendorfer et al. 2009). The excessive FFA release rates from adipose tissue into the systemic circulation lead to an increased availability of FAs for uptake by the liver, facilitating intra-hepatic lipids accumulation, and worsening hepatic IR (Boden and Shulman 2002). Furthermore, among causes of lipid accumulation in NAFLD, a key role is also played by fatty acid transporters and their regulators, fatty acid translocase/cluster differentiation protein-36 (FAT/CD36), fatty acid transport protein (FATP) and fatty acid binding proteins (FABPs), which could further enhance hepatic lipid uptake (Greco et al. 2008). FAT/CD36 is a long chain fatty acid transporter, which mediates FA uptake and intracellular transport from plasma into different cell types (skeletal muscle cells, adipocytes and hepatocytes), and it is regulated by insulin (Fig. 3.1) (Lavoie and Gauthier 2006). Different experimental approaches confirm the role of these findings. Specific activation of FAT/CD36 leads to hepatomegaly and hepatic steatosis (Zhou et al. 2008). On the other hand, deletion of FATP and FABPs reverses hepatic steatosis (Doege et al. 2006; Newberry et al. 2006).

Intrahepatic Lipid Synthesis The *de novo* lipogenesis (DNL), which is under the control of several transcription factors in response to feeding and fasting conditions, converts carbohydrates into saturated fatty acids (SFAs), and it is increased after consumption of a meal. DNL plays an important role in promoting hepatic fat accumulation, appearing significantly increased in subjects with NAFLD compared to healthy subjects (Aarsland and Wolfe 1998; Korchak 1962). Thus DNL is responsible of 25 % of TGs stored in the liver of NAFLD patients and it is still unclear the importance of this pathway in the pathogenesis of fatty liver. Intriguingly DNL can be detected not only in hyperinsulinemic condition but also in healthy subjects through a high-carbohydrate meal; acute fructose ingestion increases DNL and post-prandial TG levels (Parks et al. 2008). Insulin and glucose both drive DNL, through the respective activation of sterol regulatory element binding proteins (SREBPs) and carbohydrate response element-binding protein (ChREBP), which transcriptionally induce most genes involved in FA synthesis (Browning and Horton 2004).

SREBPs are transcription factors that control FA, TG, and cholesterol synthesis, and play a key role in the homeostasis of lipid metabolism (Fig. 3.1). Three forms have been identified; SREBP-1c and SREBP-2 are the predominant form in the rodent and human liver. SREBP-1c activates genes involved in FAs and TGs synthesis, whereas SREBP-2 mediates cholesterol synthesis; both 1c and 2 inhibit the expression of PEPCK when there is a high uptake of carbohydrates, lowering blood glucose that is used for lipogenesis (Sozio and Crabb 2008). SREBP-1c is the predominant isoform in the liver, and its expression is regulated by nutrients and nutrition-related signals, resulting suppressed during fasting and induced in feeding state. Nutrients such as carbohydrates and SFAs induce SREBP-1c, whereas polyunsaturated fatty acids (PUFAs) are inhibitors of SREBP-1c. SREBP-1c is also regulated by insulin (promoter) and glucagon (inhibitor). Insulin and hyperglycemia induce increased SREBP-1c expression, as shown in obese, insulin resistant and hyperinsulinemic conditions (Caballero et al. 2009; Higuchi et al. 2008; Nakamuta et al. 2005a; Shimomura et al. 2000; Tamura and Shimomura 2005). SREBP-1c is also inhibited by protein kinase A (PKA) and by adenosine monophosphate-activated protein kinase (AMPK), which are activated during energy-producing processes (Foretz et al. 1999). Overexpression of SREBPs results in increased lipogenesis and in the development of hepatic steatosis (Shimano et al. 1996; Shimano et al. 1997), through the transcription of target genes involved in FA and TG synthesis, such as FAS, ACC, long chain fatty acid elongase, and stearoyil-CoA desaturase 1 (SCD1), which have been shown to be increased in patients with NAFLD (Foufelle and Ferre 2002; Postic and Girard 2008). ACC converts acetyl-CoA to malonyl-CoA, whereas FAS catalyzes the last step in FA biosynthesis, the conversion of malonyl-CoA into SFAs. High levels of malonyl-CoA inhibit carnitine palmitoyltransferase-1 (CPT-1), which transport FAs into mitochondria (Capeau 2008; Greco et al. 2008). SCD1 has an important role in protection of the liver from lipotoxicity, converting SFAs to monounsaturated fatty acids (MUFAs), that can be safely stored (Li et al. 2009).

In addition to the regulation of homeostasis of lipid metabolism SREBP-1c is involved in the regulation of glucose metabolism by directly suppression of insulin receptor substrate-2 (IRS-2), which plays a major role in the modulation of hepatic insulin signaling during the fasting state, resulting in inhibition of the downstream phosphatidylinositol 3-kinases/protein kinase B pathway (PI3K/Akt), and in a lack of blocking of the forkhead box protein O1 (FOXO1). In normal conditions FOXO1 is a transcription factor that activates gluconeogenic pathways through glucagon action, and is inhibited by insulin through phosphorylation by the kinase PI3K/Akt. Overexpression of SREBP-1c induces FOXO1, which activates gluconeogenesis through the expression of PEPCK and G-6-Pase (Wolfrum et al. 2004). A reduction in IRS-2 activity leads to impaired insulin sensitivity, decreased glycogen synthesis, and increased gluconeogenesis. This may activate SREBP-1c and ChREBP, thus favoring the switch from glycogen synthesis to lipogenesis (Ide et al. 2004). Therefore, chronic SREBP-1c activation has a crucial role in the hepatic lipotoxicity, and in the control of glucose and lipid metabolism, causing both IR and lipids accumulation.

ChREBP is a transcription factor involved in lipid synthesis, which is activated by high concentrations of glucose, independently of insulin; thus, it has an important role in states of hyperglycemia (Denechaud et al. 2008). ChREBP plays a key role in the conversion of carbohydrates to TGs in response to glucose excess (Fig. 3.1) (Towle et al. 1997). ChREBP is activated directly from glucose, and indirectly from insulin through the induction of glucokinase which allows the conversion of phosphorylated glucose to xylulose-5-phosphate, with subsequent activation of protein phosphatase 2A (PP2A) that dephosphorylates and activates ChREBP (Kabashima et al. 2003). ChREBP is inhibited by phosphorylation that is mediated by PKA and AMPK (Kawaguchi et al. 2001; Kawaguchi et al. 2002). Activated ChREBP induces transcription of gene involved in glycolysis (liver pyruvate kinase, LPK), gluconeogenesis (G-6-Pase), and, in synergy with SREBP, of lipogenic genes (ACC and FAS) (Iizuka and Horikawa 2008). The final result is the conversion of excess glucose to FAs (Browning and Horton 2004).

Hepatic Lipid Export After a meal, dietary lipids are transported from the gut into the systemic circulation in the form of chylomicrons and then stored in the liver, where they are assembled to form the VLDL (Fig. 3.1). VLDL are complex lipoprotein particles constituted by proteins (apolipoprotein B-100, ApoB-100), lipids (TGs or cholesterol ester), and phospholipids. Hepatic VLDL packages involve the fusion of an ApoB-100 molecule with lipid droplets of TG through the action of the microsomal transfer protein (MTP) within hepatocytes, preventing TGs accumulation (Fabbrini et al. 2010). The lipid excess is oxidized or released as VLDL. However, the hepatic ability to oxidized fat is limited; an increase in intrahepatic fat content leads to an up-regulation of oxidative mechanisms which are less efficient due to mitochondrial uncoupling (Gastaldelli et al. 2007; Sanyal et al. 2001). Unoxidized intrahepatocellular FAs are esterified to TGs, which can be incorporated into VLDL, secreted into the circulation and delivered to the peripheral tissue, or stored within the liver.

In patients with NAFLD, VLDL synthesis and export are impaired. In a state of IR the rates of VLDL secretion are increased and accelerated, with subsequent enhancement of serum levels of LDL and of atherogenic small LDL (sLDL). The delipidation chain is under the control of two enzymes; cholesteryl ester transfer protein (CETP), and hepatic lipase (HL). CETP is a protein responsible for lipid molecule interchange between lipoproteins, enriching VLDL in cholesterol, whereas HL regulates plasma concentration of atherogenic lipoprotein, such as intermediate-density lipoproteins (IDL) and sLDL. CETP activity is increased in NAFLD and may favor, together with HL, the formation of sLDL (Lucero et al. 2011). Nevertheless, while patients with NAFLD present increased rates of secretion of VLDL compared to control, this secretion rate reaches a plateau and results inadequate to compensate the degree of intracellular TG accumulation (Adiels et al. 2006; Fabbrini et al. 2008). The mechanism responsible for the inadequate increase in hepatic TG export is unknown, but it might be related to inability of the liver to secrete large VLDL particles, which cannot be secreted through the sinusoidal endothelial pores (Horton et al. 1999). In fact, secretion rate of VLDL-ApoB-100 is equal between subjects with high and low intrahepatic TGs content, with consequent production of very large TG-rich VLDL. Moreover, the hepatic synthesis of ApoB-100, which represents the rate-limiting step in the formation of VLDL, has been shown markedly altered by insulin (Dashti et al. 1989; Pullinger et al. 1989). Thus, a state of hyperinsulinemia may cause an impaired ApoB-100 synthesis and consequent accumulation of TGs within hepatocytes. Impaired secretion of VLDL may also be secondary to a decreased activity of MTP, which is essential for the formation of VLDL, packaging TGs with ApoB. Polymorphisms or high-fat diet-induced methylation of MTP are associated with a decreased MTP activity and enhanced TG accumulation in the liver (Chang et al. 2010; Namikawa et al. 2004).

Fatty Acid Oxidation Lipid accumulation determines injury of hepatocytes with impairment in mitochondrial number and/or oxidative function. Mitochondria dysfunction could affect several cellular functions within hepatocytes, both directly, through increased oxidative stress and reduced FAO, and indirectly, affecting energy-requiring processes. The intrahepatic FAO occurs mainly within mitochondria, and lesser in peroxisomes and microsomes. As mentioned before FAs can be both esterified to TGs or oxidized with the production of adenosine triphosphate (ATP). The uptake of FA into the mitochondria matrix is regulated by CPT-1 and -2. CPT-1 levels have been found decreased in patients with NAFLD, consequentially to the inhibition by high levels of malonyl-CoA, suggesting a reduction in mitochondrial FAO (Nakamuta et al. 2005a). However, although FAO appears to be increased in liver of both NAFLD and NASH patients (Sanyal et al. 2001), this increase is not sufficient *per se* to prevent TG accumulation. Overall, contrasting results have been reported about FAO in NAFLD, so further studies are needed (Higuchi et al. 2008).

Mitochondrial dysfunction results in increased production of reactive oxygen species (ROS) and, thus, oxidative stress. ROS can activates stress responses leading to increased activity of mitogen activate protein kinase (MAPK) and c-Jun N-terminal kinase (JNK), two kinases that inhibit IRS1 and IRS2, with consequent decreased activation of PI3K.

PI3K activation is involved in the translocation of glucose transporter type 4 (GLUT4) to the plasma membrane, that facilitate glucose uptake and its utilization in muscle and adipose tissue. Therefore, inhibition of PI3K results in reduced glucose uptake in the organs, and in exacerbated IR. The excess of FFAs, due to impaired FAO, can also trigger the endoplasmic reticulum (ER) stress (Kharroubi et al. 2004; Ozcan et al. 2004), with consequent induction of the uncoupled protein response that initiates an adaptive response, worsening hepatocyte IR (Wei et al. 2006). The unfolded protein response (UPR) activates an adaptive response that results in increased levels of pro-apoptotic proteins and induction of SREBP-1c and SREBP-2 (Sozio and Crabb 2008).

3.3 Lipid Sensing Nuclear Receptors as Modulators of Hepatic Metabolism

Nuclear receptors (NRs) are transcription factors that act as intracellular sensors for endocrine hormones and dietary lipids. NRs are key players in the coordination of the development, metabolism, circadian rhythms, cell growth and differentiation. NRs function as ligand-dependent transcription factors transducing different signals into modulation of gene transcription (Browning and Horton 2004; Mangelsdorf et al. 1995; Shulman and Mangelsdorf 2005). NRs show considerable specificity in their activation and tissue-specific expression (Bookout et al. 2006), and they can work as monomers, homodimers, and heterodimers (Mangelsdorf et al. 1995) . In the human genome 48 NRs have been identified, while in rodents there are 49 (Mangelsdorf et al. 1995). NRs are characterized by a conserved modular structure including a ligand-independent activation function domain (AF-1), a ligand-binding domain (LBD), a DNA-binding domain (DBD, binding to a specific hormone response element), and a region involved in the transcriptional activation of target-gene expression (Lanz et al. 1999; Mangelsdorf et al. 1995; McKenna and O'Malley 2002; Onate et al. 1998). In the absence of ligand, the NR is bound to transcriptional co-repressor complexes that cause chromatin condensation and gene silencing. Ligand binding to NRs induces a change in NR three-dimensional conformation, which results in increased affinity and recruitment of tissue-specific co-regulators, leading to the activation of the transcription machinery (Lonard and O'Malley 2007; Mangelsdorf et al. 1995; McKenna and O'Malley 2001). We will focus on the lipid-sensing NRs, namely the peroxisome proliferator-activated receptors (PPARs), the liver X receptors (LXRs), and the farnesoid X receptor (FXR), which all form heterodimers with the retinoid X receptors (RXR) and are active players in the control of mechanisms involved in the pathogenesis of MetS, and NAFLD (hepatic lipid uptake and export, intrahepatic lipid synthesis, FAO, carbohydrate metabolism, and inflammation), and in the development of their complications.

3.3.1 Liver X Receptor

LXRs are key members of the nuclear receptors family, which are recognized as important regulators of cholesterol metabolism, lipid biosynthesis, regulation of the storage and oxidation of dietary fat, and of glucose homeostasis (Kalaany et al. 2005; Steffensen and Gustafsson 2004). Increased levels of LXR have been reported in patients with NAFLD, with direct correlation to the severity of NAFLD (Ai and Chen 2007; Higuchi et al. 2008). To date two LXRs genes, LXR-α and -β, have been described with different tissue distribution; LXR-β is ubiquitously expressed, whereas LXR-α is mainly expressed in the liver, adipose tissue, macrophages, small intestine, kidneys, and gonads (Repa and Mangelsdorf 2000). LXRs are activated by their endogenous ligands, the oxidized cholesterol derivatives, referred to as oxysterols (Janowski et al. 1996; Willy et al. 1995).

LXRs regulate cholesterol homeostasis through the control of bile acid (BA) production (in rodents), cholesterol influx, transport and efflux, and impact energy homeostasis through the regulation of glucose metabolism (in rodents). LXRs positively regulate some aspects of BA metabolism, such as the transcription of cholesterol 7α hydroxylase (CYP7A1), which represent a rate-limiting step in BA synthesis, and the transcription of the BA-conjugating enzyme UDP-glucuronosyltransferase 1–3 (UGT1A3), which facilitate BA elimination (Barbier et al. 2009; Peet et al. 1998).

The synthesis and secretion of BAs are essential for dietary lipid absorption and for removal the excess of cholesterol from the liver. LXRs also modulate cholesterol efflux through the regulation of several members of the ATP-binding cassette (ABC) transporters.

In macrophages, LXRs activation stimulates the expression of ABCA1 and ABCG1, which mediate the efflux of cholesterol and phospholipids to lipoprotein, whereas in liver stimulates ABCG5 and ABCG8, which eliminate hepatic cholesterol through its secretion into the bile (Chawla et al. 2001a). LXR also regulates intestinal cholesterol absorption through the negative regulation of the intestinal sterol transporter Niemann-Pick C1-like 1 (NPC1L1) (Duval et al. 2006) and facilitates uptake and storage of cholesterol by the liver through the expression of CEPT (Luo et al. 2001).

LXRs also regulate FA metabolism through the expression of SREBP-1c which is, as mentioned before, a master regulator of FA synthesis (Repa et al. 2000). Furthermore, LXRs directly up-regulate enzymes involved in FA synthesis and in FA influx, such as FAS, ACC, FAT/CD36 (Cha and Repa 2007; Repa et al. 2000; Tontonoz and Mangelsdorf 2003a; Zhou et al. 2008). These pro-lipogenic effects limit the use of synthetic LXR agonists for therapeutically use (Cha and Repa 2007; Joseph et al. 2002; Zhang et al. 2003). On the other hand, LXR promotes FAO in skeletal muscle and adipocytes by stimulating pyruvate dehydrogenase kinase 4 (PDK4), which suppress glycolysis in favor of FAO (Kase et al. 2005; Stenson et al. 2009). LXR is also involved in the regulation of glucose metabolism through the suppression of hepatic gluconeogenesis by inhibiting PEPCK, G-6-Pase, and peroxisome proliferator-activated receptor-gamma coactivators 1alpha (PGC-1α) and by the up-regulation of GLUT4 in adipose tissue (Cao et al. 2003; Laffitte et al. 2003). These functions suggest that LXR plays a key role in cholesterol homeostasis, lipogenesis, and also in response to nutritional state (Fig. 3.1).

3.3.2 Farsenoid X Receptor

FXR-α is the master regulator of bile acids physiology, and acts as bile acid sensor that protect cells and organs from BAs toxicity. FXR-α is activated by cholic acid (CA) and chenodeoxycholic acid (CDCA) (Chawla et al. 2001b; Makishima et al. 1999; Parks et al. 1999; Wang et al. 1999). Two FXR genes have been identified: FXR-α and FXR-β. FXR-β constitutes a pseudo-gene in humans and primates (Otte et al. 2003). FXR-α is highly expressed in liver and intestine where it modulates the expression of genes involved in bile acid (BA) conjugation, efflux from the liver, and re-uptake of the reabsorbed BA by the enterohepatic circulation (Wagner et al. 2008; Zhang and Edwards 2008).

FXR-α also mediates the reverse cholesterol transport and efflux, and up-regulates phospholipids transfer proteins and several apolipoproteins to induce overall reduction of plasma lipids; treatment with FXR-α ligands reduces serum levels of VLDL,

TGs and HDL (Watanabe et al. 2004a). Moreover, FXR-α contributes to the regulation of lipid- and carbohydrate-induced changes during the fast/feed shift, by inducing genes involved in lipoprotein/clearance of TG-rich lipoprotein and in insulin sensitivity/signaling, and repressing hepatic genes involved in DNL (Fig. 3.1) (Lee et al. 2006; Zhang et al. 2004).

CDCA showed the ability of lowering plasma TG levels in patients with cholesterol gallstones and hypertriglyceridemia (Bell et al. 1973; Carulli et al. 1981; Leiss and Von 1982; Miller and Nestel 1974). FXR lowers circulating VLDL by different mechanisms. FXR induces the expression of short heterodimer partner (SHP), that inhibits the liver receptor homolog-1 (LRH-1)—mediated activation of SREBP-1c, reducing DNL, and modulating the MTP-mediated packaging of TGs with ApoB-100 in VLDL, thus decreasing VLDL secretion (Bilz et al. 2006; Gutierrez et al. 2006; Hirokane et al. 2004; Kast et al. 2001; Lambert et al. 2003; Ma et al. 2006; Shih et al. 2006; Sinal et al. 2000; Thomas et al. 2008; Vacca et al. 2011; Wagner et al. 2011; Watanabe et al. 2004b; Zhang et al. 2006a; Zhang et al. 2006b).

FXR also induces the LPL activity by decreasing the hepatic expression of LPL inhibitors (e.g. ApoC-III), while inducing the LPL cofactors apolipoprotein C-II (directly) and apolipoprotein A-V (ApoAV, mechanism PPARα-mediated), thus increasing the VLDL clearance (Bilz et al. 2006; Gutierrez et al. 2006; Hirokane et al. 2004; Kast et al. 2001; Lambert et al. 2003; Ma et al. 2006; Pineda et al. 2003; Savkur et al. 2005a; Shih et al. 2006; Sinal et al. 2000; Vacca et al. 2011; Wagner et al. 2011; Watanabe et al. 2004b; Zhang et al. 2006a; Zhang et al. 2006b). FXR activation also decreases total serum cholesterol levels, reduces the expression of apolipoprotein A-I (ApoA-I), and increases HDL remodeling (Angelin et al. 1978; Bateson et al. 1978; Claudel et al. 2002; Lambert et al. 2003; Leiss and Von 1982; Shepherd et al. 1979; Urizar et al. 2000; Watanabe et al. 2004b). Additionally, FXR decreases LDL-C plasma levels, enhancing the expression of the LDLr (Nakahara et al. 2002). In turn, the inhibition of CYP7A1 in liver results in increased hepatocyte cholesterol and oxysterols levels, that activates LXR, with a subsequent ubiquitination of the LDLr (Tontonoz and Mangelsdorf 2003b; Zelcer et al. 2009). Long-term CDCA-mediated FXR activation reduces plasma LDL-C levels (Cariou et al. 2006; Schoenfield and Lachin 1981; Wang et al. 2006; Watanabe et al. 2004b; Zhang et al. 2006a).

Conversely, FXR deficiency in rodents is associated with a massive cholesterol and TG deposition in the liver as well as elevated circulating FFA levels (Ma et al. 2006; Sinal et al. 2000; Zhang et al. 2006a; Zhang et al. 2006b), and BA-sequestrating agent treatment results in increased plasma TG and VLDL levels in dyslipidemic patients (Angelin et al. 1978; Beil et al. 1982; Crouse 1987). FXR-α also modulates glucose metabolism in SHP-dependent manner. SHP regulates energy and glucose homeostasis through loss of repression of PGC-1α; stimulates hepatic gluconeogenesis, inhibits glycolysis, increases energy expenditure, improves glucose uptake in skeleton muscle through GLUT4 and promotes insulin secretion in pancreatic β-cells (Claudel et al. 2005; Vacca et al. 2011; Wagner et al. 2011; Wang et al. 2005). In addition activation of FXR results in increased phosphorylation of IRS-1 and IRS-2, which increase insulin sensitivity and decrease glucose plasma levels

(Zhang et al. 2004). Finally FXR suppresses glycolysis and promotes FAO through the up-regulation of hepatic PDK4 (Savkur et al. 2005b).

3.3.3 Peroxisome Proliferator-Activated Receptor

PPARs are members of the NRs superfamily that act as "fatty acid sensors" to control metabolic programs and to regulate energy homeostasis (Tontonoz and Spiegelman 2008). Three PPARs isoforms have been identified in mammals, PPAR-α, -γ, -β/δ, with tissue-specific distribution and different effects in terms of gene expression. PPAR-α is predominantly expressed in tissues where active FA catabolism occurs (e.g. liver, brown fat, kidney, heart, and skeletal muscle), while PPAR-γ is mainly expressed in adipose tissue, pancreatic β-cells, with lower levels in the liver, cardiac and skeletal muscle. PPAR-α and -γ are also expressed in vascular endothelium, vascular smooth muscle, and macrophages/foam cells. PPAR-β/δ is ubiquitously expressed in many tissues. PPARs can accommodate a wide diversity of natural and synthetic compounds, including native and modified (oxidized and nitrated) FAs, eicosanoids, derivatives of PUFAs, fibrates and thiazolidinediones (TZDs) (Forman et al. 1995; Gottlicher et al. 1992; Kliewer et al. 1992; Kliewer et al. 1994; Nolte et al. 1998; Sohda et al. 1990; Spiegelman 1998).

In the liver, PPAR-α is a master transcriptional regulator of genes involved in hepatic production and uptake of FA, mitochondrial and peroxisomal β-oxidation of FA, inflammation, and vascular function (Bernal-Mizrachi et al. 2003; Koo et al. 2004; Reddy and Hashimoto 2001). PPAR-α lowers serum levels of TGs by various mechanisms; by inducing the LPL activity through the inhibition of hepatic ApoC-III (LPL inhibitor) and the activation of ApoAV (LPL cofactor), by enhancing the expression of gene involved in FA uptake through the activation of FAT/CD36, by increasing the VLDL-ApoB-100 catabolism and the FAO via activation of CPT-1, acyl-CoA dehydrogenase and cytochrome P450-ω-hydroxylase (Schoonjans et al. 1996; Staels et al. 1995). Activated PPAR-α also increases the hepatic expression of ApoA-I and apolipoprotein A-II (ApoA-II), which raise HDL cholesterol levels, and promote HDL mediated cholesterol efflux from macrophages through the induction of ABCA1 and the reduction of CETP levels (Chinetti et al. 2001; Shah et al. 2010).

PPAR-α may play a role in NAFLD; PPAR-α null mice fed with high fat diet develop fatty liver (Kersten et al. 1999; Lee et al. 1995). Moreover, PPAR-α is activated by intrahepatic FAs that have been synthesized *de novo* to maintain glucose and lipid homeostasis (Chakravarthy et al. 2005). On the basis of this knowledge pharmacological activation of PPAR-α could be beneficial for restoring the homeostasis of lipid metabolism and inflammation, and, thus, for hepatic steatosis. The fibrates are utilized in clinical practice for their hypolipidemic properties, especially in the treatment of hypertriglyceridemia and mixed hyperlipidemia.

The fibrates also improve liver steatosis in mouse models of NAFLD; however these findings are not fully confirmed in humans, probably due to species-differences between rodents and humans (Harano et al. 2006; Kallwitz et al. 2008). In fact, although fibrates ameliorate glucose and lipid profile in patients with NASH, weak effects on liver function test and histology have been confirmed (Athyros et al. 2006; Fernandez-Miranda et al. 2008; Laurin et al. 1996; Nakamuta et al. 2005b).

PPAR-γ regulates genes involved in adipocyte differentiation and proliferation, FA uptake and storage, lipolysis, inflammation, and insulin signaling. Activated PPAR-γ increases insulin sensitivity and promotes FA uptake and storage into adipose tissue, through the modulation of several genes such as LPL, FATP, FABP, FAT/CD36 and PEPCK, which reduce plasma levels of FFA and redirect FA from the liver to the adipose tissue (Kallwitz et al. 2008; Ribon et al. 1998; Shi et al. 2006; Wu et al. 1998). PPAR-γ activation in adipose tissue appears to be protective against hepatic steatosis through the induction of genes that participate in TG accumulation in adipocytes, which leads to a decreased storage of FAs in the liver. In addition, PPAR-γ activation in adipocytes increases the production of adiponectin, which stimulates FAO and glucose utilization, and improves insulin sensitivity by its activation of AMPK (You et al. 2005). The protective role of PPAR-γ is confirmed in humans with a dominant negative mutation of PPAR-γ, who develop NAFLD and MetS (Savage et al. 2003). PPAR-γ is a potential therapeutic target for IR and hepatic steatosis; thiazolidinediones (TZDs) are a class of PPAR-γ agonist widely used in type 2 diabetes patients which restore cell sensitivity to insulin through a direct effects on lipid metabolism in adipose tissue and by secondary effects on lipid and glucose metabolism in liver and skeletal muscle (Jiang et al. 2002; Way et al. 2001). TZDs promote a different adipocytes differentiation and the transcription of different sets of genes involved in FFA flux, which increase their uptake and storage in subcutaneous adipose tissue, rather than visceral tissue. The result is a reduction of serum FAs and TGs, and a decreased of IR and blood glucose levels. Moreover PPAR-γ is believed to increase glucose uptake into liver and skeletal muscle cells by the expression and translocation to the cell surface of the glucose transporter type 1 (GLUT1) and GLUT4 (Kramer et al. 2001). PPAR-γ agonist may also restore insulin sensitivity through the reduction of TNF-α (Cabrero et al. 2002) and the induction of adiponectin expression (Maeda et al. 2001). TZDs have been evaluated in NASH patients, with promising but not fully supported results (Chinetti et al. 2001; Mayerson et al. 2002; Sanyal et al. 2010).

PPAR-β/δ recognizes dietary FAs carried by VLDL, and regulates directly genes involved in glucose and FA homeostasis, and indirectly the inflammatory responses (Kersten et al. 2000). PPAR-β/δ regulates genes involved in FA transport and oxidation in adipocytes, such as long-chain acyl-CoA dehydrogenase, CPT-1, long-chain acyl-CoA synthetase. Overexpression of PPAR-β/δ in adipose tissue protects from diet-induced hypertriglyceridemia, obesity and hepatic steatosis (Wang et al. 2003). PPAR-β/δ is also involved in the modulation of lipoprotein metabolism; in animal models of atherosclerosis, PPAR-β/δ agonist GW501516 determines an increase of HDL levels, and a decreased of both LDL and sLDL (Leibowitz et al. 2000; Oliver Jr et al. 2001; Takata et al. 2008; Tanaka et al. 2003; Wallace et al. 2005). Although in some pilot studies GW501516 treatment has promoted a substantial increase in HDL (Riserus et al. 2008; Sprecher et al. 2007), the physiological relevance of

PPAR β/δ-mediated HDL-raising effect remains still elusive in humans. Future insights may come from newer agonists, such as KD-3010 and MBX-8025, currently in the pre-clinical phases of research (Billin 2008).

3.4 Liver External Factors Influencing Hepatic Function

The adipose tissue is a complex and potent endocrine organ, which expresses and secretes factors with important endocrine functions, such as cytokines, adipokines, and a series of hormonal peptides. In the presence of excess of visceral adipose tissue it has been observed a low-grade state of inflammation that results in increased production of FFA and pro-inflammatory cytokines, such as tumor necrosis factors-α (TNF-α) and interleukin-1, and a decreased production of adipokines. Systemic oxidative stress can also determine alterations in the production of adipokines. Among these proteins adiponectin has been found to play an important role in FFA metabolism (Xu et al. 2003). Plasma adiponectin concentrations have been found lower in patients with NAFLD, type 2 diabetes mellitus, IR, obesity and other conditions associated with MetS, when compared to age- and sex-matched controls, and inversely correlated with IR and hepatic steatosis (Bugianesi et al. 2005b; Hui et al. 2004; Shklyaev et al. 2003). Thus, low adiponectin levels may be involved in the worsening of both steatosis and IR (Bugianesi et al. 2005a). The main target of adiponectin action is the liver, where stimulates FAO and glucose utilization by its activation of AMPK (You et al. 2005), and PPAR-α. Adiponectin also inhibits hepatic glucose production through the inhibition of the expression of G-6-Pase (Capeau 2008). Moreover, the catabolism of VLDL and HDL is regulated by adiponectin; abnormal levels decrease the catabolism of VLDL and increase catabolism of HDL.

3.5 Deranged Metabolism is not Enough to Justify Progression of Hepatic Damage: Is the "Two Hits" Model Still Actual?

Only a minority of patients with NAFLD develop NASH. The factors responsible for the switch from NAFLD to NASH remain elusive. However, the proposed "two hits" model, although simplistic, provides a pathophysiological rationale to the progression of liver damage (Day and James 1998). According to the two mechanisms or toxic "hit" of NAFLD etiopathogenesis, the first mechanism is the reversible increase of the intrahepatic TG content, which leads to metabolic and molecular alteration that sensitize liver to the second hit, referred as oxidative stress and cytokines induce liver damage. As mentioned before, IR is believed to be a condition strongly associated to the pathogenesis of NAFLD, which involves multiple sites; the muscle, with decreased glucose uptake and utilization, the adipose tissue, with impaired suppression of adipose tissue lipolysis and increased release of glycerol and non-esterified fatty acids (NEFAs) into the systemic circulation, and the liver, with overproduction of glucose and increased DNL. In addition to increased efflux of FFAs from adipose tissue to the liver (Lewis et al. 2002) and DNL, another source of hepatic TG is represented by dietary intake; all these factors promote hepatic TG accumulation, which culminates in NAFLD (Donnelly et al. 2005).

However, in a minority of patients, a second hit, capable of inducing necrosis, inflammation and fibrosis is required for the development of NASH (Day and James 1998; James and Day 1999). The process by which NAFLD turns into NASH remains uncertain despite this double-hit hypothesis. Indeed, the progression of liver damage may depend on a complex interplay between exogenic factors (e.g. drugs, moderate amounts of alcohol) and genetic predisposition (similar to those responsible for the MetS) (Day 2006; Petta et al. 2009). Candidates for the second hit can be represented by oxidative stress, lipid peroxidation, and increased production of ROS, with subsequent inflammation and fibrosis (McCullough 2006). In patients with NASH, structural abnormalities in the mitochondria are more frequently observed compared to NAFLD, with a decreased activity of the mitochondrial respiratory chain complexes and an inefficient ATP production. Consequent lipotoxicity from SFAs, abnormal production of adipokines, such as increased levels of leptin (pro-inflammatory and pro-fibrotic), and decreased levels of adiponectin (anti-inflammatory and insulin-sensitising), increased production of pro-inflammatory cytokines, such as TNF-α, interleukin-6 and -1β, have been all evocated in causing the progression of hepatocyte injury (Dowman et al. 2010).

3.6 Epidemiological Studies Identify NAFLD as a Travel Companion of the Metabolic Syndrome

The prevalence of fatty liver is rapidly growing, tied closely to the enhanced prevalence of obesity, and it has become an emerging health problem and an increasingly common cause of cirrhosis in the economically developed nations of the world. The real prevalence remains unknown due to the absence of accurate and non-invasive diagnostic measures. It is estimated that approximately 20 % of the general population has NAFLD, whereas 3 % has NASH (Falck-Ytter et al. 2001); these prevalence rates are increased in certain subpopulations such as patients with obesity and type 2 diabetes. Recent epidemiological data have shown that fatty liver is closely associated with one or more variables of the MetS. NAFLD association with the MetS is higher than in general population; approximately 50–100 % of subjects with NASH are overweight, 50–60 % are hypertensive, and 50–60 % have dyslipidemias (Bacon et al. 1994). Autopsy studies on NAFLD patients have shown that steatosis occurs in 36 % of lean patients, and in 72 % of obese patients. Furthermore, NASH is present in up to 3 % of lean patients, and up to 19 % of obese patients (Wanless and Lentz 1990b). The prevalence of NAFLD may be also different in relation to the ethnic group. A population study on 2.287 multiethnic individuals using magnetic resonance imaging (MRI) to assess the presence of NAFLD has shown that prevalence was greater in whites (33 %) and Hispanics (45 %) compared to blacks (24 %) (Browning et al.

2004; Ruhl and Everhart 2003). This difference may be due to ethnic differences in lipid metabolism, excess adiposity, and IR (Browning et al. 2004; Romeo et al. 2008). Prevalence of NASH rises with age, and is greater between 50 and 60 years (Angulo et al. 1999), with a higher prevalence in females (60–83 %) (Ludwig et al. 1980; Matteoni et al. 1999). However recent epidemiological data have shown an increase in cases of NASH among children and among adult males. Browning et al. have shown that the fatty liver in the white subpopulation prevails in males (42 %) than women (24 %) (Browning et al. 2004). Age > 45 years represents an independent predictor factor for the risk of fibrosis and of progression of liver damage (Angulo et al. 1999; Matteoni et al. 1999). Probably age is associated to the duration of the liver injury and with the exposition to other factors which increase the risk of fibrosis.

3.7 Clinical Presentation of Hepatic Disease in the Metabolic Syndrome

The most common feature of hepatic disease in the MetS is NAFLD, which is believed one of the most frequent causes of abnormal liver enzymes. NAFLD is a silent disease, with a non-specific clinical presentation, and serum markers of hepatic damage often within the normal ranges. The assessment of steatosis is instrumental, while the diagnosis of NAFLD is made after exclusion of all other causes of fatty liver. NAFLD is mainly asymptomatic; right upper abdominal quadrant discomfort, unexplained hepatomegaly and chronic fatigue can be rarely recognized. Fatty liver (hyperechogenic liver at ultrasound scan, US, and/or hepatomegaly) is detected incidentally in case of imaging performed for other reasons, or when abnormal serum liver function tests (elevated alanine aminotransferase, ALT, and gamma-glutamyltransferase, GGT) are detected (Reid and Sanyal 2004). In this case NASH is strongly suspected. NASH is characterized by ALT usually higher than aspartate aminotransferase (AST) and, when AST:ALT ratio is greater than 1.0 it is suspicious of the progression through fibrosis and cirrhosis (Angulo et al. 1999); in this case albumin levels and bilirubinaemia should also be taken in account (Oh et al. 2008). GGT is also frequently elevated and has been proposed as marker of increased mortality (Ghouri et al. 2010; Haring et al. 2009). However, the presence of normal serum liver enzymes does not exclude the presence of NASH (Abrams et al. 2004). When NAFLD and NASH are diagnosed, a complete metabolic assessment of patient should be performed to investigate the presence of MetS and increased cardiovascular risk. Therefore, increased fasting serum glucose, serum insulin, triglyceridemia and cholesterol are common features in patients with NAFLD/NASH. On the other hand, individuals with obesity, diabetes and hypertriglyceridemia, and MetS should always be screened (liver enzymes/US) for NAFLD/NASH. Many imaging options can be used for the diagnosis of NAFLD, such as US, computed tomography (CT), MRI, and different non-invasive scoring systems for histology prediction (Joy et al. 2003; Saadeh et al. 2002). US is the easiest and cheapest diagnostic technique, documenting an increased echodensity of the liver (e.g. "brilliant liver"). However, the sensitivity of US is low, since hepatic steatosis can be detected when the lipid accumulation involves more

Table 3.2 Brunt scoring system for NASH. (Brunt et al. 1999)

Grade2	Steatosis	Balloning of hepatocytes	Degree of inflammation
1	$\leq$ 33 %	Minimal	Lobular mild-moderate, portal absent-mild
2	33–66 %	Present	Lobular mild-moderate, portal mild-moderate
3	$\geq$ 66 %	Marked	Lobular severe, portal mild-moderate

Table 3.3 Brunt scoring for fibrosis

Stage	Fibrosis
1	Focal or extensive perisinusoidal fibrosis
2	Focal or extensive perisinusoidal and periportal fibrosis
3	Bridging fibrosis
4	Extensive bridging fibrosis, cirrhosis

than 30 % of the hepatic tissue, and US cannot assess the presence and/or degree of inflammation and fibrosis (Harrison et al. 2003). On the other hand, the costs of CT and MRI and their inability into distinguish simple steatosis, from steatohepatitis, and/or fibrosis, do not motivate their common use. Several non-invasive diagnostic techniques and scoring systems have been developed with the aim of increasing the non-invasive assessment of liver fibrosis and/or the presence of NASH. This is the case of the elastography (FibroScan) that, assessing the strength (elasticity) of the liver, can recognize the increased parenchymal echo texture and vascular blurring (Sanyal 2002), and of different mathematical algorithms proposed as surrogate markers of NASH progression. Some examples are the APRI index (aspartate transaminase AST-to platelet ratio index) (Cales et al. 2009; Shaheen and Myers 2007; Shaheen and Myers 2008), the Fibro-Test (Poynard et al. 2007) (combining haptoglobin, α2-macroglobulin, ApoA-I, GGT, and total bilirubin to predict inflammation, steatosis and fibrosis in NAFLD), the Hepascore (predicting the extent of liver fibrosis or cirrhosis in adults through the combination of bilirubin, GGT, hyaluronic acid, α2-macroglobulin) (Stauber and Lackner 2007), and the Bard score (body mass index, AST/ALT ratio, presence of diabetes), which is a simple scoring system used to identify NAFLD in patients with a low risk of advanced disease (Harrison et al. 2008). All these surrogate markers may support the clinical decisions but, to date, cannot replace the liver biopsy. Liver biopsy represents the gold standard for the diagnosis, providing the degree of steatosis, the presence of necroinflammatory lesions and of fibrosis, which are the distinctive features of NASH from simple steatosis (Brunt 2004). Brunt scoring system is the most accepted score for grading the histological degree of NASH, and is based on four main histological lesions: steatosis, hepatocyte ballooning, inflammation, and fibrosis (Brunt et al. 1999). See Table 3.2 and 3.3 for details.

Table 3.4 Pharmacological treatment of NAFLD

Weight loss agents	Orlistat
Insulin sensitizing agents	Metformin
	Thiazolidinediones
Lipid-lowering agents	Statins
	Fibrate
	Omega-3 fatty acids
	Probucol
Anti-oxidants	Vitamin E
	Vitamin C
	Probiotics
Hepatoprotective agents	Ursodeoxycholic acid
	Pentoxyfylline

3.8 From Pathophysiology to Treatment

Current treatments for MetS and NAFLD are focused on weight loss, amelioration of the metabolic balance, reduction of cardiovascular risk, and limiting disease progression. Due to a poor compliance of the patients, the success rates of this treatment are not satisfactory; lifestyle changes are very difficult in an "obeso-genic" environment. The first line approach to patients with MetS and NASH are lifestyle changes, namely diet and increased aerobic physical exercise, in order to lose weight and improve insulin sensitivity (Neuschwander-Tetri 2009). Weight loss represents the mainstay in the non-pharmacological treatment in MetS and NAFLD patients. Weight loss should not be fast in order to prevent the risk of progression of liver damage (Grattagliano et al. 2000; James and Day 1998). Patients should lose 5–10 % of their baseline weight over 6–12 months, at a rate of 0.5–1.5 kg/week (Okita et al. 2001). Gradual weight loss and increased physical activity in obese patients with liver disease result in improvement of serum insulin, markers of hepatic disease progression (liver enzymes, steatosis, inflammation, and fibrosis), and in quality of life (Andersen et al. 1991; Dixon et al. 2004; Hickman et al. 2004). In the treatment of MetS and NAFLD there is no diet that can be considered a "gold standard". The best dietary strategy probably consists in eating less, mainly healthy foods. It should be limited the consumption of foods enriched in SFAs and carbohydrates at high glycemic index, while it should be encouraged the consumption of vegetables, legumes, fruit, soy derivatives, and fish. Alcohol, smoke, and potentially hepatotoxic medications should be also avoided. The validated pharmacological approaches in the treatment of MetS and NAFLD are summarized in Table 3.4. In the management of NAFLD, the only pharmacological approaches that showed certain levels of success are antioxidants and drugs improving IR (McClain et al. 2004; Sanyal et al. 2001).

3.9 Conclusions

The liver is the metabolic workhouse responsible for the maintenance of the whole body lipid and glucose homeostasis. Liver impairment may have its origin in hyperinsulinemia and hyperglycemia, which lead to hepatic IR that could affect peripheral sites; alternatively peripheral decreased insulin sensitivity may cause both adipose and hepatic IR. Fatty liver and IR are strongly associated, and both increase the prevalence of CVD. Future clinical and basic research should address the molecular mechanisms that guide this vicious cycle in order to provide novel therapeutic measures for MetS.

References

Third Report of the National Cholesterol Education Program (NCEP) Expert Panel on Detection, Evaluation, and Treatment of High Blood Cholesterol in Adults (Adult Treatment Panel III) (2002) final report. Circulation 106(25):3143–3421 (PM:12485966)

Aarsland A, Wolfe RR (1998) Hepatic secretion of VLDL fatty acids during stimulated lipogenesis in men. J Lipid Res 39(6):1280–1286 (PM:9643360)

Abrams GA, Kunde SS, Lazenby AJ, Clements RH (2004) Portal fibrosis and hepatic steatosis in morbidly obese subjects: a spectrum of nonalcoholic fatty liver disease. Hepatology 40(2):475–483 (PM:15368453)

Adiels M, Taskinen MR, Packard C, Caslake MJ, Soro-Paavonen A, Westerbacka J, Vehkavaara S, Hakkinen A, Olofsson SO, Yki-Jarvinen H, Boren J (2006) Overproduction of large VLDL particles is driven by increased liver fat content in man. Diabetologia 49(4):755–765 (PM:16463046)

Ai ZL, Chen DF (2007). The significance and effects of liver X receptor alpha in nonalcoholic fatty liver disease in rats. Zhonghua Gan Zang Bing Za Zhi 15(2):127–130 (PM:17362639)

Altomonte J, Cong L, Harbaran S, Richter A, Xu J, Meseck M, Dong HH (2004) Foxo1 mediates insulin action on apoC-III and triglyceride metabolism. J Clin Invest 114(10):1493–1503 (PM:15546000)

Andersen T, Gluud C, Franzmann MB, Christoffersen P (1991) Hepatic effects of dietary weight loss in morbidly obese subjects. J Hepatol 12(2):224–229 (PM:2051001)

Angelin B, Einarsson K, Hellstrom K, Leijd B (1978) Effects of cholestyramine and chenodeoxycholic acid on the metabolism of endogenous triglyceride in hyperlipoproteinemia. J Lipid Res 19(8):1017–1024 (PM:731123)

Angulo P, Keach JC, Batts KP, Lindor KD (1999) Independent predictors of liver fibrosis in patients with nonalcoholic steatohepatitis. Hepatology 30(6):1356–1362 (PM:10573511)

Athyros VG, Mikhailidis DP, Didangelos TP, Giouleme OI, Liberopoulos EN, Karagiannis A, Kakafika AI, Tziomalos K, Burroughs AK, Elisaf MS (2006) Effect of multifactorial treatment on non-alcoholic fatty liver disease in metabolic syndrome: a randomised study. Curr Med Res.Opin 22(5):873–883 (PM:16709309)

Bacon BR, Farahvash MJ, Janney CG, Neuschwander-Tetri BA (1994) Nonalcoholic steatohepatitis: an expanded clinical entity. Gastroenterology 107(4):1103–1109 (PM:7523217)

Balkau B, Charles MA (1999) Comment on the provisional report from the WHO consultation. European Group for the Study of Insulin Resistance (EGIR). Diabet Med 16(5):442–443 (PM:10342346)

Barbier O, Trottier J, Kaeding J, Caron P, Verreault M (2009) Lipid-activated transcription factors control bile acid glucuronidation. Mol Cell Biochem 326(1–2):3–8 (PM:19130183)

Bateson MC, Maclean D, Evans JR, Bouchier IA (1978) Chenodeoxycholic acid therapy for hypertriglyceridaemia in men. Br J Clin Pharmacol 5(3):249–254 (PM:656270)

Beil U, Crouse JR, Einarsson K, Grundy SM (1982) Effects of interruption of the enterohepatic circulation of bile acids on the transport of very low density-lipoprotein triglycerides. Metabolism 31(5):438–444 (PM:7078426)

Bell GD, Lewis B, Petrie A, Dowling RH (1973) Serum lipids in cholelithiasis: effect of chenodeoxycholic acid therapy. Br Med J 3(5879):520–523 (PM:4741607)

Bernal-Mizrachi C, Weng S, Feng C, Finck BN, Knutsen RH, Leone TC, Coleman T, Mecham RP, Kelly DP, Semenkovich CF (2003) Dexamethasone induction of hypertension and diabetes is PPAR-alpha dependent in LDL receptor-null mice. Nat Med 9(8):1069–1075 (PM:12847522)

Billin AN (2008) PPAR-beta/delta agonists for Type 2 diabetes and dyslipidemia: an adopted orphan still looking for a home. Expert Opin Investig Drugs 17(10):1465–1471 (PM:18808307)

Bilz S, Samuel V, Morino K, Savage D, Choi CS, Shulman GI (2006) Activation of the farnesoid X receptor improves lipid metabolism in combined hyperlipidemic hamsters. Am J Physiol Endocrinol Metab 290(4):E716–E722 (PM:16291572)

Bischof MG, Bernroider E, Krssak M, Krebs M, Stingl H, Nowotny P, Yu C, Shulman GI, Waldhausl W, Roden M (2002) Hepatic glycogen metabolism in type 1 diabetes after long-term near normoglycemia. Diabetes 51(1):49–54 (PM:11756322)

Boden G (2003) Effects of free fatty acids on gluconeogenesis and glycogenolysis. Life Sci 72(9):977–988 (PM:12495777)

Boden G, Shulman GI (2002) Free fatty acids in obesity and type 2 diabetes: defining their role in the development of insulin resistance and beta-cell dysfunction. Eur J Clin Invest 32(Suppl 3):14–23 (PM:12028371)

Bookout AL, Jeong Y, Downes M, Yu RT, Evans RM, Mangelsdorf DJ (2006) Anatomical profiling of nuclear receptor expression reveals a hierarchical transcriptional network. Cell 126(4):789–799 (PM:16923397)

Brown MS, Goldstein JL (1986) A receptor-mediated pathway for cholesterol homeostasis. Science 232(4746):34–47 (PM:3513311)

Browning JD, Horton JD (2004) Molecular mediators of hepatic steatosis and liver injury. J Clin Invest 114(2):147–152 (PM:15254578)

Browning JD, Szczepaniak LS, Dobbins R, Nuremberg P, Horton JD, Cohen JC, Grundy SM, Hobbs HH (2004) Prevalence of hepatic steatosis in an urban population in the United States: impact of ethnicity. Hepatology 40(6):1387–1395 (PM:15565570)

Brunt EM (2004) Nonalcoholic steatohepatitis. Semin Liver Dis 24(1):3–20 (PM:15085483)

Brunt EM, Janney CG, Di Bisceglie AM, Neuschwander-Tetri BA, Bacon BR (1999) Nonalcoholic steatohepatitis: a proposal for grading and staging the histological lesions. Am J Gastroenterol 94(9):2467–2474 (PM:10484010)

Bugianesi E (2005) Review article: steatosis, the metabolic syndrome and cancer. Aliment Pharmacol Ther 22(Suppl 2) 40–43 (PM:16225471)

Bugianesi E, Gastaldelli A, Vanni E, Gambino R, Cassader M, Baldi S, Ponti V, Pagano G, Ferrannini E, Rizzetto M (2005a) Insulin resistance in non-diabetic patients with non-alcoholic fatty liver disease: sites and mechanisms. Diabetologia 48(4):634–642 (PM:15747110)

Bugianesi E, Pagotto U, Manini R, Vanni E, Gastaldelli A, de Iasio R, Gentilcore E, Natale S, Cassader M, Rizzetto M, Pasquali R, Marchesini G (2005b) Plasma adiponectin in nonalcoholic fatty liver is related to hepatic insulin resistance and hepatic fat content, not to liver disease severity. J Clin Endocrinol Metab 90(6):3498–3504 (PM:15797948)

Caballero F, Fernandez A, De Lacy AM, Fernandez-Checa JC, Caballeria J, Garcia-Ruiz C (2009) Enhanced free cholesterol, SREBP-2 and StAR expression in human NASH. J Hepatol 50(4):789–796 (PM:19231010)

Cabrero A, Laguna JC, Vazquez M (2002) Peroxisome proliferator-activated receptors and the control of inflammation. Curr Drug Targets Inflamm Allergy 1(3):243–248 (PM:14561188)

Caldwell SH, Oelsner DH, Iezzoni JC, Hespenheide EE, Battle EH, Driscoll CJ (1999) Cryptogenic cirrhosis: clinical characterization and risk factors for underlying disease. Hepatology 29(3):664–669 (PM:10051466)

Cales P, Laine F, Boursier J, Deugnier Y, Moal V, Oberti F, Hunault G, Rousselet MC, Hubert I, Laafi J, Ducluzeaux PH, Lunel F (2009) Comparison of blood tests for liver fibrosis specific or not to NAFLD. J Hepatol 50(1):165–173 (PM:18977552)

Cao G, Liang Y, Broderick CL, Oldham BA, Beyer TP, Schmidt RJ, Zhang Y, Stayrook KR, Suen C, Otto KA, Miller AR, Dai J, Foxworthy P, Gao H, Ryan TP, Jiang XC, Burris TP, Eacho PI, etgen

GJ (2003) Antidiabetic action of a liver X receptor agonist mediated by inhibition of hepatic gluconeogenesis. J Biol Chem 278(2):1131–1136 (PM:12414791)

Capeau J (2008) Insulin resistance and steatosis in humans. Diabetes Metab 34(6 Pt 2):649–657 (PM:19195626)

Cariou B, van Harmelen K, Duran-Sandoval D, van Dijk TH, Grefhorst A, Abdelkarim M, Caron S, Torpier G, Fruchart JC, Gonzalez FJ, Kuipers F, Staels B (2006) The farnesoid X receptor modulates adiposity and peripheral insulin sensitivity in mice. J Biol Chem 281(16):11039–11049 (PM:16446356)

Caron S, Verrijken A, Mertens I, Samanez CH, Mautino G, Haas JT, Duran-Sandoval D, Prawitt J, Francque S, Vallez E, Muhr-Tailleux A, Berard I, Kuipers F, Kuivenhoven JA, Biddinger SB, Taskinen MR, Van GL, Staels B (2011) Transcriptional activation of apolipoprotein CIII expression by glucose may contribute to diabetic dyslipidemia. Arterioscler Thromb Vasc Biol 31(3):513–519 (PM:21183731)

Carulli N, Ponz de Leon M, Podda M, Zuin M, Strata A, Frigerio G, Digrisolo A (1981) Chenodeoxycholic acid and ursodeoxycholic acid effects in endogenous hypertriglyceridemias. A controlled double-blind trial. J Clin Pharmacol 21(10):436–442 (PM:7031100)

Cha JY, Repa JJ (2007) The liver X receptor (LXR) and hepatic lipogenesis. The carbohydrate-response element-binding protein is a target gene of LXR. J Biol Chem 282(1):743–751 (PM:17107947)

Chakravarthy MV, Pan Z, Zhu Y, Tordjman K, Schneider JG, Coleman T, Turk J, Semenkovich CF (2005) "New" hepatic fat activates PPARalpha to maintain glucose, lipid, and cholesterol homeostasis. Cell Metab 1(5):309–322 (PM:16054078)

Chang X, Yan H, Fei J, Jiang M, Zhu H, Lu D, Gao X (2010) Berberine reduces methylation of the MTTP promoter and alleviates fatty liver induced by a high-fat diet in rats. J Lipid Res, 51(9):2504–2515 (PM:20567026)

Chawla A, Boisvert WA, Lee CH, Laffitte BA, Barak Y, Joseph SB, Liao D, Nagy L, Edwards PA, Curtiss LK, Evans RM, Tontonoz P (2001a) A PPAR gamma-LXR-ABCA1 pathway in macrophages is involved in cholesterol efflux and atherogenesis. Mol Cell 7(1):161–171 (PM:11172721)

Chawla A, Repa JJ, Evans RM, Mangelsdorf DJ (2001b) Nuclear receptors and lipid physiology: opening the X-files. Science 294(5548):1866–1870 (PM:11729302)

Chen M, Breslow JL, Li W, Leff T (1994) Transcriptional regulation of the apoC-III gene by insulin in diabetic mice: correlation with changes in plasma triglyceride levels. J Lipid Res 35(11):1918–1924 (PM:7868970)

Chinetti G, Lestavel S, Bocher V, Remaley AT, Neve B, Torra IP, Teissier E, Minnich A, Jaye M, Duverger N, Brewer HB, Fruchart JC, Clavey V, Staels B (2001) PPAR-alpha and PPAR-gamma activators induce cholesterol removal from human macrophage foam cells through stimulation of the ABCA1 pathway. Nat Med 7(1):53–58 (PM:11135616)

Claudel T, Staels B, Kuipers F (2005) The Farnesoid X receptor: a molecular link between bile acid and lipid and glucose metabolism. Arterioscler Thromb Vasc Biol 25(10):2020–2030 (PM:16037564)

Claudel T, Sturm E, Duez H, Torra IP, Sirvent A, Kosykh V, Fruchart JC, Dallongeville J, Hum DW, Kuipers F, Staels B (2002) Bile acid-activated nuclear receptor FXR suppresses apolipoprotein A-I transcription via a negative FXR response element. J Clin Invest 109(7):961–971 (PM:11927623)

Cohn JS, Patterson BW, Uffelman KD, Davignon J, Steiner G (2004). Rate of production of plasma and very-low-density lipoprotein (VLDL) apolipoprotein C-III is strongly related to the concentration and level of production of VLDL triglyceride in male subjects with different body weights and levels of insulin sensitivity. J Clin Endocrinol Metab 89(8):3949–3955 (PM:15292332)

Crouse JR III (1987) Hypertriglyceridemia: a contraindication to the use of bile acid binding resins. Am J Med 83(2):243–248 (PM:3618626)

Dashti N, Williams DL, Alaupovic P (1989) Effects of oleate and insulin on the production rates and cellular mRNA concentrations of apolipoproteins in HepG2 cells. J Lipid Res 30(9):1365–1373 (PM:2689548)

Day CP (2006) From fat to inflammation. Gastroenterology 130(1):207–210 (PM:16401483)

Day CP, James OF (1998) Steatohepatitis: a tale of two "hits"? Gastroenterology 114(4):842–845 (PM:9547102)

Denechaud PD, Bossard P, Lobaccaro JM, Millatt L, Staels B, Girard J, Postic C (2008) ChREBP, but not LXRs, is required for the induction of glucose-regulated genes in mouse liver. J Clin Invest 118(3):956–964 (PM:18292813)

Dixon JB, Bhathal PS, Hughes NR, O'Brien PE (2004) Nonalcoholic fatty liver disease: improvement in liver histological analysis with weight loss. Hepatology 39(6):1647–1654 (PM:15185306)

Doege H, Baillie RA, Ortegon AM, Tsang B, Wu Q, Punreddy S, Hirsch D, Watson N, Gimeno RE, Stahl A (2006) Targeted deletion of FATP5 reveals multiple functions in liver metabolism: alterations in hepatic lipid homeostasis. Gastroenterology 130(4):1245–1258 (PM:16618416)

Donnelly KL, Smith CI, Schwarzenberg SJ, Jessurun J, Boldt MD, Parks EJ (2005) Sources of fatty acids stored in liver and secreted via lipoproteins in patients with nonalcoholic fatty liver disease. J Clin Invest 115(5):1343–1351 (PM:15864352)

Dowman JK, Tomlinson JW, Newsome PN (2010) Pathogenesis of non-alcoholic fatty liver disease. QJM 103(2):71–83 (PM:19914930)

Duval C, Touche V, Tailleux A, Fruchart JC, Fievet C, Clavey V, Staels B, Lestavel S (2006) Niemann-Pick C1 like 1 gene expression is down-regulated by LXR activators in the intestine. Biochem Biophys Res Commun 340(4):1259–1263 (PM:16414355)

Fabbrini E, Mohammed BS, Magkos F, Korenblat KM, Patterson BW, Klein S (2008) Alterations in adipose tissue and hepatic lipid kinetics in obese men and women with nonalcoholic fatty liver disease. Gastroenterology 134(2):424–431 (PM:18242210)

Fabbrini E, Sullivan S, Klein S (2010) Obesity and nonalcoholic fatty liver disease: biochemical, metabolic, and clinical implications. Hepatology 51(2):679–689 (PM:20041406)

Falck-Ytter Y, Younossi ZM, Marchesini G, McCullough AJ (2001) Clinical features and natural history of nonalcoholic steatosis syndromes. Semin Liver Dis 21(1):17–26 (PM:11296693)

Fernandez-Miranda C, Perez-Carreras M, Colina F, Lopez-Alonso G, Vargas C, Solis-Herruzo JA (2008) A pilot trial of fenofibrate for the treatment of non-alcoholic fatty liver disease. Dig Liver Dis 40(3):200–205 (PM:18261709)

Foretz M, Pacot C, Dugail I, Lemarchand P, Guichard C, Le Lièpvre X, Berthelier-Lubrano C, Spiegelman B, Kim JB, Ferre P, Foufelle F (1999) ADD1/SREBP-1c is required in the activation of hepatic lipogenic gene expression by glucose. Mol Cell Biol 19(5):3760–3768 (PM:10207099)

Forman BM, Tontonoz P, Chen J, Brun RP, Spiegelman BM, Evans RM (1995) 15-Deoxy-delta 12, 14-prostaglandin J2 is a ligand for the adipocyte determination factor PPAR gamma. Cell 83(5):803–812 (PM:8521497)

Foufelle F, Ferre P (2002) New perspectives in the regulation of hepatic glycolytic and lipogenic genes by insulin and glucose: a role for the transcription factor sterol regulatory element binding protein-1c. Biochem J 366(Pt 2):377–391 (PM:12061893)

Gastaldelli A, Cusi K, Pettiti M, Hardies J, Miyazaki Y, Berria R, Buzzigoli E, Sironi AM, Cersosimo E, Ferrannini E, Defronzo RA (2007) Relationship between hepatic/visceral fat and hepatic insulin resistance in nondiabetic and type 2 diabetic subjects. Gastroenterology 133(2):496–506 (PM:17681171)

Gastaldelli A, Toschi E, Pettiti M, Frascerra S, Quinones-Galvan A, Sironi AM, Natali A, Ferrannini E (2001) Effect of physiological hyperinsulinemia on gluconeogenesis in nondiabetic subjects and in type 2 diabetic patients. Diabetes 50(8):1807–1812 (PM:11473042)

Ghouri N, Preiss D, Sattar N (2010) Liver enzymes, nonalcoholic fatty liver disease, and incident cardiovascular disease: a narrative review and clinical perspective of prospective data. Hepatology 52(3):1156–1161 (PM:20658466)

Ginsberg HN, Brown WV (2011) Apolipoprotein CIII: 42 years old and even more interesting. Arterioscler Thromb Vasc Biol 31(3):471–473 (PM:21325666)
Gottlicher M, Widmark E, Li Q, Gustafsson JA (1992) Fatty acids activate a chimera of the clofibric acid-activated receptor and the glucocorticoid receptor. Proc Natl Acad Sci U S A 89(10):4653–4657 (PM:1316614)
Grattagliano I, Vendemiale G, Caraceni P, Domenicali M, Nardo B, Cavallari A, Trevisani F, Bernardi M, Altomare E (2000) Starvation impairs antioxidant defense in fatty livers of rats fed a choline-deficient diet. J Nutr 130(9):2131–2136 (PM:10958803)
Greco D, Kotronen A, Westerbacka J, Puig O, Arkkila P, Kiviluoto T, Laitinen S, Kolak M, Fisher RM, Hamsten A, Auvinen P, Yki-Jarvinen H (2008) Gene expression in human NAFLD. Am J Physiol Gastrointest Liver Physiol 294(5):G1281–G1287 (PM:18388185)
Grundy SM, Cleeman JI, Daniels SR, Donato KA, Eckel RH, Franklin BA, Gordon DJ, Krauss RM, Savage PJ, Smith SC Jr, Spertus JA, Costa F (2005) Diagnosis and management of the metabolic syndrome: an American Heart Association/National Heart, Lung, and Blood Institute Scientific Statement. Circulation 112(17):2735–2752 (PM:16157765)
Guettier JM, Georgopoulos A, Tsai MY, Radha V, Shanthirani S, Deepa R, Gross M, Rao G, Mohan V (2005) Polymorphisms in the fatty acid-binding protein 2 and apolipoprotein C-III genes are associated with the metabolic syndrome and dyslipidemia in a South Indian population. J Clin Endocrinol Metab 90(3):1705–1711 (PM:15598690)
Gutierrez A, Ratliff EP, Andres AM, Huang X, McKeehan WL, Davis RA (2006) Bile acids decrease hepatic paraoxonase 1 expression and plasma high-density lipoprotein levels via FXR-mediated signaling of FGFR4. Arterioscler Thromb Vasc Biol 26(2):301–306 (PM:16284190)
Hamaguchi M, Kojima T, Takeda N, Nagata C, Takeda J, Sarui H, Kawahito Y, Yoshida N, Suetsugu A, Kato T, Okuda J, Ida K, Yoshikawa T (2007) Nonalcoholic fatty liver disease is a novel predictor of cardiovascular disease. World J Gastroenterol 13(10):1579–1584 (PM:17461452)
Harano Y, Yasui K, Toyama T, Nakajima T, Mitsuyoshi H, Mimani M, Hirasawa T, Itoh Y, Okanoue T (2006) Fenofibrate, a peroxisome proliferator-activated receptor alpha agonist, reduces hepatic steatosis and lipid peroxidation in fatty liver Shionogi mice with hereditary fatty liver. Liver Int 26(5):613–620 (PM:16762007)
Haring R, Wallaschofski H, Nauck M, Dorr M, Baumeister SE, Volzke H (2009) Ultrasonographic hepatic steatosis increases prediction of mortality risk from elevated serum gamma-glutamyl transpeptidase levels. Hepatology 50(5):1403–1411 (PM:19670414)
Harrison SA, Oliver D, Arnold HL, Gogia S, Neuschwander-Tetri BA (2008) Development and validation of a simple NAFLD clinical scoring system for identifying patients without advanced disease. Gut 57(10):1441–1447 (PM:18390575)
Harrison SA, Torgerson S, Hayashi PH (2003) The natural history of nonalcoholic fatty liver disease: a clinical histopathological study. Am J Gastroenterol 98(9):2042–2047 (PM:14499785)
Hickman IJ, Jonsson JR, Prins JB, Ash S, Purdie DM, Clouston AD, Powell EE (2004) Modest weight loss and physical activity in overweight patients with chronic liver disease results in sustained improvements in alanine aminotransferase, fasting insulin, and quality of life. Gut 53(3):413–419 (PM:14960526)
Higuchi N, Kato M, Shundo Y, Tajiri H, Tanaka M, Yamashita N, Kohjima M, Kotoh K, Nakamuta M, Takayanagi R, Enjoji M (2008) Liver X receptor in cooperation with SREBP-1c is a major lipid synthesis regulator in nonalcoholic fatty liver disease. Hepatol Res 38(11):1122–1129 (PM:18684130)
Hirokane H, Nakahara M, Tachibana S, Shimizu M, Sato R (2004) Bile acid reduces the secretion of very low density lipoprotein by repressing microsomal triglyceride transfer protein gene expression mediated by hepatocyte nuclear factor-4. J Biol Chem 279(44):45685–45692 (PM:15337761)
Horton JD, Shah NA, Warrington JA, Anderson NN, Park SW, Brown MS, Goldstein JL (2003) Combined analysis of oligonucleotide microarray data from transgenic and knockout mice identifies direct SREBP target genes. Proc Natl Acad Sci U S A 100(21):12027–12032 (PM:14512514)

Horton JD, Shimano H, Hamilton RL, Brown MS, Goldstein JL (1999) Disruption of LDL receptor gene in transgenic SREBP-1a mice unmasks hyperlipidemia resulting from production of lipid-rich VLDL. J Clin Invest 103(7):1067–1076 (PM:10194480)

Hui JM, Hodge A, Farrell GC, Kench JG, Kriketos A, George J (2004) Beyond insulin resistance in NASH: TNF-alpha or adiponectin? Hepatology 40(1):46–54 (PM:15239085)

Ide T, Shimano H, Yahagi N, Matsuzaka T, Nakakuki M, Yamamoto T, Nakagawa Y, Takahashi A, Suzuki H, Sone H, Toyoshima H, Fukamizu A, Yamada N (2004) SREBPs suppress IRS-2-mediated insulin signalling in the liver. Nat Cell Biol 6(4):351–357 (PM:15048126)

Iizuka K, Horikawa Y (2008) ChREBP: a glucose-activated transcription factor involved in the development of metabolic syndrome. Endocr J 55(4):617–624 (PM:18490833)

James O, Day C (1999) Non-alcoholic steatohepatitis: another disease of affluence. Lancet 353(9165):1634–1636 (PM:10335777)

James OF, Day CP (1998) Non-alcoholic steatohepatitis (NASH): a disease of emerging identity and importance. J Hepatol 29(3):495–501 (PM:9765002)

Janowski BA, Willy PJ, Devi TR, Falck JR, Mangelsdorf DJ (1996) An oxysterol signalling pathway mediated by the nuclear receptor LXR alpha. Nature 383(6602):728–731 (PM:8878485)

Jiang G, Dallas-Yang Q, Li Z, Szalkowski D, Liu F, Shen X, Wu M, Zhou G, Doebber T, Berger J, Moller DE, Zhang BB (2002) Potentiation of insulin signaling in tissues of Zucker obese rats after acute and long-term treatment with PPARgamma agonists. Diabetes 51(8):2412–2419 (PM:12145152)

Jong MC, Hofker MH, Havekes LM (1999) Role of ApoCs in lipoprotein metabolism: functional differences between ApoC1, ApoC2, and ApoC3. Arterioscler Thromb Vasc Biol 19(3):472–484 (PM:10073946)

Joseph SB, Laffitte BA, Patel PH, Watson MA, Matsukuma KE, Walczak R, Collins JL, Osborne TF, Tontonoz P (2002) Direct and indirect mechanisms for regulation of fatty acid synthase gene expression by liver X receptors. J Biol Chem 277(13):11019–11025 (PM:11790787)

Joy D, Thava VR, Scott BB (2003). Diagnosis of fatty liver disease: is biopsy necessary? Eur J Gastroenterol Hepatol 15(5):539–543 (PM:12702913)

Kabashima T, Kawaguchi T, Wadzinski BE, Uyeda K (2003) Xylulose 5-phosphate mediates glucose-induced lipogenesis by xylulose 5-phosphate-activated protein phosphatase in rat liver. Proc Natl Acad Sci U S A 100(9):5107–5112 (PM:12684532)

Kalaany NY, Gauthier KC, Zavacki AM, Mammen PP, Kitazume T, Peterson JA, Horton JD, Garry DJ, Bianco AC, Mangelsdorf DJ (2005) LXRs regulate the balance between fat storage and oxidation. Cell Metab 1(4):231–244 (PM:16054068)

Kallwitz ER, McLachlan A, Cotler SJ (2008) Role of peroxisome proliferators-activated receptors in the pathogenesis and treatment of nonalcoholic fatty liver disease. World J Gastroenterol 14(1):22–28 (PM:18176957)

Kase et, Wensaas AJ, Aas V, Hojlund K, Levin K, Thoresen GH, Beck-Nielsen H, Rustan AC, Gaster M (2005) Skeletal muscle lipid accumulation in type 2 diabetes may involve the liver X receptor pathway. Diabetes 54(4):1108–1115 (PM:15793250)

Kast HR, Nguyen CM, Sinal CJ, Jones SA, Laffitte BA, Reue K, Gonzalez FJ, Willson TM, Edwards PA (2001) Farnesoid X-activated receptor induces apolipoprotein C-II transcription: a molecular mechanism linking plasma triglyceride levels to bile acids. Mol Endocrinol 15(10):1720–1728 (PM:11579204)

Kawaguchi T, Osatomi K, Yamashita H, Kabashima T, Uyeda K (2002) Mechanism for fatty acid "sparing" effect on glucose-induced transcription: regulation of carbohydrate-responsive element-binding protein by AMP-activated protein kinase. J Biol Chem 277(6):3829–3835 (PM:11724780)

Kawaguchi T, Takenoshita M, Kabashima T, Uyeda K (2001) Glucose and cAMP regulate the L-type pyruvate kinase gene by phosphorylation/dephosphorylation of the carbohydrate response element binding protein. Proc Natl Acad Sci U S A 98(24):13710–13715 (PM:11698644)

Kersten S, Desvergne B, Wahli W (2000) Roles of PPARs in health and disease. Nature 405(6785):421–424 (PM:10839530)

Kersten S, Seydoux J, Peters JM, Gonzalez FJ, Desvergne B, Wahli W (1999) Peroxisome proliferator-activated receptor alpha mediates the adaptive response to fasting. J Clin Invest 103(11):1489–1498 (PM:10359558)

Kharroubi I, Ladriere L, Cardozo AK, Dogusan Z, Cnop M, Eizirik DL (2004) Free fatty acids and cytokines induce pancreatic beta-cell apoptosis by different mechanisms: role of nuclear factor-kappaB and endoplasmic reticulum stress. Endocrinology 145(11):5087–5096 (PM:15297438)

Kliewer SA, Forman BM, Blumberg B, Ong ES, Borgmeyer U, Mangelsdorf DJ, Umesono K, Evans, RM (1994) Differential expression and activation of a family of murine peroxisome proliferator-activated receptors. Proc Natl Acad Sci U S A 91(15):7355–7359 (PM:8041794)

Kliewer SA, Umesono K, Noonan DJ, Heyman RA, Evans RM (1992) Convergence of 9-cis retinoic acid and peroxisome proliferator signalling pathways through heterodimer formation of their receptors. Nature 358(6389):771–774 (PM:1324435)

Koo SH, Satoh H, Herzig S, Lee CH, Hedrick S, Kulkarni R, Evans RM, Olefsky J, Montminy M (2004) PGC-1 promotes insulin resistance in liver through PPAR-alpha-dependent induction of TRB-3. Nat Med 10(5):530–534 (PM:15107844)

Korchak HM (1962) Regulation of hepatic lipogenesis. Tufts Folia Med 8:134–143 (PM:14034742)

Korenblat KM, Fabbrini E, Mohammed BS, Klein S (2008) Liver, muscle, and adipose tissue insulin action is directly related to intrahepatic triglyceride content in obese subjects. Gastroenterology 134(5):1369–1375 (PM:18355813)

Kozlitina J, Boerwinkle E, Cohen JC, Hobbs HH (2011) Dissociation between APOC3 variants, hepatic triglyceride content and insulin resistance. Hepatology 53(2):467–474 (PM:21274868)

Kraegen EW, Clark PW, Jenkins AB, Daley EA, Chisholm DJ, Storlien LH (1991) Development of muscle insulin resistance after liver insulin resistance in high-fat-fed rats. Diabetes 40(11):1397–1403 (PM:1936601)

Kramer D, Shapiro R, Adler A, Bush E, Rondinone CM (2001) Insulin-sensitizing effect of rosiglitazone (BRL-49653) by regulation of glucose transporters in muscle and fat of Zucker rats. Metabolism 50(11):1294–1300 (PM:11699047)

Laffitte BA, Chao LC, Li J, Walczak R, Hummasti S, Joseph SB, Castrillo A, Wilpitz DC, Mangelsdorf DJ, Collins JL, Saez E, Tontonoz P (2003) Activation of liver X receptor improves glucose tolerance through coordinate regulation of glucose metabolism in liver and adipose tissue. Proc Natl Acad Sci U S A 100(9):5419–5424 (PM:12697904)

Lambert G, Amar MJ, Guo G, Brewer HB Jr, Gonzalez FJ, Sinal CJ (2003) The farnesoid X-receptor is an essential regulator of cholesterol homeostasis. J Biol Chem 278(4):2563–2570 (PM:12421815)

Lanz RB, McKenna NJ, Onate SA, Albrecht U, Wong J, Tsai SY, Tsai MJ, O'Malley BW (1999) A steroid receptor coactivator, SRA, functions as an RNA and is present in an SRC-1 complex. Cell 97(1):17–27 (PM:10199399)

Larter CZ, Chitturi S, Heydet D, Farrell GC (2010) A fresh look at NASH pathogenesis. Part 1: the metabolic movers. J Gastroenterol Hepatol 25(4):672–690 (PM:20492324)

Laurin J, Lindor KD, Crippin JS, Gossard A, Gores GJ, Ludwig J, Rakela J, McGill DB (1996) Ursodeoxycholic acid or clofibrate in the treatment of non-alcohol-induced steatohepatitis: a pilot study. Hepatology 23(6):1464–1467 (PM:8675165)

Lavoie JM, Gauthier MS (2006) Regulation of fat metabolism in the liver: link to non-alcoholic hepatic steatosis and impact of physical exercise. Cell Mol Life Sci 63(12):1393–1409 (PM:16649140)

Lee FY, Lee H, Hubbert ML, Edwards PA, Zhang Y (2006) FXR, a multipurpose nuclear receptor. Trends Biochem Sci 31(10):572–580 (PM:16908160)

Lee SS, Pineau T, Drago J., Lee EJ, Owens JW, Kroetz DL, Fernandez-Salguero PM, Westphal H, Gonzalez FJ (1995) Targeted disruption of the alpha isoform of the peroxisome proliferator-activated receptor gene in mice results in abolishment of the pleiotropic effects of peroxisome proliferators. Mol Cell Biol 15(6):3012–3022 (PM:7539101)

Leibowitz MD, Fievet C, Hennuyer N, Peinado-Onsurbe J, Duez H, Bergera J, Cullinan CA, Sparrow CP, Baffic J, Berger GD, Santini C, Marquis RW, Tolman RL, Smith RG, Moller DE, Auwerx J

(2000) Activation of PPARdelta alters lipid metabolism in db/db mice. FEBS Lett 473(3):333–336 (PM:10818235)

Leiss O, Von BK (1982) Different effects of chenodeoxycholic acid and ursodeoxycholic acid on serum lipoprotein concentrations in patients with radiolucent gallstones. Scand J Gastroenterol 17:(5) 587–592 (PM:7178821)

Lewis GF, Carpentier A, Adeli K, Giacca A (2002) Disordered fat storage and mobilization in the pathogenesis of insulin resistance and type 2 diabetes. Endocr Rev 23(2):201–229 (PM:11943743)

Li ZZ, Berk M, McIntyre TM, Feldstein AE (2009) Hepatic lipid partitioning and liver damage in nonalcoholic fatty liver disease: role of stearoyl-CoA desaturase. J Biol Chem 284(9):5637–5644 (PM:19119140)

Lonard DM, O'Malley BW (2007) Nuclear receptor coregulators: judges, juries, and executioners of cellular regulation. Mol Cell 27(5):691–700 (PM:17803935)

Lucero D, Zago V, Lopez GI, Graffigna M, Lopez GH, Fainboim H, Miksztowicz V, Rosso LG, Belli S, Levalle O, Berg G, Brites F, Wikinski R, Schreier L (2011) Does non-alcoholic fatty liver impair alterations of plasma lipoproteins and associated factors in metabolic syndrome? Clin Chim Acta 412(7–8) 587–592 (PM:21159313)

Ludwig J, Viggiano TR, McGill DB, Oh BJ (1980) Nonalcoholic steatohepatitis: Mayo Clinic experiences with a hitherto unnamed disease. Mayo Clin Proc 55(7):434–438 (PM:7382552)

Luo Y, Liang CP, Tall AR (2001) The orphan nuclear receptor LRH-1 potentiates the sterol-mediated induction of the human CETP gene by liver X receptor. J Biol Chem 276(27):24767–24773 (PM:11331284)

Ma K, Saha PK, Chan L, Moore DD (2006) Farnesoid X receptor is essential for normal glucose homeostasis. J Clin Invest 116(4):1102–1109 (PM:16557297)

Maeda N, Takahashi M, Funahashi T, Kihara S, Nishizawa H, Kishida K, Nagaretani H, Matsuda M, Komuro R, Ouchi N, Kuriyama H, Hotta K, Nakamura T, Shimomura I, Matsuzawa Y (2001) PPARgamma ligands increase expression and plasma concentrations of adiponectin, an adipose-derived protein. Diabetes 50(9):2094–2099 (PM:11522676)

Makishima M, Okamoto AY, Repa JJ, Tu H, Learned RM, Luk A, Hull MV, Lustig KD, Mangelsdorf DJ, Shan B (1999) Identification of a nuclear receptor for bile acids. Science 284(5418):1362–1365 (PM:10334992)

Mangelsdorf DJ, Thummel C, Beato M, Herrlich P, Schutz G, Umesono K, Blumberg B, Kastner P, Mark M, Chambon P, Evans RM (1995) The nuclear receptor superfamily: the second decade. Cell 83(6):835–839 (PM:8521507)

Marchesini G, Brizi M, Bianchi G, Tomassetti S, Bugianesi E, Lenzi M, McCullough AJ, Natale S, Forlani G, Melchionda N (2001) Nonalcoholic fatty liver disease: a feature of the metabolic syndrome. Diabetes 50(8):1844–1850 (PM:11473047)

Marrero JA, Fontana RJ, Su GL, Conjeevaram HS, Emick DM, Lok AS (2002) NAFLD may be a common underlying liver disease in patients with hepatocellular carcinoma in the United States. Hepatology 36(6):1349–1354 (PM:12447858)

Matteoni CA, Younossi ZM, Gramlich T, Boparai N, Liu YC, McCullough AJ (1999) Nonalcoholic fatty liver disease: a spectrum of clinical and pathological severity. Gastroenterology 116(6):1413–1419 (PM:10348825).

Mayerson AB, Hundal RS, Dufour S, Lebon V, Befroy D, Cline GW, Enocksson S, Inzucchi SE, Shulman GI, Petersen KF (2002) The effects of rosiglitazone on insulin sensitivity, lipolysis, and hepatic and skeletal muscle triglyceride content in patients with type 2 diabetes. Diabetes 51(3):797–802 (PM:11872682)

McClain CJ, Mokshagundam SP, Barve SS, Song Z, Hill DB, Chen T, Deaciuc I (2004) Mechanisms of non-alcoholic steatohepatitis. Alcohol 34(1):67–79 PM:15670668

McCullough AJ (2006) Pathophysiology of nonalcoholic steatohepatitis. J Clin Gastroenterol 40(Suppl 1):S17–S29 (PM:16540762)

McKenna NJ, O'Malley BW (2001) Nuclear receptors, coregulators, ligands, and selective receptor modulators: making sense of the patchwork quilt. Ann N Y Acad Sci 949:3–5 (PM:11795367)

McKenna NJ, O'Malley BW (2002) Minireview: nuclear receptor coactivators—an update. Endocrinology 143(7):2461–2465 (PM:12072374)

Miller M, Rhyne J, Chen H, Beach V, Ericson R, Luthra K, Dwivedi M, Misra A (2007) APOC3 promoter polymorphisms C-482 T and T-455°C are associated with the metabolic syndrome. Arch Med Res 38(4):444–451 (PM:17416293)

Miller NE, Nestel PJ (1974) Triglyceride-lowering effect of chenodeoxycholic acid in patients with endogenous hypertriglyceridaemia. Lancet 2(7886):929–931 (PM:4138590)

Mittendorfer B, Magkos F, Fabbrini E, Mohammed BS, Klein S (2009) Relationship between body fat mass and free fatty acid kinetics in men and women. Obesity (Silver.Spring) 17(10):1872–1877 (PM:19629053)

Musso G, Gambino R, Bo S, Uberti B, Biroli G, Pagano G, Cassader M (2008) Should nonalcoholic fatty liver disease be included in the definition of metabolic syndrome? A cross-sectional comparison with Adult Treatment Panel III criteria in nonobese nondiabetic subjects. Diabetes Care 31(3):562–568 (PM:18056890)

Musunuru K (2010) Atherogenic dyslipidemia: cardiovascular risk and dietary intervention. Lipids 45(10):907–914 (PM:20524075)

Nakahara M, Fujii H, Maloney PR, Shimizu M, Sato R (2002) Bile acids enhance low density lipoprotein receptor gene expression via a MAPK cascade-mediated stabilization of mRNA. J Biol Chem 277(40):37229–37234 (PM:12149270)

Nakamuta M, Kohjima M, Morizono S, Kotoh K, Yoshimoto T, Miyagi I, Enjoji M (2005a) Evaluation of fatty acid metabolism-related gene expression in nonalcoholic fatty liver disease. Int J Mol Med 16(4):631–635 (PM:16142397)

Nakamuta M, Morizono S, Soejima Y, Yoshizumi T, Aishima S, Takasugi S, Yoshimitsu K, Enjoji M, Kotoh K, Taketomi A, Uchiyama H, Shimada M, Nawata H, Maehara Y (2005b) Short-term intensive treatment for donors with hepatic steatosis in living-donor liver transplantation. Transplantation 80(5):608–612 (PM:16177634)

Namikawa C, Shu-Ping Z, Vyselaar JR, Nozaki Y, Nemoto Y, Ono M, Akisawa N, Saibara T, Hiroi M, Enzan H, Onishi S (2004) Polymorphisms of microsomal triglyceride transfer protein gene and manganese superoxide dismutase gene in non-alcoholic steatohepatitis. J Hepatol 40(5):781–786 (PM:15094225)

Neuschwander-Tetri BA (2009) Lifestyle modification as the primary treatment of NASH. Clin Liver Dis 13(4):649–665 (PM:19818311)

Newberry EP, Xie Y, Kennedy SM, Luo J, Davidson NO (2006) Protection against Western diet-induced obesity and hepatic steatosis in liver fatty acid-binding protein knockout mice. Hepatology 44(5):1191–1205 (PM:17058218)

Nolte RT, Wisely GB, Westin S, Cobb JE, Lambert MH, Kurokawa R, Rosenfeld MG, Willson TM, Glass CK, Milburn MV (1998) Ligand binding and co-activator assembly of the peroxisome proliferator-activated receptor-gamma. *Nature*, 395, (6698) 137–143 (PM:9744270)

Oh MK, Winn J, Poordad F (2008) Review article: diagnosis and treatment of non-alcoholic fatty liver disease. Aliment Pharmacol Ther 28(5):503–522 (PM:18532991)

Okita M, Hayashi M, Sasagawa T, Takagi K, Suzuki K, Kinoyama S, Ito T, Yamada G (2001) Effect of a moderately energy-restricted diet on obese patients with fatty liver. Nutrition 17(7–8):542–547 (PM:11448571)

Oliver WR Jr, Shenk JL, Snaith MR, Russell CS, Plunket KD, Bodkin NL, Lewis MC, Winegar DA, Sznaidman ML, Lambert MH, Xu HE, Sternbach DD, Kliewer SA, Hansen BC, Willson TM (2001) A selective peroxisome proliferator-activated receptor delta agonist promotes reverse cholesterol transport. Proc Natl Acad Sci U S A 98(9):5306–5311 (PM:11309497)

Olufadi R, Byrne CD (2008) Clinical and laboratory diagnosis of the metabolic syndrome. J Clin Pathol 61(6):697–706 (PM:18505888)

Onate SA, Boonyaratanakornkit V, Spencer TE, Tsai SY, Tsai MJ, Edwards DP, O'Malley BW (1998) The steroid receptor coactivator-1 contains multiple receptor interacting and activation domains that cooperatively enhance the activation function 1 (AF1) and AF2 domains of steroid receptors. J Biol Chem 273(20):12101–12108 (PM:9575154)

Otte K, Kranz H, Kober I, Thompson P, Hoefer M, Haubold B, Remmel B, Voss H, Kaiser C, Albers M, Cheruvallath Z, Jackson D, Casari G, Koegl M, Paabo S, Mous J, Kremoser C, Deuschle U (2003) Identification of farnesoid X receptor beta as a novel mammalian nuclear receptor sensing lanosterol. Mol Cell Biol 23(3):864–872 (PM:12529392)

Ozcan U, Cao Q, Yilmaz E, Lee AH, Iwakoshi NN, Ozdelen E, Tuncman G, Gorgun C, Glimcher LH, Hotamisligil GS (2004) Endoplasmic reticulum stress links obesity, insulin action, and type 2 diabetes. Science 306(5695):457–461 (PM:15486293)

Parks DJ, Blanchard SG, Bledsoe RK, Chandra G, Consler TG, Kliewer SA, Stimmel JB, Willson TM, Zavacki AM, Moore DD, Lehmann JM (1999) Bile acids: natural ligands for an orphan nuclear receptor. Science 284(5418):1365–1368 (PM:10334993)

Parks EJ, Skokan LE, Timlin MT, Dingfelder CS (2008) Dietary sugars stimulate fatty acid synthesis in adults. J Nutr 138(6):1039–1046 (PM:18492831)

Peet DJ, Turley SD, Ma W, Janowski BA, Lobaccaro JM, Hammer RE, Mangelsdorf DJ (1998) Cholesterol and bile acid metabolism are impaired in mice lacking the nuclear oxysterol receptor LXR alpha. Cell 93(5):693–704 (PM:9630215)

Petersen KF, Dufour S, Hariri A, Nelson-Williams C, Foo JN, Zhang XM, Dziura J, Lifton RP, Shulman GI (2010) Apolipoprotein C3 gene variants in nonalcoholic fatty liver disease. N. Engl. J Med 362(12):1082–1089 (PM:20335584)

Petta S, Muratore C, Craxi A (2009) Non-alcoholic fatty liver disease pathogenesis: the present and the future. Dig Liver Dis 41(9):615–625 (PM:19223251)

Pineda Torra I, Claudel T, Duval C, Kosykh V, Fruchart JC, Staels B (2003) Bile acids induce the expression of the human peroxisome proliferator-activated receptor alpha gene via activation of the farnesoid X receptor. Mol. Endocrinol 17(2):259–272 (PM:12554753)

Pollin TI, Damcott CM, Shen H, Ott SH, Shelton J, Horenstein RB, Post W, McLenithan JC, Bielak LF, Peyser PA, Mitchell BD, Miller M, O'Connell JR, Shuldiner AR (2008) A null mutation in human APOC3 confers a favorable plasma lipid profile and apparent cardioprotection. Science 322(5908):1702–1705 (PM:19074352)

Postic C, Girard J (2008) The role of the lipogenic pathway in the development of hepatic steatosis. Diabetes Metab 34(6 Pt 2):643–648 (PM:19195625)

Powell EE, Cooksley WG, Hanson R, Searle J, Halliday JW, Powell LW (1990) The natural history of nonalcoholic steatohepatitis: a follow-up study of forty-two patients for up to 21 years. Hepatology 11(1):74–80 (PM:2295475)

Poynard T, Morra R, Halfon P, Castera L, Ratziu V, Imbert-Bismut F, Naveau S, Thabut D, Lebrec D, Zoulim F, Bourliere M, Cacoub P, Messous D, Munteanu M, de Ledinghen V (2007) Meta-analyses of FibroTest diagnostic value in chronic liver disease. BMC Gastroenterol 7:40 (PM:17937811)

Pullinger CR, North JD, Teng BB, Rifici VA, Ronhild de Brito AE, Scott J (1989) The apolipoprotein B gene is constitutively expressed in HepG2 cells: regulation of secretion by oleic acid, albumin, and insulin, and measurement of the mRNA half-life. J Lipid Res 30(7):1065–1077 (PM:2677202)

Randle PJ (1998) Regulatory interactions between lipids and carbohydrates: the glucose fatty acid cycle after 35 years. Diabetes Metab Rev 14:(4):263–283 (PM:10095997)

Reddy JK, Hashimoto T (2001) Peroxisomal beta-oxidation and peroxisome proliferator-activated receptor alpha: an adaptive metabolic system. Annu Rev Nutr 21:193–230 (PM:11375435)

Reid BM, Sanyal AJ (2004) Evaluation and management of non-alcoholic steatohepatitis. Eur J Gastroenterol Hepatol 16(11):1117–1122 (PM:15489568)

Repa JJ, Liang G, Ou J, Bashmakov Y, Lobaccaro JM, Shimomura I, Shan B, Brown MS, Goldstein JL, Mangelsdorf DJ (2000) Regulation of mouse sterol regulatory element-binding protein-1c gene (SREBP-1c) by oxysterol receptors, LXRalpha and LXRbeta. Genes Dev 14(22):2819–2830 (PM:11090130)

Repa JJ, Mangelsdorf DJ (2000) The role of orphan nuclear receptors in the regulation of cholesterol homeostasis. Annu Rev Cell Dev Biol 16:459–481 (PM:11031244)

Ribon V, Johnson JH, Camp HS, Saltiel AR (1998) Thiazolidinediones and insulin resistance: peroxisome proliferatoractivated receptor gamma activation stimulates expression of the CAP gene. Proc Natl Acad Sci U S A 95(25):14751–14756 (PM:9843961)

Rinella ME, Green RM (2004) The methionine-choline deficient dietary model of steatohepatitis does not exhibit insulin resistance. J Hepatol 40(1):47–51 (PM:14672613)

Riserus U, Sprecher D, Johnson T, Olson E, Hirschberg S, Liu A, Fang Z, Hegde P, Richards D, Sarov-Blat L, Strum JC, Basu S, Cheeseman J, Fielding BA, Humphreys SM, Danoff T, Moore NR, Murgatroyd P, O'Rahilly S, Sutton P, Willson T, Hassall D, Frayn KN, Karpe F (2008) Activation of peroxisome proliferator-activated receptor (PPAR)delta promotes reversal of multiple metabolic abnormalities, reduces oxidative stress, and increases fatty acid oxidation in moderately obese men. Diabetes 57(2):332–339 (PM:18024853)

Rizki G, Arnaboldi L, Gabrielli B, Yan J, Lee GS, Ng RK, Turner SM, Badger TM, Pitas RE, Maher JJ (2006) Mice fed a lipogenic methionine-choline-deficient diet develop hypermetabolism coincident with hepatic suppression of SCD-1. J Lipid Res 47(10):2280–2290 (PM:16829692)

Roden M, Bernroider E (2003) Hepatic glucose metabolism in humans—its role in health and disease. Best Pract Res Clin Endocrinol Metab 17(3):365–383 (PM:12962691)

Romeo S, Kozlitina J, Xing C, Pertsemlidis A, Cox D, Pennacchio LA, Boerwinkle E, Cohen JC, Hobbs HH (2008) Genetic variation in PNPLA3 confers susceptibility to nonalcoholic fatty liver disease. Nat Genet 40(12):1461–1465 (PM:18820647)

Ruhl CE, Everhart JE (2003) Determinants of the association of overweight with elevated serum alanine aminotransferase activity in the United States. Gastroenterology 124(1):71–79 (PM:12512031)

Saadeh S, Younossi ZM, Remer EM, Gramlich T, Ong JP, Hurley M, Mullen KD, Cooper JN, Sheridan MJ (2002) The utility of radiological imaging in nonalcoholic fatty liver disease. Gastroenterology 123(3):745–750 (PM:12198701)

Samuel VT, Liu ZX, Qu X, Elder BD, Bilz S, Befroy D, Romanelli AJ, Shulman GI (2004) Mechanism of hepatic insulin resistance in non-alcoholic fatty liver disease. J Biol Chem 279(31):32345–32353 (PM:15166226)

Sanyal AJ (2002) AGA technical review on nonalcoholic fatty liver disease. Gastroenterology 123(5):1705–1725 (PM:12404245)

Sanyal AJ, Campbell-Sargent C, Mirshahi F, Rizzo, WB, Contos MJ, Sterling RK, Luketic VA, Shiffman ML, Clore JN (2001) Nonalcoholic steatohepatitis: association of insulin resistance and mitochondrial abnormalities. Gastroenterology 120(5):1183–1192 (PM:11266382)

Sanyal AJ, Chalasani N, Kowdley KV, McCullough A, Diehl AM, Bass NM, Neuschwander-Tetri BA, Lavine JE, Tonascia J, Unalp A, Van NM, Clark J, Brunt EM, Kleiner DE, Hoofnagle JH, Robuck PR (2010) Pioglitazone, vitamin E, or placebo for nonalcoholic steatohepatitis. N Engl J Med 362(18):1675–1685 (PM:20427778)

Savage DB, Tan GD, Acerini CL, Jebb SA, Agostini M, Gurnell M, Williams RL, Umpleby AM, Thomas EL, Bell JD, Dixon AK, Dunne F, Boiani R, Cinti S, Vidal-Puig A, Karpe F, Chatterjee VK, O'Rahilly S (2003) Human metabolic syndrome resulting from dominant-negative mutations in the nuclear receptor peroxisome proliferator-activated receptor-gamma. Diabetes 52(4):910–917 (PM:12663460)

Savkur RS, Bramlett KS, Michael LF, Burris TP (2005a) Regulation of pyruvate dehydrogenase kinase expression by the farnesoid X receptor. Biochem Biophys Res Commun 329(1):391–396 (PM:15721319)

Savkur RS, Bramlett KS, Michael LF, Burris TP (2005b) Regulation of pyruvate dehydrogenase kinase expression by the farnesoid X receptor. Biochem Biophys Res Commun 329(1):391–396 (PM:15721319)

Schattenberg JM, Wang Y, Singh R, Rigoli RM, Czaja MJ (2005) Hepatocyte CYP2E1 overexpression and steatohepatitis lead to impaired hepatic insulin signaling. J Biol Chem 280(11):9887–9894 (PM:15632182)

Schoenfield LJ, Lachin JM (1981) Chenodiol (chenodeoxycholic acid) for dissolution of gallstones: the National Cooperative Gallstone Study. A controlled trial of efficacy and safety. Ann Intern Med 95(3):257–282 (PM:7023307)

Schoonjans K, Staels B, Auwerx J (1996) Role of the peroxisome proliferator-activated receptor (PPAR) in mediating the effects of fibrates and fatty acids on gene expression. J Lipid Res 37(5):907–925 (PM:8725145)

Shah A, Rader DJ, Millar JS (2010) The effect of PPAR-alpha agonism on apolipoprotein metabolism in humans. Atherosclerosis 210(1):35–40 (PM:20005515)

Shaheen AA, Myers RP (2007) Diagnostic accuracy of the aspartate aminotransferase-to-platelet ratio index for the prediction of hepatitis C-related fibrosis: a systematic review. Hepatology 46(3):912–921 (PM:17705266)

Shaheen AA, Myers RP (2008) Systematic review and meta-analysis of the diagnostic accuracy of fibrosis marker panels in patients with HIV/hepatitis C coinfection. HIV.Clin Trials 9(1):43–51 (PM:18215981)

Shepherd J, Packard CJ, Morgan HG, Third JL, Stewart JM, Lawrie TD (1979) The effects of cholestyramine on high density lipoprotein metabolism. Atherosclerosis 33(4):433–444 (PM:228682)

Shi H, Cave B, Inouye K, Bjorbaek C, Flier JS (2006) Overexpression of suppressor of cytokine signaling 3 in adipose tissue causes local but not systemic insulin resistance. Diabetes 55(3):699–707 (PM:16505233)

Shih DM, Kast-Woelbern HR, Wong J, Xia YR, Edwards PA, Lusis AJ (2006) A role for FXR and human FGF-19 in the repression of paraoxonase-1 gene expression by bile acids. J Lipid Res 47(2):384–392 (PM:16269825)

Shimano H, Horton JD, Hammer RE, Shimomura I, Brown MS, Goldstein JL (1996) Overproduction of cholesterol and fatty acids causes massive liver enlargement in transgenic mice expressing truncated SREBP-1a. J Clin Invest 98(7):1575–1584 (PM:8833906)

Shimano H, Horton JD, Shimomura I, Hammer RE, Brown MS, Goldstein JL (1997) Isoform 1c of sterol regulatory element binding protein is less active than isoform 1a in livers of transgenic mice and in cultured cells. J Clin Invest 99(5):846–854 (PM:9062341)

Shimomura I, Matsuda M, Hammer RE, Bashmakov Y, Brown MS, Goldstein JL (2000) Decreased IRS-2 and increased SREBP-1c lead to mixed insulin resistance and sensitivity in livers of lipodystrophic and ob/ob mice. Mol Cell 6(1):77–86 (PM:10949029)

Shklyaev S, Aslanidi G, Tennant M, Prima V, Kohlbrenner E, Kroutov V, Campbell-Thompson M, Crawford J, Shek EW, Scarpace PJ, Zolotukhin S (2003) Sustained peripheral expression of transgene adiponectin offsets the development of diet-induced obesity in rats. Proc Natl Acad Sci U S A 100(24):14217–14222 (PM:14617771)

Shulman AI, Mangelsdorf DJ (2005) Retinoid x receptor heterodimers in the metabolic syndrome. N Engl J Med 353(6):604–615 (PM:16093469)

Sinal CJ, Tohkin M, Miyata M, Ward JM, Lambert G, Gonzalez FJ (2000) Targeted disruption of the nuclear receptor FXR/BAR impairs bile acid and lipid homeostasis Cell 102(6):731–744 (PM:11030617)

Sohda T, Momose Y, Meguro K, Kawamatsu Y, Sugiyama Y, Ikeda H (1990) Studies on antidiabetic agents. Synthesis and hypoglycemic activity of 5-[4-(pyridylalkoxy)benzyl]-2,4-thiazolidinediones. Arzneimittelforschung 40(1):37–42 (PM:2339998)

Sozio M, Crabb DW (2008) Alcohol and lipid metabolism. Am J Physiol Endocrinol Metab 295(1):E10–E16 (PM:18349117)

Spiegelman BM (1998) PPAR-gamma: adipogenic regulator and thiazolidinedione receptor. Diabetes 47(4):507–514 (PM:9568680)

Sprecher DL, Massien C, Pearce G, Billin AN, Perlstein I, Willson TM, Hassall DG, Ancellin N, Patterson SD, Lobe DC, Johnson TG (2007) Triglyceride:high-density lipoprotein cholesterol effects in healthy subjects administered a peroxisome proliferator activated receptor delta agonist. Arterioscler Thromb Vasc Biol 27(2):359–365 (PM:17110604)

Staels B, Vu-Dac N, Kosykh VA, Saladin R, Fruchart JC, Dallongeville J, Auwerx J (1995) Fibrates downregulate apolipoprotein C-III expression independent of induction of peroxisomal acyl coenzyme A oxidase. A potential mechanism for the hypolipidemic action of fibrates. J Clin Invest 95(2):705–712 (PM:7860752)

Stauber RE, Lackner C (2007) Noninvasive diagnosis of hepatic fibrosis in chronic hepatitis C. World J Gastroenterol 13(32):4287–4294 (PM:17708599)

Steffensen KR, Gustafsson JA (2004) Putative metabolic effects of the liver X receptor (LXR). Diabetes 53(Suppl 1) S36–S42 (PM:14749264)

Stenson BM, Ryden M, Steffensen KR, Wahlen K, Pettersson AT, Jocken JW, Arner P, Laurencikiene J (2009) Activation of liver X receptor regulates substrate oxidation in white adipocytes. Endocrinology 150(9):4104–4113 (PM:19556420)

Sundaram M, Zhong S, Bou KM, Links PH, Zhao Y, Iqbal J, Hussain MM, Parks RJ, Wang Y, Yao Z (2010) Expression of apolipoprotein C-III in McA-RH7777 cells enhances VLDL assembly and secretion under lipid-rich conditions. J Lipid Res 51(1):150–161 (PM:19622837)

Takata Y, Liu J, Yin F, Collins AR, Lyon CJ, Lee CH, Atkins AR, Downes M, Barish GD, Evans RM, Hsueh WA, Tangirala RK (2008) PPARdelta-mediated antiinflammatory mechanisms inhibit angiotensin II-accelerated atherosclerosis. Proc Natl Acad Sci U S A 105(11):4277–4282 (PM:18337495)

Tamura S, Shimomura I (2005) Contribution of adipose tissue and de novo lipogenesis to nonalcoholic fatty liver disease. J Clin Invest 115(5):1139–1142 (PM:15864343)

Tanaka T, Yamamoto J, Iwasaki S, Asaba H, Hamura H, Ikeda Y, Watanabe M, Magoori K, Ioka RX, Tachibana K, Watanabe Y, Uchiyama Y, Sumi K, Iguchi H, Ito S, Doi T, Hamakubo T, Naito M, Auwerx J, Yanagisawa M, Kodama T, Sakai, J (2003) Activation of peroxisome proliferator-activated receptor delta induces fatty acid beta-oxidation in skeletal muscle and attenuates metabolic syndrome. Proc Natl Acad Sci U S A 100(26):15924–15929 (PM:14676330)

Teli MR, James OF, Burt AD, Bennett MK, Day CP (1995) The natural history of nonalcoholic fatty liver: a follow-up study. Hepatology 22(6):1714–1719 (PM:7489979)

Thomas C, Pellicciari R, Pruzanski M, Auwerx J, Schoonjans K (2008) Targeting bile-acid signalling for metabolic diseases. Nat Rev Drug Discov 7(8):678–693 (PM:18670431)

Tontonoz P, Mangelsdorf DJ (2003a) Liver X receptor signaling pathways in cardiovascular disease. Mol Endocrinol 17(6):985–993 (PM:12690094)

Tontonoz P, Mangelsdorf DJ (2003b) Liver X receptor signaling pathways in cardiovascular disease. Mol Endocrinol 17(6):985–993 (PM:12690094)

Tontonoz P, Spiegelman BM (2008) Fat and beyond: the diverse biology of PPARgamma. Annu Rev Biochem 77:289–312 (PM:18518822)

Towle HC, Kaytor EN, Shih HM (1997) Regulation of the expression of lipogenic enzyme genes by carbohydrate. Annu Rev Nutr 17:405–433 (PM:9240934)

Tulenko TN, Sumner AE (2002) The physiology of lipoproteins. J Nucl Cardiol 9(6):638–649 (PM:12466789)

Unger RH (2003) Lipid overload and overflow: metabolic trauma and the metabolic syndrome. Trends Endocrinol Metab 14(9):398–403 (PM:14580758)

Urizar NL, Dowhan DH, Moore DD (2000) The farnesoid X-activated receptor mediates bile acid activation of phospholipid transfer protein gene expression. J Biol Chem 275(50):39313–39317 (PM:10998425)

Vacca M, Degirolamo C, Mariani-Costantini R, Palasciano G, Moschetta A (2011) Lipid-sensing nuclear receptors in the pathophysiology and treatment of the metabolic syndrome. Wiley Interdiscip Rev Syst Biol Med 3(5):562–587 (PM:21755605)

Wagner M, Zollner G, Trauner M (2008) Nuclear bile acid receptor farnesoid X receptor meets nuclear factor-kappaB: new insights into hepatic inflammation. Hepatology 48(5):1383–1386 (PM:18972560)

Wagner M, Zollner G, Trauner M (2011) Nuclear receptors in liver disease. Hepatology 53(3):1023–1034 (PM:21319202)

Wallace JM, Schwarz M, Coward P, Houze J, Sawyer JK, Kelley KL, Chai A, Rudel LL (2005) Effects of peroxisome proliferator-activated receptor alpha/delta agonists on HDL-cholesterol in vervet monkeys. J Lipid Res 46(5):1009–1016 (PM:15716581)

Wang H, Chen J, Hollister K, Sowers LC, Forman BM (1999) Endogenous bile acids are ligands for the nuclear receptor FXR/BAR. Mol Cell 3(5):543–553 (PM:10360171)

Wang L, Liu J, Saha P, Huang J, Chan L, Spiegelman B, Moore DD(2005) The orphan nuclear receptor SHP regulates PGC-1alpha expression and energy production in brown adipocytes. Cell Metab 2(4):227–238 (PM:16213225)

Wang Y, Jones PJ, Woollett LA, Buckley DD, Yao L, Granholm NA, Tolley EA, Heubi JE (2006) Effects of chenodeoxycholic acid and deoxycholic acid on cholesterol absorption and metabolism in humans. Transl Res 148(1):37–45 (PM:16887497)

Wang YX, Lee CH, Tiep S, Yu RT, Ham J, Kang H, Evans RM (2003) Peroxisome-proliferator-activated receptor delta activates fat metabolism to prevent obesity. Cell 113(2):159–170 (PM:12705865)

Wanless IR, Lentz JS (1990a) Fatty liver hepatitis (steatohepatitis) and obesity: an autopsy study with analysis of risk factors. Hepatology 12(5):1106–1110 (PM:2227807)

Wanless IR, Lentz JS. (1990b) Fatty liver hepatitis (steatohepatitis) and obesity: an autopsy study with analysis of risk factors. Hepatology 12(5):1106–1110 (PM:2227807)

Watanabe M, Houten SM, Wang L, Moschetta A, Mangelsdorf DJ, Heyman RA, Moore DD, Auwerx J (2004a) Bile acids lower triglyceride levels via a pathway involving FXR, SHP, and SREBP-1c. J Clin Invest 113(10):1408–1418 (PM:15146238)

Watanabe M, Houten SM, Wang L, Moschetta A, Mangelsdorf DJ, Heyman RA, Moore DD, Auwerx J (2004b) Bile acids lower triglyceride levels via a pathway involving FXR, SHP, and SREBP-1c. JClin Invest 113(10) 1408–1418:(PM:15146238)

Way JM, Harrington WW, Brown KK, Gottschalk WK, Sundseth SS, Mansfield TA, Ramachandran RK, Willson TM, Kliewer SA (2001) Comprehensive messenger ribonucleic acid profiling reveals that peroxisome proliferator-activated receptor gamma activation has coordinate effects on gene expression in multiple insulin-sensitive tissues. Endocrinology 142(3):1269–1277 (PM:11181544)

Wei Y, Wang D, Topczewski F, Pagliassotti MJ (2006) Saturated fatty acids induce endoplasmic reticulum stress and apoptosis independently of ceramide in liver cells. Am J Physiol Endocrinol Metab 291(2):E275–E281 (PM:16492686)

Westerbacka J, Kolak M, Kiviluoto T, Arkkila P, Siren J, Hamsten A, Fisher RM, Yki-Jarvinen H (2007) Genes involved in fatty acid partitioning and binding, lipolysis, monocyte/macrophage recruitment, and inflammation are overexpressed in the human fatty liver of insulin-resistant subjects. Diabetes 56(11):2759–2765 (PM:17704301)

Willy PJ, Umesono K, Ong ES, Evans RM, Heyman RA, Mangelsdorf DJ (1995) LXR, a nuclear receptor that defines a distinct retinoid response pathway. Genes Dev 9(9):1033–1045 (PM:7744246)

Wolfrum C, Asilmaz E, Luca E, Friedman JM, Stoffel M (2004) Foxa2 regulates lipid metabolism and ketogenesis in the liver during fasting and in diabetes. Nature 432(7020):1027–1032 (PM:15616563)

Wu Z, Xie Y, Morrison RF, Bucher NL, Farmer SR (1998) PPARgamma induces the insulin-dependent glucose transporter GLUT4 in the absence of C/EBPalpha during the conversion of 3T3 fibroblasts into adipocytes. J Clin Invest 101(1):22–32 (PM:9421462)

Xu A, Wang Y, Keshaw H, Xu LY, Lam KS, Cooper GJ (2003) The fat-derived hormone adiponectin alleviates alcoholic and nonalcoholic fatty liver diseases in mice. J Clin Invest 112(1):91–100 (PM:12840063)

Yao Z, Wang, Y (2012) Apolipoprotein C-III and hepatic triglyceride-rich lipoprotein production. Curr Opin Lipidol 23(3):206–212 (PM:22510806)

You M, Considine RV, Leone TC, Kelly DP, Crabb DW (2005) Role of adiponectin in the protective action of dietary saturated fat against alcoholic fatty liver in mice. Hepatology 42(3):568–577 (PM:16108051)

Zelcer N, Hong C, Boyadjian R, Tontonoz P (2009) LXR regulates cholesterol uptake through Idol-dependent ubiquitination of the LDL receptor. Science 325(5936):100–104 (PM:19520913)

Zhang Y, Castellani LW, Sinal CJ, Gonzalez FJ, Edwards PA (2004) Peroxisome proliferator-activated receptor-gamma coactivator 1alpha (PGC-1alpha) regulates triglyceride metabolism by activation of the nuclear receptor FXR. Genes Dev 18(2):157–169 (PM:14729567)

Zhang Y, Edwards PA (2008) FXR signaling in metabolic disease. FEBS Lett 582(1):10–18 (PM:18023284)

Zhang Y, Lee FY, Barrera G, Lee H., Vales C, Gonzalez FJ, Willson TM, Edwards PA (2006a) Activation of the nuclear receptor FXR improves hyperglycemia and hyperlipidemia in diabetic mice. Proc Natl Acad Sci U S A 103(4):1006–1011 (PM:16410358)

Zhang Y, Wang X, Vales C, Lee FY, Lee H, Lusis AJ, Edwards PA (2006b) FXR deficiency causes reduced atherosclerosis in Ldlr-/- mice. Arterioscler Thromb Vasc Biol, 26(10):2316–2321 (PM:16825595)

Zhang Y, Yin L, Hillgartner FB (2003) SREBP-1 integrates the actions of thyroid hormone, insulin, cAMP, and medium-chain fatty acids on ACCalpha transcription in hepatocytes. J Lipid Res 44(2):356–368 (PM:12576518)

Zhou J, Febbraio M, Wada T, Zhai Y, Kuruba R, He J, Lee JH, Khadem S, Ren S, Li S, Silverstein RL, Xie W (2008) Hepatic fatty acid transporter Cd36 is a common target of LXR, PXR, and PPARgamma in promoting steatosis. Gastroenterology 134(2):556–567 PM:18242221

Zoltowska M, Ziv E, Delvin E, Stan S, Bar-On H, Kalman R, Levy E (2001) Circulating lipoproteins and hepatic sterol metabolism in Psammomys obesus prone to obesity, hyperglycemia and hyperinsulinemia. Atherosclerosis 157(1):85–96 (PM:11427207)

Chapter 4
Role of Adipose Tissue in the Pathogenesis and Treatment of Metabolic Syndrome

David Sanchez-Infantes and Jacqueline M. Stephens

Abstract Adipocytes are highly specialized cells that play a major role in energy homeostasis in vertebrate organisms. Excess adipocyte size or number is a hallmark of obesity, which is currently a global epidemic. Obesity is not only the primary disease of fat cells, but also a major risk factor for the development of Type 2 diabetes, cardiovascular disease, hypertension, and metabolic syndrome (MetS). Today, adipocytes and adipose tissue are no longer considered passive participants in metabolic pathways. In addition to storing lipid, adipocytes are highly insulin sensitive cells that have important endocrine functions. Altering any one of these functions of fat cells can result in a metabolic disease state and dysregulation of adipose tissue can profoundly contribute to MetS. For example, adiponectin is a fat specific hormone that has cardio-protective and anti-diabetic properties. Inhibition of adiponectin expression and secretion are associated with several risk factors for MetS. For this purpose, and several other reasons documented in this chapter, we propose that adipose tissue should be considered as a viable target for a variety of treatment approaches to combat MetS.

Keywords Metabolic syndrome · Adipocyte · Adipose tissue · Adipokines

4.1 Metabolic Syndrome

"MetS" has been traditionally defined as a condition whose major features consists of obesity, insulin resistance, the development of Type 2 diabetes (T2DM), and accelerated cardiovascular disease (Chen et al. 1999; Isomaa et al. 2001; Grundy 2004; Grundy et al. 2005; Grundy 2007; Lorenzo et al. 2007; Bruce and Byrne 2009). The development and appearance of other traditional risk factors, including hypertension, dyslipidemia, and non-traditional risk factors such as inflammation and coagulopathy, are also associated with this condition (Chen et al. 1999; Isomaa et al. 2001; Grundy 2004; Grundy et al. 2005; Grundy 2007; Lorenzo et al. 2007;).

J. M. Stephens (✉) · D. Sanchez-Infantes
Pennington Biomedical Research Center,
Louisiana State University, Baton Rouge 70808, LA, USA
e-mail: jsteph1@lsu.edu

M. Orešič, A. Vidal-Puig (eds.), *A Systems Biology Approach to Study Metabolic Syndrome*,
DOI 10.1007/978-3-319-01008-3_4,

Many other names have been used for this disease state including “syndrome X” “cardio-metabolic syndrome”, “insulin resistance syndrome”, and “pre-diabetes”. Regardless of the terminology used, there has been a steady rise in incidence of MetS as a result of the staggering increase in the prevalence of obesity and T2DM that has now reached epidemic proportions. In fact, a number of components of the MetS have become increasingly prevalent in children (Ford et al. 2008; Tailor et al. 2010). The MetS has been and will continue to represent one of the most important public health problems facing society today. Given the known complications for MetS, it is imperative that effective strategies are implemented to improve the underlying pathophysiologic factors that contribute to the development of this disease.

4.2 Adipocytes

Adipocytes are highly specialized cells that play a major role in energy homeostasis in vertebrate organisms. Excess adipocyte size or number is a hallmark of obesity, which is currently a global epidemic. Obesity is not only the primary disease of fat cells, but also a major risk factor for the development of T2DM, cardiovascular disease, hypertension, and metabolic syndrome. Obesity and its related disorders result in dysregulation of the mechanisms that control the expression of metabolic genes in adipocytes. Therefore, understanding adipocyte differentiation is relevant not only for understanding the pathogenesis of metabolic diseases, but also for identifying proteins or pathways which might be appropriate targets for pharmacological interventions. In the last fifteen years, significant advances towards understanding the regulatory processes involved in adipocyte differentiation and fat cell function have been made. These observations have an impact on understanding the pathogenesis and treatment of MetS, a threatening cluster of metabolic diseases that decreases life expectancy and pose an epidemic global threat.

4.3 Adipocyte Hypertrophy and Hyperplasia in MetS

A critical function of adipocytes is the storage of lipids. Adipocytes are the primary constituent of adipose tissue and the lack of adipose tissue expandability can be associated with the onset of metabolic syndrome (central obesity, dyslipidemia, glucose intolerance and hypertension) (Mlinar and Marc 2011). An increase in fat mass can occur as a result of increased intracellular lipids and greater adipocyte size (hypertrophy) and/or by increasing adipocyte number (hyperplasia) (De Ferranti and Mozaffarian 2008). Adipocyte hypertrophy, evident in both overweight and T2DM patients (Bahceci et al. 2007), was originally considered the primary manner that adipose tissue mass increased in adults. However, adipocyte hyperplasia is now known to contribute to the increased adipose tissue mass of obesity (De Ferranti and Mozaffarian 2008). Hyperplasia is dependent on the formation of new adipocytes in a process referred to as adipogenesis, a developmental event that is initiated in utero and occurs throughout life. There is evidence that the stages of adipocyte developmental are different in humans and rodents. In mice, expression of adipocyte markers can be observed as early as day 15 (with a total of 21 days of gestation time) (Das et al. 2001). In humans, the differentiation of fat tissue, as judged by the appearance of lipid-laden cells, can be observed during the second trimester of gestation (between weeks 14 and 24 of gestation) (Deng and Scherer 2010). In recent years, a goal of many investigators in both academia and industry was to identify inhibitors of adipogenesis for potential use as treatment for metabolic diseases including obesity and MetS. However, recent studies suggest that inhibitors of adipogenesis are a poor choice for amelioration of metabolic disease states. A decade ago, Danforth proposed that a failure in adipocyte differentiation could cause T2DM (Danforth 2000). As discussed later, this hypothesis is generally recognized and supported by independent lines of investigation in adipocyte biology.

Obesity is a major risk factor for comorbidities associated with MetS, including T2DM, stroke, hypertension, and cardiovascular disease (Grundy 2004). Energy imbalance leads to storage of excess energy in adipocytes, which can exhibit both hypertrophy and hyperplasia. It is known that the presence of lipid-laden adipocytes correlates with insulin resistance (Unger 2002). In addition, adipose hypertrophy and hyperplasia are associated with intracellular abnormalities of adipocyte function, including endoplasmic reticulum and mitochondrial stress. The consequences of these stressors may cause increased rates of ROS production, mitochondrial dysfunction, altered adipokines synthesis, changes in free fatty acids, and the presence of inflammatory mediators (De Ferranti and Mozaffarian 2008). The factors regulating adipose hypertrophy and hyperplasia are not clearly understood, but circulating insulin and glucocorticoid concentrations both appear to stimulate preadipocyte differentiation (Avram et al. 2007).

It is important to point out that one of the key causal components of MetS is the accumulation of "ectopic" fat (Bruce and Byrne 2009). In particular, an increase in visceral fat is associated with insulin resistance (Despres 2001). Visceral fat accumulation, rather than whole body adiposity, has been implicated in the development of diabetes, lipid disorders, hypertension, and atherosclerosis (Matsuzawa 2010). There is evidence that inflammatory adipokines can be released from this visceral fat and these inflammatory cytokines have a negative effect on the production of adiponectin (Suganami et al. 2005). In addition, an accumulation of visceral fat impairs the suppression of free fatty acids that normally occurs through adipose tissue lipolysis in response to insulin resistance (Jensen 2008). Therefore, the accumulation of visceral fat may produce an alteration of circulating LDL-cholesterol and HDL-cholesterol that is associated with impairments of endothelial function and increased cardiovascular risk (Wyne 2003).

4.4 Adipose Tissue as an Endocrine Organ

Historically, adipose tissue has been considered an energy storage depot with few interesting attributes. Due to the dramatic rise in obesity and its metabolic consequences, the study of adipose tissue has gained tremendous scientific interest in the past two decades (Rabe et al. 2008). Scientific discoveries in the last twenty years have revealed that adipocytes and adipose tissue are highly active endocrine cells/tissues that secrete several endocrine factors, termed adipokines, that comprise a new hormonal network linking adipose tissue with other tissues and organs including skeletal muscle, the adrenal cortex, various regions of the brain and the sympathetic nervous system (Ronti et al. 2006). These adipose tissue produced mediators can modulate food intake, energy balance, homeostasis, immunity, insulin sensitivity, angiogenesis, blood pressure, and glucose and lipid metabolism (Ronti et al. 2006; Ahima 2006; Deng and Scherer 2010). The changes in each individual adipokine are the result of coordinated changes of specific transcriptional programs that affect entire groups of adipocyte gene products as well as posttranslational mechanisms that affect the release and activity of these cellular mediators (Deng et al. 2010). Adipose tissue expansion that occurs in obesity can substantially alter adipokine secretion and profoundly contribute to the development of metabolic diseases (Lehr et al. 2011).

4.4.1 Adipokines Secreted by Adipose Tissue

Adipokines can be broadly defined as any hormone or cytokine produced from a cell present in adipose tissue. Adipocytes represent the largest and most abundant type of cell in adipose tissue and produce hormones, including leptin and adiponectin that are exclusively produced and secreted from fat cells. The primary function of these hormones appears to be endocrine in nature and these hormones act in distant target tissues such as the brain, liver, and skeletal muscle. There are a variety of other cell types in adipose tissue including immune cells such as macrophages and T cells. These immune cells also contribute to the production of hormonal cellular mediators. In particular, macrophages produce cytokines that appear to act in a paracrine manner on adjacent adipocytes. These inflammatory mediators are often referred to as adipocytokines, the most common of which include TNFα, IL-6, and IFNγ. As indicated below and summarized in Table 4.1, several adipokines are regulated in metabolic syndrome.

4.4.2 Leptin

The first hormone shown to be exclusively secreted from adipocytes was identified by positional cloning of the obese gene (Zhang et al. 1994). Since its identification in 1994, leptin has attracted much attention as one of the most important signals for the regulation of food intake and energy homeostasis (Friedman and Halaas 1998; Elmquist et al. 1999; Bates and Myers 2003). Leptin plays an important role in energy balance by inhibiting appetite, stimulating thermogenesis, enhancing fatty acid oxidation, modulating glucose homeostasis, and reducing body weight by decreasing fat mass (Ahima 2006). Leptin binds to its receptors in various regions of the central nervous system, including the hypothalamus and brainstem where it activates neural

Table 4.1 The function and regulation of adipokines in Metabolic Syndrome

Adipose tissue source	Adipokine	Functions	Regulation in metabolic syndrome	References
Adipocytes	Leptin	Satiety signal Increases energy expenditure Improves lipid metabolism Angiogenic activity Improves insulin sensitivity	Decreased in U.S Pima Indians with T2DM Increased in obese humans Hyperleptinemia and leptin insensitivity in Israeli sand rat Circulating leptin levels predict development of metabolic syndrome	Fox et al. 1999 Ahima 2006 Zimmet et al. 1999 Franks et al. 2005; Galletti et al. 2007
	Adiponectin	Improves insulin sensitivity Reduces inflammation and atherosclerosis	Decreased in obese, insulin resistance and T2DM subjects Decreased in rhesus monkeys with obesity and insulin resistance Low plasma adiponectin is associated with metabolic syndrome	Pajvany and Scherer 2003 Hotta et al. 2001 Hulthe et al. 2003
	Resistin	Inflammation Insulin resistance	Increased in obesity and insulin resistance rat and mouse models Plasma resistin is increased in metabolic syndrome Resistin levels are not associated with metabolic syndrome	Qatanani et al. 2009 Norata et al. 2007 Utzschneider et al. 2005
Non Adipocytes	PAI-1	Vascular homeostasis Inflammation	Increased in obese rats Increased expression and activity in metabolic syndrome	Funahashi et al. 1999 Mertens et al. 2006
	TNFα	Inhibits adipocyte differentiation Proinflammatory Induces insulin resistance	Increased in lipoatrophic adipose tissue Increased mRNA expression in adipose tissue in obese mice	Bastard et al. 2002b Hotamisligil et al. 1993
	IFNγ	Inhibits adipocyte differentiation Proinflammatory	Increased mRNA expression in adipose tissue from obese animals	Rocha et al. 2008
	IL-6	Proinflammatory Can be associated with insulin resistance	Increased in obese humans	Bruun et al. 2003; Bastard et al. 2006

PAI-1 Plasminogen Activator Inhibitor 1, *TNF-α* Tumor Necrosis Factor alpha, *IFNγ* Interferon gamma, *IL-6* Interleukin 6

pathways that decrease appetite and increase sympathetic nervous system activity and energy expenditure (Rahmouni et al. 2005; Hall et al. 2007). Blood leptin levels are related with adiposity, but this correlation is only part of its regulation. Leptin decreases quickly during fasting and produces an increase of glucocorticoids and a reduction in thyroid hormones (T4), sex hormones and growth hormone (Ronti et al. 2006). Some studies indicate that subcutaneous fat rather than visceral fat has the greatest ability to increase leptin levels (Takahashi et al. 1996; Wajchenberg 2000; Fain et al. 2004). Leptin reduces intracellular lipid levels in the skeletal muscle, liver and β-pancreatic cells (Minokoshi and Kim 2002), and improves insulin sensitivity (Liu et al. 1998; Gutierrez-Juarez et al. 2004). In fact, in patients with lipodystrophy and leptin deficiency, leptin therapy improves glucose tolerance and decreases triglycerides levels (Ronti et al. 2006). In rodents, both intravenous and intracerebroventricular administration of leptin improves peripheral insulin action (Liu et al. 1998; Gutierrez-Juarez et al. 2004). More recently, intriguing preclinical data suggests that pharmacological leptin doses may exert beneficial effects for type 1 diabetics (Wang et al. 2010).

In addition, it has been suggested that hyperleptinemia and leptin resistance could be a primary factor in the development of MetS. Individual variations in blood leptin levels have been shown to be related with factors that produce MetS (Leyva and Godsland 1998). Epidemiological data also demonstrate that hyperleptinemia and hyperinsulinemia play a synergic central role in the genesis of the components of MetS (Zimmet and Boyko 1999). Moreover, leptin may also affect the vascular system. In fact, *in vitro* and *in vivo* studies indicate that leptin has angiogenic activity (Sierra-Honigmann et al. 1998) and may contribute to the arterial thrombosis due to its interaction with platelet leptin receptor (Bodary et al. 2002). Collectively, these studies indicate that leptin therapy is limited to a subset of patients with lipodystrophy and may be a future treatment for Type 1 diabetes. Given that leptin resistance is common in obesity and evidence that leptin may be an important modulator of the vascular system, it is highly unlikely that leptin will be viable therapeutic for patients with MetS.

4.4.3 Adiponectin

The adiponectin gene is located in chromosome 3q27 at a locus susceptible to T2DM, MetS and coronary heart disease (Yamauchi et al. 2003a). Adiponectin is secreted by adipocytes and has been shown to have anti-diabetic and anti-atherogenic properties (Yamauchi et al. 2003b). Native adiponectin exists as homotrimers that form dimers of trimers (hexamers) and high molecular weight (HMW) complexes (Kadowaki and Yamauchi 2005). Several studies have demonstrated that the HMW multimer form of adiponectin is the most active form of this hormone (Waki et al. 2003; Ouchi et al. 2004; Kobayashi et al. 2004). Four mutations in the adiponectin globular domain have been identified. The most frequent mutation is associated with low adiponectin blood levels and individuals with this mutation have some characteristics of metabolic syndrome including hypertension, hyperlipidemia, diabetes and/or atherosclerosis (Kondo et al. 2002).

Adiponectin levels in plasma are 5–30 μg/ml, and these levels tend to be lower in men than in women. Although adiponectin is only secreted from adipocytes, blood levels are lower in obese subjects as compared to non-obese subjects. In general, this is the only adipokine whose expression is decreased in obesity as the majority of other adipose tissue hormones have increased levels in obesity that correlate with increased body mass (Arita et al. 1999). A large number of studies have demonstrated a correlation between insulin sensitivity and high adiponectin levels in humans and experimental animals (Yamauchi et al. 2001; Maeda et al. 2001; Combs et al. 2001; Asensio-López et al. 2011). Circulating levels of adiponectin are decreased in obesity, insulin resistance and T2DM (Pajvani and Scherer 2003). Low adiponectin levels have also been observed in hypertension subjects, and the lowest levels can be found in patients with coronary heart disease. The incidence of death in cardiovascular disease is higher in patients with low adiponectin levels (Szmitko et al. 2007).

As indicated above, the HMW form of adiponectin appears to be the most active form. This complex has the most potent insulin-sensitizing activity of all the adiponectin complexes (Pajvani et al. 2004), and it can suppress hepatic glucose production (Waki et al. 2003). Although the mechanisms implicated in adiponectin action are largely unknown, there is evidences that phosphorylation of 5′-adenosine monophosphate—activated protein kinase may play a role in its action (Waki et al. 2003; Ouchi et al. 2004). Administration of adiponectin to rodents has been shown to increase the tyrosine phosphorylation of insulin receptor and result in increased insulin sensitivity (Berg et al. 2001; Combs et al. 2001). Similar studies have been validated in humans (Yamauchi et al. 2001; Stefan et al. 2002). In rhesus monkeys that spontaneously develop obesity and insulin resistance, adiponectin blood levels decrease proportionally to the progression of insulin resistance during the development of T2DM (Hotta et al. 2001). A negative correlation between adiponectin levels and body weight, and a positive correlation between adiponectin levels and glucose uptake were established in these monkeys. In addition, the decrease in the adiponectin levels occurred before hyperglycemia. Adiponectin does not appear to have a direct action on insulin secretion, as insulin levels do not increase after adiponectin injection. In addition, adiponectin injection could improve insulin sensitivity and hyperglycemia in mice with obesity, diabetes and lipodistrophy (Yamauchi et al. 2001). However, mice lacking insulin receptors in adipocytes have an elevation in plasma adiponectin (Bluher et al. 2004) and clinical studies demonstrate that patients with antibodies against insulin receptors showed a significant increase in serum adiponectin (Semple et al. 2006). These observations suggest that loss of insulin signaling may enhance adiponectin secretion (Deng et al. 2010). Despite these observations, adiponectin appears to be a potential therapeutic for MetS as high levels are associated with insulin sensitivity and cardiovascular health.

4.4.4 Resistin

Resistin was identified as a gene whose expression was suppressed by the insulin sensitizing drugs, thiazolidinediones (Steppan et al. 2001). In this study, the researchers suggested that this novel adipocyte expressed protein, could be a link between obesity and insulin resistance, and for this reason, they called it "resistin" (Steppan et al. 2001). Resistin acts to increase blood glucose and insulin levels and decrease glucose tolerance. In obese mice, immunoneutralization of resistin results in decreased blood glucose levels and improved insulin sensitivity and support the hypothesis that resistin induces insulin resistance and/or contributes to inhibiting insulin sensitivity (Ronti et al. 2006). In mice, resistin is produced in adipocytes. However, in humans, resistin is produced in macrophages. Despite the differences in the source of this hormone, the function of resistin to promote insulin resistance is independent of it source. An elegant transgenic model was produced that expressed human resistin in mouse macrophages, rather than adipocytes. The modulation and function of the hormone was similar in these mice despite the macrophage specific expression that was employed to mimic expression of the human resistin gene (Qatanani et al. 2009). Not surprisingly, TZDs have been shown to inhibit resistin gene expression in human macrophages (Patel et al. 2003; Lehrke et al. 2004) and a decrease in serum resistin levels has been observed in rodents as well as humans (Jung et al. 2005; Kamin et al. 2005). The signaling mechanism(s) by which resistin impairs insulin action are just beginning to be unraveled and the receptor for this adipokine is still unknown. There is evidence to suggest that resistin inhibits the AMP-activated kinase (AMPK) in liver and muscle (Satoh et al. 2004; Banerjee et al. 2004; Muse et al. 2004) and it has also been shown to activate SOCS-3 in mouse adipose tissue (Steppan et al. 2005).

In humans, resistin expression is abundant in monocytes/macrophages, which play an important role in inflammation and atherosclerosis (Patel et al. 2003; Savage et al. 2001). A correlation between circulating resistin levels and inflammation markers has been established and research suggests this fact is predictive of coronary atherosclerosis (Reilly et al. 2005). In addition, serum resistin levels show significant correlation with vascular inflammation (Choi et al. 2011). Given that MetS is associated with insulin resistance and cardiovascular risk, the inhibition of resistin expression and/or activity may represent a viable therapeutic intervention for metabolic syndrome.

4.4.5 Plasminogen Activator Inhibitor 1

Plasminogen Activator Inhibitor 1 (PAI-1) is a fibrinolysis system regulator that inhibits the plasminogen activator action and plasmin formation. PAI-1 is produced in adipose tissue and its expression has been implicated in inflammation that accompanies obesity-associated diseases (Theodorou G et al. 2011). An increase of PAI-1 could promote a tendency to have thrombosis and may produce the thrombotic cardiovascular disease that can occur in visceral obesity and MetS (Funahashi et al. 1999). Increased expression and activity of PAI-1 has been demonstrated in MetS (Mertens et al. 2006).

4.4.6 Inflammatory Cytokines Produces in Adipose Tissue

During chronic inflammation, adipose tissue expresses elevated levels of acute phase reactants and pro-inflammatory cytokines (Wellen and Hotamisligil 2005). The presence of infiltrating immune cells, such as macrophages and T cells is well documented and studies in the last decade suggest that these cells are modulated in conditions of obesity and T2DM. Inflammation may be one of the mechanisms by which high postprandial insulin and glucose responses increase the risk of T2DM (Brand-Miller et al. 2007). In the last decade, several pro-inflammatory cytokines have been shown to be produced in adipose tissue including TNFα, IL-6 and IFNγ.

4.4.7 Tumor Necrosis Factor α and Interferon γ

It is well established that the tumor necrosis alpha (TNFα) (Torti et al. 1985) and IFNγ (Keay and Grossberg 1980) are potent inhibitors of adipocyte differentiation. TNFα is not secreted by adipocytes but it is secreted by infiltrating macrophages that are present in adipose tissue (Mohamed-Ali et al. 1997; Xu et al. 2002). IFNγ is produced from both NK cells (O'Rourke et al. 2009) and T cells (Duffaut et al. 2009; Rocha et al. 2008; Strissel et al. 2010; Yang et al. 2010) present in adipose tissue. Of note, it is also known that both of these cytokines induce insulin resistance in adipocytes (Stephens et al. 1992; Mcgillicuddy et al. 2009). Several mechanisms have been proposed to explain the association between TNFα and obesity related insulin resistance. Chronic exposure of adipocytes to TNFα results in an inhibition of GLUT4 expression and a loss of insulin sensitive glucose uptake (Stephens et al. 1992). TNFα also results in a decrease of IRS-1 expression in adipocytes (Stephens et al. 1997) and can cause an increase of fatty acids release from adipocytes (Bruun et al. 2003). In addition, it has been demonstrated that TNFα can reduce adiponectin expression and secretion (Arita et al. 1999; Maeda et al. 2001). Collectively these studies indicate that there are several mechanisms in which TNFα can induce and promote insulin resistance in adipocytes.

Another cause for inflammation has been attributed to endoplasmic reticulum stress (ER stress) (Hotamisligil 2006). The ER is a major site for protein as well as for lipid and sterol synthesis (Schroder and Kaufman 2005; Wang et al. 2005). ER stress is increased in adipose tissue in response to both dietary and genetic obesity (Ozcan et al. 2004; Boden et al. 2008), and several studies report that it can trigger activation of several serine/threonine kinases, including c-Jun N-terminal kinase (JNK) and IκB kinase. ER stress can produce an impairment of insulin action as well as the ability to activate the synthesis and release of proinflammatory cytokines (Hotamisligil 2006; Hirosumi et al. 2002; Boden et al. 2008).

4.4.8 Interleukin 6

Interleukin-6 (IL)-6 is produced by adipocytes and other cells (fibroblasts, preadipocytes, endothelial cells, monocytes) in adipose tissue (Trayhurn and Wood

2004; Bastard et al. 2006). Approximately 30 % of circulating IL-6 is derived from white adipose tissue, with visceral fat producing higher levels of IL-6 than subcutaneous fat (Dube et al. 2008). It has been reported that plasma IL-6 concentrations correlate positively with human obesity (Trayhurn and Wood 2004; Bastard et al. 2006; Bruun et al. 2003), insulin resistance (Bastard et al. 2000 and 2002a), hypertension and hyperlipidemia (Recasens et al. 2004). It has been suggested that elevated IL-6 levels may be predictive of T2DM (Bastard et al. 2000) and this cytokine can have a negative effect on adiponectin secretion (Simons et al. 2007). IL-6 has metabolic effects that include reducing hepatic insulin sensitivity and decreasing glucose uptake in adipocytes (Eder et al. 2009). There is also evidence that IL-6 promotes lipolysis and fat oxidation (Van Hall et al. 2003; Petersen et al. 2005). IL-6 is a potent stimulator for the production of reactive oxygen and nitrogen by macrophages and monocytes, and may produce an increase in oxidative stress (Fernández-Sanchez et al. 2011). Therefore, this cytokine seems to play a key role in inflammation and alteration in glucose and lipid metabolism that can occur with metabolic syndrome.

4.5 Adipocytes as Drugs Targets for the Metabolic Syndrome

A primary goal of treatment for patients with MetS should include changes in lifestyle such as improving diet and physical activity. Adipose tissue plays a very important role in MetS, and therefore represents a key target in the treatment of this disease. There are a variety of drugs that are used in the treatment of metabolic syndrome. For example, drugs that lower blood pressure are commonly used. In addition, statins are used to reduce cholesterol levels. However, the beneficial effects of these drugs are not mediated by primary effects on adipocytes. However, there is one class of drugs that is used for treatment of metabolic syndrome that largely acts by improving adipocyte development and fat cell function.

The glitazones or thiazolidinediones (TZDs) are oral drugs used to improve insulin sensitivity and glucose tolerance in T2DM (Wang et al. 2010; Quinn et al. 2008). TZDs exert beneficial effects on several components of the MetS and cardiovascular risk markers (Sarafidis et al. 2010). In addition, TZDs have been reported to improve cardiovascular prognosis in patients with insulin resistance (Reasner 2002). These drugs are agonists of the nuclear receptor PPARγ, a transcription factor that regulates the expression of many genes in adipocytes. As indicated above, insulin resistance is often accompanied by an increase on pro-inflammatory cytokines that are derived from adipose tissue macrophages and act in a paracrine manner on surrounding adipocytes. TZDs have been shown to decrease the expression and release of these cytokines and other insulin resistance mediators produced in adipose tissue (free fatty acids and resistin), that results in an improvement in insulin sensitivity, particularly in muscle and liver (Pittas and Greenberg 2002; Giles and Sander 2007).

TZD administration has also been shown to increase adiponectin plasma levels in insulin resistance subjects. In fact, it has been demonstrated that these drugs can increase adiponectin transcription (Maeda et al. 2001) as well as circulating hormone levels. In normal subjects, a 14 days rosiglitazone treatment resulted in an increase in adiponectin levels (Combs et al. 2001). HMW adiponectin has been reported to have a greater increase than total adiponectin levels in patients with T2DM treated with TZDs (Pajvani et al. 2004). The precise mechanism(s) by which TZDs mediates the increased release of adiponectin is not yet clear, but may be due to the up regulation of several ER chaperons that could alleviate ER stress-induced adiponectin regulation (Wang and Scherer 2008; Liu et al. 2008; Zhou et al. 2010). In addition, TZDs may improve mitochondrial function and ROS levels that results in a favorable increase of adiponectin secretion (Wilson-Fritch et al. 2003; Wilson-Fritch et al. 2004).

It can be readily argued that PPARγ activation caused by TZDs can promote adipocyte differentiation of preadipocytes and result in an increase in fat body mass. Currently available TDZs (e.g. rosiglitazone) are associated with important side effects, such as edema and weight gain, suggesting that the investigation of alternative TZDs with better pharmacological properties is warranted (Faine et al. 2011). A recent novel synthetic compound, SR1664, has been identified that has a unique mode of binding to PPARγ (Choi et al. 2011). These researchers report that this compound completely lacks classical transcriptional agonism and inhibits the Cdk5-mediated PPARγ phosphorylation in cultured adipocytes and in insulin-resistant mice. They have observed that the non-agonist SR1664 has potent anti-diabetic activity that is not accompanied by fluid retention and weight gain that are serious side effects of many of TZDs. These recent studies suggest that designing therapeutics that effect covalent modifications of PPARγ may be potentially effective treatments for metabolic syndrome.

One treatment to target adipocytes that should not be considered for metabolic syndrome is the use of factors that inhibit adipocyte differentiation. As indicated above, proinflammatory cytokines that inhibit adipocyte development also cause insulin resistance (Fig. 4.1). In fact, most inhibitors of adipogenesis are associated with insulin resistance. In addition, there are new model systems that clearly suggest that limitations in adipose tissue expansion are associated with insulin resistance. Mice that are very obese, but have unlimited adipose tissue expansion, are metabolically healthy and insulin sensitive (Kim et al. 2007). Overall, these studies support Danforth's idea and indicate that inhibition of adipogenesis is associated with insulin resistance. Emerging evidence of metabolically healthy obese individuals suggest that obesity is likely preferable to insulin resistance in terms of overall health. Collectively, these results indicate that inhibiting adipogenesis is not a viable therapeutic approach.

Another drug that has been demonstrated to be effective in the treatment of the metabolic syndrome is metformin (N, N-dimethylimidodicarbonimidic diamide hydrochloride), a biguanide developed from galegine that is found in *Galega Officinalis*. Despite its use in T2DM treatment for more than 50 years, the mechanism of metformin action is not entirely clear. In the last decade, a link between metformin and AMP- activated protein kinase (AMPK) activity has been proposed (Zhou et al. 2001; Shaw et al. 2005). AMPK is considered as a master switch in regulating glucose and lipid metabolism. In adipose tissue, activated AMPK inhibits deposition of fat, and enhances breakdown and burning of stored fat, resulting in a reduction of body weight. These data suggest that AMPK may be a key player in the development of new treatments for obesity, T2DM and the metabolic syndrome (Misra 2008). In fact, metformin exerts cardioprotective actions via AMPK and increases the

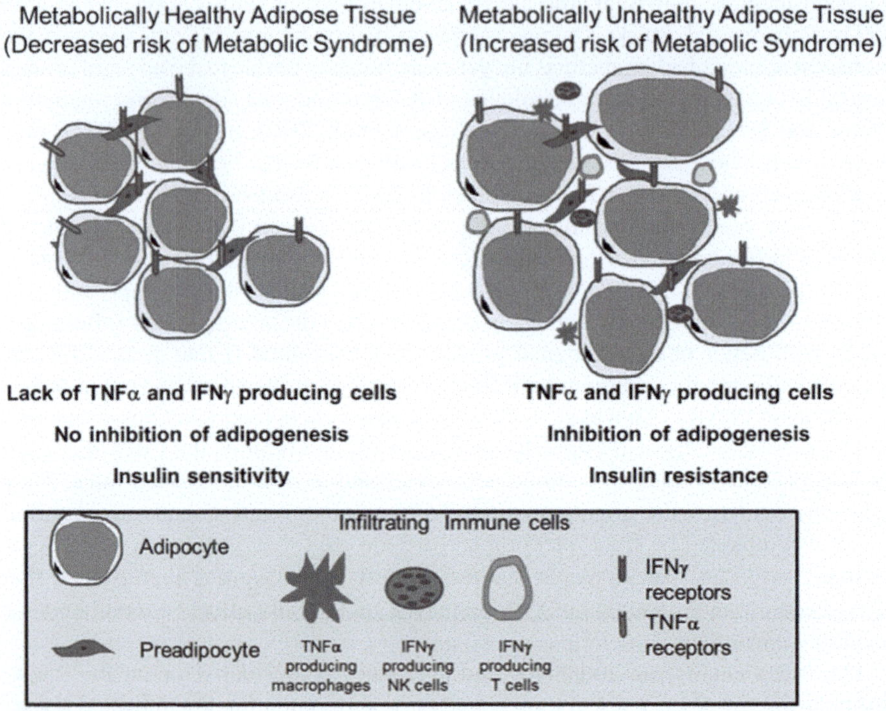

Fig. 4.1 Characteristics of adipose tissue in conditions of increased and decreased risk of Metabolic Syndrome. Adipose tissue of comprised of various cell types including adipocytes, preadipocytes, and infiltrated immune cells. Healthy adipose tissue does not contain immune cells that produce proinflammatory cytokines such as TNFα and IFNγ. Healthy adipocytes are also insulin sensitive and do no have impairments in adipogenesis. Adipose Tissue that is metabolically unhealthy contains immune cells that produce IFNγ and TNFα that act in a paracrine manner on adipocytes to promote insulin resistance and inhibit adipocyte development

expression of adiponectin and its receptors (adipoR1 and adipoR2) in skeletal muscle and adipose tissue (Asensio-López et al. 2011). Moreover, metformin activates AMPK and reduces acetyl-CoA carboxylase protein levels in human adipose tissue (Boyle et al. 2011). These results suggest that its effects on fat cells may mediate some of the beneficial effect of metformin to improve glucose homeostasis.

Although it is a more challenging solution, particularly for those that are economically disadvantaged, proper nutrition should be considered as an option to combat metabolic syndrome. There is some evidence of functional ingredients having the capability to modulate parameters that are altered in metabolic syndrome. Adipose tissue can be a target for these ingredients, particularly if they are capable of regulating adipokine levels via regulation of diet. It is known that the consumption of a high-carbohydrate and low-fiber meal results in a transient decrease in adiponectin levels and dietary fiber intake can increase the levels of this adipocyte specific hormone (Galisteo et al. 2005). Prolonged feeding of a 3.5 % *Plantago ovata* husk-supplemented diet in Zucker rats resulted in decreased body weight gain, reduced hyperinsulinemia and dyslipidemia, and restored plasma adiponectin level. In humans, a cross sectional analysis in women with T2DM revealed that the intake of cereal fiber and fruit-fiber was significantly associated with increasing plasma adiponectin levels (Qi et al. 2005). If the intake of functional foods may induce an increase in the adiponectin levels, this approach could produce health benefits associated with insulin action and glucose metabolism, but may also reduce the inflammatory state (Yannakoulia et al. 2008; Nilsson et al. 2008). Adiponectin has been shown to reverse the deleterious effects of TNFα and other cytokines in cultured cells (Ouchi et al. 1999; Okamoto et al. 2002). Perhaps the postprandial inflammatory response produced after high-fat, high-carbohydrate meal in obese individuals (Patel et al. 2007), could be improved with dietary fiber and modulation of pro-inflammatory cytokine levels (Esposito et al. 2003).

Dietary fiber can also decrease leptin levels, improving the development of leptin resistance due to increased adiposity. In fact, it has been demonstrated that in conditions of high-sugar/low-fiber that it takes longer to observe a decrease in glucose and leptin levels as compared to conditions of low-sugar/high-fiber (Spruijt-Metz et al. 2009). Moreover, a reduction in circulating leptin was reported after supplementation with fungal chitosan in a high dietary fiber treatment that was associated with lower fat mass development (Neyrinck et al. 2009). Accordingly, functional ingredients such as particular dietary fibers could be an important tool to modulate glucose and lipid metabolism and reduce inflammation to prevent the development of metabolic syndrome.

4.6 Concluding Remarks

Metabolic Syndrome is a threat to industrialized nations that is usually accompanied by abdominal obesity. Although there are metabolically healthy obese individuals, obesity is a substantial risk factor for poor metabolic control and altered glucose and lipid homeostasis. Today, adipocytes and adipose tissue are no longer considered passive participants in metabolic pathways. In addition to storing lipid, adipocytes are highly insulin sensitive cells that have profound endocrine functions. One hormone that is produced exclusively in adipocytes, adiponectin, has cardio-protective and anti-diabetic properties. For this reason, and several other reasons documented in this chapter, we propose that adipose tissue should be considered as a viable target for pharmacotherapeutic approaches to combat metabolic syndrome.

References

Ahima RS (2006) Adipose tissue as an endocrine organ. Obesity 14(5):242–249

Arita Y, Kihara S, Ouchi N, Takahashi M, Maeda K, Miyagawa J, Hotta K, Shimomura I, Nakamura T, Miyaoka K, Kuriyama H, Nishida M, Yamashita S, Okubo K, Matsubara K, Muraguchi M, Ohmoto Y, Funahashi T, Matsuzawa Y (1999) Paradoxical decrease of an adipose-specific protein, adiponectin, in obesity. Biochem Biophys Res Commun 257:79–83

Asensio-López MC, Lax A, Pascual-Figal DA, Valdés M, Sánchez-Más J (2011) Metformin protects against doxorubicin-induced cardiotoxicity: involvement of the adiponectin cardiac system. Free Radic Biol Med 51(10):1861–1871

Avram MM, Avram AS, James WD (2007) Subcutaneous fat in normal and diseased states 3. Adipogenesis: from stem cell to fat cell. J Am Acad Dermatol 56:472–492

Bahceci M, Gokalp D, Bahceci S, Tuzcu A, Atmaca S, Arikan S (2007) The correlation between adiposity and adiponectin, tumor necrosis factor alpha, interleukin-6 and high sensitivity C-reactive protein levels. Is adipocyte size associated with inflammation in adults?. J Endocrinol Invest 30:210–214

Banerjee RR, Rangwala SM, Shapiro JS, Rich AS, Rhoades B, Qi Y, Wang J, Rajala MW, Pocai A, Scherer PE, Steppan CM, Ahima RS, Obici S, Rossetti L, Lazar MA (2004) Regulation of fasted blood glucose by resistin. Science 303:1195–1198

Bastard JP, Jardel C, Bruckert E, Blondy P, Capeau J, Laville M, Vidal H, Hainque B, (2000) Elevated levels of interleukin 6 are reduced in serum and subcutaneous adipose tissue of obese women after weight loss. J Clin Endocrinol Metab 85:3338–3342

Bastard JP, Maachi M, Van Nhieu JT, Jardel C, Bruckert E, Grimaldi A, Robert JJ, Capeau J, Hainque B (2002a) Adipose tissue IL-6 content correlates with resistance to insulin activation of glucose uptake both in vivo and in vitro. J Clin Endocrinol Metab 87:2084–2089

Bastard JP, Caron M, Vidal H, Jan V, Auclair M, Vigouroux C, Luboinski J, Laville M, Maachi M, Girard PM, Rozenbaum W, Levan P, Capeau J (2002b) Association between altered expression of adipogenic factor SREBP1 in lipoatrophic adipose tissue from HIV-1-infected patients and abnormal adipocyte differentiation and insulin resistance. Lancet 359(9311):1026–1031

Bastard JP, Maachi M, Lagathu C, Kim MJ, Caron M, Vidal H, Capeau J, Feve B, (2006) Recent advances in the relationship between obesity, inflammation and insulin resistance. Eur Cytokine Netw 17:4–12

Bates SH, Myers MG (2003) The role of leptin receptor signaling in feeding and neuroendocrine function. Trends Endocrinol Metab 14:447–452

Berg AH, Combs TP, Du X, Brownlee M, Scherer PE (2001) The adipocyte-secreted protein Acrp30 enhances hepatic insulin action. Nat Med 7:947–953

Bluher M, Wilson-Fritch L, Leszyk J, Laustsen PG, Corvera S, Kahn CR (2004) Role of insulin action and cell size on protein expression patterns in adipocytes. J Biol Chem 279:31902–31909

Bodary PF, Westrick RJ, Wickenheiser KJ, Shen Y, Eitzman DT (2002) Effect of leptin on arterial thrombosis following vascular injury in mice. JAMA 287:1706–1709

Boden G, Duan X, Homko C, Molina EJ, Song W, Perez O, Cheung P, Merali S (2008) Increase in endoplasmic reticulum stress-related proteins and genes in adipose tissue of obese, insulin-resistant individuals. Diabetes 57:2438–2444

Boyle JG, Logan PJ, Jones GC, Small M, Sattar N, Connell JM, Cleland SJ, Salt IP (2011) AMP-activated protein kinase is activated in adipose tissue of individuals with type 2 diabetes treated with metformin: a randomised glycaemia-controlled crossover study. Diabetologia 54(7):1799–1809

Brand-Miller J, Dickinson S, Barclay A, Celermajer D (2007) The glycemic index and cardiovascular disease risk. Curr Atheroscler Rep 9:479–485

Bruce KD, Byrne CD (2009) The metabolic syndrome: common origins of a multifactorial disorder. Postgrad Med J 85(1009):614–621

Bruun JM, Verdich C, Toubro S, Astrup A, Richelsen B (2003) Association between measures of insulin sensitivity and circulating levels of interleukin-8, interleukin-6 and tumor necrosis factor-α. Effect of weight loss in obese men. Eur J Endocrinol 148:535–542

Chen W, Srinivasan SR, Elkasabany A, Berenson GS (1999) Cardiovascular risk factors clustering features of insulin resistance syndrome (Syndrome X) in a biracial (Black-White) population of children, adolescents, and young adults: the Bogalusa Heart Study. Am J Epidemiol 150:667–674

Choi JH, Banks AS, Kamenecka TM, Busby SA, Chalmers MJ, Kumar N, Kuruvilla DS, Shin Y, He Y, Bruning JB, Marciano DP, Cameron MD, Laznik D, Jurczak MJ, Schürer SC, Vidović D,

Shulman GI, Spiegelman BM, Griffin PR (2011) Antidiabetic actions of a non-agonist PPARγ ligand blocking Cdk5-mediated phosphorylation. Nature 477(7365):477–481

Combs TP, Berg AH, Obici S, Scherer PE, Rossetti L (2001) Endogenous glucose production is inhibited by the adipose-derived protein Acrp30. J Clin Invest 108:1875–1881

Danforth E Jr (2000) Failure of adipocyte differentiation causes type II diabetes mellitus? Nat Genet 26:13

Das K, Lin Y, Widen E, Zhang Y, Scherer PE (2001) Chromosomal localization, expression pattern, and promoter analysis of the mouse gene encoding adipocyte-specific secretory protein Acrp30. Biochem Biophys Res Commun 280(4):1120–1129

De Ferranti S, Mozaffarian D (2008) The perfect storm: obesity, adipocyte dysfunction, and metabolic consequences. Clin Chem 54(6):945–955

Deng Y, Scherer PE (2010) Adipokines as novel biomarkers and regulators of the metabolic syndrome. Ann N Y Acad Sci 1212:E1–E19

Despres JP (2001) Health consequences of visceral obesity. Ann Med 3:534–541

Dube MG, Torto R, Kalra SP (2008) Increased leptin expression selectively in the hypothalamus suppresses inflammatory markers CRP and IL-6 in leptin-deficient diabetic obese mice. Peptides 29:593–598

Duffaut C, Zakaroff-Girard A, Bourlier V, Decaunes P, Maumus M, Chiotasso P, Sengenès C, Lafontan M, Galitzky J, Bouloumié A (2009) Interplay between human adipocytes and T lymphocytes in obesity: CCL20 as an adipochemokine and T lymphocytes as lipogenic modulators. Arterioscler Thromb Vasc Biol 29:1608–1614

Eder K, Baffy N, Falus A, Fulop AK (2009) The major inflammatory mediator interleukin-6 and obesity. Inflamm Res 58:727–736

Elmquist JK, Elias CF, Saper CB (1999) From lesions to leptin: hypothalamic control of food intake and body weight. Neuron 22:221–232

Esposito K, Nappo F, Giugliano F, Di Palo C, Ciotola M, Barbieri M, Paolisso G, Giugliano D (2003) Meal modulation of circulating interleukin 18 and adiponectin concentrations in healthy subjects and in patients with type 2 diabetes mellitus. Am J Clin Nutr 78:1135–1140

Fain JN, Madan AK, Hiler ML, Cheema P, Bahouth SW (2004) Comparison of the release of adipokines by adipose tissue, adipose tissue matrix, and adipocytes from visceral and subcutaneous abdominal adipose tissues of obese humans. Endocrinol 145:2273–2282

Faine LA, Rudnicki M, César FA, Heras BL, Boscá L, Souza ES, Hernandes MZ, Galdino SL, Lima MC, Pitta IR, Abdalla DS (2011) Anti-inflammatory and antioxidant properties of a new arylidene-thiazolidinedione in macrophages. Curr Med Chem 18(22):3351–3360

Fernández-Sánchez A, Madrigal-Santillán E, Bautista M, Esquivel-Soto J, Morales-González A, Esquivel-Chirino C, Durante-Montiel I, Sánchez-Rivera G, Valadez-Vega C, Morales-González JA (2011) Inflammation, oxidative stress, and obesity. Int J Mol Sci 12(5):3117–3132

Ford ES, Li C, Zhao G, Pearson WS, Mokdad AH (2008) Prevalence of the metabolic syndrome among U.S. adolescents using the definition from the International Diabetes Federation. Diabetes Care 31(3):587–589

Fox C, Esparza J, Nicolson M, Bennett PH, Schulz LO, Valencia ME, Ravussin E (1999) Plasma leptin concentrations in Pima Indians living in drastically different environments. Diabetes Care 22(3):413–417

Franks PW, Brage S, Luan J, Ekelund U, Rahman M, Farooqi IS, Halsall I, O'Rahilly S, Wareham NJ (2005) Leptin predicts a worsening of the features of the metabolic syndrome independently of obesity. Obes Res 13(8):1476–1484

Friedman JM, Halaas JL (1998) Leptin and the regulation of body weight in mammals. Nature 395:763–770

Funahashi T, Nakamura T, Shimomura I, Maeda K, Kuriyama H, Takahashi M, Arita Y, Kihara, S, Matsuzawa Y (1999) Role of adipocytokines on the pathogenesis of atherosclerosis in visceral obesity. Internal Medicine 38(2):202–206

Galisteo M, Sánchez M, Vera R, González M, Anguera A, Duarte J, Zarzuelo A (2005) A diet supplemented with husks of plantago ovata reduces the development of endothelial dysfunction,

hypertension, and obesity by affecting adiponectin and TNF-alpha in obese zucker rats. J Nutr 135:2399–2404
Galletti F, Barbato A, Versiero M, Iacone R, Russo O, Barba G, Siani A, Cappuccio FP, Farinaro E, Della Valle E, Strazzullo P (2007) Circulating leptin levels predict the development of metabolic syndrome in middle-aged men: an 8-year follow-up study. J Hypertens 25(8):1671–1677
Giles TD, Sander GE (2007) Effects of thiazolidindiones on blood pressure. Curr Hypertens Rep 9(4):332–337
Grundy SM (2004) Obesity, metabolic syndrome, and cardiovascular disease. J Clin Endocrinol Metab 89(6):2595–2560
Grundy SM (2007) Metabolic syndrome: a multiplex cardiovascular risk factor. J Clin Endocrinol Metab 92(2):399–404
Grundy SM, Cleeman JI, Daniels SR, Donato KA, Eckel RH, Franklin BA, Gordon DJ, Krauss RM, Savage PJ, Smith SC, Spertus JA, Costa F (2005) Diagnosis and management of the metabolic syndrome: an American Heart Association/National Heart, Lung, and Blood Institute Scientific Statement. Circulation 112:2735–2752
Gutierrez-Juarez R, Obici S, Rossetti L (2004) Melanocortin-independent effects of leptin on hepatic glucose fluxes. J Biol Chem 279:49704–49715
Hall JE, da Silva AA, Brandon E, Stec DE, Ying Z, Jones DW (2007) Pathophysiology of obesity hypertension and target organ injury. In: Lip GYP, Hall JE (Eds) Comprehensive Hypertension. Elsevier, New York 447–468
Hirosumi J, Tuncman G, Chang L, Gorgun CZ, Uysal KT, Maeda K, Karin M, Hotamisligil GS (2002) A central role for JNK in obesity and insulin resistance. Nature 420:333–336
Hotamisligil GS (2006) Inflammation and metabolic disorders. Nature 444:860–867
Hotamisligil GS, Shargill NS, Spiegelman BM (1993) Adipose expression of tumor necrosis factor-alpha: direct role in obesity-linked insulin resistance. Science 259(5091):87–91
Hotta K, Funahashi T, Bodkin NL, Ortmeyer HK, Arita Y, Hansen BC, Matsuzawa Y (2001) Circulating concentrations of the adipocyte protein adiponectin are decreased in parallel with reduced insulin sensitivity during the progression to type 2 diabetes in rhesus monkeys. Diabetes 50:1126–1133
Hulthe J, Hultén LM, Fagerberg B (2003) Low adipocyte-derived plasma protein adiponectin concentrations are associated with the metabolic syndrome and small dense low-density lipoprotein particles: atherosclerosis and insulin resistance study. Metabolism 52(12):1612–1614
Isomaa B, Almgren P, Tuomi T, Forsén B, Lahti K, Nissén M, Taskinen MR, Groop L (2001) Cardiovascular morbidity and mortality associated with the metabolic syndrome. Diabetes Care 24:683–689
Jensen MD (2008) Role of body fat distribution and the metabolic complications of obesity. J Clin Endocrinol Metab 93(11 Suppl 1):57–63
Jung HS, Youn BS, Cho YM, Yu KY, Park HJ, Shin CS, Kim SY, Lee HK, Park KS (2005) The effects of rosiglitazone and metformin on the plasma concentrations of resistin in patients with type 2 diabetes mellitus. Metabolism 54:314–320
Kadowaki T, Yamauchi T (2005) Adiponectin and adiponectin receptors. Endoc.r Rev 26:439–451
Kamin D, Hadigan C, Lehrke M, Mazza S, Lazar MA, Grinspoon S (2005) Resistin levels in HIV-infected patients with lipoatrophy decrease in response to rosiglitazone. J Clin Endocrinol Metab 90:3423–3426
Keay S, Grossberg SE (1980). Interferon inhibits the conversion of 3T3-L1 mouse fibroblasts into adipocytes. Proc Natl Acad Sci USA 77(7):4099–4103
Kim JY, van de Wall E, Laplante M, Azzara A, Trujillo ME, Hofmann SM, Schraw T, Durand JL, Li H, Li G, Jelicks LA, Mehler MF, Hui DY, Deshaies Y, Shulman GI, Schwartz GJ, Scherer PE (2007) Obesity-associated improvements in metabolic profile through expansion of adipose tissue. J Clin Invest 117(9):2621–2637
Kobayashi H, Ouchi N, Kihara S, Walsh K, Kumada M, Abe Y, Funahashi T, Matsuzawa Y (2004) Selective suppression of endothelial cell apoptosis by the high molecular weight form of adiponectin. Circ Res 94:e27–31

Kondo H, Shimomura I, Matsukawa Y, Kumada M, Takahashi M, Matsuda M, Ouchi N, Kihara S, Kawamoto T, Sumitsuji S, Funahashi T, Matsuzawa Y (2002) Association of adiponectin mutation with type 2 diabetes. Diabetes 51:2325–2328

Lehr S, Hartwig S, Lamers D, Famulla S, Mueller S, Hanisch FG, Cuvelier C, Ruige J, Eckardt K, Ouwens DM, Sell H, Eckel J (2011) Identification and validation of novel adipokines released from primary human adipocytes. Mol Cell Proteomics 26 Sep

Lehrke M, Reilly MP, Millington SC, Iqbal N, Rader DJ, Lazar MA (2004). An inflammatory cascade leading to hyperresistinemia in humans. PLoS Med 1:e45

Leyva F, Godsland IF (1998) Hyperleptinemia as a component of a metabolic syndrome of cardiovascular risk. Arterioscler Throm Vasc Biol 8:928–933

Liu L, Karkanias GB, Morales JC, Hawkins M, Barzilai N, Wang J, Rossetti L (1998) Intracerebroventricular leptin regulates hepatic but not peripheral glucose fluxes. J Biol Chem 273:31160–31167

Liu M, Zhou L, Xu A, Lam KS, Wetzel MD, Xiang R, Zhang J, Xin X, Dong LQ, Liu F (2008) A disulfide-bond a oxidoreductase-like protein (DsbA-L) regulates adiponectin multimerization. Proc Natl Acad Sci USA 105:18302–18307

Lorenzo C, Williams K, Hunt KJ, Haffner SM (2007) The National Cholesterol Education Program-Adult Treatment Panel III, International Diabetes Federation, and World Health Organization definitions of the metabolic syndrome as predictors of incident cardiovascular disease and diabetes. Diabetes Care 30(1):8–13

Maeda N, Takahashi M, Funahashi T, Kihara S, Nishizawa H, Kishida K, Nagaretani H, Matsuda M, Komuro R, Ouchi N, Kuriyama H, Hotta K, Akamura T, Shimomura I, Matsuzawa Y (2001) PPAR gamma ligands increase expression and plasma concentrations of adiponectin an adipose derived protein. Diabetes 50:2094–2099

Matsuzawa Y (2010) Establishment of a concept of visceral fat syndrome and discovery of adiponectin. Proc Jpn Acad Ser B Phys Biol Sci 86:131–141

McGillicuddy FC, Chiquoine EH, Hinkle CC, Kim RJ, Shah R, Roche HM, Smyth EM, Reilly MP (2009) Interferon gamma attenuates insulin signaling, lipid storage, and differentiation in human adipocytes via activation of the JAK/STAT pathway. J Biol Chem 284(46):31936–31944

Mertens I, Verrijken A, Michiels JJ, Van der Planken M, Ruige JB, Van Gaal LF (2006) Among inflammation and coagulation markers, PAI-1 is a true component of the metabolic syndrome. Int J Obes (Lond) 30(8):1308–1314

Minokoshi Y, Kim YB (2002). Leptin stimulates fatty acid oxidation by activating AMP-activated protein kinasi. Nature 415:339–343

Misra P (2008) AMP activated protein kinase: a next generation target for total metabolic control. Expert Opin Ther Targets 12(1):91–100

Mlinar B, Marc J (2011) Review: new insights into adipose tissue dysfunction in insulin resistance. Clin Chem Lab Med Sep 6. [Epub ahead of print]

Mohamed-Ali V, Goodrick S, Rawesh A, Katz DR, Miles JM, Yudkin JS, Klein S, Coppack SW (1997) Subcutaneous adipose tissue releases interleukin-6, but not tumor necrosis factor-alpha, in vivo. J Clin Endocrinol Metab 82:4196–4200

Muse ED, Obici S, Bhanot S, Monia BP, MacKay RA, Rajala MW, Scherer PE, Rossetti L (2004) Role of resistin in diet-induced hepatic insulin resistance. J Clin Invest 114:232–239

Neyrinck AM, Bindels LB, De Backer F, Pachikian BD, Cani PD, Delzenne NM (2009) Dietary supplementation with chitosan derived from mushrooms changes adipocytokine profile in diet-induced obese mice, a phenomenon linked to its lipid-lowering action. Int Immunopharmacol 9:767–773

Nilsson AC, Ostman EM, Holst JJ, Björck IM (2008) Including indigestible carbohydrates in the evening meal of healthy subjects improves glucose tolerance, lowers inflammatory markers, and increases satiety after a subsequent standardized breakfast. J Nutr 138:732–739

Norata GD, Ongari M, Garlaschelli K, Raselli S, Grigore L, Catapano AL (2007) Plasma resistin levels correlate with determinants of the metabolic syndrome. Eur J Endocrinol 156(2):279–284

O'Rourke RW, Metcalf MD, White AE, Madala A, Winters BR, Maizlin II, Jobe BA, Roberts CT Jr, Slifka MK, Marks DL (2009) Depot-specific differences in inflammatory mediators and a role for NK cells and IFN-gamma in inflammation in human adipose tissue. Int J Obes (Lond) 33:978–990

Okamoto Y, Kihara S, Ouchi N, Nishida M, Arita Y, Kumada M, Ohashi K, Sakai N, Shimomura I, Kobayashi H, Terasaka N, Inaba T, Funahashi T, Matsuzawa Y (2002) Adiponectin reduces atherosclerosis in apolipoprotein E-deficient mice. Circulation 106(22):2767–2770

Ouchi N, Kihara S, Arita Y, Maeda K, Kuriyama H, Okamoto Y, Hotta K, Nishida M, Takahashi M, Nakamura T, Yamashita S, Funahashi T, Matsuzawa Y (1999) Novel modulator for endothelial adhesion molecules: adipocyte-derived plasma protein adiponectin. Circulation 100(25):2473–2476

Ouchi N, Kobayashi H, Kihara S, Kumada M, Sato K, Inoue T, Funahashi T, Walsh K (2004) Adiponectin stimulates angiogenesis by promoting cross-talk between AMP-activated protein kinase and Akt signaling in endothelial cells. J Biol Chem 279:1304–1309

Ozcan U, Cao Q, Yilmaz E, Lee AH, Iwakoshi NN, Ozdelen E, Tuncman G, Görgün C, Glimcher LH, Hotamisligil GS (2004) Endoplasmic reticulum stress links obesity, insulin action, and type 2 diabetes. Science 306:457–461

Pajvani UB, Scherer PE (2003) Adiponectin: systemic contributor to insulin sensitivity. Curr Diab Rep 3(3):207–213

Pajvani UB, Hawkins M, Combs TP, Rajala MW, Doebber T, Berger JP, Wagner JA, Wu M, Knopps A, Xiang AH, Utzschneider KM, Kahn SE, Olefsky JM, Buchanan TA, Scherer PE (2004) Complex distribution, not absolute amount of adiponectin, correlates with thiazolidinedione-mediated improvement in insulin sensitivity. J Biol Chem 279:12152–12162

Patel L, Buckels AC, Kinghorn IJ, Murdock PR, Holbrook JD, Plumpton C, Macphee CH, Smith SA (2003) Resistin is expressed in human macrophages and directly regulated by PPARγ activators. Biochem Biophys Res Commun 300:472–476

Patel C, Ghanim H, Ravishankar S, Sia CL, Viswanathan P, Mohanty P, Dandona P (2007) Prolonged reactive oxygen species generation and nuclear factor-κB activation following a highfat high-carbohydrate meal in the obese. J Clin Endocrinol Metab 92:4476–4479

Petersen EW, Carey AL, Sacchetti M, Steinberg GR, Macaulay SL, Febbraio MA, Pedersen BK (2005) Acute IL-6 treatment increases fatty acid turnover in elderly humans in vivo and in tissue culture in vitro. Am J Physiol Endocrinol Metab 288:155–162

Pittas AG, Greenberg AS (2002) Thiazolidinediones in the treatment of diabetes. Expert Opin Pharmacother 3:529–540

Qatanani M, Szwergold NR, Greaves DR, Ahima RS, Lazar MA (2009) Macrophage-derived human resistin exacerbates adipose tissue inflammation and insulin resistance in mice. J Clin Invest 119(3):531–539

Qi L, Rimm E, Liu S, Rifai N, Hu FB (2005) Dietary glycemic index, glycemic load, cereal fiber, and plasma adiponectin concentration in diabetic men. Diabetes Care 28:1022–1028

Quinn CE, Hamilton PK, Lockhart CJ, McVeigh GE (2008) Thiazolidinediones: effects on insulin resistance and the cardiovascular system. Br J Pharmacol 153(4):636–645

Rabe K, Lehrke M, Parhofer KG, Broedl UC (2008) Adipokines and insulin resistance. Mol Med 14:741–751

Rahmouni K, Correia ML, Haynes WG, Mark AL (2005) Obesity-associated hypertension: new insights into mechanisms. Hypertension 45:9–14

Reasner CA (2002) Where thiazolinediones will fit. Diabetes Metab Res Rev 82:530–535

Recasens M, Ricart W, Fernández-Real JM (2004) Obesity and inflammation. Rev Med Univ Navarra 48:49–54

Reilly MP, Lehrke M, Wolfe ML, Rohatgi A, Lazar MA, Rader DJ (2005) Resistin is an inflammatory marker of atherosclerosis in humans. Circulation 111(7):932–939

Rocha VZ, Folco EJ, Sukhova G, Shimizu K, Gotsman I, Vernon AH, Libby P (2008) Interferon-gamma, a Th1 cytokine, regulates fat inflammation: a role for adaptive immunity in obesity. Circ Res 103:467–476

Ronti T, Lupattelli G, Mannarino E (2006) The endocrine function of adipose tissue: an update. Clin Endocrinol 64:355–365

Sarafidis PA, Georgianos PI, Lasaridis AN (2010) PPAR-γ agonism for cardiovascular and renal protection. Cardiovasc Ther Sep 15. doi:10.1111/j.1755-5922.2010.00222.x

Satoh H, Nguyen MT, Miles PD, Imamura T, Usui I, Olefsky JM (2004) Adenovirus-mediated chronic "hyper-resistinemia" leads to in vivo insulin resistance in normal rats. J Clin Invest 114:224–231

Savage DB, Setter CP, Klenk ES, Segal DG, Vidal-Puig A, Considine RV, O'Rahilly S (2001) Resistin/fizz3 expression in relation to obesity and peroxisome proliferator-activated receptor-γ action in humans. Diabetes 50:2199–2202

Schroder M, Kaufman RJ (2005) The mammalian unfolded protein response. Annu Rev Biochem 74:739–789

Semple RK, Soos MA, Luan J, Mitchell CS, Wilson JC, Gurnell M, Cochran EK, Gorden P, Chatterjee VK, Wareham NJ, O'Rahilly S (2006) Elevated plasma adiponectin in humans with genetically defective insulin receptors. J Clin Endocrinol Metab 91:3219–3223

Shaw RJ, Lamia KA, Vasquez D, Koo SH, Bardeesy N, Depinho RA, Montminy M, Cantley LC (2005) The kinase LKB1 mediates glucose homeostasis in liver and therapeutic effects of metformin. Science 310:1642–1646

Sierra-Honigmann MR, Nath AK, Murakami C, García-Cardena G, Papapetropoulos A, Sessa WC, Madge LA, Schechner JS, Schwabb MB, Polverini PJ, Flores-Riveros JR (1998) Biological action of leptin as an angiogenic factor. Science 281:1683–1686

Simons PJ, van den Pangaart PS, Aerts JM, Boon L (2007) Pro-inflammatory delipidizing cytokines reduce adiponectin secretion from human adipocytes without affecting adiponectin oligomerization. J Endocrinol 192(2):289–299

Spruijt-Metz D, Belcher B, Anderson D, Lane CJ, Chou CP, Salter-Venzon D, Davis JN, Hsu YW, Neuhouser ML, Richey JM, McKenzie TL, McClain A, Goran MI, Weigensberg MJ (2009) High-Sugar/Low-Fiber Meal Compared with a Low-Sugar/High-Fiber Meal Leads to Higher Leptin and Physical Activity Levels in Overweight Latina Females. J Am Diet Assoc 109:1058–1063

Stefan N, Vozarova B, Funahashi T, Matsuzawa Y, Weyer C, Lindsay RS, Youngren JF, Havel PJ, Pratley RE, Bogardus C, Tataranni PA (2002). Plasma adiponectin concentration is associated with skeletal muscle insulin receptor tyrosine phosphorylation and low plasma concentration precedes a decrease in whole-body insulin sensitivity in humans. Diabetes 51:1884–1888

Stephens JM, Carter BZ, Pekala PH, Malter JS (1992) Tumor necrosis factor alpha-induced glucose transporter (GLUT-1) mRNA stabilization in 3T3-L1 preadipocytes. Regulation by the adenosine-uridine binding factor. J Biol Chem 267(12):8336–8341

Stephens JM, Lee J, Pilch PF (1997) Tumor necrosis factor-alpha-induced insulin resistance in 3T3-L1 adipocytes is accompanied by a loss of insulin receptor substrate-1 and GLUT4 expression without a loss of insulin receptor-mediated signal transduction. J Biol Chem 272(2):971–976

Steppan CM, Bailey ST, Bhat S, Brown EJ, Banerjee RR, Wright CM, Patel HR, Ahima RS, Lazar MA (2001) The hormone resistin links obesity to diabetes. Nature 409:307–312

Steppan CM, Wang J, Whiteman EL, Birnbaum MJ, Lazar MA (2005) Activation of SOCS-3 by resistin. Mol Cell Biol 25(4):1569–1575

Strissel KJ, DeFuria J, Shaul ME, Bennett G, Greenberg AS, Obin MS (2010) T-cell recruitment and Th1 polarization in adipose tissue during diet-induced obesity in C57BL/6 mice. Obesity (Silver Spring) 18:1918–1925

Suganami T, Nishida J, Ogawa Y (2005) A paracrine loop between adipocytes and macrophages aggravates inflammatory changes: role of free fatty acids and tumor necrosis factor alpha. Arterioscler Thromb Vasc Biol, 25:2062–2068

Szmitko PE, Teoh H, Stwart DJ, Verma S (2007) Adiponectin and cardiovascular disease: state of the art? Am J Physiol Heart Circ Physiol 292(4):H1655–H1663

Tailor AM, Peeters PH, Norat T, Vineis P, Romaguera D (2010) An update on the prevalence of the metabolic syndrome in children and adolescents. Int J Pediatr Obes 5(3):202–13

Takahashi M, Funahashi T, Shimomura I, Miyaoka K, Matsuzawa Y (1996) Plasma leptin and body fat distribution. Horm Metab Res 28:751–752

Theodorou G, Lampidonis AD, Laliotis GP, Bizelis I, Politis I (2011) Expression of plasminogen activator-related genes in the adipose tissue of lactating dairy sheep in the early post-weaning period. J Anim Physiol Anim Nutr (Berl) doi:10.1111/j.1439-0396.2011.01154.x

Torti FM, Dieckmann B, Beutler B, Cerami A, Ringold GM (1985) A macrophage factor inhibits adipocyte gene expression: an in vitro model of cachexia. Science 229(4716):867–869

Trayhurn P, Wood IS (2004) Adipokines: inflammation and the pleiotropic role of white adipose tissue. Br J Nutr 92:347–355

Unger RH (2002) "Lipotoxic diseases". Annu Rev Med 53:319–336

Utzschneider KM, Carr DB, Tong J, Wallace TM, Hull RL, Zraika S, Xiao Q, Mistry JS, Retzlaff BM, Knopp RH, Kahn SE (2005) Resistin is not associated with insulin sensitivity or the metabolic syndrome in humans. Diabetologia 48(11):2330–2333

Van Hall G, Steensberg A, Sacchetti M, Fischer C, Keller C, Schjerling P, Hiscock N, Møller K, Saltin B, Febbraio MA, Pedersen BK 2003. Interleukin-6 stimulates lipolysis and fat oxidation in humans. J Clin Endocrinol Metab 88:3005–3010

Wajchenberg BL (2000) Subcutaneous and visceral adipose tissue: their relation to the metabolic syndrome. Endocr Rev 21:697–738

Waki H, Yamauchi T, Kamon J, Ito Y, Uchida S, Kita S, Hara K, Hada Y, Vasseur F, Froguel P, Kimura S, Nagai R, Kadowaki T (2003) Impaired multimerization of human adiponectin mutants associated with diabetes: molecular structure and multimer formation of adiponectin. J Biol Chem 278:40352–40363

Wang H, Kouri G, Wollheim CB (2005) ER stress and SREBP-1 activation are implicated in β-cell glucolipotoxicity. J Cell Sci 118:3905–3915

Wang AP, Li X, Zheng Y, Liu BL, Huang G, Yan X, Liu Z, Zhou Z (2010) Thiazolidinediones protect mouse pancreatic β-cells directly from cytokine-induced cytotoxicity through PPARγ-dependent mechanisms. Acta Diabetol [Epub ahead of print]

Wang MY, Chen L, Clark GO, Lee Y, Stevens RD, Ilkayeva OR, Wenner BR, Bain JR, Charron MJ, Newgard CB, Unger RH (2010) Leptin therapy in insulin-deficient type I diabetes. Proc Natl Acad Sci USA 107(11):4813–4819

Wang ZV, Scherer PE (2008) DsbA-L is a versatile player in adiponectin secretion. Proc Natl Acad Sci USA 105:18077–18078

Wellen KE, Hotamisligil GS (2005). Inflammation, stress, and diabetes. J Clin Invest 115:1111–1119

Wilson-Fritch L, Burkart A, Bell G, Mendelson K, Leszyk J, Nicoloro S, Czech M, Corvera S (2003) Mitochondrial biogenesis and remodeling during adipogenesis and in response to the insulin sensitizer rosiglitazone. Mol Cell Biol 23:1085–1094

Wilson-Fritch L, Nicoloro S, Chouinard M, Lazar MA, Chui PC, Leszyk J, Straubhaar J, Czech MP, Corvera S (2004) Mitochondrial remodeling in adipose tissue associated with obesity and treatment with rosiglitazone. J Clin Invest 114:1281–1289

Wyne KL (2003) Free fatty acids and type 2 diabetes mellitus. Am J Med 115(Suppl) 8A:29–36

Xu H, Uysal KT, Becherer JD, Arner P, Hotamisligil GS (2002) Altered tumor necrosis factor-alpha (TNF-alpha) processing in adipocytes and increased expression of transmembrane TNF-alpha in obesity. Diabetes 51:1876–1883

Yamauchi T, Kamon J, Waki H, Terauchi Y, Kubota N, Hara K, Mori Y, Ide T, Murakami K, Tsuboyama-Kasaoka N, Ezaki O, Akanuma Y, Gavrilova O, Vinson C, Reitman ML, Kagechika H, Shudo K, Yoda M, Nakano Y, Tobe K, Nagai R, Kimura S, Tomita M, Froguel P, Kadowaki T (2001) The fat derived hormone adiponectin reverses insulin resistance associated with both lipoatrophy and obesity. Nat Med 7:941–946

Yamauchi T, Hara K, Kubota N, Terauchi Y, Tobe K, Froguel P, Nagai R, Kadowaki T (2003a) Dual roles of adiponectin/Acrp30 in vivo as an anti-diabetic and anti-atherogenic adipokine. Curr Drug Targets Immune Endocr Metabol Disord 3(4):243–54

Yamauchi T, Kamon J, Waki H, Imai Y, Shimozawa N, Hioki K (2003b) Globular adiponectin Protected ob/ob mice from diabetes and ApoE-deficient mice from atherosclerosis. J Biol Chem 278(4):2461–2468

Yang H, Youm YH, Vandanmagsar B, Ravussin A, Gimble JM, Greenway F, Stephens JM, Mynatt RL, Dixit VD (2010) Obesity increases the production of proinflammatory mediators from adipose tissue T cells and compromises TCR repertoire diversity: implications for systemic inflammation and insulin resistance. J Immunol 185:1836–1845

Yannakoulia M, Yiannakouris N, Melistas L, Kontogianni MD, Malagaris I, Mantzoros CS (2008) A dietary pattern characterized by high consumption of whole-grain cereals and low-fat dairy products and low consumption of refined cereals is positively associated with plasma adiponectin levels in healthy women. Metabolism 57:824–830

Zhang Y, Proenca R, Maffei M, Barone M, Leopold L, Friedman JM (1994) Positional cloning of the mouse obese gene and its human homologue. Nature 372:425–432

Zhou G, Myers R, Li Y, Chen Y, Shen X, Fenyk-Melody J, Wu M, Ventre J, Doebber T, Fujii N, Musi N, Hirshman MF, Goodyear LJ, Moller DE (2001). Role of AMP-activated protein kinase in mechanism of metformin action. J Clin Invest 108:1167–1174

Zhou L, Liu M, Zhang J, Chen H, Dong LQ, Liu F (2010) DsbA-L alleviates endoplasmic reticulum stress-induced adiponectin downregulation. Diabetes 59(11):2809–2816

Zimmet P, Boyko EJ, Collier GR, de Courten M (1999) Etiology of the metabolic syndrome: potential role of insuline resistance, leptin resistance and other players. Ann NY Acad Sci 892:25–44

Chapter 5
The Beta Cell in Metabolic Syndrome

Bader Zarrouki, Ghislaine Fontés, Meriem Semache, Julie Amyot and Vincent Poitout

Abstract The pancreatic beta cell is equipped with a highly sophisticated machinery to precisely sense the metabolic status of the organism and secrete the exactly appropriate amount of insulin to maintain blood glucose levels in a very narrow range. When the Metabolic Syndrome develops, insulin resistance imposes an additional burden on the beta cell, which then hypersecretes insulin to meet the demand. In the majority of individuals, the beta cell can sustain this additional workload and maintain normoglycemia. In a subset of predisposed individuals however, this compensatory response eventually fails and diabetes develops. Once diabetes is established, beta-cell function continues to deteriorate over time. The molecular and cellular mechanisms underlying beta-cell failure are not fully understood, although several hypotheses have been proposed. Amongst these, glucolipotoxicity; defective mitochondrial metabolism and oxidative stress; inflammation; amyloid deposits, disruption of autophagic flux; endoplasmic reticulum stress; beta-cell dedifferentiation and exhaustion from chronic hypersecretion probably contribute to some extent, perhaps at various stages of the disease progression and differently between individuals. Thus, beta-cell failure is likely mediated by a number of interrelated and complex mechanisms, which is reflected in the inability of the current therapeutic options to significantly slow down disease progression. This provides a strong argument for early interventions aimed at preventing the functional demise of pancreatic beta cells in the Metabolic Syndrome.

Keywords Beta-cell · Glucolipotoxicity · Insulin · Islet · Metabolic syndrome

---

V. Poitout (✉) · B. Zarrouki · G. Fontés · M. Semache · J. Amyot
Montreal Diabetes Research Center, CRCHUM—Technopôle Angus,
2901 Rachel Est, Montréal, H1W 4A4, QC, Canada
e-mail: vincent.poitout@umontreal.ca

V. Poitout · B. Zarrouki · G. Fontés
Department of Medicine, University of Montreal, Montreal, QC, Canada

V. Poitout · M. Semache · J. Amyot
Department of Biochemistry, University of Montreal, Montreal, QC, Canada

V. Poitout
Department of Nutrition, University of Montreal, Montreal, QC, Canada

M. Orešič, A. Vidal-Puig (eds.), *A Systems Biology Approach to Study Metabolic Syndrome*,
DOI 10.1007/978-3-319-01008-3_5,

5.1 Introduction

The dramatic increase in the incidence of diabetes, which is expected to affect more than 550 million people worldwide by 2030, represents a tremendous economical and social burden (Whiting et al. 2011). Obesity is a major risk factor for type 2 diabetes (T2D), which develops in a subset of genetically predisposed individuals (Lusis et al. 2008) when insulin secretion becomes insufficient to adequately compensate for insulin resistance. Genetic evidence supporting a critical role for the beta cell in the pathogenesis of T2D lies in the observation that the vast majority of genes identified in genome-wide association studies for T2D are related to beta-cell development or function (Herder and Roden 2011). Nonetheless, the combined contribution of identified polymorphisms to the risk of developing diabetes remains less than 10 % (Herder and Roden 2011; Lusis et al. 2008), illustrating the polygenic nature of this complex disease and the major role of environmental and/or epigenetic factors. In addition, several longitudinal studies in humans have shown that beta-cell function declines over time following the initial diagnosis (1995, 1998), and have suggested that the beta-cell secretory capacity is already significantly impaired at the time of diagnosis (Holman 1998). Therefore, identification of the early environmental factors that impair beta-cell function and promote its demise is a key prerequisite to any therapeutic strategy aimed at preserving insulin secretion and preventing T2D in the context of the Metabolic Syndrome (MetS). In fact, all metabolic perturbations that define the MetS, namely hypertriglyceridemia, reduced HDL cholesterol and fasting hyperglycemia (Alberti et al. 2009), have been shown to contribute to beta-cell dysfunction and failure.

The purpose of this chapter is to review our current understanding of the factors underlying the progressive deterioration of beta-cell function in the context of the MetS. We have particularly emphasized glucolipotoxicity, a major focus of interest in our laboratory, but have also attempted to present recent findings supporting the contribution of mitochondrial dysfunction, inflammation, autophagy, endoplasmic reticulum (ER) stress, beta-cell dedifferentiation and exhaustion in beta-cell failure. Not mentioned in this chapter, albeit clearly important, are the epigenetic changes that occur in the beta cell in response to the foetal or neonatal environment and that predispose to subsequent failure in the face of the MetS. These have been extensively reviewed elsewhere (Nolan et al. 2011) and we will only focus here on the environmental factors proposed to contribute to beta-cell failure in the adult, once the MetS is established.

5.2 Physiological Regulation of Pancreatic Beta-cell Function

The pancreatic beta cell is the key cell in the body that produces and secretes insulin, the only hypoglycaemic hormone. Beta cells are located within the islets of Langerhans of the pancreas and represent approximately 80 % of the islet mass (which itself represents 2–3 % of the total pancreatic mass), the remaining 20 % being accounted

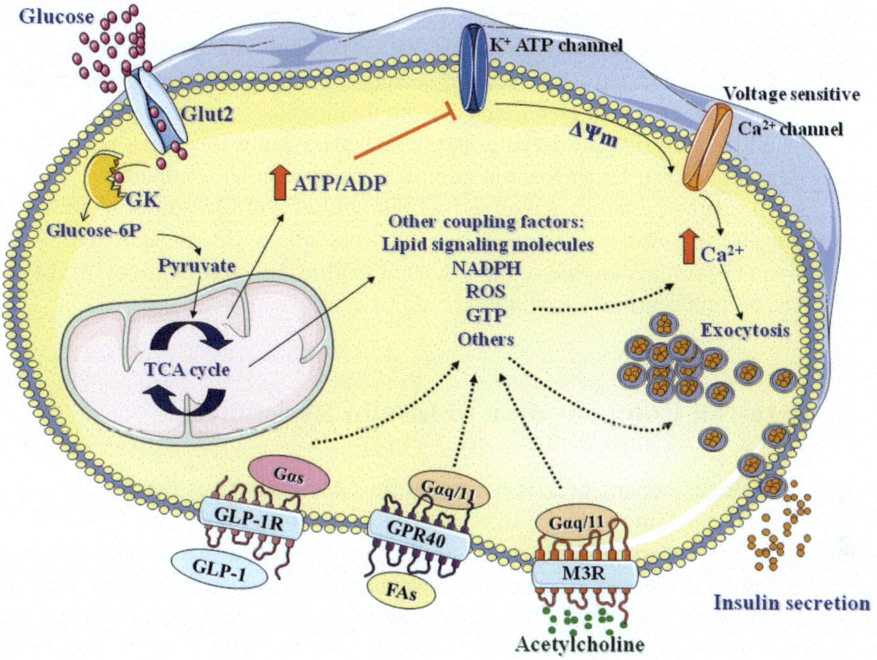

Fig. 5.1 Regulation of insulin secretion. The increase in *ATP/ADP* ratio following glucose metabolism leads to the closure of the ATP-sensitive *K + channel*, membrane depolarization, opening of voltage-sensitive calcium channels, calcium influx, and insulin exocytosis. This pathway is further amplified by other intracellular signals. Several extracellular molecules potentiate insulin secretion through activation of *GPCRs*. These receptors are coupled to downstream signalling pathways leading to an increase in cAMP levels and/or intracellular calcium that synergize to amplify insulin secretion. *GLP-1* Glucagon-like peptide-1. *FAs* Fatty acids. *M3R* Muscarinic receptor 3. This figure was produced using Servier Medical Art

for by glucagon-secreting alpha cells; somatostatin-secreting delta cells; and pancreatic polypeptide-secreting PP cells. Under normal circumstances, blood glucose levels are maintained within a very narrow range despite large variations in food intake and energy expenditure. This is achieved by a very tight regulation of insulin secretion from the beta cell, which continuously adjusts its output in response to metabolic, hormonal, and neuronal signals (Fig. 5.1). Amongst these, glucose is the major regulator of insulin release. The beta cell is equipped with a glucose-sensing apparatus consisting of the glucose transporter GLUT2 and the high-Km, low-affinity isoform of hexokinase glucokinase, the kinetic properties of which ensure that intracellular glucose concentrations rapidly equilibrate with extracellular levels within the physiological and supraphysiological range of glycemia (approximately 5.5–16.7 mM). Following its uptake and phosphorylation, glucose (in the form of glucose-6-phosphate) enters glycolysis to form pyruvate, which is then metabolized into the Krebs cycle to generate ATP. Acceleration of glucose metabolism results in an increase in intracellular ATP/ADP ratio, closure of the ATP-sensitive K + channel, membrane depolarization, opening of voltage-sensitive calcium channels, calcium influx, and insulin exocytosis (Henquin 2000). This main—or triggering—pathway of coupling between glucose and insulin secretion is complemented by additional, amplifying pathways which essentially potentiate the secretory response to calcium influx via additional glucose-derived metabolites, the nature of which remains debated (Henquin 2009). However, the amplifying pathway is not operative without the triggering pathway being activated first (Henquin 2000). In addition, a number of nutrient and non-nutrient secretagogues can potentiate glucose-induced insulin secretion via G-protein coupled receptors, such as glucagon-like peptide-1 (GLP-1), fatty acids, and muscarinic stimulation (Fig. 5.1).

5.3 Beta-cell Compensation to Insulin Resistance

Insulin resistance, a characteristic feature of the MetS, increases the burden on the beta cell to produce more insulin to compensate for its decreased efficiency. In humans, this is believed to occur via two complementary mechanisms: enhanced insulin secretion and increased beta-cell mass. It has been appreciated for a long time that in obese humans, insulin secretion is considerably enhanced to maintain normoglycemia (Reaven et al. 1993). More recently, Butler et al. (Butler et al. 2003) have shown that 1) beta-cell mass is increased in obese non-diabetic vs. lean individuals; and 2) beta-cell mass is decreased in both obese individuals with impaired fasting glucose and T2D patients. These observations suggest that part of the beta-cell compensation to insulin resistance involves an increase in mass that, as T2D occurs and progresses, subsequently declines.

Whether the expansion in beta-cell mass results from proliferation of existing beta cells or neogenesis from ductal or acinar cells remains an intense topic of debate. While beta-cell replication clearly plays a major role in mice (Dor et al. 2004), evidence suggests that beta-cell proliferation rates are extremely low in adult humans (Meier et al. 2008) and that neogenesis from ductal cells contributes to newly formed beta-cells (Inada et al. 2008). In addition, a recent study in primates (Saisho et al. 2011) showed that neogenesis is the major supplier of newly formed adult beta-cells (80 %).

The signalling and molecular mechanisms underlying beta-cell compensation to insulin resistance in the MetS are essentially unknown; however, experimental evidence points to an important role for insulin signalling in both the liver and the beta cell. Perturbation of insulin signalling in the liver results in insulin resistance (Michael et al. 2000), which in turn is proposed to drive beta-cell hyperplasia, as demonstrated for example by the large increase in beta-cell mass upon deletion of the insulin receptor specifically in the liver (Escribano et al. 2009). Beta-cell proliferation in response to liver insulin resistance is abolished by pancreatic vagotomy, pointing to an important role of neuronal signals in mediating this response (Imai et al. 2008). Alternatively, it has been suggested that liver-derived circulating factors could play a role in beta-cell adaptation to insulin resistance (El Ouaamari et al. 2013).

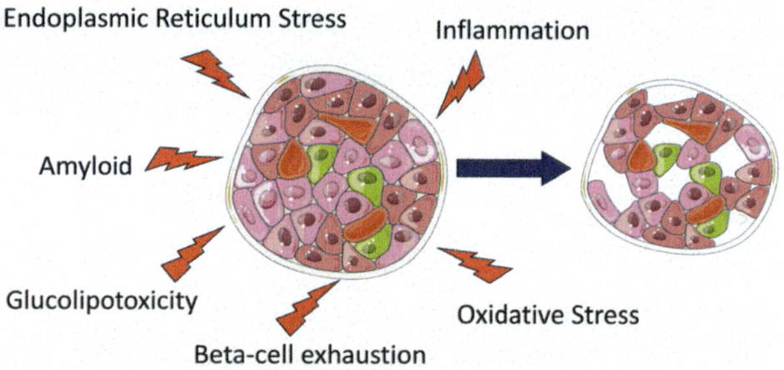

Fig. 5.2 Beta-cell dysfunction and failure in type 2 diabetes. The progressive decline in beta-cell function and mass is a result of interrelated mechanisms occurring at various stages in a subset of predisposed individuals. The relative contribution of each of these mechanisms is still not fully understood. This figure was produced using Servier Medical Art

The initial observation that beta-cell specific deletion of the insulin receptor leads to a decrease in beta-cell mass and defective glucose-induced insulin secretion (Kulkarni et al. 1999) provided the first demonstration for a role of insulin signalling in the beta cell in glucose homeostasis. This has been confirmed by subsequent studies examining the consequences of beta-cell specific deletion of genes encoding proteins in the insulin signalling pathway (reviewed in Kasuga 2006), such as insulin receptor substrate-2 (IRS-2) (Choudhury et al. 2005; Kubota et al. 2004), phosphatidyl-inositol-3 kinase (PI3K) (Kaneko et al. 2010), and 3-phosphoinositide-dependent kinase 1 (PDK-1) (Hashimoto et al. 2006). In humans, Bouche et al. (Bouche et al. 2010) have shown that preexposure of healthy subjects to exogenous insulin potentiates endogenous insulin secretion in response to glucose independently from changes in circulating free fatty acid levels (Lopez et al. 2011), supporting a role for insulin signalling in beta-cell function. This effect depends on both the prevailing levels of insulin sensitivity and the circulating insulin levels, but its relatively small magnitude makes it unlikely to contribute significantly to beta-cell compensation in the MetS (Mari et al. 2011).

5.4 Beta-cell Decompensation and Failure

In a subset of genetically predisposed individuals, the mechanisms described above are insufficient to adequately compensate for insulin resistance, diabetes occurs, and beta-cell function continues to deteriorate over time. A number of different phenomena have been proposed to contribute to beta-cell failure (Fig. 5.2). We consider it unlikely that any of these phenomena alone is sufficient to explain beta-cell failure in T2D, and they all probably contribute, to a variable extent depending on the individuals and the stage of the disease, to beta-cell dysfunction. Most of the studies described below have been performed in vitro or in rodent models, and whether or not the findings are relevant to human T2D remains, in the majority of the cases, to be directly demonstrated. Despite these limitations, significant advances have been made in the past few years towards a better understanding of beta-cell dysfunction in T2D.

5.4.1 *Glucolipotoxicity*

The MetS provides a situation in which dyslipidemia (lipotoxicity) and hyperglycemia (glucotoxicity) are simultaneously present and synergistically alter the function of multiple organs, including the beta cell (Poitout and Robertson 2008). Accordingly, a large number of ex vivo and in vivo studies in rodents have shown that excessive fatty acids levels, in the presence of high glucose, have detrimental effects on beta-cell function and, in some cases, viability. Investigations in humans are admittedly less conclusive, in part due to the inherent difficulty in designing clinical studies to examine the long-term effects of nutrient excess on glucose homeostasis. Nonetheless, Carpentier et al. (Carpentier et al. 1999) have shown that insulin secretion is increased upon acute lipid infusion in healthy subjects but diminished when the infusion is prolonged to 48 h, although the latter finding was not confirmed in other studies (Boden et al. 1995; Boden and Chen 1999). Interestingly, inhibition of insulin secretion by prolonged lipid infusion appears specific to glucose, since arginine-induced secretion was not affected (Carpentier et al. 2001). Finally, the susceptibility to the inhibitory effects of lipid infusion on insulin secretion appears to be related both to the family history of T2D (Kashyap et al. 2003) and to the presence of mild hyperglycemia (Carpentier et al. 2010), the latter being consistent with the concept of glucolipotoxicity.

A large number of in vitro and preclinical studies have attempted to identify the mechanisms underlying glucolipotoxicity. Convincing evidence has been presented in favour of metabolic dysfunction (Prentki and Madiraju 2008), ER stress (Cnop et al. 2008), oxidative stress (Oprescu et al. 2007), and inflammation (Boni-Schnetzler et al. 2009) as playing a role in glucolipotoxicity of the beta-cell. This list of potential mechanisms illustrates the fact that, somewhat surprisingly, the precise molecular and cellular mechanisms of glucolipotoxicity remain debated. We believe that this is due to a number of confounding factors and differences in the experimental conditions, perhaps the most important being whether or not glucolipotoxicity is studied under conditions where significant cell death is observed. This is dependent upon the islet species as well as the concentration and nature of the fatty acids used. For example, a 24 h exposure of human islets to elevated glucose and palmitate is sufficient to observe apoptosis (El-Assaad et al. 2003), whereas cell death is not observed in rat islets after 72 h of culture under similar conditions (Kelpe et al. 2003; Moore et al. 2004; Fontes et al. 2009). Second, the fatty-acid potency is determined by the very small fraction of the total concentration that is unbound to albumin, which itself is a function of the molar ratio of fatty acids to albumin. For example, a total concentration of 0.5 mM palmitate complexed to bovine serum albumin at a molar ratio of 5:1 corresponds to an unbound concentration in the range of 20 nM (Poitout et al. 2010). Since the mode of preparation and fatty acid to albumin molar ratio vary significantly between studies, the "biologically active" fraction varies as well and with it the functional effects observed. Third, all fatty acids were not created equal and, for example, saturated fatty acids are proapoptotic while unsaturated fatty acids are not and are even protective against the effects of saturated fatty acids (Maedler et al. 2003). Thus, the greater the ability of beta-cells to desaturate fatty acids the better they survive a lipotoxic environment (Busch et al. 2005).

Another variable to consider when interpreting in vitro studies is whether the effects of lipids were examined with (glucolipotoxicity) or without (lipotoxicity) the concomitant presence of high glucose. In our experience, under conditions in which cell viability is not affected, the deleterious effects of fatty acids on beta-cell function are only observed when glucose levels are simultaneously elevated both in vitro (Jacqueminet et al. 2000; Briaud et al. 2001) and in vivo (Briaud et al. 2002; Hagman et al. 2008).

Finally, it is important to point out that the various functional effects of fatty acids have different underlying mechanism. Fatty-acid inhibition of insulin secretion is observed with both unsaturated and saturated long-chain fatty acids (Moore et al. 2004) and involves perturbations of the late stages of insulin exocytosis (Olofsson et al. 2007) by disrupting the association between calcium channels and secretory granules (Hoppa et al. 2009; Olofsson et al. 2007). In contrast, fatty-acid inhibition of insulin gene expression is restricted to saturated fatty acids such as palmitate (Moore et al. 2004), because it is mediated by de novo synthesis of ceramide (Kelpe et al. 2003) for which only palmitate can serve as a substrate. Palmitate inhibition of the insulin gene is a transcriptional effect involving altered binding of pancreas-duodenum homeobox-1 (Pdx-1) and mammalian homologue of avian MafA/L-Maf (MafA) to the insulin gene promoter (Hagman et al. 2005). A newly identified player in glucolipotoxicity of the beta cell is the serine-threonine kinase Per-Arnt-Sim kinase (PAS kinase). PAS kinase is an environmental sensor which was shown to play an important role in glucose regulation of insulin promoter activity (da Silva Xavier et al. 2004). In islets we have observed that palmitate blocks glucose-induced PASK expression and that overexpression of PASK protects from palmitate inhibition of the insulin gene (Fontes et al. 2009). Thus, our current working hypothesis is that palmitate, through de novo ceramide synthesis, blocks PASK expression and thereby alters the expression and/or activity of Pdx-1 and MafA, resulting in decreased insulin gene expression (Fig. 5.3). The precise mechanisms whereby PASK regulates Pdx-1 and MafA function are currently under investigation in our laboratory. Recent observations in humans support an important role for PASK in beta-cell function. First, PASK expression is reduced in islets from T2D patients (da Silva Xavier et al. 2010). Second, a spontaneous mutation in the gene encoding for PASK identified in young-onset diabetes in humans results in increased kinase activity and enhanced basal insulin secretion and gene expression (Semplici et al. 2011).

Recent efforts in our laboratory were aimed to examine whether the defect in insulin gene expression upon glucoliptoxic conditions was also observed in vivo. We established a rat model of chronic infusions of glucose + Intralipid (a triglyceride emulsion co-infused with heparin to raise circulating fatty-acid levels), and showed that after 72 h of infusion insulin mRNA was reduced in islets, associated

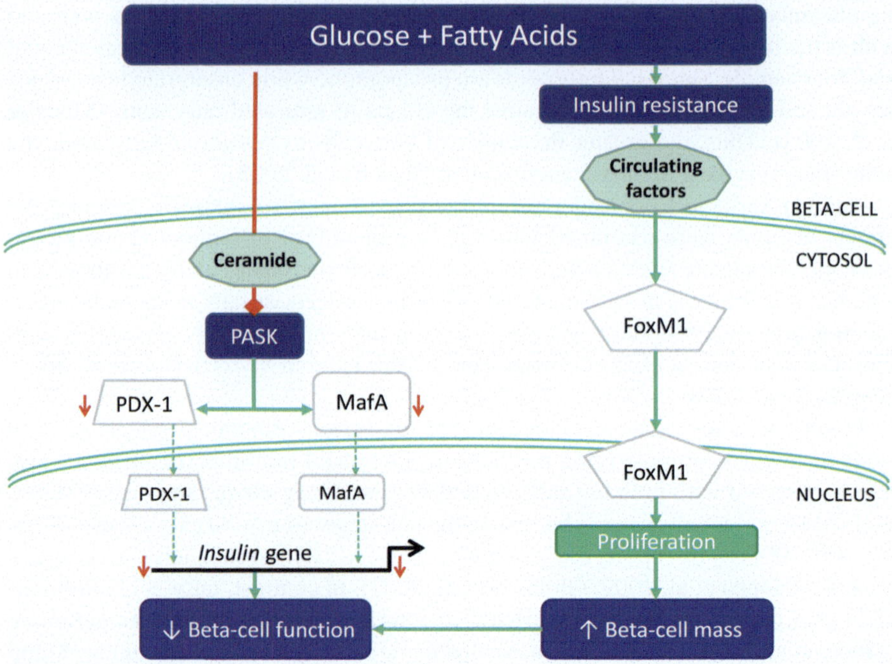

Fig. 5.3 Working model regarding the mechanisms of glucolipotoxicity. The impact of the synergistic effects of chronically elevated glucose and fatty acids is mediated at the cellular level by different mechanisms. Our current hypothesis is that the beta-cell response to glucolipotoxicity in vivo involves a detrimental arm and an adaptive arm. The detrimental arm is mediated by the intracellular generation of ceramide from palmitate, inhibition of *PASK* expression, and inhibition of the binding of *Pdx-1* and *MafA* to their cognate sequences on the insulin gene promoter resulting in a decrease in insulin gene expression. The adaptive arm is mediated by insulin resistance, probably at the liver, which through as yet unidentified circulating factors induces beta-cell proliferation via the *FoxM1* signalling pathway. It is also possible that the increase in beta-cell mass, if it produces functionally immature beta cells, indirectly contributes to defective beta-cell function

with nuclear exclusion of Pdx-1, thus recapitulating our previous ex vivo findings (Fontes et al. 2010; Hagman et al. 2008). Further, in 6-month-old rats a continuous infusion of glucose + Intralipid for 72 h induces insulin resistance and beta-cell dysfunction, characterized by defective glucose-induced insulin secretion and proinsulin biosynthesis (Fontes et al. 2010). Interestingly, this functional defect occurs despite a marked increase in beta-cell mass, largely due to beta-cell proliferation. Since these rats also develop insulin resistance in response to the infusion, these results are consistent with the possibility mentioned above that insulin resistance drives beta-cell proliferation in this model. Accordingly, a recent transcriptomic analysis showed that the most highly regulated pathways in islets from rats infused with glucose + Intralipid are related to cell proliferation, including an increase in the expression of the transcription factor FoxM1 and most of its targets (Zarrouki, Fontés, Poitout, unpublished results). Since FoxM1is necessary for beta-cell proliferation in a number of experimental conditions (Zhang et al. 2010; Ackermann Misfeldt et al. 2008;

Zhang et al. 2006) and is sufficient to induce beta-cell replication (Davis et al. 2010), our results are consistent with the possibility that in older rats subjected to chronic nutrient excess, insulin resistance promotes beta-cell proliferation via the FoxM1 pathway (Fig. 5.3). Thus, observations in this model indicate that the beta-cell response to metabolic stress includes a deleterious response (defective insulin gene expression and insulin secretion) as well as an adaptive response (increased beta-cell proliferation).

Intracellular accumulation of cholesterol has emerged in recent years as a novel pathogenic mechanism of beta-cell dysfunction. Indeed, dyslipidemia associated with the MetS is characterized by hypertriglyceridemia, decreased level of the anti-atherogenic lipoprotein HDL and increased level of the small and dense LDL pro-atherogenic lipoproteins (Alberti et al. 2009). Importantly, alterations of the lipid profile occurs long before the occurrence of T2D and is considered as an independent risk factor for the disease (von Eckardstein et al. 2000). In T2D patients, infusion of physiological levels of reconstituted HDL acutely restores beta-cell function in the absence of any improvement in insulin sensitivity (Drew et al. 2009). In addition, in vitro experiments have shown that HDL prevents beta-cell death induced by cytokines and elevated glucose levels while enhancing insulin secretion (Rutti et al. 2009). In contrast, in vitro treatment with LDL inhibits insulin secretion through the LDL receptor (Kruit et al. 2010; Rutti et al. 2009). Furthermore, oxidized LDL, which are increased in T2D patients (Colas et al. 2010), promote beta-cell apoptosis (Abderrahmani et al. 2007).

Intracellular cholesterol content is determined by cholesterol synthesis, uptake and efflux, and perturbations of any of these processes could lead to cholesterol accumulation. The efflux of cholesterol is mediated through the ABCA1 and ABCG1 transporters.

Patients suffering from Tangier disease due to loss-of-function mutations of ABCA1 are at a high risk for developing diabetes (Villarreal-Molina et al. 2007; Villarreal-Molina et al. 2008) and have defective insulin secretion (Koseki et al. 2009; Vergeer et al. 2010). Insights into the molecular mechanisms involved were provided by studies using a beta-cell specific mouse knock-out of ABCA1. Brunham et al. (Brunham et al. 2007) showed that deletion of ABCA1 in beta cells leads to impaired glucose tolerance due to defective insulin secretion in vivo as well as in isolated islets ex vivo, due to intracellular cholesterol accumulation. In a follow-up study, the same group showed that cholesterol accumulation in beta cells alters insulin exocytosis, and that conversely acute reduction of cholesterol content restores insulin secretion (Kruit et al. 2011). Another study addressed the role of ABCG1 in mice using beta-cell specific deletion (Sturek et al. 2010). The outcome was similar to that of the beta cell ABCA1 knock-out mouse, except that total cellular cholesterol level was not affected, but insulin secretory granule cholesterol composition was altered leading to morphological changes of the granules and subsequent impairment of exocytosis. Since mice fed a high-fat diet display an increase in islet cholesterol accumulation (Peyot et al. 2010), these recent observations, supported by genetic evidence in humans, suggest that abnormalities in intracellular cholesterol homeostasis might play a role in beta-cell failure in T2D.

5.4.2 *Mitochondrial Dysfunction and Oxidative Stress*

As described in Sec 5.2, mitochondrial metabolism of glucose and other nutrient secretagogues is crucial for the triggering pathway of insulin secretion. In humans, ultrastructural analysis showed mitochondrial hypertrophy in beta cells from T2D patients (Anello et al. 2005). This is associated with a decrease in insulin secretion from T2D islets and a right-shift of the dose-response curve to glucose, without changes in insulin secretion in response to arginine or sulfonylurea (Deng et al. 2004; Fernandez-Alvarez et al. 1994). These observations suggest a defect in mitochondrial glucose metabolism, supported by the observed decrease in glucose oxidation in islets from diabetic donors (Del Guerra et al. 2005; Fernandez-Alvarez et al. 1994). In this regard, studies in animal models of T2D have led to conflicting results. On the one hand, Fex et al. (Fex et al. 2007) have shown that islets isolated from high-fat fed mice have blunted insulin secretion in response to glucose associated with a decrease in glucose oxidation. In contrast, mitochondrial metabolism of other fuels such as palmitate was increased (Fex et al. 2007). On the other hand, in a similar model Peyot et al. (Peyot et al. 2010) did not detect any alteration in glucose oxidation in islets from high-fat fed mice despite a dramatic decrease in insulin secretion. Alternatively, it is conceivable that a defect in glucose oxidation might indirectly result, at least in part, from decreased glucokinase activity and reduced glucose flux through glycolysis upstream of the Krebs cycle. For example, a recent study in islets from human T2D donors showed that the impairment in insulin secretion and mitochondrial respiration was restored by a glucokinase activator (Doliba et al. 2011).

Oxidative stress has long been considered as a potential mechanism underlying beta-cell failure in T2D (Poitout and Robertson 2008), and mitochondrial metabolism of glucose promotes ROS accumulation in beta cells (Newsholme et al. 2007; Turrens 2003). During electron transfer within complexes I and II, O_2 is subjected to a single electron reduction that produces the highly reactive superoxide anion O_2^-. The latter is converted into hydrogen peroxide (H_2O_2) by the superoxide dismutase enzymes (SOD). Lastly, catalase and/or glutathione peroxidases (GPx) convert H_2O_2 into inert H_2O and O_2. It is now well accepted that ROS acutely generated in response to glucose represent an important coupling factor for insulin secretion (Pi et al. 2007; Leloup et al. 2009) but that excessive and sustained production of ROS under glucotoxic conditions results in oxidative stress (Poitout and Robertson 2008), to which the beta-cell is particularly sensitive due to an intrinsically low complement of antioxidant enzymes (Lenzen et al. 1996; Robertson 2009; Tonooka et al. 2007).

Analysis of human T2D islets revealed high levels of oxidative stress markers compared to control islets (Del Guerra et al. 2005). In the same study, a negative correlation was found between oxidative markers and insulin secretion in islets. Although the usefulness of antioxidant therapy in T2D remains debated (Golbidi et al. 2011), Paolisso et al. (Paolisso et al. 1992) demonstrated that infusion of the antioxidant glutathione improved glucose intolerance in aging patients. A role for chronic oxidative stress in beta-cell failure is supported by a large number of studies in rodents and isolated islets. Indeed, long-term culture of human islets in high glucose concentrations promotes cell death (Federici et al. 2001). In rodents, enhancing antioxidant defense by pharmacological or transgenic approaches mitigates beta-cell dysfunction (Robertson 2009). For example, in vivo studies in leptin-receptor deficient Zucker Diabetic Fatty rat and db/db mice have reported beneficial effects of the antioxidant N-acetylcysteine on blood glucose levels and beta-cell mass (Tanaka et al. 1999). The underlying mechanisms, reviewed in (Robertson 2009), are proposed to involve modulation of expression of key beta-cell genes such as insulin, Pdx-1 and MafA.

5.4.3 Inflammation

Obesity is associated with low-grade inflammation. More specifically, expansion of visceral white adipose tissue triggers a systemic inflammatory process which negatively impact glucose homeostasis and create a vicious cycle contributing to the onset of diabetes (Muoio and Newgard 2008). At the level of the beta cell, interleukin-1β (IL-1β) is proposed to play a major role both in type 1 and type 2 diabetes (Donath et al. 2003). Pioneering studies by the Donath group (Boni-Schnetzler et al. 2008; Maedler et al. 2002) have shown that 1) mRNA levels of IL-1β are increased in islets from T2D patients, 2) IL-1β is produced by islets upon long-term exposure to nutrients, mainly glucose and fatty acids, and 3) in a vicious cycle, IL-1β promotes its own production in islet through an autocrine/paracrine loop involving its own receptor (IL-1R). Importantly, IL-1β amplifies the inflammatory process by enhancing islet production of other inflammatory molecules such as MCP-1 and IL-8 (Boni-Schnetzler et al. 2008). The latter are potent chemoattractants for macrophages and other immune cells, and indeed macrophage infiltration is observed in islets from humans T2D pancreatic sections (Ehses et al. 2007). Compared to other tissues, beta cells express very high levels of IL-1R, making it more sensitive to the deleterious effect of IL-1β (Boni-Schnetzler et al. 2008). An important proof of concept that inhibition of IL-1β signalling might be beneficial for beta-cell function was provided by the demonstration that an IL-1R antagonist (IL1-Ra) not only prevents hyperglycemia in the diabetic Goto-Kakizaki rat (Ehses et al. 2009), but also significantly improves glycemic control in T2D patients (Larsen et al. 2007), an effect that lasts several months after withdrawal of the drug (Larsen et al. 2009). Downstream of IL-1β, a recent series of studies have demonstrated a novel role for a unique post-translational modification, namely hypusination of the enzyme eIF5a, in cytokine signalling in beta cells (Maier et al. 2010; Robbins et al. 2010).

Another cytokine proposed to modulate islet function in the metabolic syndrome is IL-6. Circulating levels of IL-6 are increased in obesity and are predictive of T2D (Spranger et al. 2003; Hoene and Weigert 2008), but recent studies point towards a protective role for IL-6 through its effects on the alpha cell (Ellingsgaard et al. 2008; Ellingsgaard et al. 2011). In response to exercise, skeletal muscle-derived IL-6 enhances glucagon like peptide-1 (GLP-1) production both from intestinal L cells and pancreatic alpha cells, which in turn acts in a paracrine manner to potentiate insulin secretion and lower glucose levels, without changes in systemic GLP-1 levels. This effect is mediated by an IL-6-induced switch in prohormones convertase expression in alpha cells which favours the synthesis of GLP-1 over that of glucagon. Accordingly, exogenous IL-6 administration in nutritional and genetic models of obesity has a similar impact on GLP-1 production from the alpha cells and the intestinal L cells and improves glucose homeostasis. In contrast, neutralization of IL-6 in these conditions worsens glycemic control in these mice.

5.4.4 Amyloid

Islet amyloid polypeptide (IAPP) is a 37 amino-acid peptide co-secreted with insulin whose physiological role remains unclear (reviewed in Haataja et al. 2008; Westermark et al. 2011). The importance of IAPP in islet pathology stems from its ability, restricted to the human isoform, to form amyloid deposits (Westermark 1977) and from initial pathological findings that these amyloid deposits are more frequent in diabetic pancreata (Westermark and Wilander 1978), an observation recently confirmed in islets from T2D patients (Butler et al. 2003). A cause-and-effect relationship between amyloid formation and beta-cell dysfunction was provided by the generation of rodent models expressing the human form of IAPP, which results in hyperglycemia and diabetes (Butler et al. 2004; Janson et al. 1996; Verchere et al. 1996). In rats overexpressing human IAPP, the expansion of beta-cell mass that is normally observed in wild-type animals does not occur (Matveyenko et al. 2009), suggesting a possible role for amyloid deposits under conditions of nutrient excess. Consistent with this possibility, circulating IAPP levels increase in response to high-fat feeding in mice (Westermark et al. 1998). This could be due to the fact that the IAPP gene promoter shares several regulatory elements with the insulin promoter (notably Pdx-1 binding sequences) (German et al. 1992), and that glucose induces expression of IAPP via Pdx-1 (Watada et al. 1996). In addition, fatty acids promote the expression and secretion of IAPP in insulin-secreting MIN6 cells (Qi et al. 2010) as well as amyloid fibril formation in islets (Ma and Westermark 2002). These observations raise the interesting possibility that amyloid formation may be induced and contribute to beta-cell failure in glucolipotoxic conditions, although this remains to be directly tested.

The precise mechanism by which amyloid deposits are toxic to the beta-cell, and whether it is due to the formation of insoluble IAPP oligomers or mature amyloid fibrils, is still a matter of debate (Haataja et al. 2008; Zraika et al. 2010).

5.4.5 Autophagy

Recent evidence clearly shows that the process of autophagy not only has a housekeeping role in maintaining cellular homeostasis, but also is a highly regulated process that could play an important role in the beta-cell response to environmental cues, particularly nutrient excess. Autophagy enables recycling of cellular components, such as defective or damaged organelles, ubiquitinated proteins and even lipids droplets (Rabinowitz and White 2010). This takes place in autophagosomes, which fuse with lysosomes to form autophagolysosomes. This phenomenon is critical for normal cellular function, especially in long-lived cells such as neurones and beta cells, and its impairment is associated with age-related diseases such as neurodegerative disorders and diabetes (Rabinowitz and White 2010). Autophagy-related genes (Atgs) are the main regulators of autophagy, of which Atg1 regulates the initiation steps, whereas in an intermediate phase Atg7 converts the cytosolic form of the microtubule-associated protein 1 light chain 3 (LC3) into LC3-II, commonly used as a marker for autophagy. The autophagosome then interacts with the lysosome-associated membrane protein type 2A (LAMP2A), allowing formation of autophagolysosomes where the autophagosomal content is degraded by proteases and lipases (Kaushik et al. 2010). The regulation of autophagy intersects with metabolic processes at the level of the mammalian target of rapamycin (mTOR) pathway. mTOR activation, that can occur in response to fuel excess, blocks the autophagic process, and rapamycin treatment alleviates this inhibition and potentiates the autophagic flux (Kaushik et al. 2010; Rabinowitz and White 2010). The importance of autophagy in beta-cell function was initially proposed by Marsh et al. (Marsh et al. 2007), who reported that autophagy regulates insulin content. Further, deletion of the Atg7 gene specifically in beta cells leads to increased apoptosis, reduced proliferation, and impaired insulin secretion associated with a dramatic reduction in insulin granule number, a distended ER, and altered mitochondrial morphology (Ebato et al. 2008; Jung et al. 2008). Interestingly, autophagosomes are increased in beta cells in nutritional and genetic models of T2D (Ebato et al. 2008). Likewise, insulin-secreting INS1 cells exposed to fatty acids have an increased number of autophagosomes associated with impaired insulin secretion and apoptosis (Choi et al. 2009; Ebato et al. 2008). In this system, when autophagy was pharmacologically induced using rapamycin, the deleterious outcome of fatty acids was partially prevented (Choi et al. 2009). Las et al. (Las et al. 2011) have recently shown that accumulation of autophagosomes in beta-cells under lipotoxic conditions is due to their inability to fuse with lysosomes to form autophagolysosomes. Thus, accumulation of autophagosomes occurs as a result of interrupted autophagic flux, and might impair beta-cell function by triggering ER stress or increasing ROS levels due to inefficient recycling of organelles and clearing of accumulating ubiquitinated proteins. Interestingly, analyses of pancreatic sections from T2D patients have shown an increase in autophagosomes in beta cells (Masini et al. 2009).

5.4.6 Endoplasmic Reticulum Stress

As mentioned several times throughout this chapter, the initial beta-cell response to the MetS is to secrete more insulin to compensate for insulin resistance. This increased secretory flux enhances the workload on the ER, where proteins are synthesized and routed towards the secretory pathway. Conceivably, a continuous pressure on the ER might eventually overcome its capacity to properly ensure protein folding. Therefore, the unfolded protein response (UPR) and ER stress have emerged as key players in beta-cell failure and apoptosis (Eizirik et al. 2008). Accumulation of unfolded proteins activates a protective mechanism that allows deceleration of the processing flux to permit proper folding of proteins. This relies on the activation of three separate pathways. Within the lumen of the ER, the immunoglobulin heavy chain binding protein (BIP) is a chaperone that binds to and inhibits three transmembrane proteins, namely inositol-requiring enzyme 1 (IRE1), PKR-like kinase (PERK) and activating transcription factor 6 (ATF6). As unfolded proteins accumulate, BIP dissociates from PERK, which in turn becomes active and lead to attenuation of translation through eIF2a. IRE-1 promotes splicing of the XBP1 mRNA into its active form, which stimulates genes encoding for UPR chaperones. Finally, ATF6 translocates to the Golgi apparatus where it is cleaved into its active 50 kDa fragment which then translocates to the nucleus to upregulate genes encoding for protein chaperones and the ER-associated protein degradation machinery (Scheuner and Kaufman 2008). The overall purpose of this sophisticated process is to slow down protein synthesis and to enable proper folding of the accumulating proteins. However, if the pressure on the ER continues and the UPR fails to restore normal ER function, ER stress ensues and eventually leads to apoptosis (Eizirik et al. 2008).

Several studies have provided evidence for a role of the ER stress in nutrient-induced beta-cell dysfunction and death. In insulin-secreting cells, saturated fatty acids such as palmitate induce ER stress to a greater extent than unsaturated fatty acids (Karaskov et al. 2006). Palmitate was reported to activate the PERK branch of the UPR (Cnop et al. 2007; Karaskov et al. 2006), to block ER-to-Golgi protein trafficking (Preston et al. 2009), and to deplete ER calcium stores (Gwiazda et al. 2009). In vivo, Laybutt et al. (Laybutt et al. 2007) reported an increase in ER stress markers in islets from *db/db* mice and T2D patients. Although chemical chaperones improve glycemic control in several mouse models of diabetes (Ozcan et al. 2006), these beneficial effects might be due to alleviation of ER stress in peripheral tissues (since ER stress also plays a major role in insulin resistance). In humans ER stress is clearly a mechanism involved in rare cases of monogenic diabetes due, for example, to mutations in the human orthologue of PERK in Wolcott-Rallison syndrome (Delepine et al. 2000).

5.4.7 Beta-cell Exhaustion

Can hypersecretion of insulin, the initial attempt of the beta-cell to adapt to the enhanced secretory demand in insulin resistance, be itself a factor of beta-cell demise by leading, in the long-term, to beta-cell exhaustion and failure? This concept has led to the proposal that "beta-cell rest", i.e. reduction of endogenous insulin secretion, might have benefical effects not only in type 1 but also in type 2 diabetes (reviewed in Brown and Rother 2008) and might explain the so-called "secondary failure" of sulfonylureas (Pontiroli et al. 1994). Convincing evidence in rodents suggests that inhibiting insulin secretion with the K-ATP channel opener diazoxide prevents beta-cell dysfunction in several diabetic models (Sako and Grill 1990; Leahy et al. 1994; Kullin et al. 2000). In humans, beneficial effects of diazoxide treatment were reported in both type 1 (Bjork et al. 1996; Ortqvist et al. 2004) and type 2 diabetes (Qvigstad et al. 2004; Radtke et al. 2007), although in the latter it required the concomitant administration of insulin at bedtime (Grill et al. 2009).

Indirectly supportive of this concept is genetic evidence in humans that insulin hypersecretion might eventually lead to diabetes in the absence of insulin resistance (reviewed in Aston-Mourney et al. 2008). Subjects with persistent hyperinsulinemic hypoglycemia of infancy (PHHI), harbouring genetic mutations resulting in increased insulin secretion, are at high risk for developing diabetes (Huopio et al. 2003). The underlying mutations are in the genes encoding glucokinase (resulting in a shift to the left of the glucose dose-responsiveness (Glaser et al. 1998; Christesen et al. 2008)) or the sulfonylurea receptor 1 (SUR1) (resulting in impaired channel activity (Huopio et al. 2003)).

The precise mechanisms by which a reduction in insulin secretion might prove beneficial for beta-cell function are unknown, and may be related to metabolic deceleration and alleviation of ER stress associated with hypersecretion (Andrikopoulos 2010). Nonetheless, this concept raises the intriguing possibility that current therapies aimed at enhancing insulin secretion, while clearly beneficial in terms of glucose lowering, might also represent a "double-edge sword" and indirectly contribute to beta-cell failure.

5.4.8 Beta-cell Dedifferentiation

The exquisite ability of the pancreatic beta cell to sense nutrients and accordingly adjust insulin secretion relies on the expression of a specific set of genes, which all together define the identity of a differentiated beta cell. It has been proposed in rodents that beta-cell dedifferentiation may contribute to beta-cell failure that occurs after long period of metabolic stress, as is the case in MetS (Weir and Bonner-Weir 2004). In that regard, in experimental models of chronic hyperglycemia induced by partial pancreatectomy the expression of beta-cell specific genes is lost as beta-cell hypertrophy develops (Jonas et al. 1999; Laybutt et al. 2002). Interestingly, these effects are reversed upon restoration of normoglycemia. Similar changes in gene expression were reported in islets from partially pancreatectomized Zucker fatty rats (Delghingaro-Augusto et al. 2009). In our own studies with 6-month old rats infused with glucose and Intralipid for 72 h, we observed down-regulation of several key beta-cell genes accompanied by defective insulin biosynthesis and secretion, despite an increase in beta-cell mass and proliferation (Fontes et al. 2010). Collectively, these observations support the possibility that beta-cell failure under conditions of chronic nutrient excess results, at least in part, from a loss of beta-cell identity. Recently, Talchai et al. (Talchai et al. 2012) provided a mechanistic basis for this phenomenon. They showed that under metabolic stress, beta-cell specific deletion of the transcription factor FoxO1 leads to beta cells dedifferentiation giving rise to progenitor-like endocrine cells, a subset of which subsequently trans-differentiate into alpha cells. This leads not only to relative hypoinsulinemia but also hyperglucagonemia, which contributes to the loss of glucose homeostasis.

5.5 Conclusions

The pancreatic beta cell plays a key role not only in the maintenance of glucose homeostasis but also in the transition from the MetS to overt T2D. In the face of insulin resistance that characterizes the MetS, individuals whose beta cells are adequately equipped to compensate for the increased demand will maintain normal glucose homeostasis by hypersecreting insulin. In individuals whose beta-cells are more susceptible, because of a genetic predisposition or epigenetic modifications related to the prenatal or neonatal environment, the compensatory response eventually fails and diabetes ensues. Despite intensive investigation, the precise mechanisms underlying beta-cell failure and the continuous deterioration of its function at the cellular and molecular level remain poorly understood. As reviewed in this chapter a number of different mechanisms have been proposed, all of which probably contributing a different role in different individuals at different stages of disease progression. The complexity of the pathogenesis of T2D, which is a multifactorial syndrome rather than a single disease, its slow progression spanning several years if not decades, and methodological limitations in the study of beta-cell function in humans, most notably the lack of reliable methods to non-invasively measure beta-cell mass, all probably contribute to our relatively poor understanding of the disease process. It is clear, however, that once the defect in beta-cell function is established, none of the available therapeutic options are capable of significantly slowing down the relentless degradation of glucose homeostasis (except for bariatric surgery, which is obviously only indicated in a small proportion of subjects). Therefore, efforts must be aimed at preventing the occurrence of the metabolic syndrome and preserving beta-cell function before the point of no return is reached.

Acknowledgements Work performed in our laboratory was supported by the US National Institutes of Health (R01-DK58096 from NIDDK). B.Z. was supported by the Montreal Diabetes Research Center/ Merck Frosst post-doctoral fellowship and Eli Lilly Canada. G.F. was supported by a post-doctoral fellowship from the Canadian Diabetes Association. J.A. and M.S. were supported by doctoral fellowships from the Fonds de Recherche Québec—Santé. V.P. holds the Canada Research Chair in Diabetes and Pancreatic Beta-cell Function.

References

Abderrahmani A, Niederhauser G, Favre D, Abdelli S, Ferdaoussi M, Yang JY, Regazzi R, Widmann C, Waeber G (2007) Human high-density lipoprotein particles prevent activation of the JNK pathway induced by human oxidised low-density lipoprotein particles in pancreatic beta cells. Diabetologia 50:1304–1314

Ackermann Misfeldt A, Costa RH, Gannon M (2008) Beta-cell proliferation, but not neogenesis, following 60 % partial pancreatectomy is impaired in the absence of FoxM1. Diabetes 57:3069–3077

ALberti KG, Eckel RH, Grundy SM, Zimmet PZ, Cleeman JI, Donato KA, Fruchart JC, James WP, Loria CM, Smith SC JR (2009) Harmonizing the metabolic syndrome: a joint interim statement of the International Diabetes Federation Task Force on Epidemiology and Prevention; National Heart, Lung, and Blood Institute; American Heart Association; World Heart Federation; International Atherosclerosis Society; and International Association for the Study of Obesity. Circulation 120:1640–1645

Andrikopoulos S (2010) Obesity and type 2 diabetes: slow down!—Can metabolic deceleration protect the islet beta cell from excess nutrient-induced damage? Mol Cell Endocrinol 316:140–146

Anello M, Lupi R, Spampinato D, Piro S, Masini M, Boggi U, Del Prato S, Rabuazzo AM, Purrello F, Marchetti P (2005) Functional and morphological alterations of mitochondria in pancreatic beta cells from type 2 diabetic patients. Diabetologia 48:282–289

Aston-Mourney K, Proietto J, Morahan G, Andrikopoulos S (2008) Too much of a good thing: why it is bad to stimulate the beta cell to secrete insulin. Diabetologia 51:540–545

Bjork E, Berne C, Kampe O, Wibell L, Oskarsson P, Karlsson FA (1996) Diazoxide treatment at onset preserves residual insulin secretion in adults with autoimmune diabetes. Diabetes 45:1427–1430

Boden G, Chen X (1999) Effects of fatty acids and ketone bodies on basal insulin secretion in type 2 diabetes. Diabetes 48:577–583

Boden G, Chen X, Rosner J, Barton M (1995) Effects of a 48 h fat infusion on insulin secretion and glucose utilization. Diabetes 44:1239–1242

Boni-Schnetzler M, Boller S, Debray S, Bouzakri K, Meier DT, Prazak R, Kerr-Conte J, Pattou F, Ehses JA, Schuit FC, Donath MY (2009) Free fatty acids induce a proinflammatory response in islets via the abundantly expressed interleukin-1 receptor I. Endocrinology 150:5218–5229

Boni-Schnetzler M, Thorne J, Parnaud G, Marselli L, Ehses JA, Kerr-Conte J, Pattou F, Halban PA, Weir GC, Donath MY (2008) Increased interleukin (IL)-1beta messenger ribonucleic acid expression in beta -cells of individuals with type 2 diabetes and regulation of IL-1beta in human islets by glucose and autostimulation. J Clin Endocrinol Metab 93:4065–4074

Bouche C, Lopez X, Fleischman A, Cypess AM, O'shea S, Stefanovski D, Bergman RN, Rogatsky E, Stein DT, Kahn CR, Kulkarni RN, Goldfine AB (2010) Insulin enhances glucose-stimulated insulin secretion in healthy humans. Proc Natl Acad Sci U S A 107:4770–4775

Briaud I, Harmon JS, Kelpe CL, Segu VB, Poitout V (2001) Lipotoxicity of the pancreatic beta-cell is associated with glucose-dependent esterification of fatty acids into neutral lipids. Diabetes 50:315–321

Briaud I, Kelpe CL, Johnson LM, Tran POT, Poitout V (2002) Differential effects of hyperlipidemia on insulin secretion in islets of Langerhans from hyperglycemic vs normoglycemic rats. Diabetes 51:662–668

Brown RJ, Rother KI (2008) Effects of beta-cell rest on beta-cell function: a review of clinical and preclinical data. Pediatr Diabetes 9:14–22

Brunham LR, Kruit JK, Pape TD, Timmins JM, Reuwer AQ, Vasanji Z, Marsh BJ, Rodrigues B, Johnson JD, Parks JS, Verchere CB, Hayden MR (2007) Beta-cell ABCA1 influences insulin secretion, glucose homeostasis and response to thiazolidinedione treatment. Nat Med 13:340–347

Busch AK, Gurisik E, Cordery DV, Sudlow M, Denyer GS, Laybutt DR, Hughes WE, Biden TJ (2005) Increased fatty acid desaturation and enhanced expression of stearoyl coenzyme a desaturase protects pancreatic beta-cells from lipoapoptosis. Diabetes 54:2917–2924

Butler AE, Jang J, Gurlo T, Carty MD, Soeller WC, Butler PC (2004) Diabetes due to a progressive defect in beta-cell mass in rats transgenic for human islet amyloid polypeptide (HIP Rat): a new model for type 2 diabetes. Diabetes 53:1509–1516

Butler AE, Janson J, Bonner-Weir S, Ritzel R, Rizza RA, Butler PC (2003) Beta-cell deficit and increased beta-cell apoptosis in humans with type 2 diabetes. Diabetes 52:102–110

Carpentier A, Giacca A, Lewis GF (2001) Effect of increased plasma non-esterified fatty acids (NEFAs) on arginine-stimulated insulin secretion in obese humans. Diabetologia 44:1989–1997
Carpentier A, Mittelman SD, Lamarche B, Bergman RN, Giacca A, Lewis GF (1999) Acute enhancement of insulin secretion by FFA in humans is lost with prolonged FFA elevation. Am J Physiol 276: E1055–1066
Carpentier AC, Bourbonnais A, Frisch F, Giacca A, Lewis GF (2010) Plasma nonesterified fatty acid intolerance and hyperglycemia are associated with intravenous lipid-induced impairment of insulin sensitivity and disposition index. J Clin Endocrinol Metab 95:1256–1264
Choi SE, Lee SM, Lee YJ LI LJ, Lee SJ, Lee JH, Kim Y, Jun HS, Lee KW, Kang Y (2009) Protective role of autophagy in palmitate-induced INS-1 beta-cell death. Endocrinology 150:126–134
Choudhury AI, Heffron H, Smith MA, AL-Qassab H XU AW, Selman C, Simmgen M, Clements M, Claret M, Maccoll G, Bedford DC, Hisadome K, Diakonov I, Moosajee V, Bell JD, Speakman JR, Batterham RL, Barsh GS, Ashford ML, Withers DJ (2005) The role of insulin receptor substrate 2 in hypothalamic and beta cell function. J Clin Invest 115:940–950
Christesen HB, Tribble ND, Molven A, Siddiqui J, Sandal T, Brusgaard K, Ellard S, Njolstad PR, Alm J, Brock Jacobsen B, Hussain K, Gloyn AL (2008) Activating glucokinase (GCK) mutations as a cause of medically responsive congenital hyperinsulinism: prevalence in children and characterisation of a novel GCK mutation. Eur J Endocrinol 159:27–34
Cnop M, Igoillo-Esteve M, Cunha DA, Ladriere L, Eizirik DL (2008) An update on lipotoxic endoplasmic reticulum stress in pancreatic beta-cells. Biochem Soc Trans 36:909–915
Cnop M, Ladriere L, Hekerman P, Ortis F, Cardozo AK, Dogusan Z, Flamez D, Boyce M, Yuan J, Eizirik DL (2007) Selective inhibition of eukaryotic translation initiation factor 2 alpha dephosphorylation potentiates fatty acid-induced endoplasmic reticulum stress and causes pancreatic beta-cell dysfunction and apoptosis. J Biol Chem 282:3989–3997
Colas R, Pruneta-Deloche V, Guichardant M, Luquain-Costaz C, Cugnet-Anceau C, Moret M, Vidal H, Moulin P, Lagarde M, Calzada C (2010) Increased lipid peroxidation in LDL from type-2 diabetic patients. Lipids 45:723–731
da Silva Xavier G, Rutter J, Rutter GA (2004) Involvement of Per-Arnt-Sim (PAS) kinase in the stimulation of preproinsulin and pancreatic duodenum homeobox 1 gene expression by glucose. Proc Natl Acad Sci U S A 101:8319–8324
da Silva Xavier G, Farhan H, Kim H, Caxaria S, Johnson P, Hughes S, Bugliani M, Marselli L, Marchetti P, Birzele F, Sun G, Scharfmann R, Rutter J, Siniakowicz K, Weir G, Parker H, Reimann F, Gribble FM, Rutter GA (2010) Per-arnt-sim (PAS) domain-containing protein kinase is downregulated in human islets in type 2 diabetes and regulates glucagon secretion. Diabetologia 54(4):819–817
Davis DB, Lavine JA, Suhonen JI, Krautkramer KA, Rabaglia ME, Sperger JM, Fernandez LA, Yandell BS, Keller MP, Wang IM, Schadt EE, Attie AD (2010) FoxM1 is up-regulated by obesity and stimulates beta-cell proliferation. Mol Endocrinol 24:1822–1834
Del Guerra S, Lupi R, Marselli L, Masini M, Bugliani M, Sbrana S, Torri S, Pollera M, Boggi U, Mosca F, Del Prato S, Marchetti P (2005) Functional and molecular defects of pancreatic islets in human type 2 diabetes. Diabetes 54:727–735
Delepine M, Nicolino M, Barrett T, Golamaully M, Lathrop GM, Julier C (2000) EIF2AK3, encoding translation initiation factor 2-alpha kinase 3, is mutated in patients with Wolcott-Rallison syndrome. Nat Genet 25:406–409
Delghingaro-Augusto V, Nolan CJ, Gupta D, Jetton TL, Latour MG, Peshavaria M, Madiraju SR, Joly E, Peyot ML, Prentki M, Leahy J (2009) Islet beta cell failure in the 60 % pancreatectomised obese hyperlipidaemic Zucker fatty rat: severe dysfunction with altered glycerolipid metabolism without steatosis or a falling beta cell mass. Diabetologia 52:1122–1132
Deng S, Vatamaniuk M, Huang X, Doliba N, Lian MM, Frank A, Velidedeoglu E, Desai NM, Koeberlein B, Wolf B, Barker CF, Naji A, Matschinsky FM, Markmann JF (2004) Structural and functional abnormalities in the islets isolated from type 2 diabetic subjects. Diabetes 53:624–632
Doliba NM, Qin W, Najafi H, Liu C, Buettger CW, Sotiris J, Weik-Collins H LI CD, Stanley CA, Wilson DF, Grimsby J, Sarabu R, Naji A, Matschinsky FM (2011) Glucokinase activation

repairs defective bioenergetics of islets of Langerhans isolated from type-2-diabetics. Am J Physiol Endocrinol Metab 302(1):E87–102

Donath MY, Storling J, Maedler K, Mandrup-Poulsen T (2003) Inflammatory mediators and islet beta-cell failure: a link between type 1 and type 2 diabetes. J Mol Med (Berl) 81:455–470

Dor Y, Brown J, Martinez OI, Melton DA (2004) Adult pancreatic beta-cells are formed by self-duplication rather than stem-cell differentiation. Nature 429:41–46

Drew BG, Duffy SJ, Formosa MF, Natoli AK, Henstridge DC, Penfold SA, Thomas WG, Mukhamedova N, DE Courten B, Forbes JM, Yap FY, Kaye DM, Van Hall G, Febbraio MA, Kemp BE, Sviridov D, Steinberg GR, Kingwell BA (2009) High-density lipoprotein modulates glucose metabolism in patients with type 2 diabetes mellitus. Circulation 119:2103–2111

Ebato C, Uchida T, Arakawa M, Komatsu M, Ueno T, Komiya K, Azuma K, Hirose T, Tanaka K, Kominami E, Kawamori R, Fujitani Y, Watada H (2008) Autophagy is important in islet homeostasis and compensatory increase of beta cell mass in response to high-fat diet. Cell Metab 8:325–332

Ehses JA, Perren A, Eppler E, Ribaux P, Pospisilik JA, Maor-Cahn R, Gueripel X, Ellingsgaard H, Schneider MK, Biollaz G, Fontana A, Reinecke M, Homo-Delarche F, Donath MY (2007) Increased number of islet-associated macrophages in type 2 diabetes. Diabetes 56:2356–2370

Ehses JA, Lacraz G, Giroix MH, Schmidlin F, Coulaud J, Kassis N, Irminger JC, Kergoat M, Portha B, Homo-Delarche F, Donath MY (2009) IL-1 antagonism reduces hyperglycemia and tissue inflammation in the type 2 diabetic GK rat. Proc Natl Acad Sci U S A 106:13998–14003

Eizirik DL, Cardozo AK, Cnop M (2008) The role for endoplasmic reticulum stress in diabetes mellitus. Endocr Rev 29:42–61

EL-Assaad W, Buteau J, Peyot ML, Nolan C, Roduit R, Hardy S, Joly E, Dbaibo G, Rosenberg L, Prentki M (2003) Saturated fatty acids synergize with elevated glucose to cause pancreatic beta-cell death. Endocrinology 144:4154–4163

El Ouaamari A, Kawamori D, Dirice E, Liew CW, Shadrach JL, Hu J, Katsuta H, Hollister-Lock J, Qian WJ, Wagers AJ, Kulkarni RN (2013). Liver-derived systemic factors drive beta-cell hyperplasia in insulin-resistant states. Cell Rep 3:401–410

Ellingsgaard H, Ehses JA, Hammar EB, Van Lommel L, Quintens R, Martens G, Kerr-Conte J, Pattou F, Berney T, Pipeleers D, Halban PA, Schuit F, Donath MY (2008) Interleukin-6 regulates pancreatic alpha-cell mass expansion. Proc Natl Acad Sci U S A 105:13163–13168

Ellingsgaard H, Hauselmann I, Schuler B, Habib AM, Baggio LL, Meier DT, Eppler E, Bouzakri K, Wueest S, Muller YD, Hansen AM, Reinecke M, Konrad D, Gassmann M, Reimann F, Halban PA, Gromada J, Drucker DJ, Gribble FM, Ehses JA, Donath MY (2011) Interleukin-6 enhances insulin secretion by increasing glucagon-like peptide-1 secretion from L cells and alpha cells. Nat Med 17:1481–1489

Escribano O, Guillen C, Nevado C, Gomez-Hernandez A, Kahn CR, Benito M (2009) Beta-Cell hyperplasia induced by hepatic insulin resistance: role of a liver-pancreas endocrine axis through insulin receptor A isoform. Diabetes 58:820–828

Federici M, Hribal M, Perego L, Ranalli M, Caradonna Z, Perego C, Usellini L, Nano R, Bonini P, Bertuzzi F, Marlier LN, Davalli AM, Carandente O, Pontiroli AE, Melino G, Marchetti P, Lauro R, Sesti G, Folli F (2001) High glucose causes apoptosis in cultured human pancreatic islets of Langerhans: a potential role for regulation of specific Bcl family genes toward an apoptotic cell death program. Diabetes 50:1290–1301

Fernandez-Alvarez J, Conget I, Rasschaert J, Sener A, Gomis R, Malaisse WJ (1994) Enzymatic, metabolic and secretory patterns in human islets of type 2 (non-insulin-dependent) diabetic patients. Diabetologia 37:177–181

Fex M, Nitert MD, Wierup N, Sundler F, Ling C, Mulder H (2007) Enhanced mitochondrial metabolism may account for the adaptation to insulin resistance in islets from C57BL/6J mice fed a high-fat diet. Diabetologia 50:74–83

Fontes G, Semache M, Hagman DK, Tremblay C, Shah R, Rhodes CJ, Rutter J, Poitout V (2009) Involvement of Per-Arnt-Sim Kinase and extracellular-regulated kinases-1/2 in palmitate inhibition of insulin gene expression in pancreatic beta-cells. Diabetes 58:2048–2058

Fontes G, Zarrouki B, Hagman DK, Latour MG, Semache M, Roskens V, Moore PC, Prentki M, Rhodes CJ, Jetton TL, Poitout V (2010) Glucolipotoxicity age-dependently impairs beta cell function in rats despite a marked increase in beta cell mass. Diabetologia 53:2369–2379

German MS, Moss LG, Wang J Rutter WJ (1992) The insulin and islet amyloid polypeptide genes contain similar cell-specific promoter elements that bind identical beta-cell nuclear complexes. Mol Cell Biol 12:1777–1788

Glaser B, Kesavan P, Heyman M, Davis E, Cuesta A, Buchs A, Stanley CA, Thornton PS, Permutt MA, Matschinsky FM, Herold KC (1998) Familial hyperinsulinism caused by an activating glucokinase mutation. N Engl J Med 338:226–230

Golbidi S, Ebadi SA, Laher I (2011) Antioxidants in the treatment of diabetes. Curr diabetes rev 7:106–125

Grill V, Radtke M, Qvigstad E, Kollind M, Bjorklund A (2009) Beneficial effects of K-ATP channel openers in diabetes: an update on mechanisms and clinical experiences. Diabetes Obes Metab 11(Suppl 4):143–148

Gwiazda KS, Yang TL, Lin Y, Johnson JD (2009) Effects of palmitate on ER and cytosolic Ca + homeostasis in beta-cells. Am J Physiol Endocrinol Metab 296: E690–701

Haataja L, Gurlo T, Huang CJ, Butler PC (2008) Islet amyloid in type 2 diabetes, and the toxic oligomer hypothesis. Endocr Rev 29:303–316

Hagman DK, Hays LB, Parazzoli SD, Poitout V (2005) Palmitate inhibits insulin gene expression by altering PDX-1 nuclear localization and reducing MafA expression in isolated rat islets of Langerhans. J Biol Chem 280:32413–32418

Hagman DK, Latour MG, Chakrabarti SK, Fontes G, Amyot J, Tremblay C, Semache M, Lausier JA, Roskens V, Mirmira RG, Jetton TL, Poitout V (2008) Cyclical and alternating infusions of glucose and intralipid in rats inhibit insulin gene expression and Pdx-1 binding in islets. Diabetes 57:424–431

Hashimoto N, Kido Y, Uchida T, Asahara S, Shigeyama Y, Matsuda T, Takeda A, Tsuchihashi D, Nishizawa A, Ogawa W, Fujimoto Y, Okamura H, Arden KC, Herrera PL, Noda T, Kasuga M (2006) Ablation of PDK1 in pancreatic beta cells induces diabetes as a result of loss of beta cell mass. Nat Genet 38:589–593

Henquin JC (2000) Triggering and amplifying pathways of regulation of insulin secretion by glucose. Diabetes 49:1751–1760

Henquin JC (2009) Regulation of insulin secretion: a matter of phase control and amplitude modulation. Diabetologia 52:739–751

Herder C, Roden M (2011) Genetics of type 2 diabetes: pathophysiologic and clinical relevance. Eur J Clin Invest 41:679–692

Hoene M, Weigert C (2008) The role of interleukin-6 in insulin resistance, body fat distribution and energy balance. Obes Rev 9:20–29

Holman RR (1998) Assessing the potential for alpha-glucosidase inhibitors in prediabetic states. Diabetes Res Clin Pract 40 Suppl:S21–S25

Hoppa MB, Collins S, Ramracheya R, Hodson L, Amisten S, Zhang Q, Johnson P, Ashcroft FM, Rorsman P (2009) Chronic palmitate exposure inhibits insulin secretion by dissociation of Ca(2 +) channels from secretory granules. Cell Metab 10:455–465

Huopio H, Otonkoski T, Vauhkonen I, Reimann F, Ashcroft FM, Laakso M (2003) A new subtype of autosomal dominant diabetes attributable to a mutation in the gene for sulfonylurea receptor 1. Lancet 361:301–307

Imai J, Katagiri H, Yamada T, Ishigaki Y, Suzuki T, Kudo H, Uno K, Hasegawa Y, Gao J, Kaneko K, Ishihara H, Niijima A, Nakazato M, Asano T, Minokoshi Y, Oka Y (2008) Regulation of pancreatic beta cell mass by neuronal signals from the liver. Science 322:1250–1254

Inada A, Nienaber C, Katsuta H, Fujitani Y, Levine J, Morita R, Sharma A, Bonner-Weir S (2008) Carbonic anhydrase II-positive pancreatic cells are progenitors for both endocrine and exocrine pancreas after birth. Proc Natl Acad Sci U S A 105:19915–19919

Jacqueminet S, Briaud I, Rouault C, Reach G, Poitout V (2000) Inhibition of insulin gene expression by long-term exposure of pancreatic beta-cells to palmitate is dependent upon the presence of a stimulatory glucose concentration. Metabolism 49:532–536

Janson J, Soeller WC, Roche PC, Nelson RT, Torchia AJ, Kreutter DK, Butler PC (1996) Spontaneous diabetes mellitus in transgenic mice expressing human islet amyloid polypeptide. Proc Natl Acad Sci U S A 93:7283–7288

Jonas JC, Sharma A, Hasenkamp W, Ilkova H, Patane G, Laybutt R, Bonner-Weir S, Weir GC (1999) Chronic hyperglycemia triggers loss of pancreatic beta cell differentiation in an animal model of diabetes. J Biol Chem 274:14112–14121

Jung HS, Chung KW, Won Kim J, Kim J, Komatsu M, Tanaka K, Nguyen YH, Kang TM, Yoon KH, Kim JW, Jeong YT, Han MS, Lee MK, Kim KW, Shin J, Lee MS (2008) Loss of autophagy diminishes pancreatic beta cell mass and function with resultant hyperglycemia. Cell Metab 8:318–324

Kaneko K, Ueki K, Takahashi N, Hashimoto S, Okamoto M, Awazawa M, Okazaki Y, Ohsugi M, Inabe K, Umehara T, Yoshida M, Kakei M, Kitamura T, Luo J, Kulkarni RN, Kahn CR, Kasai H, Cantley LC, Kadowaki T (2010) Class IA phosphatidylinositol 3-kinase in pancreatic beta cells controls insulin secretion by multiple mechanisms. Cell Metab 12:619–632

karaskov E, Scott C, Zhang L, Teodoro T, Ravazzola M, Volchuk A (2006) Chronic palmitate but not oleate exposure induces endoplasmic reticulum stress, which may contribute to INS-1 pancreatic beta-cell apoptosis. Endocrinology 147:3398–3407

Kashyap S, Belfort R, Gastaldelli A, Pratipanawatr T, Berria R, Pratipanawatr W, Bajaj M, Mandarino L, Defronzo R, Cusi K (2003) A sustained increase in plasma free fatty acids impairs insulin secretion in nondiabetic subjects genetically predisposed to develop type 2 diabetes. Diabetes 52:2461–2474

Kasuga M (2006) Insulin resistance and pancreatic beta cell failure. J Clin Invest 116:1756–1760

Kaushik S, Singh R, Cuervo AM (2010) Autophagic pathways and metabolic stress. Diabetes Obes Metab 12 (Suppl 2):4–14

Kelpe CL, Moore PC, Parazzoli SD, Wicksteed B, Rhodes, CJ Poitout V (2003) Palmitate inhibition of insulin gene expression is mediated at the transcriptional level via ceramide synthesis. J Biol Chem 278:30015–30021

Koseki M, Matsuyama A, Nakatani K, Inagaki M, Nakaoka H, Kawase R, Yuasa-Kawase M, Tsubakio-Yamamoto K, Masuda D, Sandoval JC, Ohama T, Nakagawa-Toyama Y, Matsuura F, Nishida M, Ishigami M, Hirano K, Sakane N, Kumon Y, Suehiro T, Nakamura T, Shimomura I, Yamashita S (2009) Impaired insulin secretion in four Tangier disease patients with ABCA1 mutations. J Atheroscler Thromb 16:292–296

Kruit JK, Kremer PH, Dai L, Tang R, Ruddle P, DE Haan W, Brunham LR, Verchere C B, Hayden MR (2010) Cholesterol efflux via ATP-binding cassette transporter A1 (ABCA1) and cholesterol uptake via the LDL receptor influences cholesterol-induced impairment of beta cell function in mice. Diabetologia 53:1110–1119

Kruit JK, Wijesekara N, Fox JE, Dai XQ, Brunham LR, Searle GJ, Morgan GP, Costin AJ, Tang R, Bhattacharjee A, Johnson JD, Light PE, Marsh BJ, Macdonald PE, Verchere CB, Hayden MR (2011) Islet cholesterol accumulation due to loss of ABCA1 leads to impaired exocytosis of insulin granules. Diabetes 60(12):3186–3196

Kubota N, Terauchi Y, Tobe K, Yano W, Suzuki R, Ueki K, Takamoto I, Satoh H, Maki T, Kubota T, Moroi M, Okada-Iwabu M, Ezaki O, Nagai R, Ueta Y, Kadowaki T, Noda T (2004) Insulin receptor substrate 2 plays a crucial role in beta cells and the hypothalamus. J Clin Invest 114:917–927

Kulkarni RN, Bruning JC, Winnay JN, Postic C, Magnuson, MA, Kahn CR (1999) Tissue-specific knockout of the insulin receptor in pancreatic beta cells creates an insulin secretory defect similar to that in type 2 diabetes. Cell 96:329–339

Kullin M LIZ, Hansen JB, Bjork E, Sandler S, Karlsson FA (2000) K(ATP) channel openers protect rat islets against the toxic effect of streptozotocin. Diabetes 49:1131–1136

Larsen CM, Faulenbach M, Vaag A, Volund A, Ehses JA, Seifert B, Mandrup-Poulsen T, Donath MY (2007) Interleukin-1-receptor antagonist in type 2 diabetes mellitus. N Engl J Med 356:1517–1526

Larsen CM, Faulenbach M, Vaag A, Ehses JA, Donath MY, Mandrup-Poulsen T (2009) Sustained effects of interleukin-1 receptor antagonist treatment in type 2 diabetes. Diabetes Care 32:1663–1668

Las G, Sereda S, Wikstrom JD, Twig G, Shirihai OS (2011) Fatty acids suppress autophagic turnover in {beta}-cells. J Biol Chem 286(49):42534–42544

Laybutt DR, Sharma A, Sgroi DC, Gaudet J, Bonner-Weir S, Weir GC (2002) Genetic regulation of metabolic pathways in beta-cells disrupted by hyperglycemia. J Biol Chem 277:10912–10921

Laybutt DR, Preston AM, Akerfeldt MC, Kench JG, Busch AK, Biankin AV, Biden TJ (2007) Endoplasmic reticulum stress contributes to beta cell apoptosis in type 2 diabetes. Diabetologia 50:752–763

Leahy JL, Bumbalo LM, Chen C (1994) Diazoxide causes recovery of beta-cell glucose responsiveness in 90 % pancreatectomized diabetic rats. Diabetes 43:173–179

Leloup C, Tourrel-Cuzin C, Magnan C, Karaca M, Castel J, Carneiro L, Colombani AL, Ktorza A, Casteilla L, Penicaud L (2009) Mitochondrial reactive oxygen species are obligatory signals for glucose-induced insulin secretion. Diabetes 58:673–681

Lenzen S, Drinkgern J, Tiedge M (1996) Low antioxidant enzyme gene expression in pancreatic islets compared with various other mouse tissues. Free Radic Biol Med 20:463–466

Lopez X, Cypess A, Manning R, O'shea S, Kulkarni RN, Goldfine AB (2011) Exogenous insulin enhances glucose-stimulated insulin response in healthy humans independent of changes in free fatty acids. J Clin Endocrinol Metab 96(12):3811–3821

Lusis AJ, Attie AD, Reue K (2008) Metabolic syndrome: from epidemiology to systems biology. Nat Rev Genet 9:819–830

Ma Z, Westermark GT (2002) Effects of free fatty acid on polymerization of islet amyloid polypeptide (IAPP) in vitro and on amyloid fibril formation in cultivated isolated islets of transgenic mice overexpressing human IAPP. Mol Med 8:863–868

Maedler K, Sergeev P, Ris F, Oberholzer J, Joller-Jemelka HI, Spinas GA, Kaiser N, Halban PA, Donath MY (2002) Glucose-induced beta cell production of IL-1beta contributes to glucotoxicity in human pancreatic islets. J Clin Invest 110:851–860

Maedler K, Oberholzer J, Bucher P, Spinas GA, Donath MY (2003) Monounsaturated fatty acids prevent the deleterious effects of palmitate and high glucose on human pancreatic beta-cell turnover and function. Diabetes 52:726–733

Maier B, Ogihara T, Trace AP, Tersey SA, Robbins RD, Chakrabarti SK, Nunemaker CS, Stull ND, Taylor CA, Thompson JE, Dondero RS, Lewis EC, Dinarello CA, Nadler JL, Mirmira RG (2010) The unique hypusine modification of eIF5A promotes islet beta cell inflammation and dysfunction in mice. J Clin Invest 120:2156–2170

Mari A, Tura A, Natali A, Anderwald C, Balkau B, Lalic N, Walker M, Ferrannini E (2011) Influence of hyperinsulinemia and insulin resistance on in vivo beta-cell function: their role in human beta-cell dysfunction. Diabetes 60(12):3141–3147

Marsh BJ, Soden C, Alarcon C, Wicksteed BL, Yaekura K, Costin AJ, Morgan GP Rhodes CJ (2007) Regulated autophagy controls hormone content in secretory-deficient pancreatic endocrine beta-cells. Mol Endocrinol 21:2255–2269

Masini M, Bugliani M, Lupi R, Del Guerra S, Boggi U, Filipponi F, Marselli L, Masiello P, Marchetti P (2009) Autophagy in human type 2 diabetes pancreatic beta cells. Diabetologia 52:1083–1086

Matveyenko AV, Gurlo T, Daval M, Butler AE, Butler PC (2009) Successful versus failed adaptation to high-fat diet-induced insulin resistance: the role of IAPP-induced beta-cell endoplasmic reticulum stress. Diabetes 58:906–916

Meier JJ, Butler AE, Saisho Y, Monchamp T, Galasso R, Bhushan A, Rizza RA, Butler PC (2008) Beta-cell replication is the primary mechanism subserving the postnatal expansion of beta-cell mass in humans. Diabetes 57:1584–1594

Michael MD, Kulkarni RN, Postic C, Previs SF, Shulman GI, Magnuson MA, Kahn CR (2000) Loss of insulin signaling in hepatocytes leads to severe insulin resistance and progressive hepatic dysfunction. Mol Cell 6:87–97

Moore PC, Ugas MA, Hagman DK, Parazzoli SD, Poitout V (2004) Evidence against the involvement of oxidative stress in fatty acid inhibition of insulin secretion. Diabetes 53:2610–2616

Muoio DM, Newgard CB (2008) Mechanisms of disease: molecular and metabolic mechanisms of insulin resistance and beta-cell failure in type 2 diabetes. Nat Rev Mol Cell Biol 9:193–205

Newsholme P, Haber EP, Hirabara SM, Rebelato EL, Procopio J, Morgan D, Oliveira-Emilio HC, Carpinelli AR, Curi R (2007) Diabetes associated cell stress and dysfunction: role of mitochondrial and non-mitochondrial ROS production and activity. J Physiol 583:9–24

Nolan CJ, Damm P, Prentki M (2011) Type 2 diabetes across generations: from pathophysiology to prevention and management. Lancet 378:169–181

Olofsson CS, Collins S, Bengtsson M, Eliasson L, Salehi A, Shimomura K, Tarasov A, Holm C, Ashcroft F, Rorsman P (2007) Long-term exposure to glucose and lipids inhibits glucose-induced insulin secretion downstream of granule fusion with plasma membrane. Diabetes 56:1888–1897

Oprescu AI, Bikopoulos G, Naassan A, Allister EM, Tang C, Park E, Uchino H, Lewis GF, Fantus IG, Rozakis-Adcock M, Wheeler MB, Giacca A (2007) Free fatty acid-induced reduction in glucose-stimulated insulin secretion: evidence for a role of oxidative stress in vitro and in vivo. Diabetes 56:2927–2937

Ortqvist E, Bjork E, Wallensteen M, Ludvigsson J, Aman J, Johansson C, Forsander G, Lindgren F, Berglund L, Bengtsson M, Berne C, Persson B, Karlsson FA (2004) Temporary preservation of beta-cell function by diazoxide treatment in childhood type 1 diabetes. Diabetes Care 27:2191–2197

Paolisso G, Giugliano D, Pizza G, Gambardella A, Tesauro P, Varricchio M, D'onofrio F (1992) Glutathione infusion potentiates glucose-induced insulin secretion in aged patients with impaired glucose tolerance. Diabetes Care 15:1–7

Peyot ML, Pepin E, Lamontagne J, Latour MG, Zarrouki B, Lussier R, Pineda M, Jetton TL, Madiraju SR, Joly E, Prentki M (2010) Beta-cell failure in diet-induced obese mice stratified according to body weight gain: secretory dysfunction and altered islet lipid metabolism without steatosis or reduced beta-cell mass. Diabetes 59:2178–2187

Pi J, Bai Y, Zhang Q, Wong V, Floering LM, Daniel K, Reece JM, Deeney JT, Andersen ME, Corkey BE, Collins S (2007) Reactive oxygen species as a signal in glucose-stimulated insulin secretion. Diabetes 56:1783–1791

Poitout V, Amyot J, Semache M, Zarrouki B, Hagman D, Fontes G (2010) Glucolipotoxicity of the pancreatic beta cell. Biochim Biophys Acta 1801:289–298

Poitout V, Robertson RP (2008) Glucolipotoxicity: fuel excess and beta-cell dysfunction. Endocr Rev 29:351–366

Pontiroli AE, Calderara A, Pozza G (1994) Secondary failure of oral hypoglycaemic agents: frequency, possible causes, and management. Diabetes Metab Rev 10:31–43

Prentki M, Madiraju SR (2008) Glycerolipid metabolism and signaling in health and disease. Endocr Rev 29:647–676

Preston AM, Gurisik E, Bartley C, Laybutt DR, Biden TJ (2009) Reduced endoplasmic reticulum (ER)-to-Golgi protein trafficking contributes to ER stress in lipotoxic mouse beta cells by promoting protein overload. Diabetologia 52:2369–2373

Qi D, Cai K, Wang O LIZ, Chen J, Deng B, Qian L LEY (2010) Fatty acids induce amylin expression and secretion by pancreatic beta-cells. Am J Physiol Endocrinol Metab 298:E99–107

Qvigstad E, kollind M, Grill V (2004) Nine weeks of bedtime diazoxide is well tolerated and improves beta-cell function in subjects with Type 2 diabetes. Diabet Med 21:73–76

Rabinowitz JD, White E (2010) Autophagy and metabolism. Science 330:1344–1348

Radtke M, Kollind M, Qvigstad E, Grill V (2007) Twelve weeks' treatment with diazoxide without insulin supplementation in Type 2 diabetes is feasible but does not improve insulin secretion. Diabet Med 24:172–177

Reaven GM, Chen YD, Hollenbeck CB, Sheu WH, Ostrega D, Polonsky KS (1993) Plasma insulin, C-peptide, and proinsulin concentrations in obese and nonobese individuals with varying degrees of glucose tolerance. J Clin Endocrinol Metab 76:44–8

Robbins RD, Tersey SA, Ogihara T, Gupta D, Farb TB, Ficorilli J, Bokvist K, Maier B, Mirmira RG (2010) Inhibition of deoxyhypusine synthase enhances islet {beta} cell function and survival in the setting of endoplasmic reticulum stress and type 2 diabetes. J Biol Chem 285:39943–39952

Robertson RP (2009) Beta-cell deterioration during diabetes: what's in the gun? Trends Endocrinol Metab 20:388–393

Rutti S, Ehses JA, Sibler RA, Prazak R, Rohrer L, Georgopoulos S, Meier DT, Niclauss N, Berney T, Donath MY, Von Eckardstein A (2009) Low- and high-density lipoproteins modulate function, apoptosis, and proliferation of primary human and murine pancreatic beta-cells. Endocrinology 150:4521–4530

Saisho Y, Manesso E, Butler AE, Galasso R, Kavanagh K, Flynn M, Zhang L, Clark P, Gurlo T, Toffolo GM, Cobelli C, Wagner JD, Butler PC (2011) Ongoing beta-cell turnover in adult nonhuman primates is not adaptively increased in streptozotocin-induced diabetes. Diabetes 60:848–856

Sako Y, Grill VE (1990) A 48-hour lipid infusion in the rat time-dependently inhibits glucose-induced insulin secretion and ß-cell oxidation through a process likely coupled to fatty acid oxidation. Endocrinology 127:1580–1589

Scheuner D, Kaufman RJ (2008) The unfolded protein response: a pathway that links insulin demand with beta-cell failure and diabetes. Endocr Rev 29:317–333

Semplici F, Vaxillaire M, Fogarty S, Semache M, Bonnefond A, Fontes G, Philippe J, Meur G, Diraison F, Sessions RB, Rutter J, Poitout V, Froguel P, Rutter GA (2011) A human mutation within the per-ARNT-sim (PAS) domain-containing protein kinase (PASK) causes basal insulin hypersecretion. J Biol Chem 286(51):44005–44014

Spranger J, Kroke A, Mohlig M, Hoffmann K, Bergmann MM, Ristow M, Boeing H, Pfeiffer AF (2003) Inflammatory cytokines and the risk to develop type 2 diabetes: results of the prospective population-based european prospective investigation into cancer and nutrition (EPIC)-Potsdam study. Diabetes 52:812–817

Sturek JM, Castle JD, Trace AP, Page LC, Castle AM, Evans-Molina C, Parks JS, Mirmira RG, Hedrick CC (2010) An intracellular role for ABCG1-mediated cholesterol transport in the regulated secretory pathway of mouse pancreatic beta cells. J Clin Invest 120:2575–2589

Talchai C, Xuan S, Lin HV, Sussel L, Accili D (2012) Pancreatic beta cell dedifferentiation as a mechanism of diabetic beta cell failure. Cell 150:1223–1234

Tanaka Y, Gleason CE, Tran PO, Harmon JS, Robertson RP (1999) Prevention of glucose toxicity in HIT-T15 cells and Zucker diabetic fatty rats by antioxidants. Proc Natl Acad Sci U S A 96:10857–10862

Tonooka N, Oseid E, Zhou H, Harmon JS, Robertson RP (2007) Glutathione peroxidase protein expression and activity in human islets isolated for transplantation. Clin Transplant 21:767–772

Turrens JF (2003) Mitochondrial formation of reactive oxygen species. J Physiol 552:335–344

UK Prospective Diabetes Study 16 (1995) Overview of 6 years' therapy of type II diabetes: a progressive disease. UK prospective piabetes study group. Diabetes 44:1249–1258

UK Prospective Diabetes Study (1998) Intensive blood-glucose control with sulphonylureas or insulin compared with conventional treatment and risk of complications in patients with type 2 diabetes (UKPDS 33). UK prospective diabetes study (UKPDS) group. Lancet 352:837–853

Verchere CB, D'alessio DA, Palmiter RD, Weir GC, Bonner-Weir S, Baskin DG, Kahn SE (1996) Islet amyloid formation associated with hyperglycemia in transgenic mice with pancreatic ß cell expression of human islet amyloid polypeptide. Proc Natl Acad Sci U S A 93:3492–3496

Vergeer M, Brunham LR, Koetsveld J, Kruit JK, Verchere CB, Kastelein JJ, Hayden MR, Stroes ES (2010) Carriers of loss-of-function mutations in ABCA1 display pancreatic beta-cell dysfunction. Diabetes Care 33:869–874

Villarreal-Molina MT, Aguilar-Salinas CA, Rodriguez-Cruz M, Riano D, Villalobos-Comparan M, Coral-Vazquez R, Menjivar M, Yescas-Gomez P, Konigsoerg-Fainstein M, Romero-Hidalgo S,

Tusie-Luna MT, Canizales-Quinteros S (2007) The ATP-binding cassette transporter A1 R230C variant affects HDL cholesterol levels and BMI in the Mexican population: association with obesity and obesity-related comorbidities. Diabetes 56:1881–1887

Villarreal-Molina MT, Flores-Dorantes MT, Arellano-Campos O, Villalobos-Comparan M, Rodriguez-Cruz M, Miliar-Garcia A, Huertas-Vazquez A, Menjivar M, Romero-Hidalgo S, Wacher NH, Tusie-Luna MT, Cruz M, Aguilar-Salinas CA, Canizales-Quinteros S (2008) Association of the ATP-binding cassette transporter A1 R230C variant with early-onset type 2 diabetes in a Mexican population. Diabetes 57:509–513

Von Eckardstein A, Schulte H, Assmann G (2000) Risk for diabetes mellitus in middle-aged Caucasian male participants of the PROCAM study: implications for the definition of impaired fasting glucose by the American Diabetes Association. Prospective Cardiovascular Munster. J Clin Endocrinol Metab 85:3101–3108

Watada H, Kajimoto Y, Kaneto H, Matsuoka T, Fujitani Y, Miyazaki J, Yamasaki Y (1996) Involvement of the homeodomain-containing transcription factor PDX-1 in islet amyloid polypeptide gene transcription. Biochem Biophys Res Commun 229:746–751

Weir GC, Bonner-Weir S (2004) Five stages of evolving beta-cell dysfunction during progression to diabetes. Diabetes 53 (Suppl 3):S16–S21

Westermark P (1977) Amyloid of human islets of Langerhans II. Electron microscopic analysis of isolated amyloid. Virchows Arch A Pathol Anat Histol 373:161–166

Westermark P, Wilander E (1978) The influence of amyloid deposits on the islet volume in maturity onset diabetes mellitus. Diabetologia 15:417–421

Westermark GT, Leckstrom A MAZ, Westermark P (1998) Increased release of IAPP in response to long-term high fat intake in mice. Horm Metab Res 30:256–258

Westermark P, Andersson A, Westermark GT (2011) Islet amyloid polypeptide, islet amyloid, and diabetes mellitus. Physiol Rev 91:795–826

Whiting DR, Guariguata L, Weil C, Shaw J (2011) IDF diabetes atlas: global estimates of the prevalence of diabetes for 2011 and 2030. Diabetes Res Clin Pract 94:311–321

Zhang H, Ackermann AM, Gusarova GA, Lowe D, Feng X, Kopsombut UG, Costa RH, Gannon M (2006) The FoxM1 transcription factor is required to maintain pancreatic beta-cell mass. Mol Endocrinol 20:1853–1866

Zhang H, Zhang J, Pope CF, Crawford LA, Vasavada RC, Jagasia SM, Gannon M (2010) Gestational diabetes mellitus resulting from impaired beta-cell compensation in the absence of FoxM1, a novel downstream effector of placental lactogen. Diabetes 59:143–152

Zraika S, Hull RL, Verchere CB, Clark A, Potter KJ, Fraser PE, Raleigh DP, Kahn SE (2010) Toxic oligomers and islet beta cell death: guilty by association or convicted by circumstantial evidence? Diabetologia 53:1046–1056

Chapter 6
The Skeletal Muscle in Metabolic Syndrome

Antonio Zorzano, David Sebastián and Montserrat Romero

Abstract Skeletal muscle metabolism shows a considerable number of alterations in the Metabolic Syndrome (MetS), these most probably linked to the development of insulin resistance. The muscle of obese subjects and type 2 diabetic patients shows lower capacity to take up glucose, to convert it into glycogen, and to oxidize it by mitochondria. In addition, muscle in these individuals shows a higher capacity to take up fatty acids, mainly as a result of increased lipid availability. Muscle shows impaired insulin signaling, and most data point to altered IRS protein activity as a result of enhanced phosphorylation in serine residues by protein serine kinases. Inflammation, excessive lipid availability, and oxidative stress have been proposed as potential causes to explain muscle insulin resistance in the MetS. A major question pending in the field is the identification of the primary causes of this disorder and the identification of the major pathways responsible for the development of insulin resistance and for the metabolic alterations present in skeletal muscle in humans.

Keywords Glucose metabolism · Glycogen · Insulin resistance · Lipid metabolism · Metabolic flexibility · Muscle

6.1 Skeletal Muscle Shows Insulin Resistance in the Metabolic Syndrome

Skeletal muscle plays a crucial role in the regulation of whole-body glucose homeostasis. Approximately 70–80 % of ingested glucose is taken up by skeletal muscle and is either stored as glycogen or oxidized for energy, or to a lesser extent stored as fat (DeFronzo et al. 1981; DeFronzo et al. 1985; Nuutila et al. 1992). Studies using

A. Zorzano (✉) · D. Sebastián · M. Romero
Institute for Research in Biomedicine (IRB Barcelona),
C/Baldiri Reixac 10, 08028 Barcelona, Spain
e-mail: antonio.zorzano@irbbarcelona.org

Departament de Bioquímica i Biologia Molecular, Facultat de Biologia,
Universitat de Barcelona, Av. Diagonal 645, 08028 Barcelona, Spain

CIBER de Diabetes y Enfermedades Metabólicas Asociadas (CIBERDEM),
Instituto de Salud Carlos III, Barcelona, Spain

M. Orešič, A. Vidal-Puig (eds.), *A Systems Biology Approach to Study Metabolic Syndrome,*
DOI 10.1007/978-3-319-01008-3_6,

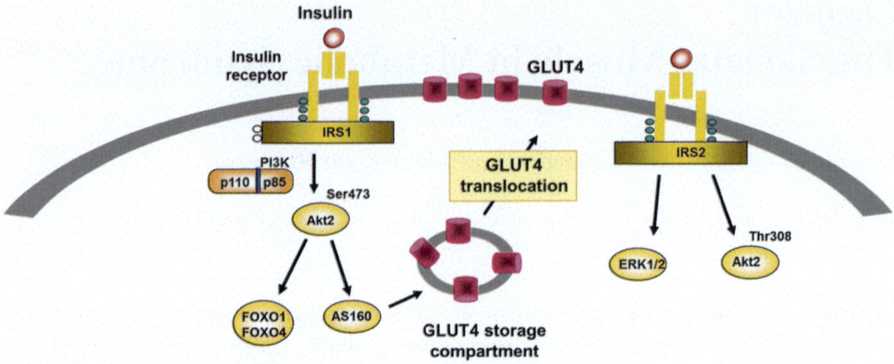

Fig. 6.1 Insulin signaling pathways in human muscle cells

euglycemic-insulin clamps suggest that non-oxidative glucose metabolism is the major pathway for glucose disposal in healthy subjects (DeFronzo et al. 1981). Moreover, during a hyperglycemic-hyperinsulinemic clamp, skeletal muscle accounts for most glucose uptake in healthy humans and over 80 % of this polysaccharide is stored as muscle glycogen (Shulman et al. 1990). After a standard meal, however, muscle glycogen synthesis accounts for approximately 30 % of the glucose ingested (Woerle et al. 2003).

A fundamental defect in the pathogenesis of type 2 diabetes mellitus is insulin resistance. In the postprandial state, insulin-mediated glucose uptake and utilization is greatly impaired in type 2 diabetic patients (Eriksson et al. 1992; DeFronzo et al. 1985; Shulman et al. 1990), and in glucose-tolerant first-degree relatives (Eriksson et al. 1992; Henriksen et al. 1994). From *in vivo* tracer studies, the defect in whole-body glucose uptake in these patients has been localized to the non-oxidative pathway for glucose metabolism (DeFronzo et al. 1981; Bonadonna et al. 1996; Shulman et al. 1990; Groop et al. 1989).

In addition, insulin resistance affects skeletal muscle in obesity and in type 2 diabetes, as assessed by *in vitro* preparations of human skeletal muscle. Thus, insulin-stimulated skeletal muscle glucose transport is impaired in type 2 diabetes and in obesity (Dohm et al. 1988; Goodyear et al. 1995). This impairment is a consequence of a reduced recruitment of GLUT4 glucose transporters to the cell surface of muscle fibers in type 2 diabetic patients (Ryder et al. 2000).

6.2 Alterations in Muscle Insulin Signaling in the Metabolic Syndrome

The critical components of insulin signaling in human muscle cells are shown in Fig. 6.1. Insulin signaling via the cell-surface receptor is linked to the phosphorylation of IRSs, namely, IRS1 and IRS2 (White 2002). Insulin signaling can be transduced

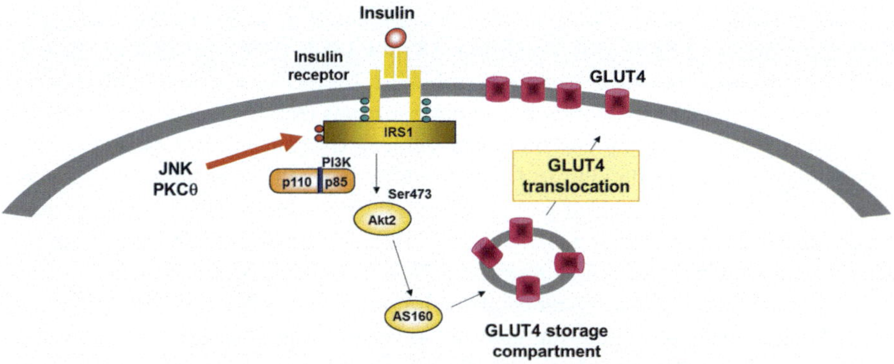

Fig. 6.2 Major metabolic alterations detected in insulin-resistant muscles

along metabolic pathways involving phosphatidylinositol 3-kinase (PI3K) and Akt or gene regulatory/mitogenic pathways including mitogen-activated protein kinases (MAPKs), where each of these steps constitutes a critical node in the regulation of metabolic and gene events controlling insulin sensitivity (Taniguchi et al. 2006). Further complexity comes from the expression of multiple isoforms of substrates and protein kinases at each of these nodes. Thus, in human myotubes, IRS1 signals to Akt, whereas IRS2 signals to MAPK (Fig. 6.1). These observations thus indicate the existence of isoform specificity in the transmission between nodes (Bouzakri et al. 2006).

Insulin signaling is negatively regulated through the activation of cell-surface receptor signaling pathways that are responsive to cytokines such as TNF-α or IL-6 (produced by macrophages that infiltrate adipose tissue in obesity) or through excessive availability of glucose or lipid (Taube et al. 2009; Chibalin et al. 2008; Schenk et al. 2008). Negative regulation of the insulin cascade occurs through serine phosphorylation events at the level of IRS isoforms and perhaps of the insulin receptor (Boura-Halfon and Zick 2009; Pirola et al. 2004; Draznin 2006).

There are many lines of evidence indicating that insulin signaling is deficient in muscle of obese and type 2 diabetes subjects (Fig. 6.2). Results from studies of insulin receptor phosphorylation in the skeletal muscle of type 2 diabetic subjects are contradictory. Insulin receptor phosphorylation appears to be normal or reduced in non-obese type 2 diabetics (Nolan et al. 1994; Krook et al. 2000; Maegawa et al. 1991; Klein et al. 1995; Arner et al. 1987). Similarly, insulin receptor phosphorylation has been reported to be normal or decreased in obese non-diabetic subjects (Goodyear et al. 1995; Nolan et al. 1994; Arner et al. 1987). Type 2 diabetic subjects have impaired insulin-stimulated tyrosine phosphorylation of IRS-1 in skeletal muscle (Fig. 6.2). This impairment is not a consequence of decreased IRS-1 protein expression (Krook et al. 2000; Bjornholm et al. 1997). A similar defect is observed at the level of PI3K in type 2 diabetic muscle (Bjornholm et al. 1997; Krook et al. 2000; Kim et al. 1999). Reduced insulin-stimulated tyrosine phosphorylation of IRS-1 and of PI3K has been also reported in skeletal muscle of obese subjects (Kim et al. 1999; Goodyear et al. 1995).

Insulin-stimulated Akt phosphorylation is impaired in skeletal muscle of type 2 diabetic patients under *in vitro* conditions (Krook et al. 1998). In addition, insulin-stimulated phosphorylation of the Akt substrate AS160 (also named TBC1D4) is defective in skeletal muscle of type 2 diabetic subjects (Karlsson et al. 2005). AS160 is a Rab GTPase-activating protein that has emerged as a candidate Akt effector for controlling GLUT4 trafficking in fat and muscle cells (Sano et al. 2003; Zeigerer et al. 2004; Bruss et al. 2005; Eguez et al. 2005; Thong et al. 2007). In this respect, recent data indicate that mice with a AS160/TBC1D4-Thr649Ala knock-in mutation are glucose-intolerant and show altered GLUT4 trafficking (Chen et al. 2011).

It has been proposed that the reduced insulin signaling via the IRS-1/PI3K pathway is attributable to the phosphorylation of IRS proteins in specific serine residues. This notion is supported by two observations: a) serine phosphorylation of IRS proteins reduces the capacity of IRS proteins to attract PI3K (Qiao et al. 1999; Qiao et al. 2002; Aguirre et al. 2002; Birnbaum 2001; White 2003); and b) serine phosphorylation of IRS-1 can also lead to an accelerated degradation of IRS-1 protein (Shah et al. 2004). Serine phosphorylation of IRS proteins can occur in response to a number of intracellular serine kinases, such as p70 S6 kinase (S6K1 kinase), c-Jun NH2-terminal kinase (JNK), protein kinase C, and IKK (Um et al. 2004; Pende et al. 2000; Hirosumi et al. 2002; Gao et al. 2004; Nguyen et al. 2005; Yuan et al. 2001; Gao et al. 2002; Li et al. 2004; Bell et al. 2000; Kim et al. 2004). In this regard, increased activities of PKC and JNK, have been found in skeletal muscle of obese and type 2 diabetic subjects (Itani et al. 2001; Bandyopadhyay et al. 2005), thus supporting a potential role of these serine kinases in the pathogenesis of insulin resistance (Fig. 6.2).

6.3 Glucose Metabolism in Muscle During Metabolic Syndrome

Insulin stimulates glucose uptake in skeletal muscle by inducing the translocation of GLUT4 to the plasma membrane and therefore glucose transport into the cell. Glucose is then phosphorylated by hexokinase and glucose-6-phosphate (G6P) and either used in the glycolytic pathway or incorporated into glycogen by glycogen synthase (GS).

Baseline muscle glycogen concentrations are 30 % lower in type 2 diabetic patients compared to matched controls (Carey et al. 2003; Shulman et al. 1990), and the rate of glycogen synthesis is 50 % lower in diabetic subjects during hyperglycemic-hyperinsulinemic clamps (Shulman et al. 1990). Postprandrial increments in muscle glycogen are also lower than in healthy subjects (Carey et al. 2003). Alterations in insulin-mediated glucose disposal are also present in obese non-diabetic subjects (Kahn et al. 2006; Bogardus et al. 1984a) and individuals with hypertension (Ferrannini et al. 1987), other features present in the Metabolic Syndrome.

The mechanisms underlying these alterations could affect several steps in the glucose metabolism pathway. Thus, defects in GS (Bogardus et al. 1984b; Damsbo et al. 1991; Wright et al. 1988) hexokinase (Kelley et al. 1996; Braithwaite et al. 1995;

Kruszynska et al. 1998; Bonadonna et al. 1996); and glucose transport (Rothman et al. 1992; Bonadonna et al. 1996; Zierath et al. 1996; Dohm et al. 1988) have all been implicated in the lower muscle glycogen synthesis in type 2 diabetes.

6.3.1 Glucose Transport

Activation of the insulin signaling pathway leads to the translocation of GLUT4 from intracellular vesicles to the plasma membrane and the subsequent increase in glucose uptake by skeletal muscle. Glucose transport is severely impaired in insulin-resistant individuals with type 2 diabetes or obesity (Bonadonna et al. 1996; Zierath et al. 1996; Rothman et al. 1992; Dohm et al. 1988) with normal or increased levels of GLUT4 mRNA or protein expression (Pedersen et al. 1990; Eriksson et al. 1992). These observations suggest that impaired GLUT4 translocation or the intrinsic activity of the glucose transporter is responsible for the defect in muscle glucose transport. Moreover, the reduced phosphorylation of AS160 in muscle of type 2 diabetic patients points to impaired GLUT4 recruitment to the cell surface (Karlsson et al. 2005).

6.3.2 Glucose Phosphorylation

Once glucose is inside the muscle cell, it is phosphorylated to G6P by the action of hexokinase II (HK-II). Glucose transport and phosphorylation are tightly coupled processes. In human skeletal muscle, insulin increases cytosolic activity, protein content and HK-II mRNA and this is associated with the translocation of the enzyme from the cytosol to the mitochondria. There are discrepant data as to whether glucose phosphorylation is impaired in type 2 diabetic subjects, probably because of the different methodologies used in the studies. Thus, some authors report that in type 2 diabetic patients glucose phosphorylation is impaired to a greater extent than glucose transport, suggesting that impaired HK-II activity is the rate-limiting step for insulin action (Bonadonna et al. 1993; Bonadonna et al. 1996). In keeping with this view, HK-II activity is reduced in muscle of type 2 diabetic patients (Kruszynska et al. 1998). However, other studies demonstrated that the defect in glucose transport exceeds the impairment in phosphorylation (Cline et al. 1999), suggesting that glucose transport is the rate-controlling step for insulin-stimulated glucose disposal in type 2 diabetic subjects.

6.3.3 Glycogen Synthesis

G6P can be used for glycogen synthesis or for oxidation. Glycogen synthesis is regulated by GS, a key insulin-regulated enzyme (Dent et al. 1990; Yki-Jarvinen et al. 1987). GS can be found in an active (dephosphorylated) or inactive (phosphorylated) form (Dent et al. 1990). Insulin triggers GS by stimulating a cascade of phosphorylation-dephosphorylation reactions that lead to the dephosphorylation of GS by the inhibition of GS kinase and activation of GS phosphatase or PP1. GS mRNA and protein levels are decreased in muscle of type 2 diabetic patients (Vestergaard et al. 1993; Vestergaard et al. 1991). Moreover, under basal conditions, total GS activity in type 2 diabetics is reduced, and the capacity of insulin to activate GS is severely impaired (Damsbo et al. 1991). However, the major abnormality in GS regulation in type 2 diabetes is its lack of activation by insulin, as a consequence of the impairment of the insulin signaling pathway.

In mice, it has been recently reported that a major mechanism by which insulin enhances GS activity in skeletal muscle is by activation induced by G6P (Bouskila et al. 2010). In this respect, the concentrations of G6P during a hyperglycaemic-hyperinsulinemic clamp were lower in muscles of type 2 diabetic patients or first-degree relatives compared to controls (Rothman et al. 1995). This is potentially the major mechanism that explains defective glycogen synthesis.

6.3.4 Glucose Oxidation

Approximately 90 % of glucose entering the glycolytic pathway is metabolized by aerobic glycolysis and 10 % is converted to lactate by the anaerobic glycolysis in human muscle (Thiebaud et al. 1982). The key enzymes that regulate glycolysis and glucose oxidation are phosphofructokinase (PFK) and pyruvate dehydrogenase (PDH), respectively. PFK is not regulated by insulin but PDH is a key insulin-regulated enzyme whose activity in muscle is acutely induced by insulin (Mandarino et al. 1986). In type 2 diabetic subjects, the glycolytic/glucose oxidative pathway has been shown to be impaired (Del Prato et al. 1993), with no alterations in PFK activity (Vestergaard et al. 1993) but with defects in insulin-stimulated PDH activity (Mandarino et al. 1986; Kelley et al. 1992).

6.4 Lipid Metabolism in Muscle During Metabolic Syndrome

Skeletal muscle plays a crucial role in lipid metabolism. In resting conditions, for instance, skeletal muscle fatty acid oxidation accounts for around 90 % of the energy requirements for this tissue (Dagenais et al. 1976; Kelley et al. 1993). Furthermore, lipid metabolism is essential in the physiology of skeletal muscle, such as in the metabolic switching that occurs during the transition from feeding to fasting conditions, and in adaptation to the energetic requirements of exercise (Achten et al. 2002). However, the Metabolic Syndrome is characterized by altered lipid metabolism in skeletal muscle and this change contributes to some of the pathological features present in the disease, such as insulin resistance. In this section, a brief overview of skeletal muscle lipid metabolism will be provided, followed by a description of the alterations present in the Metabolic Syndrome and discussion on how these alterations may contribute to insulin resistance.

Fatty acid metabolism in skeletal muscle is complex and depends on many factors at whole-body and at organ level. Although fatty acid availability (mainly determined by lipolysis in adipose tissue) is an important factor in the use of fatty acids by skeletal muscle, the regulation of lipid metabolism at muscular level is also relevant. The processes regulated at muscular level are as follows: fatty acid transport into the cell; fatty acid activation; fatty acid oxidation; fatty acid esterification or storage; and lipolysis of intramuscular lipids. However, we will focus on the processes that have been described to be altered in the Metabolic Syndrome.

6.4.1 Fatty Acid Transport

Plasma free fatty acids circulate bound to albumin (Campbell et al. 1964) and enter the muscle cell through either diffusion or protein-mediated transport (Abumrad et al. 1998; Hamilton 1998). The proteins responsible for the transport of fatty acids into the myocyte are the membrane-bound fatty acid transporters FAT/CD36 (fatty acid translocase/CD36, glycoprotein IV), FABPpm (membrane-bound fatty acid binding protein) and a family of fatty acid transport proteins (FATP 1 and 4) (Abumrad et al. 1998; Schaffer and Lodish 1994; Bonen et al. 1999; Schwieterman et al. 1988). FAT/CD36, FABPpm, FATP1, and FATP4 are co-expressed in skeletal muscle and all contribute to membrane transport, with recent evidence suggesting that FAT/CD36 and FATP4 are quantitatively the most important (Nickerson et al. 2009).

6.4.2 Fatty Acid Oxidation

Fatty acid oxidation takes place in the mitochondrial matrix by the coordinated action of three metabolic pathways: β-oxidation, the Krebs cycle, and electron transport chain (ETC). Long-chain fatty acyl-CoAs are transported into mitochondria by the action of the carnitine palmitoyltransferase (CPT) system, which is composed by CPT1, carnitine:acylcarnitine translocase (CACT) and CPT2 (McGarry and Brown 1997; Kerner and Hoppel 2000). The step catalyzed by CPT1 is tightly regulated allosterically by malonyl-CoA and is considered the rate-limiting step in fatty acid oxidation. Once inside the mitochondria, the fatty acid is degraded into two carbon fragments (acetyl-CoA) during β-oxidation. β-Hydroxy acyl-CoA dehydrogenase (β-HAD) is a key enzyme in β-oxidation and correlates with the rate of fatty acid oxidation (Kiens 1997). Acetyl-CoA generated during this process enters the Krebs cycle and through the electron transport chain ultimately leads to ATP generation.

6.4.3 Fatty Acid Storage

Fatty acids that are not oxidized are directed to lipid synthesis, mainly triacylglycerides (TAG) (Dyck et al. 1997), although other lipid species can be synthesized such as glycerolipids, phospholipids, sphingolipids and cholesterol esters. The flux of stored fatty acids depends on the concentration of the incoming fatty acid, the type of fatty acid (saturated or unsaturated), the muscle fiber type, the hormonal milieu and the energy requirements of the muscle (Watt and Hoy 2012).

6.4.4 Alterations in Skeletal Muscle Lipid Metabolism in the Metabolic Syndrome

Several alterations in lipid metabolism during obesity and type 2 diabetes have been reported in skeletal muscle. These alterations may contribute to insulin resistance but the precise mechanisms involved are unknown.

Intramuscular accumulation of lipid and its metabolites has been demonstrated in human (Boden et al. 1991; Itani et al. 2002) and in animal models of obesity (Bruce et al. 2009; Yu et al. 2002). Although TAGs are increased in muscle in obesity and type 2 diabetes, it is believed that other more reactive lipid species are responsible for lipid-induced insulin resistance, such as diacylglycerol (DAG) and ceramides. Several studies have revealed an association between high intramuscular lipids (IMCLs) and elevated levels of DAG and/or ceramide in obese insulin-resistant subjects (Adams et al. 2004; Moro et al. 2009; Thrush et al. 2009). Furthermore, DAG and ceramides are increased in skeletal muscle in obese subjects and contribute to insulin resistance (Adams et al. 2004; Itani et al. 2002).

The accumulation of lipid metabolites in muscle could be explained by an increase in fatty acid uptake and/or a decrease in their oxidation. Fatty acid uptake in skeletal muscle is increased in obesity and type 2 diabetes, and fatty acid transporters may play a critical role in this increase (Luiken et al. 2001; Bonen et al. 2004). Thus, muscle FABPpm protein expression is elevated in individuals with type 2 diabetes and obesity (Bruce et al. 2003; Simoneau et al. 1999), and there is a positive correlation between the sarcolemmal content of FAT/CD36 and TG levels in skeletal muscle of obese and type 2 diabetic patients (Bonen et al. 2004; Sampson et al. 2003) and obese rats (Han et al. 2007; Hegarty et al. 2002). Further support for the functional implication of FAT/CD36 in IMCL accumulation derives from a recent study demonstrating a dramatic reduction in IMCL content after specific inhibition of FAT/CD36 in myotubes obtained from obese patients (Aguer et al. 2010).

The deregulation of fatty acid oxidation during the Metabolic Syndrome is not as clear as the increase in fatty acid uptake. Some studies have described reduced oxidation of fatty acids in obesity (Kelley et al. 1999; Kim et al. 2000) and type 2 diabetes (Kelley and Simoneau 1994) while other studies have reported unchanged oxidation rates in obese humans (Holloway et al. 2007) and type 2 diabetic patients (Boushel et al. 2007). This discrepancy may arise from the different methodology used to assess fatty acid oxidative capacity since some studies analyzed whole muscle fatty acid oxidation (Kelley et al. 1999; Kim et al. 2000) while others examined mitochondrial function in isolated mitochondria or normalized to mitochondrial content (Boushel et al. 2007; de Feyter et al. 2008).

Fatty acid oxidation in muscle is increased in obese and in diabetic rodents (de Feyter et al. 2008; Holloway et al. 2009), thus indicating that these animal models are not similar to humans. In this regard, whole-body fatty acid oxidation assessed measuring the respiratory quotient demonstrated that obese humans show increased values compared to lean subjects, suggesting reduced fatty acid oxidation (Kelley et al. 1999), while obese mice have a decreased value, which would indicate increased fatty acid oxidation (Turner et al. 2007; Koves et al. 2008). It has been proposed that the increase in this process does not match a subsequent rise in the TCA flux, and then products of incompletely oxidized fatty acids, such as acylcarnitine derivatives, accumulate (Koves et al. 2008). The accumulation of these by-products is linked to impaired insulin action, although the mechanism has not been elucidated (Koves et al. 2008).

New technologies based on magnetic resonance spectroscopy assessment of *in vivo* mitochondrial function in humans have revealed compromised mitochondrial metabolism in insulin-resistant offspring of type 2 diabetic patients and in obesity and type 2 diabetes (Petersen et al. 2004; Befroy et al. 2007; Schrauwen-Hinderling et al. 2007). These findings suggest that fatty acid oxidation is impaired in these conditions. Moreover, the activity of some enzymes involved in fatty acid oxidation, such as CPT1, β-hydroxyacyl CoA dehydrogenase, citrate synthase and cytochrome oxidase, is decreased in obesity (Simoneau et al. 1999; Kim et al. 2000), and the expression of genes involved in oxidative capacity has also been shown to be reduced in skeletal muscle of obese and diabetic patients and of obese mice (Petersen et al. 2004; Sparks et al. 2005; Mootha et al. 2003; Patti et al. 2003).

6.5 Metabolic Inflexibility in the Metabolic Syndrome

Metabolic flexibility is defined in general as the ability of an organ, tissue or cell to adjust fuel oxidation to fuel availability. In the context of skeletal muscle, metabolic flexibility refers to the capacity of the tissue to switch from predominantly lipid oxidation and high rates of fatty acid uptake during fasting conditions to the suppression of lipid oxidation and increased glucose uptake, oxidation and storage under insulin-stimulated conditions (Kelley and Mandarino 2000). In healthy individuals, skeletal muscle displays substantial metabolic flexibility, which is lost in obese and type 2 diabetic subjects. Under basal fasting conditions, glucose uptake and oxidation are normal or even increased in obese subjects compared to lean subjects. Fatty acid uptake is also normal, but fatty acid oxidation is lower and its storage is elevated. In fed state (operationally achieved by the hyperinsulinemic-euglycemic clamp), glucose metabolism is not increased in obese subjects as it is in lean individuals, and although fatty acid uptake is comparable between lean and obese subjects, fatty acid oxidation

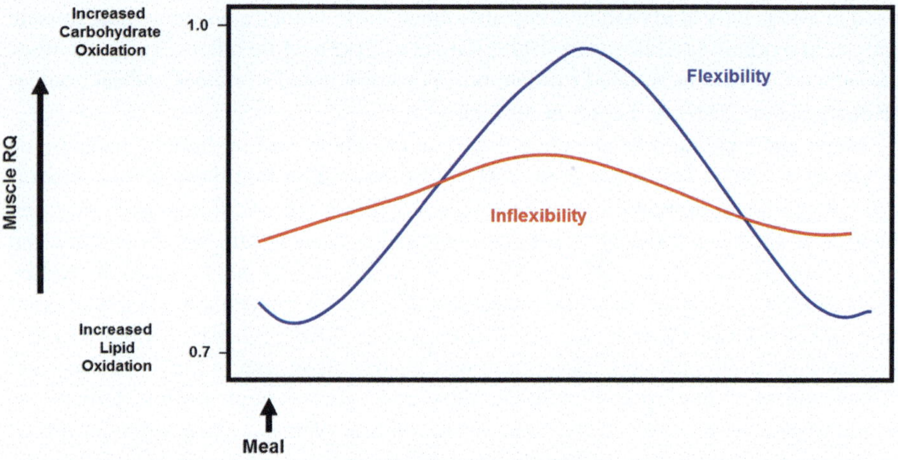

Fig. 6.3 Model for substrate oxidation during metabolic flexibility and metabolic inflexibility in skeletal muscle. A respiratory quotient (*RQ*) close to 0.7 indicates high lipid oxidation whereas a RQ close to 1 indicates high carbohydrate oxidation. After a meal, skeletal muscle switches from lipid oxidation (which is the predominant fuel used during fasting) to glucose oxidation, and this is reflected in the increase in RQ. During metabolic inflexibility, fasting lipid oxidation is lower (higher value of RQ) and after a meal the switch from lipid to glucose oxidation is impaired (lower value of RQ)

in much higher in the latter (Kelley et al. 1999) (Fig. 6.3). Therefore, obese and type 2 diabetic subjects show metabolic inflexibility in skeletal muscle and this alteration can contribute to the accumulation of lipids and reduced insulin-stimulated glucose disposal seen in skeletal muscle in obesity and type 2 diabetes.

Metabolic flexibility is influenced by both uptake and oxidation of glucose and plasma-derived fatty acids. In general, in insulin-resistant conditions there is an alteration of fatty acid utilization, and the impairments are present most clearly under stimulated conditions. Thus, lower leg muscle rates of fatty acid oxidation during fasting are related to decreased glucose storage during insulin stimulation in obese women (Colberg et al. 1995); the suppression of fat oxidation during insulin stimulation is defective in obese men and women (Kelley et al. 1999); fuel switching after a meal is impaired in type 2 diabetes and glucose intolerant subjects (Corpeleijn et al. 2008; Kelley and Simoneau 1994); and the increase in fat oxidation is blunted during exercise in type 2 diabetic patients (Blaak et al. 2000).

It has been proposed that in order to understand how lipids accumulate in muscle and cause insulin resistance, it is relevant to assess metabolic flexibility to high-fat diets (Galgani et al. 2008). Most studies addressing metabolic flexibility in the context of the Metabolic Syndrome and insulin resistant conditions show that the capacity to switch from fat to carbohydrate oxidation is generally impaired during a hyperinsulinemic clamp. However, this metabolic inflexibility is mostly the consequence of defective cellular glucose uptake. Indeed, after controlling for insulin-stimulated glucose disposal rate, metabolic flexibility is not altered in obesity regardless of the presence of type 2 diabetes.

Several factors are determinant in metabolic flexibility, such as fatty acid availability and the capacity for fatty acid uptake and oxidation in muscle. Furthermore, there is increasing evidence that defects in substrate switching cluster with disturbances in mitochondrial content and/or function (Ritov et al. 2005; Petersen et al. 2004; Simoneau and Kelley 1997), although the causal link between metabolic flexibility, insulin resistance and mitochondrial dysfunction remains to be established (Morino et al. 2006).

6.6 Mitochondrial Metabolism in Muscle

In skeletal muscle, the disruption of mitochondrial biology is evident in insulin-resistant subjects years before they develop diabetes (Petersen et al. 2004; Befroy et al. 2007; Patti et al. 2003). These individuals show a reduction in mitochondrial activity in skeletal muscle characterized by a decrease in mitochondrial oxidative and phosphorylation activity, and also increased fat accumulation in muscle and liver (Petersen et al. 2003; Petersen et al. 2004). Skeletal muscle of type 2 diabetic patients also shows a reduction in the activity of the Krebs cycle and of the respiratory chain (Kelley et al. 2002). In keeping with these observations, plasma levels of lactate are enhanced and the systemic appearance of lactate is also increased in these patients (Consoli et al. 1990; Cusi et al. 1996).

Several mechanisms are implicated in alterations of mitochondrial activity in insulin-resistant conditions, namely changes in mitochondrial density and morphology or alterations in mitochondrial function. Decreased mitochondrial mass has been demonstrated in muscle of insulin-resistant offspring of type 2 diabetic patients (Morino et al. 2005). In this regard, skeletal muscle of obese and type 2 diabetic subjects shows a lower mitochondrial DNA content (Ritov et al. 2005), and the latter present reduced citrate synthase activity (Kelley et al. 2002). In addition, several lines of evidence indicate that mitochondrial size and network are reduced in obesity and type 2 diabetes (Bach et al. 2003; Kelley et al. 2002; Toledo et al. 2006).

These alterations in mitochondrial morphology may indicate that mitochondrial dynamics is altered in insulin-resistant states. In this regard, repression of the mitochondrial fusion protein Mnf2, at mRNA and protein level, has been reported in the skeletal muscle of obese and type 2 diabetes subjects (Bach et al. 2003; Bach et al. 2005). This observation may be relevant as Mnf2-loss-of-function reduces glucose oxidation, mitochondrial membrane potential, oxygen consumption and mitochondrial proton leak in cultured cells (Bach et al. 2003; Pich et al. 2005; Chen et al. 2005). In connection with these observations, Mingrone and collaborators showed that, after bariatric surgery and massive weight loss, obese patients show induced expression of genes encoding for mitochondrial proteins, like Mnf2, associated with an amelioration of insulin sensitivity (Mingrone et al. 2005), thereby suggesting a positive correlation between Mnf2 levels and insulin sensitivity. However, under similar conditions, type 2 diabetic patients did not show any increase in the expression of these genes in response to bariatric surgery (Hernandez-Alvarez et al. 2009).

In addition to Mnf2, other proteins that participate in mitochondrial fusion and fission might play a regulatory role in mitochondrial metabolism in the context of obesity and type 2 diabetes. Thus, the expression of one of the mitochondrial proteases that is invoked in Optic atrophy 1 protein (OPA1) processing, PARL, is also reduced in an animal model of diabetes (Walder et al. 2005). In humans, a positive linear correlation has been reported between PARL mRNA levels and insulin sensitivity, as assessed by the hyperinsulinemic-euglycemic clamp. In addition, one study found the polymorphic variant Leu262Val to be associated with increased plasma insulin concentrations in one study (Walder et al. 2005), while this observation was not replicated in another (Fawcett et al. 2006). The precise mechanisms of mitochondrial dynamics regulation by PARL are far from clear, and no data on OPA1 protein abundance are available for obesity or type 2 diabetes. However, these data suggest the occurrence of multiple alterations in mitochondrial fusion in insulin-resistant states.

Regarding mitochondrial fission, Jheng and collaborators showed that skeletal muscle of both genetic and diet-induced obese mice present smaller and shorter mitochondria and increased expression of the mitochondrial fission components dynamin-related protein (Drp1) and Fis1. In addition, palmitate induced mitochondrial fragmentation and increased Drp1 and Fis1 proteins in differentiated C2C12 muscle cells. Interestingly, acute inhibition of mitochondrial fission with Mdivi-1 compound improved muscle insulin signaling and reduced plasma insulin levels in ob/ob mice (Jheng et al. 2012).

There is some discrepancy as to whether the muscle of type 2 diabetes patients shows alterations in mitochondrial function that depend or not on changes in mitochondrial mass. Thus, some studies report that this muscle shows alterations in mitochondrial function that persist after correction by mitochondrial mass (Mogensen et al. 2007; Ritov et al. 2005). However, in others, differences in electron transport chain or in oxygen consumption were not detected (Kelley et al. 2002; Boushel et al. 2007). One of the mechanisms proposed to explain the alterations in mitochondrial function in type 2 diabetes is a reduction in the expression of genes encoding for oxidative phosphorylation (Mootha et al. 2003; Patti et al. 2003). In this context, several genetic studies have suggested that a common polymorphism of the peroxisome proliferator-activated receptor γ coactivator-1α (PGC-1α) gene (Gly482Ser) is associated with the risk of type 2 diabetes (Barroso et al. 2006; Ek et al. 2001; Hara et al. 2002). In turn, the muscle of patients with type 2 diabetes and offspring of insulin-resistant subjects with this disease shows a decrease in the expression of the nuclear co-activators PGC-1α and PGC-1β (Patti et al. 2003; Mootha et al. 2003). The reduced activity of these co-activators may explain, at least in part, the defective expression of genes encoding respiratory chain subunits and the lower mitochondrial biogenesis that occurs in muscle in type 2 diabetes (Mootha et al. 2004). In this regard, since the decrease of PGC-1α and PGC-1β expression in skeletal muscle (Mootha et al. 2003; Patti et al. 2003) is paralleled by Mnf2 repression (Bach et al. 2003; Bach et al. 2005), PGC proteins were proposed to be implicated in Mfn2 transcription. In support of this notion, PGC-1α overexpression enhances Mnf2, at mRNA and protein levels, and also stimulates the transcriptional activity of the human Mnf2 gene promoter through transactivation of estrogen related-receptor α (ERRα) in muscle cells (Soriano et al. 2006). PGC-1β also induces Mnf2 expression in C2C12 myoblasts and myotubes and enhances ERRα-dependent Mnf2 transcription (Liesa et al. 2008). On the basis of these results, it is feasible that alterations in the PGC-1α/Mnf2 and PGC-1β/Mnf2 axes confer susceptibility to the development of metabolic alterations in cells.

In humans, there is some evidence of a parallelism between mitochondrial function and insulin sensitivity. In this regard, several intervention studies indicate that, after physical exercise and weight loss, obese/overweight and type 2 diabetic subjects show an increase in mitochondrial size. These subjects had an activation of electron chain transport, citrate synthase and succinate dehydrogenase in muscle, in parallel to ameliorated insulin sensitivity (Toledo et al. 2006; Menshikova et al. 2005; Menshikova et al. 2007; Toledo et al. 2007). However, other clinical studies indicate that the amelioration of insulin resistance can occur in the absence of changes in mitochondrial function. Thus, obese/overweight subjects after dietary restriction for 16 weeks show improved insulin sensitivity without changes in mitochondrial function (Toledo et al. 2008). In addition, the same results were observed in type 2 diabetic patients after treatment for 8 weeks with rosiglitazone (Schrauwen-Hinderling et al. 2008).

The observation of a correlation between insulin resistance and mitochondrial dysfunction in some human studies has led to the notion that mitochondrial dysfunction is a potential cause of insulin resistance. However, animal studies have shown that not all alterations in muscle mitochondrial activity produce insulin resistance. In this regard, ablation of the PGC-1α or PGC-1β gene in all tissues or in a tissue-specific manner in mice causes mitochondrial alterations in the absence of muscle insulin resistance (Handschin et al. 2007; Lelliott et al. 2006; Leone et al. 2005; Lin et al. 2004; Sonoda et al. 2007; Vianna et al. 2006). Moreover, apoptosis inducing factor (AIF) or transcription factor A mitochondrial (Tfam) knock-out mice show reduced respiratory chain activity in muscle tissues, without alterations in sensitivity to insulin (Pospisilik et al. 2007; Wredenberg et al. 2006). In contrast, some animal models display parallel modifications in mitochondrial activity and in insulin sensitivity. Thus, muscle-specific peroxisome proliferator activated receptor (PPAR) β knock-out mouse shows a decrease in the abundance of oxidative muscle fibers and reduced expression of genes involved in the mitochondrial respiratory chain before developing obesity and insulin resistance (Schuler et al. 2006). In addition, Wang and collaborators reported that transgenic mice overexpressing active PPARδ in skeletal muscle show a higher number of oxidative muscle fibers and are resistant to obesity and diabetes in response to a high fat diet (Wang et al. 2004). A tentative conclusion from all these data is that mitochondrial dysfunction is not sufficient to induce insulin resistance.

An alternative view is that insulin resistance has a negative impact on mitochondria. Thus, in myoblasts, insulin regulates the expression of genes encoding the subunits I and IV of cytochrome c oxidase (Pawlikowska et al. 2006), and chemical inhibition of PI3K results in the repression of citrate synthase expression (Kraft et al. 2006). Moreover, studies performed in humans reveal that insulin signaling has a direct impact on mitochondrial function in humans. Thus, administration of insulin to non-diabetic insulin-sensitive subjects caused an increase in the mitochondrial ATP production rate accompanied by stimulation in muscle mitochondrial protein synthesis, effects that were not seen in diabetic patients (Stump et al. 2003; Boirie et al. 2001; Halvatsiotis et al. 2002). Accordingly, a mosaic animal model with a 98 % reduction of insulin receptors displays enlarged mitochondria that show poorly organized cristae and shows reduced expression of PPARγ, PGC-1α and uncoupling protein 1 (UCP1) (Kitamura et al. 2004).

In summary, current information indicates that insulin signaling cascades are crucial for mitochondrial function and that disruption of mitochondrial function is not sufficient to induce insulin resistance. This view does not exclude the possibility that specific alterations of mitochondrial function are crucial in the pathogenesis of muscle insulin resistance in obesity or type 2 diabetes.

6.7 Molecular Mechanisms of Muscle Insulin Resistance and of Defective Metabolism in the Metabolic Syndrome

The main causes that trigger insulin resistance in skeletal muscle in the Metabolic Syndrome are not well known. The hypotheses with greatest experimental support regarding muscle insulin resistance are that it: a) is a consequence of chronic low-grade inflammation; or b) is a consequence of an excess of lipid availability.

The inflammation hypothesis proposes the occurrence of macrophage activation in adipose depots secondary to obesity. Under lean conditions, adipocytes secrete factors, such as interleukin (IL)-13, that promote alternative activation of macrophages. Alternatively, activated macrophages secrete anti-inflammatory mediators, such as IL-10, and may secrete insulin-sensitizing factors. Obesity induces changes in adipocyte metabolism and gene expression, resulting in increased lipolysis and the release of pro-inflammatory free fatty acids and factors that recruit and activate macrophages, such as monocyte chemotactic protein-1 (MCP-1) and tumor necrosis factor α (TNFα). Activated M1 macrophages produce large amounts of pro-inflammatory mediators, such as TNFα, IL-1β, and resistin, which act on adipocytes to induce an insulin-resistant state. This establishes a positive feedback loop that further amplifies inflammation and insulin resistance. As a result, adipose depots release a number of pro-inflammatory cytokines such as IL-6, IL-1 or TNF, which may reach muscle cells and promote insulin resistance (Schenk et al. 2008; Olefsky and Glass 2010). In keeping with this view, obese and type 2 diabetes subjects show increased circulating levels of pro-inflammatory cytokines such as IL-6, IL-8, and retinol-binding protein. In addition, circulating levels of non-esterified fatty acids are also increased as a result of a higher rate of lipolysis in adipose tissue.

There is also evidence that increased macrophage accumulation occurs in skeletal muscle in high-fat diet-induced obesity. Just as in the adipose tissue, these skeletal muscle pro-inflammatory macrophages are generally positive for F4/80, CD11b, and CD11c and exhibit an M1 macrophage-like phenotype (Nguyen et al. 2007; Patsouris et al. 2008). Histological studies indicate that these macrophages are localized mainly in intermuscular adipose depots, which accumulate within skeletal muscle in obesity. This finding raises the possibility that these skeletal muscle macrophages locally contribute to insulin resistance through paracrine mechanisms. However, this is only a hypothesis and it has not been demonstrated. In obesity, the number of macrophages in skeletal muscle is lower than in adipose tissue or also lower than the number of Kuppfer cells in liver.

The lipotoxic hypothesis is based on the observation that insulin resistance is associated with elevated levels of free fatty acids in the bloodstream. Several observations support this hypothesis. The single best predictor for the presence of insulin resistance in young, lean offspring of type 2 diabetic patients is the accumulation of fat inside muscle cells. This accumulation blocks an intracellular chain of events that normally triggers glucose transport into the cell. The buildup of DAGs inside muscle or liver cells causes the shutdown of the insulin signaling pathway. In the muscle cells, they achieve this by inhibiting the translocation of a protein, GLUT4, to the cell membrane, where it would normally work to pump glucose into the cell. Insulin-stimulated glucose transport no longer works efficiently, and the cell is insulin-resistant.

Lipotoxicity can occur in muscle when the body does not store fat in adipose tissue. This may occur due to various reasons: because adipose cells are already hypertrophic and cannot grow; because there is no capacity of adipose depots to generate enough adipose cells; or because lipid metabolism in muscle is dysfunctional and generates lipids that are toxic to cells so the cells become insulin-resistant.

As mentioned in a previous section, mitochondrial dysfunction has been proposed as a cause of insulin resistance, mainly on the basis of correlational observations. More recent data indicate that conditions/agents, such as TNF, dexamethasone, and chronic exposure to insulin or palmitate, which produce insulin resistance in muscle or adipose cells models share mitochondrial superoxide stress, which precedes insulin resistance (Houstis et al. 2006; Hoehn et al. 2009). Thus, treatment of cells with anti-oxidant compounds restores the capacity of cells to respond to insulin. Moreover, treatment of insulin-resistant obese ob/ob mice or mice on a high-fat diet with anti-oxidants showed improved glucose and insulin tolerance (Anderson et al. 2009; Hoehn et al. 2009; Houstis et al. 2006). In all, these data suggest a causal role for muscle mitochondrial reactive oxygen species (ROS) generation in insulin resistance.

6.8 Future Prospects

A major pending question in the field is to determine the primary causes of muscle insulin resistance in susceptible individuals or in rodents subjected to insulin-resistant challenges as well as to identify the elements of the pathway that trigger its development. In particular, it will be relevant to determine whether the primary causes of muscle insulin resistance are related to the activation of pro-inflammatory pathways or whether they are more associated with alterations in lipid metabolism in skeletal muscle, independent of inflammation. On the basis of the proposed role of specific mitochondrial dysfunction or mitochondrially generated ROS as a cause of insulin resistance, it will be pertinent to determine whether insulin resistant subjects show early alterations in these parameters. In this respect, and given the evidence reported to date, it may be of interest to determine whether changes in the activity of proteins involved in mitochondrial dynamics impair insulin sensitivity in muscle or other insulin-sensitive tissues.

Acknowledgements We thank Ms. T. Yates for editorial support. This study was supported by research grants from the MEC (SAF2008-03803), Grant 2009SGR915 from the "*Generalitat de Catalunya*", CIBERDEM ("*Instituto de Salud Carlos III*"), FP7-European Commission (MITIN, HEALTH-F4-2008-223450), and INTERREG IV-B-SUDOE-FEDER (DIOMED, SOE1/P1/E178). A.Z. was the recipient of a Science Intensification Award from the University of Barcelona.

References

Abumrad N, Harmon C, Ibrahimi A (1998) Membrane transport of long-chain fatty acids: evidence for a facilitated process. J Lipid Res 39:2309–2318

Achten J, Gleeson M, Jeukendrup AE (2002) Determination of the exercise intensity that elicits maximal fat oxidation. Med Sci Sports Exerc 34:92–97

Adams JM 2nd, Pratipanawatr T, Berria R, Wang E, Defronzo RA, Sullards MC, Mandarino LJ (2004) Ceramide content is increased in skeletal muscle from obese insulin-resistant humans. Diabetes 53:25–31

Aguer C, Mercier J, Man CY, Metz L, Bordenave S, Lambert K, Jean E, Lantier L, Bounoua L, Brun JF, Raynaud de ME, Andreelli F, Foretz M, Kitzmann M (2010) Intramyocellular lipid accumulation is associated with permanent relocation ex vivo and in vitro of fatty acid translocase (FAT)/CD36 in obese patients. Diabetologia 53:1151–1163

Aguirre V, Werner ED, Giraud J, Lee YH, Shoelson SE, White MF (2002) Phosphorylation of Ser307 in insulin receptor substrate-1 blocks interactions with the insulin receptor and inhibits insulin action. J Biol Chem 277:1531–1537

Anderson EJ, Lustig ME, Boyle KE, Woodlief TL, Kane DA, Lin CT, Price JW 3rd, Kang L, Rabinovitch PS, Szeto HH, Houmard JA, Cortright RN, Wasserman DH, Neufer PD (2009) Mitochondrial H2O2 emission and cellular redox state link excess fat intake to insulin resistance in both rodents and humans. J Clin Invest 119:573–581

Arner P, Pollare T, Lithell H, Livingston JN (1987) Defective insulin receptor tyrosine kinase in human skeletal muscle in obesity and type 2 (non-insulin-dependent) diabetes mellitus. Diabetologia 30:437–440

Bach D, Pich S, Soriano FX, Vega N, Baumgartner B, Oriola J, Daugaard JR, Lloberas J, Camps M, Zierath JR, Rabasa-Lhoret R, Wallberg-Henriksson H, Laville M, Palacin M, Vidal H, Rivera F, Brand M, Zorzano A (2003) Mitofusin-2 determines mitochondrial network architecture and mitochondrial metabolism. A novel regulatory mechanism altered in obesity. J Biol Chem 278:17190–17197

Bach D, Naon D, Pich S, Soriano FX, Vega N, Rieusset J, Laville M, Guillet C, Boirie Y, Wallberg-Henriksson H, Manco M, Calvani M, Castagneto M, Palacin M, Mingrone G, Zierath JR, Vidal H, Zorzano A (2005) Expression of Mfn2, the Charcot-Marie-Tooth neuropathy type 2A gene, in human skeletal muscle: effects of type 2 diabetes, obesity, weight loss, and the regulatory role of tumor necrosis factor alpha and interleukin-6. Diabetes 54:2685–2693

Bandyopadhyay GK, Yu JG, Ofrecio J, Olefsky JM (2005) Increased p85/55/50 expression and decreased phosphotidylinositol 3-kinase activity in insulin-resistant human skeletal muscle. Diabetes 54:2351–2359

Barroso I, Luan J, Sandhu MS, Franks PW, Crowley V, Schafer AJ, O'rahilly S, Wareham NJ (2006) Meta-analysis of the Gly482Ser variant in PPARGC1A in type 2 diabetes and related phenotypes. Diabetologia 49:501–505

Befroy DE, Petersen KF, Dufour S, Mason GF, de Graaf RA, Rothman DL, Shulman GI (2007) Impaired mitochondrial substrate oxidation in muscle of onsulin-resistant offspring of type 2 diabetes patients. Diabetes 56:1376–1381

Bell KS, Schmitz-Peiffer C, Lim-Fraser M, Biden TJ, Cooney GJ, Kraegen EW (2000) Acute reversal of lipid-induced muscle insulin resistance is associated with rapid alteration in PKC-theta localization. Am J Physiol Endocrinol Metab 279:E1196–1201

Birnbaum MJ (2001) Turning down insulin signaling. J Clin Invest 108:655–659

Bjornholm M, Kawano Y, Lehtihet M, Zierath JR (1997) Insulin receptor substrate-1 phosphorylation and phosphatidylinositol 3-kinase activity in skeletal muscle from NIDDM subjects after in vivo insulin stimulation. Diabetes 46:524–527

Blaak EE, Aggel-Leijssen DP van, Wagenmakers AJ, Saris WH, Baak MA van (2000) Impaired oxidation of plasma-derived fatty acids in type 2 diabetic subjects during moderate-intensity exercise. Diabetes 49:2102–2107

Boden G, Jadali F, White J, Liang Y, Mozzoli M, Chen X, Coleman E, Smith C (1991) Effects of fat on insulin-stimulated carbohydrate metabolism in normal men. J Clin Invest 88:960–966

Bogardus C, Lillioja S, Mott D, Reaven GR, Kashiwagi A, Foley JE (1984a) Relationship between obesity and maximal insulin-stimulated glucose uptake in vivo and in vitro in Pima Indians. J Clin Invest 73:800–805

Bogardus C, Lillioja S, Stone K, Mott D (1984b) Correlation between muscle glycogen synthase activity and in vivo insulin action in man. J Clin Invest 73:1185–1190

Boirie Y, Short KR, Ahlman B, Charlton M, Nair KS (2001) Tissue-specific regulation of mitochondrial and cytoplasmic protein synthesis rates by insulin. Diabetes 50:2652–2658

Bonadonna RC, Del Prato S, Saccomani MP, Bonora E, Gulli G, Ferrannini E, Bier D, Cobelli C, Defronzo RA (1993) Transmembrane glucose transport in skeletal muscle of patients with non-insulin-dependent diabetes. J Clin Invest 92:486–494

Bonadonna RC, Del Prato S, Bonora E, Saccomani MP, Gulli G, Natali A, Frascerra S, Pecori N, Ferrannini E, Bier D, Cobelli C, Defronzo RA (1996) Roles of glucose transport and glucose phosphorylation in muscle insulin resistance of NIDDM. Diabetes 45:915–925

Bonen A, Miskovic D, Kiens B (1999) Fatty acid transporters (FABPpm, FAT, FATP) in human muscle. Can J Appl Physiol 24:515–523

Bonen A, Parolin ML, Steinberg GR, Calles-Escandon J, Tandon NN, Glatz JF, Luiken JJ, Heigenhauser GJ, Dyck DJ (2004) Triacylglycerol accumulation in human obesity and type 2 diabetes is associated with increased rates of skeletal muscle fatty acid transport and increased sarcolemmal FAT/CD36. FASEB J 18:1144–1146

Boura-Halfon S, Zick Y (2009) Phosphorylation of IRS proteins, insulin action, and insulin resistance. Am J Physiol Endocrinol Metab 296:E581–591

Boushel R, Gnaiger E, Schjerling P, Skovbro M, Kraunsoe R, Dela F (2007) Patients with type 2 diabetes have normal mitochondrial function in skeletal muscle. Diabetologia 50:790–796

Bouskila M, Hunter RW, Ibrahim AF, Delattre L, Peggie M, Diepen JA van, Voshol PJ, Jensen J, Sakamoto K (2010) Allosteric regulation of glycogen synthase controls glycogen synthesis in muscle. Cell Metab 12:456–466

Bouzakri K, Zachrisson A, Al-Khalili L, Zhang BB, Koistinen HA, Krook A, Zierath JR (2006) siRNA-based gene silencing reveals specialized roles of IRS-1/Akt2 and IRS-2/Akt1 in glucose and lipid metabolism in human skeletal muscle. Cell Metab 4:89–96

Braithwaite SS, Palazuk B, Colca JR, Edwards CW 3rd, Hofmann C (1995) Reduced expression of hexokinase II in insulin-resistant diabetes. Diabetes 44:43–48

Bruce CR, Anderson MJ, Carey AL, Newman DG, Bonen A, Kriketos AD, Cooney GJ, Hawley JA (2003) Muscle oxidative capacity is a better predictor of insulin sensitivity than lipid status. J Clin Endocrinol Metab 88:5444–5451

Bruce CR, Hoy AJ, Turner N, Watt MJ, Allen TL, Carpenter K, Cooney GJ, Febbraio MA, Kraegen EW (2009) Overexpression of carnitine palmitoyltransferase-1 in skeletal muscle is sufficient to enhance fatty acid oxidation and improve high-fat diet-induced insulin resistance. Diabetes 58:550–558

Bruss MD, Arias EB, Lienhard GE, Cartee GD (2005) Increased phosphorylation of Akt substrate of 160 kDa (AS160) in rat skeletal muscle in response to insulin or contractile activity. Diabetes 54:41–50

Campbell J, Martucci AD, Green GR (1964) Plasma albumin as an acceptor of free fatty acids. Biochem J 93:183–189

Carey PE, Halliday J, Snaar JE, Morris PG, Taylor R (2003) Direct assessment of muscle glycogen storage after mixed meals in normal and type 2 diabetic subjects. Am J Physiol Endocrinol Metab 284:E688–694

Cline GW, Petersen KF, Krssak M, Shen J, Hundal RS, Trajanoski Z, Inzucchi S, Dresner A, Rothman DL, Shulman GI (1999) Impaired glucose transport as a cause of decreased insulin-stimulated muscle glycogen synthesis in type 2 diabetes. N Engl J Med 341:240–246

Colberg SR, Simoneau JA, Thaete FL, Kelley DE (1995) Skeletal muscle utilization of free fatty acids in women with visceral obesity. J Clin Invest 95:1846–1853

Consoli A, Nurjhan N, Reilly JJ Jr, Bier DM, Gerich JE (1990) Mechanism of increased gluconeogenesis in noninsulin-dependent diabetes mellitus. Role of alterations in systemic, hepatic, and muscle lactate and alanine metabolism. J Clin Invest 86:2038–2045

Corpeleijn E, Mensink M, Kooi ME, Roekaerts PM, Saris WH, Blaak EE (2008) Impaired skeletal muscle substrate oxidation in glucose-intolerant men improves after weight loss. Obesity (Silver Spring) 16:1025–1032

Cusi K, Consoli A, Defronzo RA (1996) Metabolic effects of metformin on glucose and lactate metabolism in noninsulin-dependent diabetes mellitus. J Clin Endocrinol Metab 81:4059–4067

Chen H, Chomyn A, Chan DC (2005) Disruption of fusion results in mitochondrial heterogeneity and dysfunction. J Biol Chem 280:26185–26192

Chen S, Wasserman DH, Mackintosh C, Sakamoto K (2011) Mice with AS160/TBC1D4-Thr649Ala knockin mutation are glucose intolerant with reduced insulin sensitivity and altered GLUT4 trafficking. Cell Metab 13:68–79

Chibalin AV, Leng Y, Vieira E, Krook A, Bjornholm M, Long YC, Kotova O, Zhong Z, Sakane F, Steiler T, Nylen C, Wang J, Laakso M, Topham MK, Gilbert M, Wallberg-Henriksson H, Zierath JR (2008) Downregulation of diacylglycerol kinase delta contributes to hyperglycemia-induced insulin resistance. Cell 132:375–386

Dagenais GR, Tancredi RG, Zierler KL (1976) Free fatty acid oxidation by forearm muscle at rest, and evidence for an intramuscular lipid pool in the human forearm. J Clin Invest 58:421–431

Damsbo P, Vaag A, Hother-Nielsen O, Beck-Nielsen H (1991) Reduced glycogen synthase activity in skeletal muscle from obese patients with and without type 2 (non-insulin-dependent) diabetes mellitus. Diabetologia 34:239–245

DeFronzo RA, Jacot E, Jequier E, Maeder E, Wahren J, Felber JP (1981) The effect of insulin on the disposal of intravenous glucose. Results from indirect calorimetry and hepatic and femoral venous catheterization. Diabetes 30:1000–1007

DeFronzo RA, Gunnarsson R, Bjorkman O, Olsson M, Wahren J (1985) Effects of insulin on peripheral and splanchnic glucose metabolism in noninsulin-dependent (type II) diabetes mellitus. J Clin Invest 76:149–155

Del Prato S, Bonadonna RC, Bonora E, Gulli G, Solini A, Shank M, Defronzo RA (1993) Characterization of cellular defects of insulin action in type 2 (non-insulin-dependent) diabetes mellitus. J Clin Invest 91:484–494

Dent P, Lavoinne A, Nakielny S, Caudwell FB, Watt P, Cohen P (1990) The molecular mechanism by which insulin stimulates glycogen synthesis in mammalian skeletal muscle. Nature 348:302–308

Dohm GL, Tapscott EB, Pories WJ, Dabbs DJ, Flickinger EG, Meelheim D, Fushiki T, Atkinson SM, Elton CW, Caro JF (1988) An in vitro human muscle preparation suitable for metabolic studies. Decreased insulin stimulation of glucose transport in muscle from morbidly obese and diabetic subjects. J Clin Invest 82:486–494

Draznin B (2006) Molecular mechanisms of insulin resistance: serine phosphorylation of insulin receptor substrate-1 and increased expression of p85alpha: the two sides of a coin. Diabetes 55:2392–2397

Dyck DJ, Peters SJ, Glatz J, Gorski J, Keizer H, Kiens B, Liu S, Richter EA, Spriet LL, Vusse GJ van der, Bonen A (1997) Functional differences in lipid metabolism in resting skeletal muscle of various fiber types. Am J Physiol 272:E340–351

Eguez L, Lee A, Chavez JA, Miinea CP, Kane S, Lienhard GE, Mcgraw TE (2005) Full intracellular retention of GLUT4 requires AS160 Rab GTPase activating protein. Cell Metab 2:263–272

Ek J, Andersen G, Urhammer SA, Gaede PH, Drivsholm T, Borch-Johnsen K, Hansen T, Pedersen O (2001) Mutation analysis of peroxisome proliferator-activated receptor-gamma coactivator-1 (PGC-1) and relationships of identified amino acid polymorphisms to Type II diabetes mellitus. Diabetologia 44:2220–2226

Eriksson J, Koranyi L, Bourey R, Schalin-Jantti C, Widen E, Mueckler M, Permutt AM, Groop LC (1992) Insulin resistance in type 2 (non-insulin-dependent) diabetic patients and their relatives is not associated with a defect in the expression of the insulin-responsive glucose transporter (GLUT-4) gene in human skeletal muscle. Diabetologia 35:143–147

Fawcett KA, Wareham NJ, Luan J, Syddall H, Cooper C, O'rahilly S, Day IN, Sandhu MS, Barroso I (2006) PARL Leu262Val is not associated with fasting insulin levels in UK populations. Diabetologia 49:2649–2652

Ferrannini E, Buzzigoli G, Bonadonna R, Giorico MA, Oleggini M, Graziadei L, Pedrinelli R, Brandi L, Bevilacqua S (1987) Insulin resistance in essential hypertension. N Engl J Med 317:350–357

Feyter HM de, Lenaers E, Houten SM, Schrauwen P, Hesselink MK, Wanders RJ, Nicolay K, Prompers JJ (2008) Increased intramyocellular lipid content but normal skeletal muscle mitochondrial oxidative capacity throughout the pathogenesis of type 2 diabetes. FASEB J 22:3947–3955

Galgani JE, Moro C, Ravussin E (2008) Metabolic flexibility and insulin resistance. Am J Physiol Endocrinol Metab 295:E1009–1017

Gao Z, Hwang D, Bataille F, Lefevre M, York D, Quon MJ, Ye J (2002) Serine phosphorylation of insulin receptor substrate 1 by inhibitor kappa B kinase complex. J Biol Chem 277:48115–48121

Gao Z, Zhang X, Zuberi A, Hwang D, Quon MJ, Lefevre M, Ye J (2004) Inhibition of insulin sensitivity by free fatty acids requires activation of multiple serine kinases in 3T3-L1 adipocytes. Mol Endocrinol 18:2024–2034

Goodyear LJ, Giorgino F, Sherman LA, Carey J, Smith RJ, Dohm GL (1995) Insulin receptor phosphorylation, insulin receptor substrate-1 phosphorylation, and phosphatidylinositol 3-kinase activity are decreased in intact skeletal muscle strips from obese subjects. J Clin Invest 95:2195–2204

Groop LC, Bonadonna RC, Delprato S, Ratheiser K, Zyck K, Ferrannini E, Defronzo RA (1989) Glucose and free fatty acid metabolism in non-insulin-dependent diabetes mellitus. Evidence for multiple sites of insulin resistance. J Clin Invest 84:205–213

Halvatsiotis P, Short KR, Bigelow M, Nair KS (2002) Synthesis rate of muscle proteins, muscle functions, and amino acid kinetics in type 2 diabetes. Diabetes 51:2395–2404

Hamilton JA (1998) Fatty acid transport: difficult or easy? J Lipid Res 39:467–481

Han XX, Chabowski A, Tandon NN, Calles-Escandon J, Glatz JF, Luiken JJ, Bonen A (2007) Metabolic challenges reveal impaired fatty acid metabolism and translocation of FAT/CD36 but not FABPpm in obese Zucker rat muscle. Am J Physiol Endocrinol Metab 293:E566–575

Handschin C, Choi CS, Chin S, Kim S, Kawamori D, Kurpad AJ, Neubauer N, Hu J, Mootha VK, Kim YB, Kulkarni RN, Shulman GI, Spiegelman BM (2007) Abnormal glucose homeostasis in skeletal muscle-specific PGC-1alpha knockout mice reveals skeletal muscle-pancreatic beta cell crosstalk. J Clin Invest 117:3463–3474

Hara K, Tobe K, Okada T, Kadowaki H, Akanuma Y, Ito C, Kimura S, Kadowaki T (2002) A genetic variation in the PGC-1 gene could confer insulin resistance and susceptibility to Type II diabetes. Diabetologia 45:740–743

Hegarty BD, Cooney GJ, Kraegen EW, Furler SM (2002) Increased efficiency of fatty acid uptake contributes to lipid accumulation in skeletal muscle of high fat-fed insulin-resistant rats. Diabetes 51:1477–1484

Henriksen JE, Alford F, Handberg A, Vaag A, Ward GM, Kalfas A, Beck-Nielsen H (1994) Increased glucose effectiveness in normoglycemic but insulin-resistant relatives of patients with non-insulin-dependent diabetes mellitus. A novel compensatory mechanism. J Clin Invest 94:1196–1204

Hernandez-Alvarez MI, Chiellini C, Manco M, Naon D, Liesa M, Palacin M, Mingrone G, Zorzano A (2009) Genes involved in mitochondrial biogenesis/function are induced in response to bilio-pancreatic diversion in morbidly obese individuals with normal glucose tolerance but not in type 2 diabetic patients. Diabetologia 52:1618–1627

Hirosumi J, Tuncman G, Chang L, Gorgun CZ, Uysal KT, Maeda K, Karin M, Hotamisligil GS (2002) A central role for JNK in obesity and insulin resistance. Nature 420:333–336

Hoehn KL, Salmon AB, Hohnen-Behrens C, Turner N, Hoy AJ, Maghzal GJ, Stocker R, Remmen H van, Kraegen EW, Cooney GJ, Richardson AR, James DE (2009) Insulin resistance is a cellular antioxidant defense mechanism. Proc Natl Acad Sci U S A 106:17787–17792

Holloway GP, Thrush AB, Heigenhauser GJ, Tandon NN, Dyck DJ, Bonen A, Spriet LL (2007) Skeletal muscle mitochondrial FAT/CD36 content and palmitate oxidation are not decreased in obese women. Am J Physiol Endocrinol Metab 292:E1782–1789

Holloway GP, Benton CR, Mullen KL, Yoshida Y, Snook LA, Han XX, Glatz JF, Luiken JJ, Lally J, Dyck DJ, Bonen A (2009) In obese rat muscle transport of palmitate is increased and is channeled to triacylglycerol storage despite an increase in mitochondrial palmitate oxidation. Am J Physiol Endocrinol Metab 296:E738–747

Houstis N, Rosen ED, Lander ES (2006) Reactive oxygen species have a causal role in multiple forms of insulin resistance. Nature 440:944–948

Itani SI, Pories WJ, Macdonald KG, Dohm GL (2001) Increased protein kinase C theta in skeletal muscle of diabetic patients. Metabolism 50:553–557

Itani SI, Ruderman NB, Schmieder F, Boden G (2002) Lipid-induced insulin resistance in human muscle is associated with changes in diacylglycerol, protein kinase C, and IkappaB-alpha. Diabetes 51:2005–2011

Jheng HF, Tsai PJ, Guo SM, Kuo LH, Chang CS, Su IJ, Chang CR, Tsai YS (2012) Mitochondrial fission contributes to mitochondrial dysfunction and insulin resistance in skeletal muscle. Mol Cell Biol 32:309–319

Kahn SE, Hull RL, Utzschneider KM (2006) Mechanisms linking obesity to insulin resistance and type 2 diabetes. Nature 444:840–846

Karlsson HK, Zierath JR, Kane S, Krook A, Lienhard GE, Wallberg-Henriksson H (2005) Insulin-stimulated phosphorylation of the Akt substrate AS160 is impaired in skeletal muscle of type 2 diabetic subjects. Diabetes 54:1692–1697

Kerner J, Hoppel C (2000) Fatty acid import into mitochondria. Biochim Biophys Acta 1486:1–17

Kelley DE, Simoneau JA (1994) Impaired free fatty acid utilization by skeletal muscle in non-insulin-dependent diabetes mellitus. J Clin Invest 94:2349–2356

Kelley DE, Mandarino LJ (2000) Fuel selection in human skeletal muscle in insulin resistance: a reexamination. Diabetes 49:677–683

Kelley DE, Mokan M, Mandarino LJ (1992) Intracellular defects in glucose metabolism in obese patients with NIDDM. Diabetes 41:698–706

Kelley DE, Mokan M, Simoneau JA, Mandarino LJ (1993) Interaction between glucose and free fatty acid metabolism in human skeletal muscle. J Clin Invest 92:91–98

Kelley DE, Mintun MA, Watkins SC, Simoneau JA, Jadali F, Fredrickson A, Beattie J, Theriault R (1996) The effect of non-insulin-dependent diabetes mellitus and obesity on glucose transport and phosphorylation in skeletal muscle. J Clin Invest 97:2705–2713

Kelley DE, Goodpaster B, Wing RR, Simoneau JA (1999) Skeletal muscle fatty acid metabolism in association with insulin resistance, obesity, and weight loss. Am J Physiol 277:E1130–1141
Kelley DE, He J, Menshikova EV, Ritov VB (2002) Dysfunction of mitochondria in human skeletal muscle in type 2 diabetes. Diabetes 51:2944–2950
Kiens B (1997) Effect of endurance training on fatty acid metabolism: local adaptations. Med Sci Sports Exerc 29:640–645
Kim JY, Hickner RC, Cortright RL, Dohm GL, Houmard JA (2000) Lipid oxidation is reduced in obese human skeletal muscle. Am J Physiol Endocrinol Metab 279:E1039–1044
Kim JK, Fillmore JJ, Sunshine MJ, Albrecht B, Higashimori T, Kim DW, Liu ZX, Soos TJ, Cline GW, O'brien WR, Littman DR, Shulman GI (2004) PKC-theta knockout mice are protected from fat-induced insulin resistance. J Clin Invest 114:823–827
Kim YB, Nikoulina SE, Ciaraldi TP, Henry RR, Kahn BB (1999) Normal insulin-dependent activation of Akt/protein kinase B, with diminished activation of phosphoinositide 3-kinase, in muscle in type 2 diabetes. J Clin Invest 104:733–741
Kitamura T, Kitamura Y, Nakae J, Giordano A, Cinti S, Kahn CR, Efstratiadis A, Accili D (2004) Mosaic analysis of insulin receptor function. J Clin Invest 113:209–219
Klein HH, Vestergaard H, Kotzke G, Pedersen O (1995) Elevation of serum insulin concentration during euglycemic hyperinsulinemic clamp studies leads to similar activation of insulin receptor kinase in skeletal muscle of subjects with and without NIDDM. Diabetes 44:1310–1317
Koves TR, Ussher JR, Noland RC, Slentz D, Mosedale M, Ilkayeva O, Bain J, Stevens R, Dyck JR, Newgard CB, Lopaschuk GD, Muoio DM (2008) Mitochondrial overload and incomplete fatty acid oxidation contribute to skeletal muscle insulin resistance. Cell Metab 7:45–56
Kraft CS, Lemoine CM, Lyons CN, Michaud D, Mueller CR, Moyes CD (2006) Control of mitochondrial biogenesis during myogenesis. Am J Physiol Cell Physiol 290:C1119–1127
Krook A, Roth RA, Jiang XJ, Zierath JR, Wallberg-Henriksson H (1998) Insulin-stimulated Akt kinase activity is reduced in skeletal muscle from NIDDM subjects. Diabetes 47:1281–1286
Krook A, Bjornholm M, Galuska D, Jiang XJ, Fahlman R, Myers MG Jr, Wallberg-Henriksson H, Zierath JR (2000) Characterization of signal transduction and glucose transport in skeletal muscle from type 2 diabetic patients. Diabetes 49:284–292
Kruszynska YT, Mulford MI, Baloga J, Yu JG, Olefsky JM (1998) Regulation of skeletal muscle hexokinase II by insulin in nondiabetic and NIDDM subjects. Diabetes 47:1107–1113
Lelliott CJ, Medina-Gomez G, Petrovic N, Kis A, Feldmann HM, Bjursell M, Parker N, Curtis K, Campbell M, Hu P, Zhang D, Litwin SE, Zaha VG, Fountain KT, Boudina S, Jimenez-Linan M, Blount M, Lopez M, Meirhaeghe A, Bohlooly YM, Storlien L, Stromstedt M, Snaith M, Oresic M, Abel ED, Cannon B, Vidal-Puig A (2006) Ablation of PGC-1beta results in defective mitochondrial activity, thermogenesis, hepatic function, and cardiac performance. PLoS Biol 4:e369
Leone TC, Lehman JJ, Finck BN, Schaeffer PJ, Wende AR, Boudina S, Courtois M, Wozniak DF, Sambandam N, Bernal-Mizrachi C, Chen Z, Holloszy JO, Medeiros DM, Schmidt RE, Saffitz JE, Abel ED, Semenkovich CF, Kelly DP (2005) PGC-1alpha deficiency causes multi-system energy metabolic derangements: muscle dysfunction, abnormal weight control and hepatic steatosis. PLoS Biol 3:e101
Li Y, Soos TJ, Li X, Wu J, Degennaro M, Sun X, Littman DR, Birnbaum MJ, Polakiewicz RD (2004) Protein kinase C Theta inhibits insulin signaling by phosphorylating IRS1 at Ser(1101). J Biol Chem 279:45304–45307
Liesa M, Borda-D'agua B, Medina-Gomez G, Lelliott CJ, Paz JC, Rojo M, Palacin M, Vidal-Puig A, Zorzano A (2008) Mitochondrial fusion is increased by the nuclear coactivator PGC-1beta. PLoS One 3:e3613
Lin J, Wu PH, Tarr PT, Lindenberg KS, St-Pierre J, Zhang CY, Mootha VK, Jager S, Vianna CR, Reznick RM, Cui L, Manieri M, Donovan MX, Wu Z, Cooper MP, Fan MC, Rohas LM, Zavacki AM, Cinti S, Shulman GI, Lowell BB, Krainc D, Spiegelman BM (2004) Defects in adaptive energy metabolism with CNS-linked hyperactivity in PGC-1alpha null mice. Cell 119:121–135

Luiken JJ, Arumugam Y, Dyck DJ, Bell RC, Pelsers MM, Turcotte LP, Tandon NN, Glatz JF, Bonen A (2001) Increased rates of fatty acid uptake and plasmalemmal fatty acid transporters in obese Zucker rats. J Biol Chem 276:40567–40573

Maegawa H, Shigeta Y, Egawa K, Kobayashi M (1991) Impaired autophosphorylation of insulin receptors from abdominal skeletal muscles in nonobese subjects with NIDDM. Diabetes 40:815–819

Mandarino LJ, Madar Z, Kolterman OG, Bell JM, Olefsky JM (1986) Adipocyte glycogen synthase and pyruvate dehydrogenase in obese and type II diabetic subjects. Am J Physiol 251:E489–496

Mcgarry JD, Brown NF (1997) The mitochondrial carnitine palmitoyltransferase system. From concept to molecular analysis. Eur J Biochem 244:1–14

Menshikova EV, Ritov VB, Ferrell RE, Azuma K, Goodpaster BH, Kelley DE (2007) Characteristics of skeletal muscle mitochondrial biogenesis induced by moderate-intensity exercise and weight loss in obesity. J Appl Physiol 103:21–27

Menshikova EV, Ritov VB, Toledo FG, Ferrell RE, Goodpaster BH, Kelley DE (2005) Effects of weight loss and physical activity on skeletal muscle mitochondrial function in obesity. Am J Physiol Endocrinol Metab 288:E818–825

Mingrone G, Manco M, Calvani M, Castagneto M, Naon D, Zorzano A (2005) Could the low level of expression of the gene encoding skeletal muscle mitofusin-2 account for the metabolic inflexibility of obesity? Diabetologia 48:2108–2114

Mogensen M, Sahlin K, Fernstrom M, Glintborg D, Vind BF, Beck-Nielsen H, Hojlund K (2007) Mitochondrial respiration is decreased in skeletal muscle of patients with type 2 diabetes. Diabetes 56:1592–1599

Mootha VK, Lindgren CM, Eriksson KF, Subramanian A, Sihag S, Lehar J, Puigserver P, Carlsson E, Ridderstrale M, Laurila E, Houstis N, Daly MJ, Patterson N, Mesirov JP, Golub TR, Tamayo P, Spiegelman B, Lander ES, Hirschhorn JN, Altshuler D, Groop LC (2003) PGC-1alpha-responsive genes involved in oxidative phosphorylation are coordinately downregulated in human diabetes. Nat Genet 34:267–273

Mootha VK, Handschin C, Arlow D, Xie X, St Pierre J, Sihag S, Yang W, Altshuler D, Puigserver P, Patterson N, Willy PJ, Schulman IG, Heyman RA, Lander ES, Spiegelman BM (2004) Erralpha and Gabpa/b specify PGC-1alpha-dependent oxidative phosphorylation gene expression that is altered in diabetic muscle. Proc Natl Acad Sci U S A 101:6570–6575

Morino K, Petersen KF, Dufour S, Befroy D, Frattini J, Shatzkes N, Neschen S, White MF, Bilz S, Sono S, Pypaert M, Shulman GI (2005) Reduced mitochondrial density and increased IRS-1 serine phosphorylation in muscle of insulin-resistant offspring of type 2 diabetic parents. J Clin Invest 115:3587–3593

Morino K, Petersen KF, Shulman GI (2006) Molecular mechanisms of insulin resistance in humans and their potential links with mitochondrial dysfunction. Diabetes 55(Suppl 2):9–15

Moro C, Galgani JE, Luu L, Pasarica M, Mairal A, Bajpeyi S, Schmitz G, Langin D, Liebisch G, Smith SR (2009) Influence of gender, obesity, and muscle lipase activity on intramyocellular lipids in sedentary individuals. J Clin Endocrinol Metab 94:3440–3447

Nguyen MT, Satoh H, Favelyukis S, Babendure JL, Imamura T, Sbodio JI, Zalevsky J, Dahiyat BI, Chi NW, Olefsky JM (2005) JNK and tumor necrosis factor-alpha mediate free fatty acid-induced insulin resistance in 3T3-L1 adipocytes. J Biol Chem 280:35361–35371

Nguyen MT, Favelyukis S, Nguyen AK, Reichart D, Scott PA, Jenn A, Liu-Bryan R, Glass CK, Neels JG, Olefsky JM (2007) A subpopulation of macrophages infiltrates hypertrophic adipose tissue and is activated by free fatty acids via Toll-like receptors 2 and 4 and JNK-dependent pathways. J Biol Chem 282:35279–35292

Nickerson JG, Alkhateeb H, Benton CR, Lally J, Nickerson J, Han XX, Wilson MH, Jain SS, Snook LA, Glatz JF, Chabowski A, Luiken JJ, Bonen A (2009) Greater transport efficiencies of the membrane fatty acid transporters FAT/CD36 and FATP4 compared with FABPpm and FATP1 and differential effects on fatty acid esterification and oxidation in rat skeletal muscle. J Biol Chem 284:16522–16530

Nolan JJ, Freidenberg G, Henry R, Reichart D, Olefsky JM (1994) Role of human skeletal muscle insulin receptor kinase in the in vivo insulin resistance of noninsulin-dependent diabetes mellitus and obesity. J Clin Endocrinol Metab 78:471–477

Nuutila P, Koivisto VA, Knuuti J, Ruotsalainen U, Teras M, Haaparanta M, Bergman J, Solin O, Voipio-Pulkki LM, Wegelius U et al (1992) Glucose-free fatty acid cycle operates in human heart and skeletal muscle in vivo. J Clin Invest 89:1767–1774

Olefsky JM, Glass CK (2010) Macrophages, inflammation, and insulin resistance. Annu Rev Physiol 72:219–246

Patsouris D, Li PP, Thapar D, Chapman J, Olefsky JM, Neels JG (2008) Ablation of CD11c-positive cells normalizes insulin sensitivity in obese insulin resistant animals. Cell Metab 8:301–309

Patti ME, Butte AJ, Crunkhorn S, Cusi K, Berria R, Kashyap S, Miyazaki Y, Kohane I, Costello M, Saccone R, Landaker EJ, Goldfine AB, Mun E, Defronzo R, Finlayson J, Kahn CR, Mandarino LJ (2003) Coordinated reduction of genes of oxidative metabolism in humans with insulin resistance and diabetes: Potential role of PGC1 and NRF1. Proc Natl Acad Sci U S A 100:8466–8471

Pawlikowska P, Gajkowska B, Hocquette JF, Orzechowski A (2006) Not only insulin stimulates mitochondriogenesis in muscle cells, but mitochondria are also essential for insulin-mediated myogenesis. Cell Prolif 39:127–145

Pedersen O, Bak JF, Andersen PH, Lund S, Moller DE, Flier JS, Kahn BB (1990) Evidence against altered expression of GLUT1 or GLUT4 in skeletal muscle of patients with obesity or NIDDM. Diabetes 39:865–870

Pende M, Kozma SC, Jaquet M, Oorschot V, Burcelin R, Le Marchand-Brustel Y, Klumperman J, Thorens B, Thomas G (2000) Hypoinsulinaemia, glucose intolerance and diminished beta-cell size in S6K1-deficient mice. Nature 408:994–997

Petersen KF, Befroy D, Dufour S, Dziura J, Ariyan C, Rothman DL, Dipietro L, Cline GW, Shulman GI (2003) Mitochondrial dysfunction in the elderly: possible role in insulin resistance. Science 300:1140–1142

Petersen KF, Dufour S, Befroy D, Garcia R, Shulman GI (2004) Impaired mitochondrial activity in the insulin-resistant offspring of patients with type 2 diabetes. N Engl J Med 350:664–671

Pich S, Bach D, Briones P, Liesa M, Camps M, Testar X, Palacin M, Zorzano A (2005) The Charcot-Marie-Tooth type 2A gene product, Mfn2, up-regulates fuel oxidation through expression of OXPHOS system. Hum Mol Genet 14:1405–1415

Pirola L, Johnston AM, Obberghen E van (2004) Modulation of insulin action. Diabetologia 47: 170–184

Pospisilik JA, Knauf C, Joza N, Benit P, Orthofer M, Cani PD, Ebersberger I, Nakashima T, Sarao R, Neely G, Esterbauer H, Kozlov A, Kahn CR, Kroemer G, Rustin P, Burcelin R, Penninger JM (2007) Targeted deletion of AIF decreases mitochondrial oxidative phosphorylation and protects from obesity and diabetes. Cell 131(3):476–491

Qiao LY, Goldberg JL, Russell JC, Sun XJ (1999) Identification of enhanced serine kinase activity in insulin resistance. J Biol Chem 274:10625–10632

Qiao LY, Zhande R, Jetton TL, Zhou G, Sun XJ (2002) In vivo phosphorylation of insulin receptor substrate 1 at serine 789 by a novel serine kinase in insulin-resistant rodents. J Biol Chem 277:26530–26539

Ritov VB, Menshikova EV, He J, Ferrell RE, Goodpaster BH, Kelley DE (2005) Deficiency of subsarcolemmal mitochondria in obesity and type 2 diabetes. Diabetes 54:8–14

Rothman DL, Shulman RG, Shulman GI (1992) 31P nuclear magnetic resonance measurements of muscle glucose-6-phosphate. Evidence for reduced insulin-dependent muscle glucose transport or phosphorylation activity in non-insulin-dependent diabetes mellitus. J Clin Invest 89: 1069–1075

Rothman DL, Magnusson I, Cline G, Gerard D, Kahn CR, Shulman RG, Shulman GI (1995) Decreased muscle glucose transport/phosphorylation is an early defect in the pathogenesis of non-insulin-dependent diabetes mellitus. Proc Natl Acad Sci U S A 92:983–987

Ryder JW, Yang J, Galuska D, Rincon J, Bjornholm M, Krook A, Lund S, Pedersen O, Wallberg-Henriksson H, Zierath JR, Holman GD (2000) Use of a novel impermeable biotinylated photolabeling reagent to assess insulin- and hypoxia-stimulated cell surface GLUT4 content in skeletal muscle from type 2 diabetic patients. Diabetes 49:647–654

Sampson MJ, Davies IR, Braschi S, Ivory K, Hughes DA (2003) Increased expression of a scavenger receptor (CD36) in monocytes from subjects with Type 2 diabetes. Atherosclerosis 167:129–134

Sano H, Kane S, Sano E, Miinea CP, Asara JM, Lane WS, Garner CW, Lienhard GE (2003) Insulin-stimulated phosphorylation of a Rab GTPase-activating protein regulates GLUT4 translocation. J Biol Chem 278:14599–14602

Schaffer JE, Lodish HF (1994) Expression cloning and characterization of a novel adipocyte long chain fatty acid transport protein. Cell 79:427–436

Schenk S, Saberi M, Olefsky JM (2008) Insulin sensitivity: modulation by nutrients and inflammation. J Clin Invest 118:2992–3002

Schrauwen-Hinderling VB, Kooi ME, Hesselink MK, Jeneson JA, Backes WH, Echteld CJ van, Engelshoven JM van, Mensink M, Schrauwen P (2007) Impaired in vivo mitochondrial function but similar intramyocellular lipid content in patients with type 2 diabetes mellitus and BMI-matched control subjects. Diabetologia 50:113–120

Schrauwen-Hinderling VB, Mensink M, Hesselink MK, Sels JP, Kooi ME, Schrauwen P (2008) The insulin-sensitizing effect of rosiglitazone in type 2 diabetes mellitus patients does not require improved in vivo muscle mitochondrial function. J Clin Endocrinol Metab 93:2917–2921

Schuler M, Ali F, Chambon C, Duteil D, Bornert JM, Tardivel A, Desvergne B, Wahli W, Chambon P, Metzger D (2006) PGC1alpha expression is controlled in skeletal muscles by PPARbeta, whose ablation results in fiber-type switching, obesity, and type 2 diabetes. Cell Metab 4:407–414

Schwieterman W, Sorrentino D, Potter BJ, Rand J, Kiang CL, Stump D, Berk PD (1988) Uptake of oleate by isolated rat adipocytes is mediated by a 40-kDa plasma membrane fatty acid binding protein closely related to that in liver and gut. Proc Natl Acad Sci U S A 85:359–363

Shah OJ, Wang Z, Hunter T (2004) Inappropriate activation of the TSC/Rheb/mTOR/S6K cassette induces IRS1/2 depletion, insulin resistance, and cell survival deficiencies. Curr Biol 14: 1650–1656

Shulman GI, Rothman DL, Jue T, Stein P, Defronzo RA, Shulman RG (1990) Quantitation of muscle glycogen synthesis in normal subjects and subjects with non-insulin-dependent diabetes by 13 C nuclear magnetic resonance spectroscopy. N Engl J Med 322:223–228

Simoneau JA, Kelley de (1997) Altered glycolytic and oxidative capacities of skeletal muscle contribute to insulin resistance in NIDDM. J Appl Physiol 83:166–171

Simoneau JA, Veerkamp JH, Turcotte LP, Kelley de (1999) Markers of capacity to utilize fatty acids in human skeletal muscle: relation to insulin resistance and obesity and effects of weight loss. FASEB J 13:2051–2060

Sonoda J, Mehl IR, Chong LW, Nofsinger RR, Evans RM (2007) PGC-1beta controls mitochondrial metabolism to modulate circadian activity, adaptive thermogenesis, and hepatic steatosis. Proc Natl Acad Sci U S A 104:5223–5228

Soriano FX, Liesa M, Bach D, Chan DC, Palacin M, Zorzano A (2006) Evidence for a mitochondrial regulatory pathway defined by peroxisome proliferator-activated receptor-gamma coactivator-1 alpha, estrogen-related receptor-alpha, and mitofusin 2. Diabetes 55:1783–1791

Sparks LM, Xie H, Koza RA, Mynatt R, Hulver MW, Bray GA, Smith SR (2005) A high-fat diet coordinately downregulates genes required for mitochondrial oxidative phosphorylation in skeletal muscle. Diabetes 54:1926–1933

Stump CS, Short KR, Bigelow ML, Schimke JM, Nair KS (2003) Effect of insulin on human skeletal muscle mitochondrial ATP production, protein synthesis, and mRNA transcripts. Proc Natl Acad Sci U S A 100:7996–8001

Taniguchi CM, Emanuelli B, Kahn CR (2006) Critical nodes in signalling pathways: insights into insulin action. Nat Rev Mol Cell Biol 7:85–96

Taube A, Eckardt K, Eckel J (2009) Role of lipid-derived mediators in skeletal muscle insulin resistance. Am J Physiol Endocrinol Metab 297(5): E1004–1012

Thiebaud D, Jacot E, Defronzo RA, Maeder E, Jequier E, Felber JP (1982) The effect of graded doses of insulin on total glucose uptake, glucose oxidation, and glucose storage in man. Diabetes 31:957–963

Thong FS, Bilan PJ, Klip A (2007) The Rab GTPase-activating protein AS160 integrates Akt, protein kinase C, and AMP-activated protein kinase signals regulating GLUT4 traffic. Diabetes 56:414–423

Thrush AB, Brindley DN, Chabowski A, Heigenhauser GJ, Dyck DJ (2009) Skeletal muscle lipogenic protein expression is not different between lean and obese individuals: a potential factor in ceramide accumulation. J Clin Endocrinol Metab 94:5053–5061

Toledo FG, Menshikova EV, Ritov VB, Azuma K, Radikova Z, Delany J, Kelley DE (2007) Effects of physical activity and weight loss on skeletal muscle mitochondria and relationship with glucose control in type 2 diabetes. Diabetes 56:2142–2147

Toledo FG, Menshikova EV, Azuma K, Radikova Z, Kelley CA, Ritov VB, Kelley de (2008) Mitochondrial capacity in skeletal muscle is not stimulated by weight loss despite increases in insulin action and decreases in intramyocellular lipid content. Diabetes 57:987–994

Toledo FG, Watkins S, Kelley de (2006) Changes induced by physical activity and weight loss in the morphology of intermyofibrillar mitochondria in obese men and women. J Clin Endocrinol Metab 91:3224–3227

Turner N, Bruce CR, Beale SM, Hoehn KL, So T, Rolph MS, Cooney GJ (2007) Excess lipid availability increases mitochondrial fatty acid oxidative capacity in muscle: evidence against a role for reduced fatty acid oxidation in lipid-induced insulin resistance in rodents. Diabetes 56:2085–2092

Um SH, Frigerio F, Watanabe M, Picard F, Joaquin M, Sticker M, Fumagalli S, Allegrini PR, Kozma SC, Auwerx J, Thomas G (2004) Absence of S6K1 protects against age- and diet-induced obesity while enhancing insulin sensitivity. Nature 431:200–205

Vestergaard H, Bjorbaek C, Andersen PH, Bak JF, Pedersen O (1991) Impaired expression of glycogen synthase mRNA in skeletal muscle of NIDDM patients. Diabetes 40:1740–1745

Vestergaard H, Lund S, Larsen FS, Bjerrum OJ, Pedersen O (1993) Glycogen synthase and phosphofructokinase protein and mRNA levels in skeletal muscle from insulin-resistant patients with non-insulin-dependent diabetes mellitus. J Clin Invest 91:2342–2350

Vianna CR, Huntgeburth M, Coppari R, Choi CS, Lin J, Krauss S, Barbatelli G, Tzameli I, Kim YB, Cinti S, Shulman GI, Spiegelman BM, Lowell BB (2006) Hypomorphic mutation of PGC-1beta causes mitochondrial dysfunction and liver insulin resistance. Cell Metab 4:453–464

Walder K, Kerr-Bayles L, Civitarese A, Jowett J, Curran J, Elliott K, Trevaskis J, Bishara N, Zimmet P, Mandarino L, Ravussin E, Blangero J, Kissebah A, Collier GR (2005) The mitochondrial rhomboid protease PSARL is a new candidate gene for type 2 diabetes. Diabetologia 48:459–468

Wang YX, Zhang CL, Yu RT, Cho HK, Nelson MC, Bayuga-Ocampo CR, Ham J, Kang H, Evans RM (2004) Regulation of muscle fiber type and running endurance by PPARdelta. PLoS Biol 2:e294

Watt MJ, Hoy AJ (2012) Lipid metabolism in skeletal muscle: generation of adaptive and maladaptive intracellular signals for cellular function. Am J Physiol Endocrinol Metab 302(11):E1315–1328

White MF (2002) IRS proteins and the common path to diabetes. Am J Physiol Endocrinol Metab 283:E413–422

White MF (2003) Insulin signaling in health and disease. Science 302:1710–1711

Woerle HJ, Meyer C, Dostou JM, Gosmanov NR, Islam N, Popa E, Wittlin SD, Welle SL, Gerich JE (2003) Pathways for glucose disposal after meal ingestion in humans. Am J Physiol Endocrinol Metab 284:E716–725

Wredenberg A, Freyer C, Sandstrom ME, Katz A, Wibom R, Westerblad H, Larsson NG (2006) Respiratory chain dysfunction in skeletal muscle does not cause insulin resistance. Biochem Biophys Res Commun 350(1):202–207

Wright KS, Beck-Nielsen H, Kolterman OG, Mandarino LJ (1988) Decreased activation of skeletal muscle glycogen synthase by mixed-meal ingestion in NIDDM. Diabetes 37:436–440

Yki-Jarvinen H, Mott D, Young AA, Stone K, Bogardus C (1987) Regulation of glycogen synthase and phosphorylase activities by glucose and insulin in human skeletal muscle. J Clin Invest 80:95–100

Yu C, Chen Y, Cline GW, Zhang D, Zong H, Wang Y, Bergeron R, Kim JK, Cushman SW, Cooney GJ, Atcheson B, White MF, Kraegen EW, Shulman GI (2002) Mechanism by which fatty acids inhibit insulin activation of insulin receptor substrate-1 (IRS-1)-associated phosphatidylinositol 3-kinase activity in muscle. J Biol Chem 277:50230–50236

Yuan M, Konstantopoulos N, Lee J, Hansen L, Li ZW, Karin M, Shoelson SE (2001) Reversal of obesity- and diet-induced insulin resistance with salicylates or targeted disruption of Ikkbeta. Science 293:1673–1677

Zeigerer A, Mcbrayer MK, Mcgraw TE (2004) Insulin stimulation of GLUT4 exocytosis, but not its inhibition of endocytosis, is dependent on RabGAP AS160. Mol Biol Cell 15:4406–4415

Zierath JR, He L, Guma A, Odegoard Wahlstrom E, Klip A, Wallberg-Henriksson H (1996) Insulin action on glucose transport and plasma membrane GLUT4 content in skeletal muscle from patients with NIDDM. Diabetologia 39:1180–1189

Chapter 7
The Central Nervous System in Metabolic Syndrome

Luís Martins, Andrew J. Whittle, Rubén Nogueiras, Antonio Vidal-Puig, Carlos Diéguez and Miguel López

Abstract In the last decades obesity and its related metabolic disorders have increased at an epidemic rate in the developed and developing world. Lipid accumulation in peripheral tissues, such as heart, skeletal muscle, liver, and pancreatic β cells leads to lipotoxicity. It is an important factor contributing to type 2 diabetes, steatotic liver disease, insulin resistance and heart failure, disorders that are usually known as the Metabolic Syndrome (MetS). Recent work has demonstrated that hypothalamic sensing of circulating lipids and modulation of hypothalamic endogenous fatty acid and lipid metabolism are mechanisms modulating energy homeostasis at the whole body level. Enzymes, such as AMP-activated protein kinase (AMPK) and fatty acid synthase (FAS), plus intermediate metabolites, such as malonyl-CoA and long chain fatty acid-CoAs (LCFA-CoAs), have a key role in this neuronal network, integrating peripheral signals with classical neuropeptide-based mechanisms. One major point that needs to be addressed is whether impairment of lipid metabolism and accumulation of specific lipid species in the hypothalamus, leading to lipotoxicity, has deleterious effects on hypothalamic neurons that may contribute to MetS. In this chapter, we discuss data about hypothalamic lipid metabolism, with emphasis on the events typically associated to lipotoxicity such as endoplasmic reticulum (ER) stress. More extensive knowledge about these molecular mechanisms will be of great relevance for the treatment of obesity and metabolic syndrome.

Keywords AMP-Activated protein kinase (AMPK) · Fatty acids · Hypothalamus · Lipid metabolism · Metabolic syndrome

---

M. López · L. Martins (✉) · R. Nogueiras · C. Diéguez
Department of Physiology, CIMUS,
University of Santiago de Compostela- Instituto de Investigación Sanitaria,
15782 Santiago de Compostela, Spain
e-mail: luisricardo.jesus@usc.es

CIBER Fisiopatología de la Obesidad y Nutrición (CIBERobn),
15706 Santiago de Compostela, Spain

A. J. Whittle · A. Vidal-Puig
Metabolic Research Laboratories Level 4, NIHR Cambridge Biomedical Research Centre,
Addenbrooke's Hospital, Institute of Metabolic Science, University of Cambridge,
Box 289, Cambridge CB2 0QQ, United Kingdom
e-mail: ajv22@cam.ac.uk

M. Orešič, A. Vidal-Puig (eds.), *A Systems Biology Approach to Study Metabolic Syndrome*,
DOI 10.1007/978-3-319-01008-3_7,

7.1 Energy Balance and the Obesity Epidemic

Increasing prevalence of obesity is now one of the main health issues in both developed and developing countries, causing enormous economic problems for these societies (Farooqi and O'Rahilly 2005; Flier 2004; Friedman 2003; Medina-Gomez and Vidal-Puig 2005). Given that previous knowledge of the physiopathology of obesity is likely to reveal novel pharmacological targets, much effort is focused on identifying and characterizing the molecular mechanisms controlling energy balance and particularly feeding behaviour (Fruhbeck 2008; Medina-Gomez and Vidal-Puig 2005). The reasons for the increasing prevalence of obesity are multifaceted, but social and environmental factors together with genetic predisposition can lead overall to positive energy balance. The altered energy homeostasis that elicits obesity is associated to insulin resistance and type 2 diabetes, fatty liver and a range of other disorders, globally known as the Metabolic Syndrome (Cota et al. 2007; Farooqi and O'Rahilly 2005; Flier 2004; Fruhbeck 2008; Medina-Gomez and Vidal-Puig 2005; Plum et al. 2006; Sarafidis and Nilsson 2006), as well as several types of cancer (Calle and Kaaks 2004). Recent investigations have shown that housekeeping metabolic pathways, such as *de novo* lipogenesis, play an important role in the regulation of energy balance. By responding to hormonal and nutrient signals in peripheral tissues and also in the central nervous system (CNS) these housekeeping mechanisms can in turn modify energy balance (Carling et al. 2008; Kahn et al. 2005; Lage et al. 2008; Ruderman et al. 2003).

7.2 The Hypothalamus is a Key Brain Region Regulating Energy Balance

The hypothalamus is the brain region located below the thalamus, comprising the major portion of the ventral diencephalon, and regulates an enormous number of homeostatic functions; among them, the regulation of endocrine axes and energy balance are of particular importance. The hypothalamus is organized in anatomically-defined neuronal clusters, called nuclei, which form interconnected neuronal circuits via axonal projections (Fig. 7.1; Elmquist et al. 2005; Gao and Horvath 2007; López et al. 2007a, b, 2009; Morton et al. 2006). These nuclei respond to changes in energy status by altering the expression of specific neurotransmitters/neuromodulators that result in changes in energy intake and expenditure (Elmquist et al. 2005; Gao and Horvath 2007; López et al. 2007a, b, 2009; Morton et al. 2006).

The arcuate nucleus (ARC) is considered the "*master hypothalamic centre*" for food intake control. Two different neuronal populations in the ARC receive peripheral nutritional/feeding signals. One set of neurons produces the orexigenic (feeding-promoting) neuropeptides agouti-related protein (AgRP) and neuropeptide Y (NPY) (Fig. 7.1). These neurons mostly project to other "*second order*" neurons located in other hypothalamic nuclei, such as the paraventricular nucleus (PVH). A second ARC population of neurons expresses the anorexigenic (feeding inhibitors)

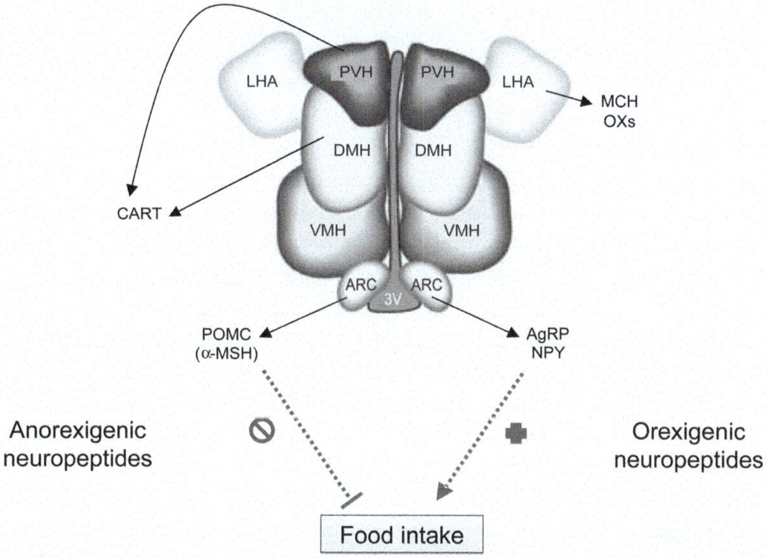

Fig. 7.1 Hypothalamic neuropeptides regulating food intake. The hypothalamus is structured in several nuclei, such as the arcuate (*ARC*), dorsomedial (*DMH*), paraventricular (*PVH*), ventromedial (*VMH*) and lateral hypothalamic area (*LHA*). Food intake is regulated at hypothalamic level by orexigenic (feeding stimulators) and anorexigenic neuropeptides (feeding inhibitors). Neuropeptide Y (*NPY*), agouti-related protein (*AgRP*), orexins (*OXs*) and melanin-concentrating hormone (*MCH*) exert orexigenic actions, whereas alpha-melanocyte stimulating hormone (*α-MSH*) and cocaine and amphetamine-regulated transcript (*CART*) are the main anorexigenic signals. The expression of these neuropeptides is under the control of central and peripheral signals

products of proopiomelanocortin (POMC), the precursor of alpha-melanocyte stimulating hormone (α-MSH) and the cocaine and amphetamine regulated transcript (CART). These neurons project their axons more broadly within the CNS, to secondary hypothalamic nuclei such as the dorsomedial nucleus (DMH), the lateral hypothalamic area (LHA) and the perifornical area, as well as the PVH. Dorsal to the ARC lies the VMH, which mainly collects projections from AgRP/NPY and CART/POMC neurons in the ARC. Furthermore, the VMH neurons project their axons to the ARC, DMH, LHA and brainstem regions such as the nucleus of the solitary tract. Hypothalamic neurons respond to peripheral metabolic signals, such as glucose, amino acids and lipids and also to hormones, such as leptin, ghrelin, adiponectin (ADPN), resistin (RSTN), glucagón-like peptide-1 and insulin, by modifying the expression and synthesis of neuropeptides. When energy intake exceeds expenditure, the expression of orexigenic neuropeptides such as AgRP and NPY diminishes. On the other hand, the expression of anorexigenic neuropeptides such as CART and POMC increases. Opposite changes happen when energy expenditure surpasses feeding (Elmquist et al. 2005; Gao and Horvath 2007; López et al. 2007a, b; Morton et al. 2006).

7.3 Lipid Sensing in the Hypothalamus

Circulating nutrients are derived either exogenously (via food intake) or endogenously (via hepatic glucose production and adipocyte lipolysis). Plasma long chain fatty acids (LCFAs) are mainly bound to albumin and cross the blood brain barrier (BBB) mostly by simple diffusion of the unbound form. A small proportion of fatty acids enter the brain via direct uptake of lipoprotein particles mediated by lipoprotein receptors on the luminal surface of the cerebrovascular endothelium (Lam et al. 2005b; Qi et al. 2002; Rapoport 2001). In general, the rate of entry of fatty acids into the brain is relative to its plasma concentration (Miller et al. 1987; Rapoport 1996). After crossing the BBB, LCFAs are esterified to LCFAs-CoAs, a reaction catalyzed by the long chain fatty acyl-CoA synthetase enzymes. This takes place inside neurons and glial cells, where the LCFAs-CoAs are metabolized for use in lipid biosynthesis and, to a much lesser extent, within oxidative pathways such as β-oxidation via CPT1 (Lam et al. 2005b; Miller et al. 1987).

Whether alterations in the lipid sensing mechanism are pathophysiological mechanisms leading to obesity and comorbidities is still unclear. However, it is worth mentioning that *1)* impairment of this central nutrient-sensing pathway is sufficient to disrupt energy homeostasis and induce obesity and *2)* the anorectic response to oleic acid (OA) depends on the nutritional state, being suppressed by short-term overfeeding (Morgan et al. 2004; Pocai et al. 2006). In fact, the hypothalamic responses to LCFAs disappear in overfed animals, which did not show changes either in AgRP or NPY after OA treatment (Morgan et al. 2004; Pocai et al. 2006). Bearing in mind these data, it is conceivable that in hyperphagic and obese states there may be a resistance to the satiety effects of LCFAs and a desensitization of the AgRP and NPY responses to circulating fatty acids, which contributes to body weight gain. Interestingly, inhibition of CPT1 activity restores lipid sensing, normalizing the hypothalamic levels of LCFA-CoAs and markedly inhibiting feeding behavior and hepatic glucose fluxes in overfed rats (Pocai et al. 2006). It is clear that in this model central inhibition of lipid oxidation is enough to restore hypothalamic lipid sensing, as well as glucose and energy homeostasis, and may be an efficient strategy for the treatment of diet-induced obesity and insulin resistance. However, before this can be carried out in a clinical setting the molecular underpinnings of this event, which are partially unsolved, must be understood. Some lines of evidence suggest that reactive oxygen species could be one of the mitochondrial effectors implicated in both the hormonal (Andrews et al. 2008) and nutrient sensing at the level of the hypothalamus (Benani et al. 2007). This is a very interesting idea but it implies the presence of mechanisms downstream of β-oxidation. An alternative hypothesis may be that excessive accumulation of LCFAs in hypothalamic neurons might be redirected to non-oxidative pathways, producing non-esterified fatty acids. This would result in the production of toxic reactive lipid species and cause ER stress and neuronal lipotoxic effects that can affect neuronal function and affect the mechanisms of energy homeostasis.

7.4 Hypothalamic Lipid Metabolism: A Housekeeping Pathway Regulating Energy Balance

Regardless of the well-established role of neuropeptides and peripheral signals in feeding, current data demonstrates that modulation of hypothalamic lipid metabolism is an important mechanism regulating food intake. Under lipogenic conditions, excess glucose in the cell is first converted to pyruvate via glycolysis in the cytoplasm. Pyruvate enters the mitochondria and is converted to acetyl-CoA and transported as citrate from mitochondria into cytoplasm. ATP citrate lyase then reconverts citrate to acetyl-CoA. The *de novo* fatty acid biosynthesis pathway comprises three key enzymes, acetyl-CoA carboxylase (ACC), FAS and malonyl-CoA decarboxylase (MCD). ACC catalyzes the carboxylation of acetyl-CoA to malonyl-CoA in an ATP-dependent manner. The synthesis step of malonyl-CoA is a reversible regulated mechanism and MCD converts malonyl-CoA back to acetyl-CoA. Acetyl-CoA and malonyl-CoA are then used as the substrates for the production of palmitate by FAS. Malonyl-CoA is an intermediary in the biosynthesis of fatty acids but also an important modulator of the balance between *de novo* lipogenesis and fatty acid oxidation. Levels of malonyl-CoA depend on the equilibrium between ACC, FAS and MCD activities. The activities of ACC and MCD are regulated by phosphorylation by AMPK. Activated AMPK phosphorylates and inhibits ACC, whilst activating MCD, thereby reducing malonyl-CoA and the flux of substrates in the fatty acid biosynthetic pathway (Carling 2004; Kahn et al. 2005; Ruderman et al. 2003). Decreased malonyl-CoA levels further stimulate carnitine palmitoyltransferase 1 (CPT1), the enzyme importing LCFAs-CoA into mitochondria. Thus, by altering the flux through the pathway, levels of malonyl-CoA can be regulated to activate or inhibit mitochondrial β-oxidation (Dowell et al. 2005; Lam et al. 2005b; López et al. 2007a; Ruderman et al. 2003). Of relevance, ACC, AMPK, CPT1, FAS and MCD mRNA and protein expression have been detected at elevated levels in the ARC, DMH, PVH and VMH hypothalamic nuclei in rodents and humans (Kim et al. 2002; López et al. 2006; Minokoshi et al. 2004; Sorensen et al. 2002), suggesting that the fatty acid biosynthetic pathway is important in these cell clusters.

Despite the fact that anatomical support suggested fatty acid metabolism was important in hypothalamic neurons, the interest in the role of fatty acid metabolism in energy balance came from the field of oncology (Kuhajda et al. 1994). The discovery that numerous tumours expressed high levels of FAS suggested that inhibition of this enzyme could be a target for cancer treatment (Kuhajda et al. 1994). Remarkably, treatments with FAS inhibitors, such as C75 and cerulenin, resulted in massive weight loss, which was associated with marked hypophagia (Loftus et al. 2000). Noteworthy, the anorectic effect of these drugs, especially C75, is mediated by the accumulation of malonyl-CoA in the hypothalamus, replicating a signal of nutrient abundance (Hu et al. 2003). In fact, simultaneous inhibition of FAS and ACC (by 5-(tetradecyloxy)-2-furoic acid, TOFA), prevents malonyl-CoA accumulation and consequently does not result in decreased food consumption (Hu et al. 2003; Loftus et al. 2000; López et al. 2006). Conversely, reduction of malonyl-CoA levels by adeno-associated virus-mediated gene transfer of MCD into the medial basal hypothalamus of rats results in increased food intake and progressive weight gain (He et al. 2006; Hu et al. 2005).

Despite the relevance of these data, they can appear contradictory in some ways: FAS inhibitors are anorectic (Hu et al. 2003; Loftus et al. 2000; López et al. 2006), implying that the activity of FAS (an anabolic enzyme) is usually involved in an orexigenic tone. On the other hand, CPT1 (a catabolic enzyme) is also orexigenic (Obici et al. 2003; Wolfgang et al. 2006). The explanation for this apparent paradox is found in the malonyl-CoA levels and the integrated nature of metabolism, illustrated by the cooperative regulation of fatty acid synthesis and oxidation. Levels of malonyl-CoA play a key role in this system by acting both as a substrate for FAS but also as a potent allosteric inhibitor of CPT1, the enzyme importing fatty acyl-CoA into the mitochondria for β-oxidation (Dowell et al. 2005; López et al. 2007a; Ruderman et al. 2003). In summary, selective inhibition of FAS leads to cytoplasmatic malonyl-CoA accumulation in hypothalamic neurons, which consequently inhibits CPT1 activity, leading to decreased feeding (López et al. 2006, 2008a, b; Wolfgang et al. 2006).

7.5 Hypothalamic AMP-Activated Protein Kinase a Cellular Sensor Regulating Whole Body Energy Balance

AMP-activated protein kinase (AMPK) is the downstream component of a kinase cascade that acts as a gauge of cellular energy levels (Carling et al. 2008; Kahn et al. 2005; Ruderman et al. 2003). Accumulating evidence has demonstrated that AMPK is also involved in the regulation of energy balance at the entire body level. AMPK is a serine/threonine protein kinase composed of a catalytic subunit (α1 or α2) and two regulatory subunits (β1, or β2 or β3 and γ1 or γ2 or γ3). AMPK is activated by phosphorylation on Thr172 of the α subunit, a process catalyzed by LKB1 or Ca + 2/calmodulin-dependent protein kinase kinase alpha and beta (CaMKKα and CaMKKβ), (Carling et al. 2008; Kahn et al. 2005; Ruderman et al. 2003). Recently, it has been reported that transforming growth factor-beta-activated kinase (TAK1) also activates AMPK (Fig. 7.2; Xie et al. 2006). Current data also point out that protein phosphatase 2C (PP2C) inactivates AMPK activity by dephosphorylation (Steinberg et al. 2006). AMPK is allosterically activated by AMP, which also inhibits PP2C, increasing phosphorylation in Thr172 and then activation of AMPK. Finally, a novel mechanism modulating AMPK independently of AMP and phosphorylation/dephosphorylation processes has been documented: cell-death-inducing like-effector A (Cidea) forms a complex with the β subunit of AMPK, which promotes an ubiquitination-mediated degradation of AMPK, reducing its activity (Qi et al. 2008). Whatever the mechanism, activated (phosphorylated) AMPK is a counter-regulatory response in many tissues to switch off ATP-consuming processes (such as fatty acid synthesis) whilst switching on catabolic processes that produce ATP (such as fatty acid β-oxidation) and restore the AMP:ATP ratio (Carling et al. 2008; Kahn et al. 2005; Ruderman et al. 2003).

Until recently, AMPK was viewed as a cellular energy gauge, modulating and being modulated by the AMP:ATP ratio. However, current data are also showing that AMPK plays a major role in hypothalamic neurons as a sensor and regulator of

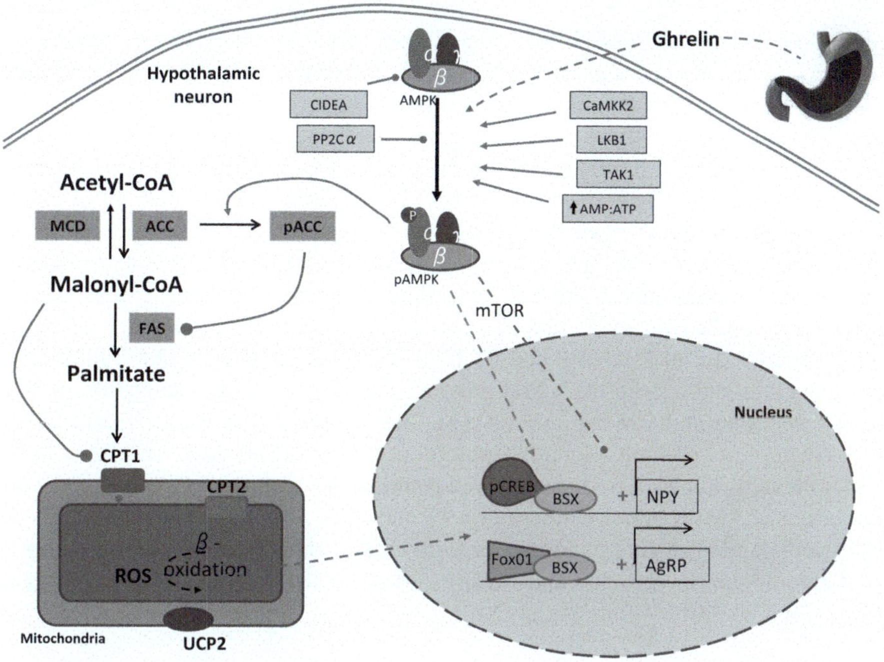

Fig. 7.2 Hypothalamic fatty acid metabolism integrates peripheral signals with neuropeptide systems. In the hypothalamus, anorectic signals inhibit AMP-activated protein kinase (*AMPK*) phosphorylation and activation, while orexigenic signals stimulate *AMPK* phosphorylation and activation. These signals are integrated and progress through alterations in the tumor suppressor *LKB1* and the Ca + 2/calmodulin-dependent protein kinase kinase 2 pathways (*CaMKK2*). In the particular case of ghrelin, activated *AMPK* in turn phosphorylates and inactivates acetyl-CoA carboxylase (*ACC*), decreasing the cytoplasmatic pool of malonyl-CoA. The net result of this action is an increase in carnitine palmitoyltransferase 1 (*CPT1*) activity and then fatty acid beta oxidation, which promotes the generation of reactive oxygen species, which are buffered by uncoupling protein 2 (*UCP2*). This mechanism is critical for ghrelin-induced electric activation of agouti-related peptide/neuropeptide Y (*AgRP/NPY*) neurons. Ghrelin-induced upregulation of *Agrp* and *Npy* genes is mediated through a mechanism involving hypothalamic homeobox domain transcription factor (*BSX*), forkhead box O1 (*FoxO1*) and the phosphorylated *cAMP* response-element binding protein (*pCREB*)

whole body energy homeostasis, linking metabolic status to classical neuropeptide/neurotransmitter systems and ultimately regulating feeding. Morphological evidence has shown that AMPK is highly expressed in several key hypothalamic nuclei, such as the ARC, PVH, VMH and the LHA (Lage et al. 2010; López et al. 2006, 2008a; McCrimmon et al. 2006; Minokoshi et al. 2004). Functional data have also demonstrated that modulation of AMPK in the hypothalamus is part of the adaptive changes observed during physiological regulation of feeding. Fasting increases AMPK activity in several hypothalamic regions, whilst refeeding inhibits it (Andersson et al. 2004; López et al. 2008a; Minokoshi et al. 2004). In line with this evidence, activation of AMPK in the hypothalamus increases feeding and body weight gain, while inhibition of hypothalamic AMPK activity promotes hypophagia and weight loss (Andersson et al. 2004; López et al. 2008a, 2010; Minokoshi et al. 2004).

Remarkably, hypothalamic AMPK has recently emerged as a new central regulator of peripheral metabolism and thermogenesis through the autonomic nervous system. Data from our group have recently revealed that AMPK in the VMH mediates the effects of thyroid hormone (triiodothyronine, T3), nicotine and bone morphogenetic protein 8b (BMP8b) (Cannon and Nedergaard 2010; López et al. 2010; Martínez de Morentin et al. 2012; Whittle et al. 2012). Specifically, central T3 overfeeding or hyperthyroidism, and nicotine decrease the activity of hypothalamic AMPK, and additionally to BMP8b, stimulate sympathetic nervous system activity and upregulate thermogenic markers in brown adipose tissue (BAT), such as uncoupling proteins 1 and 3 (UCP1 and UCP3). Worth mentioning is that inhibition of the lipogenic pathway in the VMH prevents centrally-mediated activation of BAT by T3, nicotine and BMP8b and reverses the weight loss associated with hyperthyroidism and smoking. Equally, inhibition of thyroid hormone receptors in the VMH prevents hyperthyroidism-induced leanness. Overall this evidence shows that inhibition of AMPK activity induced by T3 and nicotine and the associated increase in hypothalamic lipogenesis alongside with central BMP8b are a key regulators of BAT thermogenesis (Cannon and Nedergaard 2010; López et al. 2010; Martínez de Morentin et al. 2012; Whittle et al. 2012).

7.6 Hypothalamic AMPK in the Development of Obesity and Disease-Associated Feeding Disorders

The importance of hypothalamic AMPK in energy balance is clear but its role in the progression to obesity is vague. It has been recently demonstrated that impaired responses of AMPK to leptin might be involved in leptin resistance in obese states (Martin et al. 2006). In this sense, neuronal protein tyrosine phosphatase 1B (PTP1B) KO mice, which are resistant to diet-induced obesity due to leptin hypersensitivity, show decreased hypothalamic AMPK activity accompanied by alterations in neuropeptide expression and increased thermogenesis in BAT and muscle. Such changes collectively promote leanness and increased energy expenditure (Xue et al. 2009). Recent work has also demonstrated that disruption of kinase suppressor of RAS 2 (KSR2) *in vivo* impairs AMPK-regulated processes affecting fatty acid oxidation and thermogenesis to cause obesity (Costanzo-Garvey et al. 2009). Despite their increased adiposity, KSR2 KO mice are hypophagic and hyperactive but expend less energy than wild-type mice. In addition, hyperinsulinemic-euglycemic clamp studies reveal that KSR2 KO mice are profoundly insulin resistant. The expression of genes mediating oxidative phosphorylation is also downregulated in the adipose tissue of these mice. Overall, these data demonstrate that KSR2 KO mice are highly efficient in conserving energy, revealing a novel role for the KSR2-AMPK duo in the regulation of energy metabolism. Interestingly, KSR2 is mainly detected in the brain suggesting that these actions are probably exerted within the CNS (Costanzo-Garvey et al. 2009).

Numerous mouse models with selective deletions of AMPK subunits have been produced (for extensive review see (Kahn et al. 2005; Long and Zierath 2006)). However, the global AMPKα2 KO fed a high fat diet (HFD) (Viollet et al. 2003) and the AMPKα2 KO in POMC neurons (POMCα2KO) fed standard chow and HFD, display increased body weight and fat mass (Claret et al. 2007; Long and Zierath 2006). On the contrary, mice with a targeted ablation of AMPKα2 in AgRP neurons (AgRPα2KO) have an age-dependent lean phenotype (Claret et al. 2007). Together these evidence point to a very important role for AMPK in the ARC on feeding control. Furthermore, it has been also recently reported that AMPKβ1 KO mice either fed low fat diet or HFD have reduced food intake, reduced adiposity, and reduced total body mass without changes in metabolic rate, physical activity, adipose tissue lipolysis, or lipogenesis (Dzamko et al. 2010). The main problem in the understanding of these phenotypes is the pathway's redundancy itself. In fact, the absence of marked alterations in terms of body weight and feeding in some of the AMPK KOs is probably related to compensation by the remaining catalytic subunits, *i.e.* in the AMPKα2 KO mice, AMPKα1 is upregulated (Viollet et al. 2003). Whatever the case, additional experiments are necessary, given the fact that it has been recently demonstrated that at least the obese phenotype in POMCα2 KO mice is not due to compensatory upregulation of AMPKα1 (Claret et al. 2007). The development of other mouse models specifically targeting AMPK subunits in precise hypothalamic nuclei or even neuronal populations will help to elucidate the role of hypothalamic AMPK in: *(a)* the pathophysiology of obesity and *(b)* the physiological response to peripheral hormones and metabolites. Consistent with this, it has been recently demonstrated that the deletion of AMPKα2 in hypothalamic POMC and AgRP neurons is essential for glucose sensing but not to mediate the effects of leptin and insulin (Claret et al. 2007). Similar studies should be performed in mice with selective deletion of AMPKα1 and 2 in the VMH, which seems to be the key hypothalamic nucleus integrating AMPK's actions not just on food intake (Lage et al. 2010; López et al. 2008a) and glucose homeostasis (Fan et al. 2009; McCrimmon et al. 2004; McCrimmon et al. 2006; McCrimmon et al. 2008) but also on energy expenditure (López et al. 2010).

7.7 Endoplasmic Reticulum Stress in the Hypothalamus Leads to Impaired Energy Balance

Alterations to the function of protein folding machinery of the ER are usually known as ER stress, and its induction is one of the mechanisms induced by lipotoxicity (Basseri and Austin 2008; Eizirik et al. 2008; Hotamisligil 2008; Lin et al. 2008; Oyadomari and Mori 2004; Ron and Walter 2007; Yang and Hotamisligil 2008). The main function of the ER is the synthesis of secretory proteins and resident proteins but the ER is also involved in diverse functions such as lipid biosynthesis, metabolism, oligosaccharide synthesis, calcium storage, and drug detoxification (Eizirik et al. 2008; Lin et al. 2008; Oyadomari and Mori 2004; Ron and Hubbard 2008; Ron and Walter 2007). In eukaryotic cells, most secreted and transmembrane

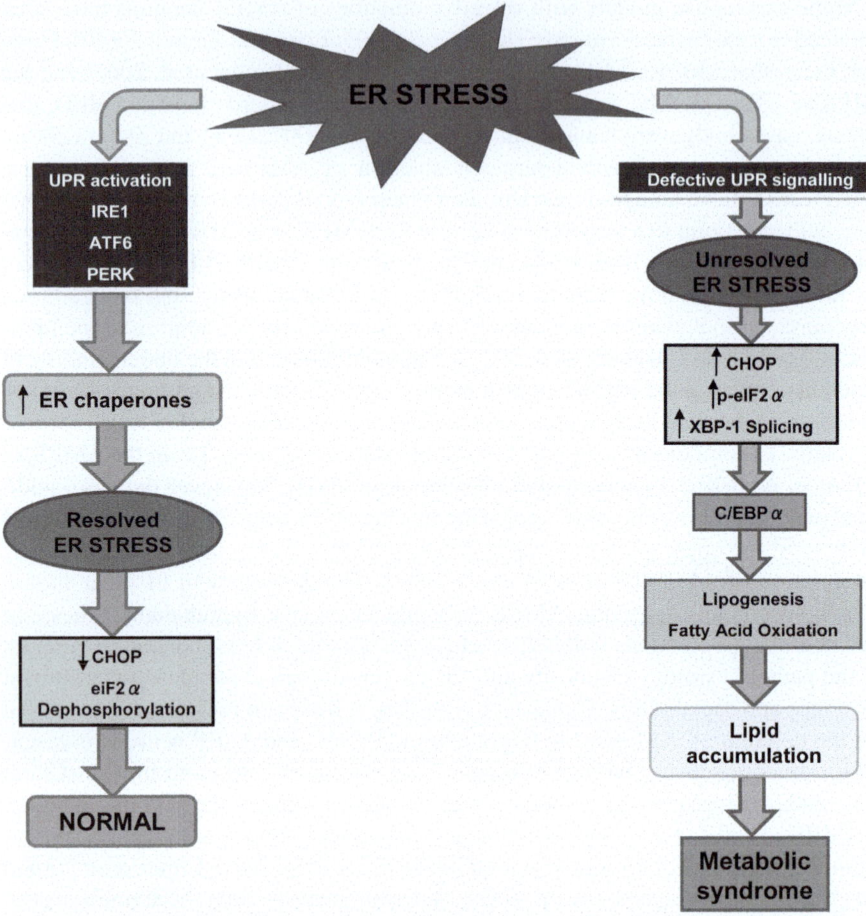

Fig. 7.3 A model illustrating how prolonged and unresolved ER stress disrupts lipid homeostasis and results in the development of metabolic syndrome. Under conditions of physiological *ER stress*, cells respond by activating the unfolded protein response (*UPR*). The subsequent increase in the expression of ER chaperones and phosphorylation of eukaryotic translation initiation factor 2 α (*eIF2α*) reduces cellular stress (*left side*). However, if *UPR* signaling is impaired (*right side*), increased *C/EBP* homologous protein (*CHOP*) expression and/or phosphorylation of *eIF2α* can result in increased expression of lipogenic genes, as well as decreased expression of genes required for fatty acid oxidation. These metabolic changes induced by ER stress can lead to the development of metabolic syndrome hallmarks, such as hepatic steatosis

proteins fold and mature in the lumen of the ER in a process kept in check by the unfolded protein response (UPR). Three different classes of UPR transducers have been identified: *(1)* inositol-requiring protein-1 (IRE1), *(2)* activating transcription factor-6 (ATF6) and *(3)* protein kinase RNA-like ER kinase (PERK) (Fig. 7.3; Eizirik et al. 2008; Lin et al. 2008; Oyadomari and Mori 2004; Ron and Hubbard 2008; Ron and Walter 2007). Activation of these pathways as part of the ER stress response, caused by accumulation of unfolded proteins in the lumen of the ER, results in a coordinated transcriptional response associated to attenuation of protein synthesis and upregulation of ER folding machinery (ER chaperones proteins, such as GRP78, GRP94 and ERp72) and degradation of irreversibly misfolded proteins (Eizirik et al. 2008; Lin et al. 2008; Oyadomari and Mori 2004; Ron and Hubbard 2008; Ron and Walter 2007). If this UPR adaptive response is not sufficient to resolve the protein-folding defect, ER dysfunction can lead to cell dysfunction and ultimately apoptotic cell death (Eizirik et al. 2008; Lin et al. 2008; Ron and Hubbard 2008; Ron and Walter 2007).

Recent evidence shows that ER stress occurs under conditions of overnutrition in both peripheral tissues and the CNS (Hosoi et al. 2008; Ozcan et al. 2009; Zhang et al. 2008). Free fatty acids (FFAs) and hyperglycemia in β cells promote activation of the UPR, leading to decreased insulin mRNA expression and inhibition of insulin signaling (Cunha et al. 2008; El Assaad et al. 2003; Lipson et al. 2006, 2008). FFAs and glucose have been involved in PERK-mediated eukaryotic translation initiation factor 2 α (eIF2α) phosphorylation and IRE1α-mediated X-box binding protein 1 (Xbp1) mRNA splicing *in vivo* and in isolated rat islets (Cnop et al. 2007; Cunha et al. 2008; El Assaad et al. 2003; Elouil et al. 2007; Lipson et al. 2006, 2008) and adipose tissue of obese rodent models and humans (Boden et al. 2008; Sharma et al. 2008). The type of lipids involved and particularly their degree of saturation seems to be relevant, with saturated fat being particularly prone to promote a greater UPR in β cells inducing Xbps, C/EBP homologous protein (CHOP, also known as GADD153) and GRP78 expression (Cnop et al. 2001; Cunha et al. 2008; El Assaad et al. 2003; Lipson et al. 2006, 2008). CHOP is a member of the C/EBP family of transcription factors (Oyadomari and Mori 2004; Ron and Habener 1992) and it is known to inhibit adipocyte differentiation in response to metabolic stress (Tang and Lane 2000). *Chop* gene induction is preferentially regulated through the PERK/eIF2α/ATF4 UPR pathway, although the IRE1α/XBP1 and ATF6α pathways have also been implicated (Eizirik et al. 2008; Lin et al. 2008; Ron and Hubbard 2008; Ron and Walter 2007). There is strong evidence demonstrating that impaired CHOP function in peripheral tissues promotes alterations in energy homeostasis. Female CHOPKO mice develop obesity; however no clear mechanisms involving either food intake or thermogenesis have been identified (Ariyama et al. 2007). HFD feeding results in obesity in female CHOPKO and heterozygous mice, although their caloric intake does not differ from that of wild-type. Fat cell area is increased in mesenteric fat but not in subcutaneous fat in CHOPKO mice fed a HFD (Ariyama et al. 2007). There is evidence that the UPR promotes CHOP expression in other systems such as β cell apoptosis and diabetes (Huang et al. 2007; Laybutt et al. 2007; Oyadomari et al. 2002; Yusta et al. 2006). Further evidence indicates that genetic ablation of CHOP provides an advantage to the β cell by preventing oxidative damage and promoting β cell survival (Song et al. 2008). In the context of the CNS induction of CHOP has been related to neuronal toxicity and death in diverse neurodegenerative diseases (Pennuto et al. 2008; Schapansky et al. 2007; Silva et al. 2005). Thus, on the basis of these data, it is tempting to speculate that CHOP could be a mediator or ER stress in the hypothalamus following selective lipid damage. Further work will be necessary to address this hypothesis.

Recent evidence also links inflammation to hypothalamic ER stress and leptin resistance. Three recent works from three independent groups have demonstrated that obesity and overnutrition elicit ER stress and initiate UPR signaling pathways in the hypothalamus, which in turn leads to inhibition of leptin receptor signaling and leptin resistance (Fig. 7.3; Hosoi et al. 2008; Ozcan et al. 2009; Zhang et al. 2008). *Zhang et al* demonstrated that IKKβ/NF-κB, a well-known mediator of metabolic inflammation, is highly expressed in hypothalamic neurons although it normally remains inactive (Zhang et al. 2008). Interestingly chronic (dietary or genetic obesity) or acute (glucose or OA exposure) overnutrition activates hypothalamic IKKβ/NF-κB through, at least in part, elevated hypothalamic ER stress leading to impaired insulin and leptin signaling through a mechanism partially mediated by suppressor of cytokine signaling (SOCS3), an inhibitor of insulin and leptin signaling. Experimental activation of the IKKβ/NF-κB signaling pathway in the hypothalamus, through virus-mediated introduction of a constitutively active IKKβ, leads to weight gain and increased food intake together with impaired insulin and leptin signaling in the hypothalamus in mice fed a HFD. Also, this work demonstrates that suppression of IKKβ/NF-κB signaling in the mediobasal hypothalamus protects against insulin and leptin resistance and obesity (Zhang et al. 2008). Finally, *Zhang et al* demonstrate that obesity-induced ER stress in the hypothalamus due to diet could also be suppressed by administration of tauroursodeoxycholic acid (TUDCA), a chemical chaperone that improves the protein-folding capacity. Such treatment resulted in reduced food intake and decreased body weight (Zhang et al. 2008). Altogether, these data show that the hypothalamic IKKβ/NF-κB program is a general neural mechanism for energy balance underlying obesity and suggest that suppressing hypothalamic IKKβ/NF-κB may represent a strategy to combat obesity and related diseases, perhaps even hypothalamic lipotoxicity. In two additional studies *Hosoi et al* and *Ozcan et al* described that ER stress clearly inhibited leptin-induced signal transducer and activator of transcription 3 phosphorylation and that ER stress-induced leptin resistance was mediated through PTP1B. However, contrary to *Zhang et al,* this was not mediated through SOCS3 (Hosoi et al. 2008; Ozcan et al. 2009). It is noteworthy that both studies showed that treatment with chemical chaperones, such as 4-phenyl butyric acid and TUDCA not only rescued ER stress in the CNS and reduced body weight but also acted as leptin sensitizing agents, significantly improving ER stress-induced leptin resistance (Hosoi et al. 2008; Ozcan et al. 2009). In accordance with these observations a recent study has shown that central administration of thapsigargin, an ER stress inducer, inhibits the anorexigenic effects of leptin and insulin (Won et al. 2009).

The kinase known as mammalian target of rapamycin (mTOR) could also be affected by ER stress. This kinase integrates different sensory inputs to regulate protein synthesis rates in individual cells, and in the last years it has been related to the regulation of food intake and body weight in the CNS (Cota et al. 2006, 2007, 2008; Martins et al. 2012; Proulx et al. 2008; Varela et al. 2012; Woods et al. 2008). Considering that both nutrients and inflammatory signaling could engage the mTOR pathway, it could have an in important role in obesity-induced ER stress in the CNS. Present data show that ER stress is a critical component of the pathologies associated with dysregulated mTOR activity, and the evidence is as follows. To begin with, in a role distinct from the regulation of energy homeostasis, mTOR modulates protein translation after ER stress induced by human cytomegalovirus (Alwine 2008) or chronic hepatitis B virus infection (Yang et al. 2009). Research from recent research has demonstrated that mTOR has a key role in the translational control of gene expression during hypoxia- or oxidative-induced ER stress in both normal and cancer cells (Endo et al. 2007; Jin et al. 2009; van den et al. 2006). In the CNS, some cues can be found in the tuberous sclerosis complex (TSC), a neurogenetic disorder caused by loss-of-function mutations in either the TSC1 or TSC2 genes. Such mutations frequently result in prominent CNS-related symptoms, including epilepsy, mental retardation, and autism spectrum disorder (Di Nardo et al. 2009). Very recent data have demonstrated that loss of TSC1 or TSC2 leads to constitutive mTOR activation in cell lines and mouse or human tumors and causes ER stress and activation of the UPR (Ozcan et al. 2008). Also, hippocampal *Tsc2*-deficient neurons and brain lysates from *Tsc1*-deficient mice show elevated ER stress, indicated by increased expression of CHOP. This effect is completely reversed by the mTOR inhibitor rapamycin both *in vitro* and *in vivo* (Di Nardo et al. 2009). This evidence indicates that ER stress modulates mTOR activity in neurons through the TSC protein complex and that ER stress is elevated in cells lacking this complex. Whether the TSC-mTOR axis mediates ER stress in the hypothalamus needs to be investigated. Recent work supports this theory, showing that hypothalamic Rip-TSC1KO mice developed hyperphagia and obesity, suggesting that hypothalamic disruption of *Tsc1* may dysregulate feeding circuits via mTOR activation. In fact, Rip-TSC1KO mice displayed increased mTOR signaling and enlarged neuron cell size in a number of hypothalamic populations, including POMC neurons. Additionally, *Tsc1* deletion with *Pomc/Cre* POMC-TSC1KO mice resulted in dysregulation of POMC neurons and hyperphagia-driven obesity. Treatment with the mTOR inhibitor, rapamycin, improved the hyperphagia, obesity, and the altered POMC neuronal morphology in developing or adult POMC-TSC1KO mice (Mori et al. 2009). This work clearly demonstrates that mTOR activation in POMC neurons blocks the catabolic function of these cells to promote nutrient intake and increased adiposity. Whether these effects are linked to specific changes in ER stress would need further research, but taking into consideration the connection between both facts, this link is quite appealing.

7.8 Conclusions and Future Challenges

In the 1950's we watched to the formulation of the *Glucostatic, Lipostatic and Aminostatic Hypotheses*, that proposed the sensing of circulating levels of metabolites such as glucose, lipids and amino acids in the CNS and that feeding was modified according to levels of these molecules (Campfield et al. 1996; MAYER 1955). It took until the 21st century to demonstrate that circulating LCFAs act as nutrient surplus signals in the hypothalamus (Lam et al. 2005a; Le Foll et al. 2009; Morgan et al. 2004; Obici et al. 2002; Pocai et al. 2005; Wang et al. 2006). The last decade's work has provided pharmacological, genetic and physiological evidence that energy balance at the whole body level is modulated by hypothalamic lipid metabolism (Andrews et al. 2008; Loftus et al. 2000; López et al. 2006, 2008b, 2010; Minokoshi et al. 2004; Obici et al. 2003; Wolfgang et al. 2007). Still, there are three main questions requiring an answer. First, can alterations in hypothalamic lipid metabolism impair energy homeostasis and lead to obesity and the Metabolic Syndrome? Second, what is the molecular nature of these alterations? And third, could hypothalamic lipotoxicity play a role in the pathophysiology of the Metabolic Syndrome? Some recent and exceptional work gave evidence suggesting that this may be the case, demonstrating that a key mechanism of lipid toxicity, the induction of ER stress, is a general neural mechanism for energy homeostasis underlying obesity and leptin resistance (Hosoi et al. 2008; Ozcan et al. 2009; Won et al. 2009; Zhang et al. 2008). However, this connection relating lipotoxicity and ER-stress in the hypothalamus needs to be confirmed. Of note, these studies also show that pharmacological manipulation of ER stress may be an ideal target for obesity treatment, acting as leptin-sensitizer (Hosoi et al. 2008; Ozcan et al. 2009; Won et al. 2009; Zhang et al. 2008). This finding may be essential for guiding the newly evolving strategies for drug design and targeting ER stress could be a useful therapeutic solution to prevent or even revert obesity and its metabolic complications in humans.

In conclusion these recent data point to hypothalamic ER stress as a key mechanism leading to some of the diverse manifestations of the metabolic syndrome. The challenge over coming years will be to validate this hypothesis and identify the molecular effectors linking central lipotoxicity, ER stress and the classical pathways involved on energy homeostasis. We believe that a wider comprehension of these mechanisms operating in the CNS will facilitate the development of more efficient drugs against obesity and related metabolic disorders.

Acknowledgements The research leading to these results has received funding from the European Community's Seventh Framework Programme (FP7/2007–2013) under grant agreement n°StG-2011-281854— the *ObERStress* project (ML). 245009—the *Neurofast* project (RN, CD and ML) and 018734— the *Hepadip* project (AVP), Xunta de Galicia (ML:10PXIB208164PR; RN:2010/14), FIS (ML:PS09/01880), MICINN (RN:RyC-2008-02219 and SAF2009-07049; CD:BFU2008-02001; ML:RyC-2007-00211). CIBER de Fisiopatología de la Obesidad y Nutrición is an initiative of ISCIII. LM is a recipient of a fellowship from Fundação para a Ciência e Tecnologia, Portugal (SFRH/BD/65379/2009). AW was supported by BBSRC. Authors declare no conflict of interest.

References

Alwine JC (2008) Modulation of host cell stress responses by human cytomegalovirus. Curr Top Microbiol Immunol 325:263–279

Andersson U, Filipsson K, Abbott CR, Woods A, Smith K, Bloom SR, Carling D, Small C (2004) AMP-activated protein kinase plays a role in the control of food intake. J Biol Chem 279(13):12005–12008

Andrews ZB, Liu ZW, Walllingford N, Erion DM, Borok E, Friedman JM, Tschop MH, Shanabrough M, Cline G, Shulman GI, Coppola A, Gao, XB, Horvath TL, Diano S (2008)

UCP2 mediates ghrelin's action on NPY/AgRP neurons by lowering free radicals. Nature 454(7206):846–851

Ariyama Y, Shimizu H, Satoh T, Tsuchiya T, Okada S, Oyadomari S, Mori M (2007) Chop-deficient mice showed increased adiposity but no glucose intolerance. Obesity (Silver Spring) 15(7): 1647–1656

Basseri S, Austin RC (2008) ER stress and lipogenesis: a slippery slope toward hepatic steatosis. Dev Cell 15(6):795–796

Benani A, Troy S, Carmona MC, Fioramonti X, Lorsignol A, Leloup C, Casteilla L, Penicaud L (2007) Role for mitochondrial reactive oxygen species in brain lipid sensing: redox regulation of food intake. Diabetes 56(1):152–160

Boden G, Duan X, Homko C, Molina EJ, Song W, Perez O, Cheung P, Merali S (2008) Increase in endoplasmic reticulum stress-related proteins and genes in adipose tissue of obese, insulin-resistant individuals. Diabetes 57(9):2438–2444

BT van den, Koritzinsky M, Wouters BG (2006) Translational control of gene expression during hypoxia. Cancer Biol Ther 5(7):749–755

Calle EE, Kaaks R (2004) Overweight, obesity and cancer: epidemiological evidence and proposed mechanisms. Nat Rev Cancer 4(8):579–591

Campfield LA, Smith FJ, Burn P (1996) The OB protein (leptin) pathway—a link between adipose tissue mass and central neural networks. Horm Metab Res 28(12):619–632

Cannon B, Nedergaard J (2010) Thyroid hormones: igniting brown fat via the brain. Nat Med 16(9):965–967

Carling D (2004) AMPK. Curr Biol 14(6):R220

Carling D, Sanders MJ, Woods A (2008) The regulation of AMP-activated protein kinase by upstream kinases. Int J Obes(Lond) 32(4):55–59

Claret M, Smith MA, Batterham RL, Selman C, Choudhury AI, Fryer LG, Clements M, Al Qassab H, Heffron H, Xu AW, Speakman R, Barsh GS, Viollet B, Vaulont S, Ashford ML, Carling D, Withers, DJ (2007) AMPK is essential for energy homeostasis regulation and glucose sensing by POMC and AgRP neurons. J Clin Invest 117(8):2325–2336

Cnop M, Hannaert JC, Hoorens A, Eizirik DL, Pipeleers DG (2001) Inverse relationship between cytotoxicity of free fatty acids in pancreatic islet cells and cellular triglyceride accumulation. Diabetes 50(8):1771–1777

Cnop M, Ladriere L, Hekerman P, Ortis F, Cardozo AK, Dogusan Z, Flamez D, Boyce M, Yuan J, Eizirik DL (2007) Selective inhibition of eukaryotic translation initiation factor 2 alpha dephosphorylation potentiates fatty acid-induced endoplasmic reticulum stress and causes pancreatic beta-cell dysfunction and apoptosis. J Biol Chem 282(6):3989–3997

Costanzo-Garvey DL, Pfluger PT, Dougherty MK, Stock JL, Boehm M, Chaika O, Fernandez MR, Fisher K, Kortum RL, Hong EG, Jun JY, Ko HJ, Schreiner A, Volle DJ, Treece T, Swift AL, Winer M, Chen D, Wu M, Leon LR, Shaw AS, McNeish J, Kim JK, Morrison DK, Tschop MH, Lewis RE (2009) KSR2 is an essential regulator of AMP kinase, energy expenditure, and insulin sensitivity. Cell Metab 10(5):366–378

Cota D, Matter EK, Woods SC, Seeley RJ (2008) The role of hypothalamic mammalian target of rapamycin complex 1 signaling in diet-induced obesity. J Neurosci 28(28):7202–7208

Cota D, Proulx K, Seeley RJ (2007) The role of CNS fuel sensing in energy and glucose regulation. Gastroenterology 132(6):2158–2168

Cota D, Proulx K, Smith KA, Kozma SC, Thomas G, Woods SC, Seeley RJ (2006) Hypothalamic mTOR signaling regulates food intake. Science 312(5775):927–930

Cunha DA, Hekerman P, Ladriere L, Bazarra-Castro A, Ortis F, Wakeham MC, Moore F, Rasschaert J, Cardozo AK, Bellomo E, Overbergh L, Mathieu C, Lupi R, Hai T, Herchuelz A, Marchetti P, Rutter GA, Eizirik DL, Cnop M (2008) Initiation and execution of lipotoxic ER stress in pancreatic beta-cells. J Cell Sci 121(Pt 14):2308–2318

Di Nardo A, Kramvis I, Cho N, Sadowski A, Meikle L, Kwiatkowski DJ, Sahin M (2009) Tuberous sclerosis complex activity is required to control neuronal stress responses in an mTOR-dependent manner. J Neurosci 29(18):5926–5937

Dowell P, Hu Z, Lane MD (2005) Monitoring energy balance: metabolites of fatty acid synthesis as hypothalamic sensors. Annu Rev Biochem 74:515–534

Dzamko N, Denderen BJ van, Hevener AL, Jorgensen SB, Honeyman J, Galic S, Chen ZP, Watt MJ, Campbell DJ, Steinberg GR, Kemp BE (2010) AMPK beta1 deletion reduces appetite, preventing obesity and hepatic insulin resistance. J Biol Chem 285(1):115–122

Eizirik DL, Cardozo AK, Cnop M (2008) The role for endoplasmic reticulum stress in diabetes mellitus. Endocr Rev 29(1):42–61

El Assaad W, Buteau J, Peyot ML, Nolan C, Roduit R, Hardy S, Joly E, Dbaibo G, Rosenberg L, Prentki M (2003) Saturated fatty acids synergize with elevated glucose to cause pancreatic beta-cell death. Endocrinology 144(9):4154–4163

Elmquist JK, Coppari R, Balthasar N, Ichinose M, Lowell BB (2005) Identifying hypothalamic pathways controlling food intake, body weight, and glucose homeostasis. J Comp Neurol 493(1):63–71

Elouil H, Bensellam M, Guiot Y, Vander MD, Pascal SM, Schuit FC, Jonas JC (2007) Acute nutrient regulation of the unfolded protein response and integrated stress response in cultured rat pancreatic islets. Diabetologia 50(7):1442–1452

Endo H, Murata K, Mukai M, Ishikawa O, Inoue M (2007) Activation of insulin-like growth factor signaling induces apoptotic cell death under prolonged hypoxia by enhancing endoplasmic reticulum stress response. Cancer Res 67(17):8095–8103

Fan X, Ding Y, Brown S, Zhou L, Shaw M, Vella MC, Cheng H, McNay EC, Sherwin RS, McCrimmon RJ (2009) Hypothalamic AMP-activated protein kinase activation with AICAR amplifies counterregulatory responses to hypoglycemia in a rodent model of type 1 diabetes. Am J Physiol Regul Integr Comp Physiol 296(6):R1702–R1708

Farooqi IS, O'Rahilly S (2005) Monogenic obesity in humans. Annu Rev Med 56:443–458

Flier JS (2004) Obesity wars: molecular progress confronts an expanding epidemic. Cell 116(2):337–350

Friedman JM (2003) A war on obesity, not the obese. Science 299(5608):856–858

Fruhbeck G (2008) Are we really tackling the obesity epidemic. Obes Metab 4(2):91–95

Gao Q, Horvath TL (2007) Neurobiology of feeding and energy expenditure. Annu Rev Neurosci 30:367–398

He W, Lam TK, Obici S, Rossetti L (2006) Molecular disruption of hypothalamic nutrient sensing induces obesity. Nat Neurosci 9(2):227–233

Hosoi T, Sasaki M, Miyahara T, Hashimoto C, Matsuo S, Yoshii M, Ozawa K (2008) Endoplasmic reticulum stress induces leptin resistance. Mol Pharmacol 74(6):1610–1619

Hotamisligil GS (2008) Inflammation and endoplasmic reticulum stress in obesity and diabetes. Int J Obes (Lond) 32(7):52–54

Hu Z, Cha SH, Chohnan S, Lane MD (2003) Hypothalamic malonyl-CoA as a mediator of feeding behavior. Proc Natl Acad Sci U.S.A 100(22):12624–12629

Hu Z, Dai Y, Prentki M, Chohnan S, Lane MD (2005) A role for hypothalamic malonyl-CoA in the control of food intake. J Biol Chem 280(48):39681–39683

Huang CJ, Lin CY, Haataja L, Gurlo T, Butler AE, Rizza RA, Butler PC (2007) High expression rates of human islet amyloid polypeptide induce endoplasmic reticulum stress mediated beta-cell apoptosis, a characteristic of humans with type 2 but not type 1 diabetes. Diabetes 56(8): 2016–2027

Jin HO, Seo SK, Woo SH, Kim ES, Lee HC, Yoo DH, An S, Choe TB, Lee SJ, Hong SI, Rhee CH, Kim JI, Park IC (2009) Activating transcription factor 4 and CCAAT/enhancer-binding protein-beta negatively regulate the mammalian target of rapamycin via Redd1 expression in response to oxidative and endoplasmic reticulum stress. Free Radic Biol Med 46(8):1158–1167

Kahn BB, Alquier T, Carling D, Hardie DG (2005) AMP-activated protein kinase: ancient energy gauge provides clues to modern understanding of metabolism. Cell Metab 1(1):15–25

Kim EK, Miller I, Landree LE, Borisy-Rudin FF, Brown P, Tihan T, Townsend CA, Witters LA, Moran TH, Kuhajda FP, Ronnett GV (2002) Expression of FAS within hypothalamic neurons: a model for decreased food intake after C75 treatment. Am J Physiol Endocrinol Metab 283(5):E867–E879

Kuhajda FP, Jenner K, Wood FD, Hennigar RA, Jacobs LB, Dick JD, Pasternack GR (1994) Fatty acid synthesis: a potential selective target for antineoplastic therapy. Proc Natl Acad Sci U.S.A 91(14):6379–6383

Lage R, Diéguez C, Vidal-Puig A, López M (2008) AMPK: a metabolic gauge regulating whole-body energy homeostasis. Trends Mol Med 14(12):539–549

Lage R, Vázquez MJ, Varela L, Saha AK, Vidal-Puig A, Nogueiras R, Diéguez C, López M (2010) Ghrelin effects on neuropeptides in the rat hypothalamus depend on fatty acid metabolism actions on BSX but not on gender. FASEB J 24(8):2670–2679

Lam TK, Pocai A, Gutierrez-Juarez R, Obici S, Bryan J, Aguilar-Bryan L, Schwartz GJ, Rossetti L (2005a) Hypothalamic sensing of circulating fatty acids is required for glucose homeostasis. Nat Med 11(3):320–327

Lam TK, Schwartz GJ, Rossetti L (2005b) Hypothalamic sensing of fatty acids. Nat.Neurosci 8(5):579–584

Laybutt DR, Preston AM, Akerfeldt MC, Kench JG, Busch AK, Biankin AV, Biden TJ (2007) Endoplasmic reticulum stress contributes to beta cell apoptosis in type 2 diabetes. Diabetologia 50(4):752–763

Le Foll C, Irani BG, Magnan C, Dunn-Meynell AA, Levin BE (2009) Characteristics and mechanisms of hypothalamic neuronal fatty acid sensing. Am.J.Physiol Regul.Integr.Comp Physiol 297(3):R655–R664

Lin JH, Walter P, Yen TS (2008) Endoplasmic reticulum stress in disease pathogenesis. Annu Rev Pathol 3:399–425

Lipson KL, Fonseca SG, Ishigaki S, Nguyen LX, Foss E, Bortell R, Rossini AA, Urano F (2006) Regulation of insulin biosynthesis in pancreatic beta cells by an endoplasmic reticulum-resident protein kinase IRE1. Cell Metab 4(3):245–254

Lipson KL, Ghosh R, Urano F (2008) The role of IRE1alpha in the degradation of insulin mRNA in pancreatic beta-cells. PLoS ONE 3(2):e1648

Loftus TM, Jaworsky DE, Frehywot GL, Townsend CA, Ronnett GV, Lane MD, Kuhajda FP (2000) Reduced food intake and body weight in mice treated with fatty acid synthase inhibitors. Science 288(5475):2379–2381

Long YC, Zierath JR (2006) AMP-activated protein kinase signaling in metabolic regulation. J Clin Invest 116(7):1776–1783

López M, Lelliott CJ, Tovar S, Kimber W, Gallego R, Virtue S, Blount M, Vázquez MJ, Finer N, Powles T, O'Rahilly S, Saha AK, Diéguez C, Vidal-Puig AJ (2006) Tamoxifen-induced anorexia is associated with fatty acid synthase inhibition in the ventromedial nucleus of the hypothalamus and accumulation of malonyl-CoA. Diabetes 55(5):1327–1336

López M, Lelliott CJ, Vidal-Puig A (2007a) Hypothalamic fatty acid metabolism: a housekeeping pathway that regulates food intake. Bioessays 29(3):248–261

López M, Tovar S, Vázquez MJ, Williams LM, Diéguez C (2007b) Peripheral tissue-brain interactions in the regulation of food intake. Proc Nutr.Soc 66(1):131–155

López M, Lage R, Saha AK, Pérez-Tilve D, Vázquez MJ, Varela L, Sangiao-Alvarellos S, Tovar S, Raghay K, Rodríguez-Cuenca S, Deoliveira RM, Castañeda T, Datta R, Dong JZ, Culler M, Sleeman MW, Álvarez CV, Gallego R, Lelliott CJ, Carling D, Tschop MH, Diéguez C, Vidal-Puig A (2008a) Hypothalamic fatty acid metabolism mediates the orexigenic action of ghrelin. Cell Metab 7(5):389–399

López M, Saha AK, Diéguez C, Vidal-Puig A (2008b) The AMPK-malonyl-CoA-CPT1 axis in the control of hypothalamic neuronal function-Reply. Cell Metab 8(3):176

López M, Tena-Sempere M, Diéguez C (2009) Cross-talk between orexins (hypocretins) and the neuroendocrine axes (hypothalamic-pituitary axes). Front Neuroendocrinol 31(2):113–127

López M, Varela L, Vázquez MJ, Rodríguez-Cuenca S, González CR, Velagapudi VR, Morgan DA, Schoenmakers E, Agassandian K, Lage R, de Morentin PB, Tovar S, Nogueiras R, Carling D, Lelliott C, Gallego R, Oresic M, Chatterjee K, Saha AK, Rahmouni, K, Diéguez C, Vidal-Puig A (2010) Hypothalamic AMPK and fatty acid metabolism mediate thyroid regulation of energy balance. Nat Med 16(9):1001–1008

Martin TL, Alquier T, Asakura K, Furukawa N, Preitner F, Kahn BB (2006) Diet-induced obesity alters AMP kinase activity in hypothalamus and skeletal muscle. J Biol Chem 281(28): 18933–18941

Martínez de Morentin PB, Whittle AJ, Ferno J, Nogueiras R, Diéguez C, Vidal-Puig A, López M (2012) Nicotine induces negative energy balance through hypothalamic AMP-activated protein kinase. Diabetes 61(4):807–817

Martins L, Fernández-Mallo D, Novelle MG, Vázquez MJ, Tena-Sempere M, Nogueiras R, López M Diéguez C (2012) Hypothalamic mTOR signaling mediates the orexigenic action of ghrelin. PLoS ONE 7(10):e46923

Mayer J (1955) Regulation of energy intake and the body weight: the glucostatic theory and the lipostatic hypothesis. AnnN Y Acad Sci 63(1):15–43

McCrimmon RJ, Fan X, Cheng H, McNay E, Chan O, Shaw M, Ding Y, Zhu W, Sherwin RS (2006) Activation of AMP-activated protein kinase within the ventromedial hypothalamus amplifies counterregulatory hormone responses in rats with defective counterregulation. Diabetes 55(6):1755–1760

McCrimmon RJ, Fan X, Ding Y, Zhu W, Jacob RJ, Sherwin RS (2004) Potential role for AMP-activated protein kinase in hypoglycemia sensing in the ventromedial hypothalamus. Diabetes 53(8):1953–1958

McCrimmon RJ, Shaw M, Fan X, Cheng H, Ding Y, Vella MC, Zhou L, McNay EC, Sherwin RS (2008) Key role for AMP-activated protein kinase in the ventromedial hypothalamus in regulating counterregulatory hormone responses to acute hypoglycemia. Diabetes 57(2): 444–450

Medina-Gomez G Vidal-Puig A (2005) Gateway to the metabolic syndrome. Nat Med 11(6): 602–603

Miller JC, Gnaedinger JM, Rapoport SI (1987) Utilization of plasma fatty acid in rat brain: distribution of (14C) palmitate between oxidative and synthetic pathways. J Neurochem 49(5):1507–1514

Minokoshi Y, Alquier T, Furukawa N, Kim YB, Lee A, Xue B, Mu J, Foufelle F, Ferre P, Birnbaum MJ, Stuck BJ, Kahn BB (2004) AMP-kinase regulates food intake by responding to hormonal and nutrient signals in the hypothalamus. Nature 428(6982):569–574

Morgan K, Obici S, Rossetti L (2004) Hypothalamic responses to long-chain fatty acids are nutritionally regulated. J Biol Chem 279(30):31139–31148

Mori H, Inoki K, Munzberg H, Opland D, Faouzi M, Villanueva EC, Ikenoue T, Kwiatkowski D, MacDougald OA, Myers MG, Guan KL Jr (2009) Critical role for hypothalamic mTOR activity in energy balance. Cell Metab 9(4):362–374

Morton GJ, Cummings DE, Baskin DG, Barsh GS, Schwartz MW (2006) Central nervous system control of food intake and body weight. Nature 443(7109):289–295

Obici S, Feng Z, Morgan K, Stein D, Karkanias G, Rossetti L (2002) Central administration of oleic acid inhibits glucose production and food intake. Diabetes 51(2):271–275

Obici S, Feng Z, Arduini A, Conti R, Rossetti L (2003) Inhibition of hypothalamic carnitine palmitoyltransferase-1 decreases food intake and glucose production. Nat Med 9(6):756–761

Oyadomari S, Koizumi A, Takeda K, Gotoh T, Akira S, Araki E, Mori M (2002) Targeted disruption of the Chop gene delays endoplasmic reticulum stress-mediated diabetes. J Clin Invest 109(4): 525–532

Oyadomari S Mori M (2004) Roles of CHOP/GADD153 in endoplasmic reticulum stress. Cell Death Differ 11(4):381–389

Ozcan U, Ozcan L, Yilmaz E, Duvel K, Sahin M, Manning BD, Hotamisligil GS (2008) Loss of the tuberous sclerosis complex tumor suppressors triggers the unfolded protein response to regulate insulin signaling and apoptosis. Mol Cell 29(5):541–551

Ozcan L, Ergin AS, Lu A, Chung J, Sarkar S, Nie D, Myers MG, Jr Ozcan U (2009) Endoplasmic reticulum stress plays a central role in development of leptin resistance. Cell Metab 9(1):35–51

Pennuto M, Tinelli E, Malaguti M, Del Carro U, D'Antonio M, Ron D, Quattrini A, Feltri ML, Wrabetz L (2008) Ablation of the UPR-mediator CHOP restores motor function and reduces demyelination in Charcot-Marie-Tooth 1B mice. Neuron 57(3):393–405

Plum L, Belgardt BF, Bruning JC (2006) Central insulin action in energy and glucose homeostasis. J Clin Invest 116(7):1761–1766

Pocai A, Lam TK, Gutierrez-Juarez R, Obici S, Schwartz GJ, Bryan J, Aguilar-Bryan L, Rossetti L (2005) Hypothalamic K(ATP) channels control hepatic glucose production. Nature 434(7036):1026–1031

Pocai A, Lam TK, Obici S, Gutierrez-Juarez R, Muse ED, Arduini A, Rossetti L (2006) Restoration of hypothalamic lipid sensing normalizes energy and glucose homeostasis in overfed rats. J Clin Invest 116(4):1081–1091

Proulx K, Cota D, Woods SC, Seeley RJ (2008) Fatty acid synthase inhibitors modulate energy balance via mammalian target of rapamycin complex 1 signaling in the central nervous system. Diabetes 57(12):3231–3238

Qi K, Hall M, Deckelbaum RJ (2002) Long-chain polyunsaturated fatty acid accretion in brain. Curr Opin Clin Nutr Metab Care 5(2):133–138

Qi J, Gong J, Zhao T, Zhao J, Lam P, Ye J, Li JZ, Wu J, Zhou HM, Li P (2008) Downregulation of AMP-activated protein kinase by Cidea-mediated ubiquitination and degradation in brown adipose tissue. EMBO J 27(11):1537–1548

Rapoport SI (1996) In vivo labeling of brain phospholipids by long-chain fatty acids: relation to turnover and function. Lipids 31 Suppl:S97–101

Rapoport SI (2001) In vivo fatty acid incorporation into brain phosholipids in relation to plasma availability, signal transduction and membrane remodeling. J Mol Neurosci 16(2–3):243–261

Ron D, Habener JF (1992) CHOP, a novel developmentally regulated nuclear protein that dimerizes with transcription factors C/EBP and LAP and functions as a dominant-negative inhibitor of gene transcription. Genes Dev 6(3):439–453

Ron D, Hubbard SR (2008) How IRE1 reacts to ER stress. Cell 132(1):24–26

Ron D, Walter P (2007) Signal integration in the endoplasmic reticulum unfolded protein response. Nat Rev Mol Cell Biol 8(7):519–529

Ruderman NB, Saha AK, Kraegen EW (2003) Minireview: malonyl CoA, AMP-activated protein kinase, and adiposity. Endocrinology 144(12):5166–5171

Sarafidis PA, Nilsson PM (2006) The metabolic syndrome: a glance at its history. J Hypertens 24(4):621–626

Schapansky J, Olson K, Van Der PR, Glazner G (2007) NF-kappaB activated by ER calcium release inhibits Abeta-mediated expression of CHOP protein: enhancement by AD-linked mutant presenilin 1. Exp Neurol 208(2):169–176

Sharma NK, Das SK, Mondal AK, Hackney OG, Chu WS, Kern PA, Rasouli N, Spencer HJ, Yao-Borengasser A, Elbein SC (2008) Endoplasmic reticulum stress markers are associated with obesity in nondiabetic subjects. J Clin Endocrinol Metab 93(11):4532–4541

Silva RM, Ries V, Oo TF, Yarygina O, Jackson-Lewis V, Ryu EJ, Lu PD, Marciniak SJ, Ron D, Przedborski S, Kholodilov N, Greene LA, Burke RE (2005) CHOP/GADD153 is a mediator of apoptotic death in substantia nigra dopamine neurons in an in vivo neurotoxin model of parkinsonism. J Neurochem 95(4):974–986

Song B, Scheuner D, Ron D, Pennathur S, Kaufman RJ (2008) Chop deletion reduces oxidative stress, improves beta cell function, and promotes cell survival in multiple mouse models of diabetes. J Clin Invest 118(10):3378–3389

Sorensen A, Travers MT, Vernon RG, Price NT, Barber MC (2002) Localization of messenger RNAs encoding enzymes associated with malonyl-CoA metabolism in mouse brain. Brain Res Gene Expr Patterns 1(3–4):167–173

Steinberg GR, Michell BJ, Denderen BJ van, Watt MJ, Carey AL, Fam BC, Andrikopoulos S, Proietto J, Gorgun CZ, Carling D, Hotamisligil GS, Febbraio MA, Kay TW, Kemp, BE (2006) Tumor necrosis factor alpha-induced skeletal muscle insulin resistance involves suppression of AMP-kinase signaling. Cell Metab 4(6):465–474

Tang QQ, Lane MD (2000) Role of C/EBP homologous protein (CHOP-10) in the programmed activation of CCAAT/enhancer-binding protein-beta during adipogenesis. Proc Natl Acad Sci U.S.A 97(23):12446–12450

Varela L, Martínez-Sánchez N, Gallego R, Vázquez MJ, Roa J, Gándara M, Schoenmakers E, Nogueiras R, Chatterjee K, Tena-Sempere M, Diéguez C, López M (2012) Hypothalamic mTOR pathway mediates thyroid hormone-induced hyperphagia in hyperthyroidism. J Pathol 227(2):209–222

Viollet B, Andreelli F, Jorgensen SB, Perrin C, Geloen A, Flamez D, Mu J, Lenzner C, Baud O, Bennoun M, Gomas E, Nicolas G, Wojtaszewski JF, Kahn A, Carling D, Schuit FC, Birnbaum MJ, Richter EA, Burcelin R, Vaulont S (2003) The AMP-activated protein kinase alpha2 catalytic subunit controls whole-body insulin sensitivity. J Clin Invest 111(1):91–98

Wang R, Cruciani-Guglielmacci C, Migrenne S, Magnan C, Cotero VE, Routh VH (2006) Effects of oleic acid on distinct populations of neurons in the hypothalamic arcuate nucleus are dependent on extracellular glucose levels. J Neurophysiol 95(3):1491–1498

Whittle AJ, Carobbio S, Martins L, Slawik M, Hondares E, Vázquez MJ, Morgan D, Csikasz RI, Gallego R, Rodriguez-Cuenca S, Dale M, Virtue S, Villarroya F, Cannon B, Rahmouni K, López M, Vidal-Puig A (2012) Bmp8b increases brown adipose tissue thermogenesis through both central and peripheral actions. Cell 149(4):871–885

Wolfgang MJ, Cha SH, Sidhaye A, Chohnan S, Cline G, Shulman GI, Lane MD (2007) Regulation of hypothalamic malonyl-CoA by central glucose and leptin. Proc Natl Acad Sci U S A 104(49):19285–19290

Wolfgang MJ, Kurama T, Dai Y, Suwa A, Asaumi M, Matsumoto S, Cha SH, Shimokawa T, Lane MD (2006) The brain-specific carnitine palmitoyltransferase-1c regulates energy homeostasis. Proc Natl Acad Sci U S A 103(19):7282–7287

Won JC, Jang PG, Namkoong C, Koh EH, Kim SK, Park JY, Lee KU, Kim MS (2009) Central administration of an endoplasmic reticulum stress inducer inhibits the anorexigenic effects of leptin and insulin. Obesity (Silver Spring) 17(10):1861–1865

Woods SC, Seeley RJ, Cota D (2008) Regulation of food intake through hypothalamic signaling networks involving mTOR. Annu Rev Nutr 28:295–311

Xie M, Zhang D, Dyck JR, Li Y, Zhang H, Morishima M, Mann DL, Taffet GE, Baldini A, Khoury DS, Schneider MD (2006) A pivotal role for endogenous TGF-beta-activated kinase-1 in the LKB1/AMP-activated protein kinase energy-sensor pathway. Proc Natl Acad Sci U S A 103(46):17378–17383

Xue B, Pulinilkunnil T, Murano I, Bence KK, He H, Minokoshi Y, Asakura K, Lee A, Haj F, Furukawa N, Catalano KJ, Delibegovic M, Balschi JA, Cinti S, Neel BG, Kahn BB (2009) Neuronal protein tyrosine phosphatase 1B deficiency results in inhibition of hypothalamic AMPK and isoform-specific activation of AMPK in peripheral tissues. Mol Cell Biol 29(16):4563–4573

Yang JC, Teng CF, Wu HC, Tsai HW, Chuang HC, Tsai TF, Hsu YH, Huang W, Wu LW, Su IJ (2009) Enhanced expression of vascular endothelial growth factor-A in ground glass hepatocytes and its implication in hepatitis B virus hepatocarcinogenesis. Hepatology 49(6):1962–1971

Yang L, Hotamisligil GS (2008) Stressing the brain, fattening the body. Cell 135(1):20–22

Yusta B, Baggio LL, Estall JL, Koehler JA, Holland DP, Li H, Pipeleers D, Ling Z, Drucker DJ (2006) GLP-1 receptor activation improves beta cell function and survival following induction of endoplasmic reticulum stress. Cell Metab 4(5):391–406

Zhang X, Zhang G, Zhang H, Karin M, Bai H, Cai D (2008) Hypothalamic IKKbeta/NF-kappaB and ER stress link overnutrition to energy imbalance and obesity. Cell 135(1):61–73

Chapter 8
Lipid Metabolism in Metabolic Syndrome

Jan Borén and Martin Adiels

Abstract Lipoprotein metabolism is a complex system in which abnormal concentrations of various lipoprotein particles can result from alterations in their rates of production, conversion and/or catabolism. To develop novel strategies for the prevention and treatment of dyslipidaemia, it is essential to better understand the lipid disorders in humans. Here we discuss recent advances in the field as well as show how the use of tracers labeled with stable isotopes combined with mathematical modelling provides a powerful tool for probing lipid and lipoprotein kinetics *in vivo*; thus furthering our understanding of the pathogenesis of lipid disorders.

Keywords Lipid metabolism · Liver · Tracer studies · Very low density lipopoprotein

8.1 Introduction

Cardiovascular diseases (CVD), such as stroke and ischemic heart attack, are the leading cause of death in the vast majority of the countries in the developed world, reaching well over 50 % in some countries and accounting for 30 % of all deaths worldwide (Gaziano 2005). Due to the rapid increase in obesity and sedentary lifestyle there is an increase in co-morbities such as type 2 diabetes (Zimmet et al. 2001). This will likely portend a daunting increase in CVD in the coming decades.

Changes in blood lipids, commonly known as diabetic dyslipidaemia, is an important and frequent risk factor for CVD and is a common feature of metabolic diseases thus playing an important role in the close link between metabolic diseases and CVD.

Recent studies have recognized the complex nature of diabetic dyslipidaemia, a cluster of potentially atherogenic lipid and lipoprotein abnormalities (Taskinen 2003). A better understanding of the pathophysiology behind diabetic dyslipidaemia will allow us to tailor targeted goals for the lipid management.

J. Borén (✉) · M. Adiels
Department of Molecular and Clinical Medicine, University of Gothenburg,
Wallenberg Laboratory, 413 45 Gothenburg, Sweden
e-mail: Jan.Boren@wlab.gu.se

M. Orešič, A. Vidal-Puig (eds.), *A Systems Biology Approach to Study Metabolic Syndrome*, 157
DOI 10.1007/978-3-319-01008-3_8, © Springer International Publishing Switzerland 2014

8.2 Lipoprotein Metabolism

The function of lipoprotein particles is to transport hydrophobic lipid molecules, which are not water soluble, in the blood. The general structure of a lipoprotein includes a core consisting of triglycerides and/or cholesterol esters surrounded by a surface monolayer of phospholipids, unesterified cholesterol and specific proteins (Mahley et al. 1984). Cholesterol is an important component of membranes in the cell and is also important for the synthesis of steroid hormones. The function of triglycerides is as energy source in tissue and for energy storage in adipose tissue. The protein components of the lipoprotein are known as apolipoproteins which function as cofactors for enzymes and ligands for receptors (Mahley et al. 1984). The lipoprotein particles are classified according to their density and size (Table 8.1). The density is primarily reflecting the triglyceride content of the particles with more triglycerides in the lighter and larger particles. The metabolism of lipoproteins is schematically illustrated in Fig. 8.1.

Chylomicrons Chylomicrons are triglyceride-rich lipoprotein particles synthesized in the gut during uptake of triglycerides from the food. The chylomicrons are transported with the lymph vessels into the blood. In the blood the triglycerides are catabolized by the enzyme, lipoprotein lipase (LPL), and fatty acids are released to be used as energy source in the cells, or to be stored in adipose tissue as triglycerides. The chylomicron remnant is finally taken up and catabolized by the liver. The main protein component of chylomicrons is the apolipoprotein (apo) B48.

VLDL-IDL-LDL These lipoproteins, containing the protein apoB100, represent a metabolic cascade of particles for triglyceride transport of endogenous origin. Very low density lipoprotein (VLDL) is synthesized in the liver as a triglyceride-rich lipoprotein particle. VLDL is catabolized by the enzyme LPL, and as the triglyceride content is reduced the particle becomes smaller and denser. It is first transformed into intermediate density lipoprotein (IDL), which to some extent is taken up by the liver, or further catabolized to low density lipoprotein (LDL). The transition of IDL to LDL is driven by the enzymes LPL and hepatic lipase. LDL particles are cholesterol rich particles and about 80 % of the plasma cholesterol is carried in LDL. Most of the LDL is taken up by the liver and excreted via the bile, in the form of cholesterol or after transformation to bile acids.

The liver can secrete both large triglyceride-rich $VLDL_1$ particles, and smaller more cholesterol-containing $VLDL_2$. Intravascular lipolysis of these lipoproteins gives rise to normal-size LDL that remains in the circulation for $\sim$2 days. However, when triglyceride levels are high, $VLDL_1$ particles are initially converted to large LDL particles. These large LDL particles are subsequently converted to small dense LDL (sdLDL) with a retention time of $\sim$5 days by cholesterol ester transport protein (CETP) and hepatic lipase, which both are increased in type 2, diabetes (Taskinen 2003) (Fig. 8.2). In addition to this, increased levels of $VLDL_1$ alter the composition of high density lipoproteins (HDL) ultimately leading to an increased catabolism of these particles (Rashid et al. 2003). Thus, elevation of $VLDL_1$ in diabetic dyslipidemia leads to the generation of sdLDL and increased risk for CVD.

Table 8.1 Characteristics of lipoprotein families

	Density (g/mL)	Size (nm)	TG (%)	CE (%)	PL (%)	Chol (%)	Major apolipoprotein	Minor apolipoproteins
Chylomicrons	< 0.95	80–100	90–95	2–4	2–6	1	apoB48	apoAI, apoAII, apoAIV
VLDL	0.95–1.006	30–80	50–65	8–14	12–16	4–7	apoB100	apoAI, apoCII, apoCIII, apoE
IDL	1.006–1.019	25–30	25–40	20–35	16–24	7–11	apoB100	apoCII, apoCIII, apoE
LDL	1.019–1.063	20–25	4–6	34–35	22–26	6–15	apoB100	
HDL	1.063–1.210	8–13	7	10–20	25	5	apoAI	apoAII, apoCIII, apoE, apoM

TG Content of triglycerides, *CE* cholesterol esters, *PL* phospholipids, and *chol* cholesterol

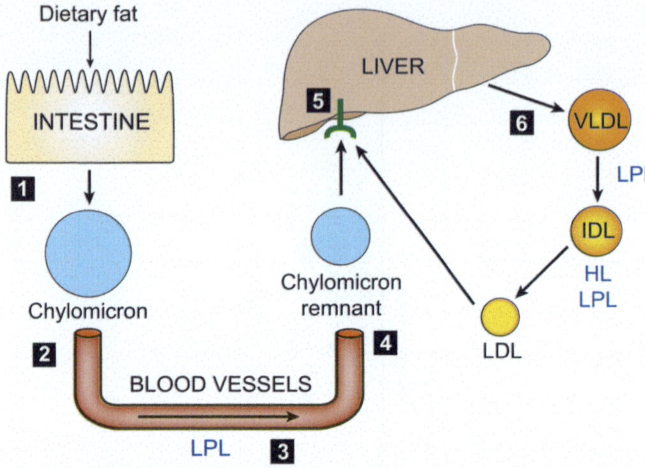

Fig. 8.1 Metabolism of apolipoprotein B (apoB)-containing lipoproteins. Dietary lipids are packed into chylomicrons (*1*). Once in the plasma (*2*), most triglyceride molecules in chylomicrons are hydrolyzed by the action of lipoprotein lipase (LPL) (*3*) resulting in smaller chylomicron remnants (*4*). The remnants are cleared from the circulation by the liver (*5*). LPL and hepatic lipase (HL) convert liver-derived VLDL to smaller intermediate-density lipoprotein (IDL) and LDL (*6*), which are removed from the circulation by the liver

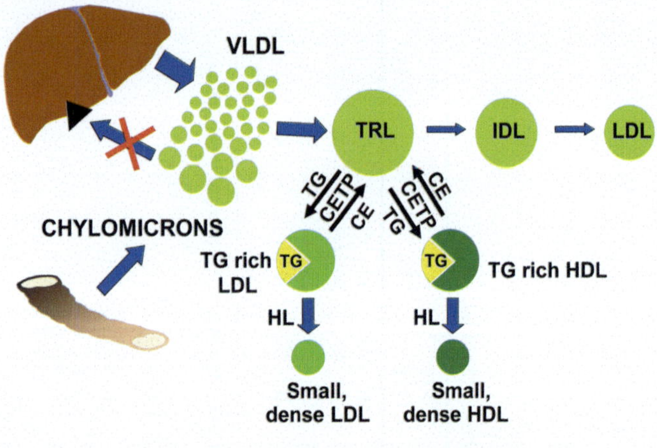

Fig. 8.2 Metabolism of triglyceride-rich lipoproteins (TRLs) and generation of sdLDL. The concentration of fasting triglycerides and the magnitude of postprandial lipoproteinemia have striking influences on LDL subclasses. The long residence time of triglyceride-rich lipoproteins (TRL) particles in circulation due to impaired triglyceride hydrolysis and delayed remnant removal results in increased exchange of core lipids, that is, triglycerides and cholesterol esters, between TRLs and LDL, mediated by cholesterol ester transfer protein (CETP). In this process, LDL particles become triglyceride-enriched. Large triglyceride-rich particles are good substrates for hepatic lipase (HL), which is commonly increased in type 2 diabetes. Taken together, the formation of small, dense LDL depends on the rate of triglyceride transfer to LDL and on the rate of its hydrolysis by HL; both processes are enhanced in diabetic dyslipidemia. In addition to this, increased levels of TRL alter the composition of HDL ultimately leading to an increased catabolism of these particles

The presence of sdLDL has been correlated with an increased risk of coronary heart disease, but this has not been shown to be fully independent of related factors such as elevated plasma triacylglycerol concentrations (Lada and Rudel 2004).

HDL These are the smallest and most dense lipoproteins with a relatively high content of phospholipids. The function of HDL is complex. The particle is secreted as a lipid-poor apoAI particle which associates with phospholipids to form a disc-shaped particle. Through uptake of cholesterol from cells and tissue, a core of cholesterol ester is formed. In the atherosclerotic plaque the cholesterol is taken up from the lipid filled macrophages. From HDL the cholesterol can be transported to the liver via two routes; HDL can either be taken up by the liver, or alternatively cholesterol can be transferred to LDL particles via the transport protein CETP.

8.3 Assembly of VLDL in the Secretory Pathway of the Cell

The assembly of VLDL involves a stepwise increase of the lipidation of the structural protein apoB100 (Olofsson and Boren 2005). It starts in the rough endoplasmic reticulum (ER) when apoB100 enters the secretory pathway co-translationally (Boren et al. 1992) and is simultaneously lipidated by the microsomal triglyceride transfer protein (MTP) (Olofsson et al. 1999; Olofsson and Boren 2005). The nascent VLDL particle is converted to a smaller $VLDL_2$ by additional lipidation (Stillemark-Billton et al. 2005). In situations with increased lipid load, the liver can also form a larger triglyceride-rich $VLDL_1$ particle. Studies have demonstrated that the secretion of $VLDL_1$ increases with increasing concentrations of liver lipids (Adiels et al. 2006b). Thus, the formation of $VLDL_1$ is regulated by the availability of lipids, and increased lipid load results in that more $VLDL_1$ particles are secreted (Adiels et al. 2005a). Thus, the assembly of $VLDL_1$ requires several steps: the formation of nascent VLDL particle, its conversion to $VLDL_2$, and further addition of triglycerides (Stillemark et al. 2000).

8.3.1 Regulators VLDL Assembly and Secretion

To date it is still unclear how the liver is able to regulate the amount of triglycerides incorporated into VLDL particles to produce either large $VLDL_1$ particles or smaller $VLDL_2$ particles. Substantial evidence indicates that the synthesis and release of $VLDL_1$ and $VLDL_2$ particles are independently regulated. Two key regulators of VLDL secretion has been shown to be fatty acids flux and insulin. An inflow of fatty acids to the liver increases the triglyceride formation which explains the influence of the fatty acids on the formation of VLDL (Adiels et al. 2008). Insulin has also been shown been shown to reduce $VLDL_1$ formation (Adiels et al. 2007). The balance between $VLDL_1$ and $VLDL_2$ is also regulated by peroxisome proliferator-activated receptor alpha (PPARα) agonists. Such agonists have a fundamental influence on the assembly of $VLDL_1$; thus, they increase the biosynthesis of apoB100 and decrease the formation of triglycerides (Lindén et al. 2002). This promotes secretion of $VLDL_2$ particles instead of the larger triglyceride-rich $VLDL_1$ particles.

8.3.2 Regulators of VLDL Turn Over

In the circulation, VLDL particles are converted to denser particles by the removal of triglycerides, primarily by lipoprotein lipase. Thus, converting the large $VLDL_1$ to the denser $VLDL_2$, and subsequently $VLDL_2$ to IDL. In turn IDL are converted to LDL by removal of triglycerides by hepatic lipase. Removal of lipoprotein from the circulation may occur by LDL-receptor mediated uptake or by heparin sulphate proteoglycans (Williams and Chen 2010).

The interaction between the lipoproteins and the receptors and lipases are facilitated by the protein composition of the lipoproteins. For example apoCIII is an inhibitor of lipoprotein lipase (Havel et al. 1973) and the apoCIII content on VLDL particles are inversely related to the turnover of the particles (Chan et al. 2006; Taskinen et al. 2011).

8.4 Dyslipidaemia is a Cardinal Feature of the Metabolic Syndrome

In type 2 diabetes, dyslipidemia is an important and common modifiable risk factor for coronary heart disease (CHD) (Turner et al. 1998). Several studies have recognised the complex nature of diabetic dyslipidemia that is a cluster of potentially atherogenic lipid and lipoprotein abnormalities (Taskinen 2003); increased plasma triglycerides and low concentrations of HDL-cholesterol (Adiels et al. 2006a), accumulation of sdLDL and excessive postprandial lipaemia. Interestingly, the different components of the diabetic dyslipidaemia are not isolated abnormalities but are metabolically closely linked to each other (Taskinen 2005) (Fig. 8.3).

Patients with a diabetic dyslipidemia have a significant increased risk of developing cardiovascular disease (CVD). Furthermore, the prognosis for CVD is much worse in the presence of type 2 diabetes. Therefore, lipid management of diabetic dyslipidaemia is of immense importance.

8.4.1 Elucidation of the Mechanism for Diabetic Dyslipidemia Requires Studies with Stable Isotopes

In order to prevent and treat diabetic dyslipidemia it is necessary to characterize the disorder to fully understand its cause. The most common way to do this is by measuring plasma lipid or apolipoprotein concentrations. However, abnormal concentrations of lipids and apolipoproteins can result from changes in the production,

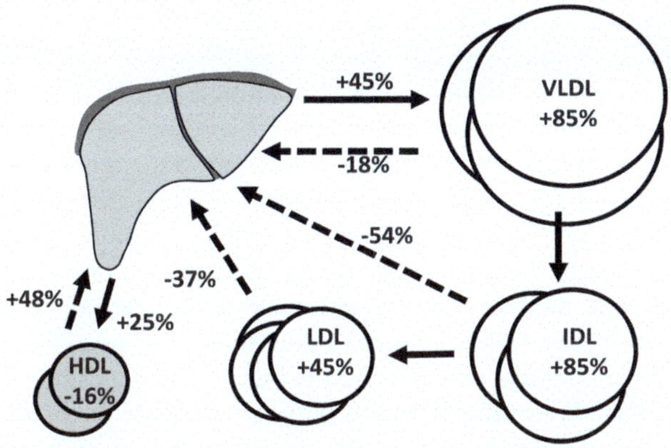

Fig. 8.3 The dysregulation of apoB and apoAI metabolism in subjects with the metabolic syndrome. ApoB-100 is the major structural protein of, VLDL, IDL and LDL. It is a large protein that is required for the assembly and secretion of apoB-containing lipoproteins. It is also a ligand for LDL receptor-mediated endocytosis of LDL and, possibly, VLDL and IDL. Subjects with the metabolic syndrome have significantly increased production (+ 45 % for VLDL, solid line) and reduced catabolism of apoB100 (− 18 %, − 54 % and − 37 % for VLDL, IDL and LDL respectively, dashed lines). This results in an increase concentration of VLDL, IDL and LDL particles. ApoAI is the major structural protein associated with HDL particles. It is important in the activation of enzymes involved in the maturation of HDL particles. In addition, a low HDL-cholesterol concentration is observed in these subjects. This is primarily due to hypercatabolism of its major structural protein apoAI (+ 48 %, dashed line) which is somewhat compensated by increased production (+ 25 %, solid line). The figure is based on studies by Watts and co-workers. (Riches et al. 1998; Chan et al. 2002; Watts et al. 2003)

conversion or catabolism of lipoprotein particles. Therefore, although static measurements and functional assays are important techniques, it is necessary to study the true unit of function (the integrated metabolic pathway) to understand the complexity of lipoprotein metabolism (Chan et al. 2004b; Chan et al. 2004a; Adiels et al. 2008).

In contrast to static measurements, kinetic experiments are of great importance in lipid research because they further the understanding of lipid metabolism *in vivo* and help to explain the pathophysiology of lipid disorders in humans (Chan et al. 2004b; Chan et al. 2004a; Adiels et al. 2008). At present, due to species specificity, no valid animal model can efficiently replace human studies to explore lipid metabolism, and the use of radioactive tracers in such studies is restricted. Therefore, stable-isotope tracer kinetic studies have become an important tool of translational research to achieve a quantitative understanding of the dynamics of metabolic processes *in vivo*.

8.5 Principles of Tracer Methodology

To study the behavior of an endogenous molecule, the tracee, the same molecule, but labeled (the tracer), is introduced into the system (usually via the bloodstream) (Chan et al. 2004b; Chan et al. 2004a). This process is termed *exogenous labeling*.

Endogenous labeling occurs when a labeled precursor of the molecule of interest is used to label this molecule (e.g., infusion of a labeled amino acid to label a protein). Ideally, the tracer can easily be detected and quantified, has the same kinetic behavior as the tracee, and does not perturb the system.

Usually, kinetic studies are performed in a steady state, where the rates of input and output for a given unlabeled tracee substance are equal and time invariant. Thus, the information provided by the tracer reflects the behavior of the tracee (Barrett et al. 2006). At various times, the amount of tracer is quantified to provide a kinetic curve. Then a mathematical model is constructed to extract all the information contained in the kinetic curve. By fitting a model to the data, it is possible to estimate the parameters of the model that characterize the flux of molecules between kinetically homogeneous pools. For example, it is possible to calculate the production, conversion or catabolism of lipoprotein particles, information that cannot be obtained by static measurements alone. Such models are usually constructed based on a priori knowledge, building *physiologically relevant* models (de Graaf et al. 2009).

The term stable isotope refers to a non-radioactive isotope of a given atom. In tracer experiments, the tracer is typically a molecule in which some atoms are replaced by stable isotopes with low natural abundance, such as ^{2}H or ^{13}C. Stable isotopes commonly used as metabolic tracers include: [$^{2}H_3$]-leucine for apolipoprotein turnover; [$^{2}H_5$]-glycerol and [^{13}C]-palmitic acid for VLDL–triglyceride metabolism; [1,2$-^{13}C$]-acetate and [^{13}C]-palmitic acid for fatty acid metabolism, and $^{2}H_2O$ for hepatic *de novo* lipogenesis.

8.6 Studies of Lipoprotein Metabolism Using Multicompartment Models

When a stable isotope tracer is injected, it is incorporated into the molecules under investigation to the same extent as the unlabelled trace molecule (by definition the tracer should have identical properties as the tracee). The fraction of labeled molecules can be estimated by measuring the enrichment of the stable isotope (tracer/tracee ratio). The tracer enables the dynamic fate of the latter to be traced through its metabolic pathways. The movement of material in the system can be described by a set of differential equations, thereby allowing quantitative information on the system to be estimated.

The tracer data generated can be analyzed using several different methodologies. One approach to analyse VLDL kinetics is to inject a bolus of tracer and determine the subsequent monoexponential slope of the decline in plasma VLDL tracer/tracer ratio. A disadvantage of this approach is that it can underestimate the true VLDL turnover rate because it does not account for recycling of the injected bolus of tracer (Patterson et al. 2002). Multicompartment modeling is a superior method to dissect the complexities of lipoprotein metabolism, and has been widely applied to systems in which material is transferred over time between compartments connected in a specific structure to permit the movement of material among the compartments.

Each compartment is assumed to represent a homogenous entity with respect to its kinetic properties. Typically, compartments represent different types of lipoprotein particles with are kinetically homogenous, while being distinct from other material in the system. Different compartments may also represent lipoprotein or molecules in different volumes (e.g., intracellular or in plasma). Usually, the model is constructed based on prior knowledge of the system. However, often data can be described by more than one model. Model selection may be carried out by examination of the fit to the experimental curves, parameter identifiability and precision of parameter estimates and by statistical tests of model results and complexity. However, the complexity of a multicompartment model is usually a compromise for what is practically possible. A very simple model may not adequately describe the kinetic heterogeneity present within the system. A model that is too complex, on the other hand, will not be supported by experimental data and, hence, will have little predictive value. Furthermore, even if the development of models is based on experimental data, several assumptions are required in order to derive the model that is to be used.

Several studies have been conducted to analyses VLDL–triglyceride turnover kinetics using stable isotopically labeled glycerol or palmitic acid tracers and mathematical modeling; however, VLDL subclasses were not analyzed and VLDL–apoB100 was not included in the models (Lemieux et al. 1999; Carpentier et al. 2002; Patterson et al. 2002). To enhance our understanding of the pathways leading to $VLDL_1$ and $VLDL_2$ and of the metabolic fate of these particles, we developed a multicompartmental mathematical model that allows the kinetics of triglycerides and apoB100 in $VLDL_1$ and $VLDL_2$ to be simultaneously assessed after a bolus injection of glycerol and leucine (Adiels et al. 2005b). Tracer/tracee curves of leucine and glycerol in $VLDL_1$ and $VLDL_2$were analyzed to determine kinetic parameters using mathematical modeling.

8.7 *In vivo* Studies Elucidates the Disturbed Lipid Metabolism in Subjects with Type 2 Diabetes and/or the Metabolic Syndrome

The model has greatly furthered our understanding of the disturbed lipoprotein metabolism in diabetic dyslipidemia. For example, the model showed a linear correlation between the $VLDL_1$ triglyceride and apoB100 production. Furthermore, the composition of the $VLDL_1$ and $VLDL_2$ particles (triglyceride:apoB100 ratio) in plasma suggests that a high level of total plasma triglycerides in achieved mainly by an increased $VLDL_1$ triglyceride pool, but this is the effect of increased number of particles rather than increased particle size. Overall, the contributions of $VLDL_2$ triglyceride pools to the total plasma TG triglyceride were less than that of the $VLDL_1$ triglyceride pools. We speculate that the $VLDL_2$ secretion reflects baseline production of triglycerides whereas the $VLDL_1$ secretion is modulated by dynamic variations in substrate fluxes. This is consistent with studies showing that insulin does not affect the direct production of $VLDL_2$ apoB100 or the lipid composition of the $VLDL_2$ particles (Malmstrom et al. 1997).

The model also revealed a significant difference in the delay times of glycerol and leucine (i.e., VLDL triglycerides and VLDL apoB100). After incorporation of glycerol into triglycerides, the completed lipoprotein enters the blood system within 21 min. In contrast, the delay time for apoB100 molecules is 33 min. These observations are consistent with a sequential assembly model of VLDL and likely provide the first evidence in man that triglyceride can be added to a primordial apoB100-containing particle in the liver (Alexander et al. 1976; Boren et al. 1994).

We continued to analyse which features of type 2 diabetes and insulin resistance correlate with $VLDL_1$ production, and revealed strong correlations with plasma glucose (Adiels et al. 2005a). By extending our study to monitor liver, intra-abdominal and subcutaneous fat as well as adiponectin, we showed that fasting insulin, plasma glucose, intra-abdominal and liver fat and the homeostasis model of assessing insulin resistance (HOMA-IR) are predictors of $VLDL_1$-apoB100 and $VLDL_1$-triglyceride production (Adiels et al. 2006b). However, in a multiple regression analysis, only liver fat and plasma glucose remained significant (Adiels et al. 2006b). Moreover, the key predictors of liver fat are intra-abdominal fat, adiponectin and plasma glucose (Adiels et al. 2006b). Further studies have demonstrated that liver fat content is closely related to triglyceride secretion in different settings and populations (Taskinen et al. 2011; Chan et al. 2010) and is a better predictor of triglyceride secretion than intra-abdominal fat (Fabbrini et al. 2009; Magkos et al. 2010).

Non-alcoholic fatty liver disease (NAFLD) is defined as fat accumulation in the liver that exceeds 5–10 % of liver weight in individuals who do not consume significant amounts of alcohol (Neuschwander-Tetri and Caldwell 2003). Recent data show that NAFLD is strongly associated with type 2 diabetes, obesity and hyperlipidaemia (Angelico et al. 2005; Clark and Diehl 2002; Youssef and McCullough 2002 ; Li et al. 2002; Marchesini et al. 2001; Marchesini et al. 2003; Bugianesi et al. 2005; Targher et al. 2005). We tested the relationship between liver fat and $VLDL_1$ suppression among subjects with a broad range of liver fat content (Adiels et al. 2007). From this study we confirmed that liver fat predicts baseline $VLDL_1$ production (Adiels et al. 2006b), and also that liver fat is associated with a lack of $VLDL_1$ suppression in response to insulin. Indeed, insulin downregulates $VLDL_1$ secretion in subjects with low liver fat levels, but fails to suppress $VLDL_1$ secretion in subjects with high liver fat levels, resulting in overproduction of $VLDL_1$ (Adiels et al. 2007). The reason for this lack of $VLDL_1$ suppression is not known. Insulin suppresses the non-esterified fatty acid (NEFA) pool to a similar extent regardless of liver fat level (Adiels et al. 2007), and thus the high $VLDL_1$ production in individuals with high levels of liver fat must be facilitated either by using a greater portion of systemic NEFA or recruiting other sources of triglycerides, such as from hepatic stores. The data suggest that liver fat content in obesity may be a better marker of metabolic derangement and CVD risk than visceral obesity *per se*.

Data on clearance of VLDL particles in metabolic diseases are less conclusive. Studies show clearance to be slightly slower (Sorensen et al. 2011) ($p = 0.08$), (Adiels et al. 2005a) (NS), (Ouguerram et al. 2003) (NS) and (Duvillard et al. 2000) ($P < 0.05$) in type 2 diabetes. ApoCIII has been shown to be inversely correlated to the rate of clearance of VLDL and $VLDL_1$ (Chan et al. 2006, Taskinen et al. 2011). But it has also been reported a relative deficiency of apoCIII in all triglyceride-rich lipoprotein species in type 2 diabetic subjects (Hiukka et al. 2005). Despite the increased plasma concentrations of apoCIII in these subjects, the concentration of apoCIII does not increase in concert with core lipids. ApoCIII also is implicated in the hepatic clearance of VLDL remnants (Gangabadage et al. 2008), which may play an important role.

8.8 Conclusions

Insight into lipid metabolism, in particular the causative relation between production, uptake and conversion of triglycerides, may be gained from *in vivo* kinetic studies. Such information may be pivotal for our understanding of dysregulation plasma lipids as well as for finding means to optimally treat dyslipidaemias.

References

Adiels M, Boren J, Caslake MJ, Stewart P, Soro A, Westerbacka J, Wennberg B, Olofsson SO, Packard C, Taskinen MR (2005a) Overproduction of $VLDL_1$ driven by hyperglycemia is a dominant feature of diabetic dyslipidemia. Arterioscler Thromb Vasc Biol 25:1697–1703

Adiels M, Packard C, Caslake MJ, Stewart P, Soro A, Westerbacka J, Wennberg B, Olofsson SO, Taskinen MR, Boren J (2005b) A new combined multicompartmental model for apolipoprotein B-100 and triglyceride metabolism in VLDL subfractions. J Lipid Res 46:58–67

Adiels M, Olofsson SO, Taskinen MR, Boren J (2006a) Diabetic dyslipidaemia. Curr Opin Lipidol 17:238–246

Adiels M, Taskinen MR, Packard C, Caslake MJ, Soro-Paavonen A, Westerbacka J, Vehkavaara S, Hakkinen A, Olofsson SO, YKI-Jarvinen H, Boren J (2006b) Overproduction of large VLDL particles is driven by increased liver fat content in man. Diabetologia 49:755–765

Adiels M, Westerbacka J, Soro-Paavonen A, Hakkinen AM, Vehkavaara S, Caslake MJ, Packard C, Olofsson SO, YKI-Jarvinen H, Taskinen MR, Boren J (2007) Acute suppression of $VLDL_1$ secretion rate by insulin is associated with hepatic fat content and insulin resistance. Diabetologia 50:2356–2365

Adiels M, Olofsson SO, Taskinen MR, Boren J (2008) Overproduction of very low-density lipoproteins is the hallmark of the dyslipidemia in the metabolic syndrome. Arterioscler Thromb Vasc Biol 28:1225–1236

Alexander CA, Hamilton RL, Havel RJ (1976) Subcellular localization of B apoprotein of plasma lipoproteins in rat liver. J Cell Biol 69:241–263

Angelico F, Del Ben M, Conti R, Francioso S, Feole K, Fiorello S, Cavallo MG, Zalunardo B, Lirussi F, Alessandri C, Violi F (2005) Insulin resistance, the metabolic syndrome, and nonalcoholic fatty liver disease. J Clin Endocrinol Metab 90:1578–1582

Barrett PH, Chan DC, Watts GF (2006) Thematic review series: patient-oriented research. Design and analysis of lipoprotein tracer kinetics studies in humans. J Lipid Res 47:1607–1619

Boren J, Graham L, Wettesten M, Scott J, White A, Olofsson SO (1992) The assembly and secretion of ApoB 100-containing lipoproteins in Hep G2 cells. ApoB 100 is cotranslationally integrated into lipoproteins. J Biol Chem 267:9858–9867

Boren J, Rustaeus S, Olofsson SO (1994) Studies on the assembly of apolipoprotein B-100- and B-48-containing very low density lipoproteins in McA-RH7777 cells. J Biol Chem 269:25879–25888

Bugianesi E, Mccullough AJ, Marchesini G (2005) Insulin resistance: a metabolic pathway to chronic liver disease. Hepatology 42:987–1000

Carpentier A, Patterson BW, Leung N, Lewis GF (2002) Sensitivity to acute insulin-mediated suppression of plasma free fatty acids is not a determinant of fasting VLDL triglyceride secretion in healthy humans. Diabetes 51:1867–1875

Clark JM, Diehl AM (2002) Hepatic steatosis and type 2 diabetes mellitus. Curr Diab Rep 2:210–215

Chan DC, Watts GF, Redgrave TG, Mori TA, Barrett PH (2002) Apolipoprotein B-100 kinetics in visceral obesity: associations with plasma apolipoprotein C-III concentration. Metabolism 51:1041–1046

Chan DC, Barrett PH, Watts GF (2004a) Lipoprotein transport in the metabolic syndrome (Part I): methodological aspects of stable isotope kinetic studies. Clin Sci (Lond) 107:221–232

Chan DC, Barrett PH, Watts GF (2004b) Lipoprotein transport in the metabolic syndrome (Part II): pathophysiological and interventional studies employing stable isotopy and modelling methods. Clin Sci (Lond)

Chan DC, Watts GF, Nguyen MN, Barrett PH (2006) Apolipoproteins C-III and A-V as predictors of Very-Low-Density Lipoprotein triglyceride and apolipoprotein B-100 Kinetics. Arterioscler Thromb Vasc Biol

Chan DC, Watts GF, Gan S, Wong AT, Ooi EM, Barrett PH (2010) Nonalcoholic fatty liver disease as the transducer of hepatic oversecretion of Very-Low-Density Lipoprotein-Apolipoprotein B-100 in obesity. Arterioscler Thromb Vasc Biol

de Graaf AA, Freidig AP, De Roos B, Jamshidi N, Heinemann M, Rullmann JA, Hall KD, Adiels M, van Ommen B (2009) Nutritional systems biology modeling: from molecular mechanisms to physiology. PLoS Comput Biol 5:e1000554

Duvillard L, Pont F, Florentin E, Galland-Jos C, Gambert P, Verges B (2000) Metabolic abnormalities of apolipoprotein B-containing lipoproteins in non-insulin-dependent diabetes: a stable isotope kinetic study. Eur J Clin Invest 30:685–694

Fabbrini E, Magkos F, Mohammed BS, Pietka T, Abumrad NA, Patterson BW, Okunade A, Klein S (2009) Intrahepatic fat, not visceral fat, is linked with metabolic complications of obesity. Proc Natl Acad Sci U S A 106:15430–15435

Gaziano TA (2005) Cardiovascular disease in the developing world and its cost-effective management. Circ 112:3547–3553

Gangabadage CS, Zdunek J, Tessari M, Nilsson S, Olivecrona G, Wijmenga SS (2008) Structure and dynamics of human apolipoprotein CIII. J Biol Chem 283:17416–17427

Havel RJ, Fielding CJ, Olivecrona T, Shore VG, Fielding PE, Egelrud T (1973) Cofactor activity of protein components of human very low density lipoproteins in the hydrolysis of triglycerides by lipoproteins lipase from different sources. BioChemistry 12:1828–1833

Hiukka A, Fruchart-Najib J, Leinonen E, Hilden H, Fruchart JC, Taskinen MR (2005) Alterations of lipids and apolipoprotein CIII in very low density lipoprotein subspecies in type 2 diabetes. Diabetologia 48:1207–1215

Lada AT, Rudel LL (2004) Associations of low density lipoprotein particle composition with atherogenicity. Curr OpinLipidol 15:19–24

Lemieux S, Patterson BW, Carpentier A, Lewis GF, Steiner G (1999) A stable isotope method using a [(2)H(5)] glycerol bolus to measure very low density lipoprotein triglyceride kinetics in humans. J Lipid Res 40:2111–2117

Li Z, Clark J, Diehl AM (2002) The liver in obesity and type 2 diabetes mellitus. Clin Liver Dis 6:867–877

Lindén D, Lindberg K, Oscarsson J, Claesson C, Asp L, Li L, Gustafsson M, Borén J, Olofsson S-O (2002) Influence of agonists to the peroxisome proliferator activated receptor a (PPARa) on the intracellular turnover and secretion of apoB-100 and apoB-48. J Biol Chem 277:23044–23053

Magkos F, Fabbrini E, Mohammed BS, Patterson BW, Klein S (2010) Increased whole-body adiposity without a concomitant increase in liver fat is not associated with augmented metabolic dysfunction. Obesity (Silver Spring) 18:1510–1515

Mahley RW, Innerarity TL, Rall SC Jr, Weisgraber KH (1984) Plasma lipoproteins: apolipoprotein structure and function. J Lipid Res 25:1277–1294

Malmstrom R, Packard CJ, Caslake M, Bedford D, Stewart P, Yki-Jarvinen H, Shepherd J, Taskinen MR (1997) Defective regulation of triglyceride metabolism by insulin in the liver in NIDDM. Diabetologia 40:454–462

Marchesini G, Brizi M, Bianchi G, Tomassetti S, Bugianesi E, Lenzi M, Mccullough AJ, Natale S, Forlani G, Melchionda N (2001) Nonalcoholic fatty liver disease: a feature of the metabolic syndrome. Diabetes 50:1844–1850

Marchesini G, Bugianesi E, Forlani G, Cerrelli F, Lenzi M, Manini R, Natale S, Vanni E, Villanova N, Melchionda N, Rizzetto M (2003) Nonalcoholic fatty liver, steatohepatitis, and the metabolic syndrome. Hepatology 37:917–923

Neuschwander-Tetri BA, Caldwell SH (2003) Nonalcoholic steatohepatitis: summary of an AASLD Single Topic Conference. Hepatology 37:1202–1219

Olofsson S-O, Asp L, Borén J (1999) The assembly and secretion of apolipoprotein B-containing lipoproteins. Curr Opin Lipidol 10:341–346

Olofsson SO, Boren J (2005) Apolipoprotein B: a clinically important apolipoprotein which assembles atherogenic lipoproteins and promotes the development of atherosclerosis. J Int Medi 258:395–410

Ouguerram K, Magot T, Zair Y, Marchini JS, Charbonnel B, Laouenan H, Krempf M (2003) Effect of atorvastatin on apolipoprotein B100 containing lipoprotein metabolism in type-2 diabetes. J Pharmacol Exp Ther 306:332–337

Patterson BW, Mittendorfer B, Elias N, Satyanarayana R, Klein S (2002) Use of stable isotopically labeled tracers to measure very low density lipoprotein-triglyceride turnover. J Lipid Res 43:223–233

Rashid S, Watanabe T, Sakaue T, Lewis GF (2003) Mechanisms of HDL lowering in insulin resistant, hypertriglyceridemic states: the combined effect of HDL triglyceride enrichment and elevated hepatic lipase activity. Clin Biochem 36:421–429

Riches FM, Watts GF, Naoumova RP, Kelly JM, Croft KD, Thompson GR (1998) Hepatic secretion of very-low-density lipoprotein apolipoprotein B-100 studied with a stable isotope technique in men with visceral obesity. Int J Obes Relat Metab Disord 22:414–423

Sorensen LP, Andersen IR, Sondergaard E, Gormsen LC, Schmitz O, Christiansen JS, Nielsen S (2011) Basal and insulin mediated VLDL-triglyceride kinetics in type 2 diabetic men. Diabetes 60:88–96

Stillemark P, Borén J, Andersson M, Larsson T, Rustaeus S, Karlsson K-A, Olofsson S-O (2000) The assembly and secretion of apolipoprotein-B48-containing very low density lipoproteins in McA-RH7777 cells. J Biol Chem 275:10506–10513

Stillemark-Billton P, Beck C, Boren J, Olofsson SO (2005) Relation of the size and intracellular sorting of apoB to the formation of VLDL 1 and VLDL 2. J Lipid Res 46:104–114

Targher G, Bertolini L, Poli F, Rodella S, Scala L, Tessari R, Zenari L, Falezza G (2005) Nonalcoholic fatty liver disease and risk of future cardiovascular events among type 2 diabetic patients. Diabetes 54:3541–3546

Taskinen MR (2003) Diabetic dyslipidaemia: from basic research to clinical practice. Diabetologia 46:733–749

Taskinen MR (2005) Type 2 diabetes as a lipid disorder. Curr Mol Med 5:297–308

Taskinen MR, Adiels M, Westerbacka J, Soderlund S, Kahri J, Lundbom N, Lundbom J, Hakkarainen A, Olofsson SO, Orho-Melander M, Boren J (2011) Dual metabolic defects are required to produce hypertriglyceridemia in obese subjects. Arterioscler Thromb Vasc Biol 31:2144–2150

Turner RC, Millns H, Neil HA, Stratton IM, Manley SE, Matthews DR, Holman RR (1998) Risk factors for coronary artery disease in non-insulin dependent diabetes mellitus: United Kingdom Prospective Diabetes Study (UKPDS: 23). BMJ 316:823–828

Watts GF, Barrett PH, Ji J, Serone AP, Chan DC, Croft KD, Loehrer F, Johnson AG (2003) Differential regulation of lipoprotein kinetics by atorvastatin and fenofibrate in subjects with the metabolic syndrome. Diabetes 52:803–811

Williams KJ, Chen K (2010) Recent insights into factors affecting remnant lipoprotein uptake. Curr Opin Lipidol 21:218–228

Youssef W, Mccullough AJ (2002) Diabetes mellitus, obesity, and hepatic steatosis. Semin Gastrointest Dis 13:17–30

Zimmet P, Alberti KG, Shaw J (2001) Global and societal implications of the diabetes epidemic. Nature 414:782–787

Chapter 9
Gut Microbiota in Metabolic Syndrome

Fredrik Bäckhed

Abstract The human gut is colonized by trillions of bacteria that complement our own human genes with 150-fold more microbial genes. Acting at the intersection between diet and host biology the gut microbiota can be considered a metabolic organ that affects host physiology. Here we explore evidence that supports the hypothesis that there is a dynamic interrelationship between our diet and gut microbiota that plays an important role in nutrition as well as modulating energy balance. Furthermore, the gut microbiota is altered in obesity and germ-free mice have reduced adiposity and are protected against diet-induced obesity. Accordingly, the gut microbiota may be considered an important environmental factor that contributes to obesity and metabolic diseases.

Keywords Gut microbiota · Metabolism · Metagenome · Obesity

9.1 Introduction

We humans are much alike and our genomes are $>99\,\%$ identical and we all have approximately the same human cellular composition. However, despite this similarity we differ drastically in our microbial communities. The greatest variation is between body sites (Costello et al. 2009). For example, the difference in microbial communities between a person's mouth and gut is comparable to the difference in microbial communities that reside in soil and seawater. Even within a body site, the differences between people are not subtle and the gut communities can differ by 80–90 % at the species level (Turnbaugh et al. 2009; Costello et al. 2009).

The human gut microbiota is established at birth when our sterile gut is colonized by bacteria from our mothers and the environment (Reinhardt et al. 2009, Dominguez-Bello et al. 2010). This initial microbiota develops into a complex ecosystem in a predictable fashion determined by internal (e.g., oxygen depletion

F. Bäckhed (✉)
Sahlgrenska Center for Cardiovascular and Metabolic Research/Wallenberg Laboratory, Department of Molecular and Clinical Medicine, Sahlgrenska University Hospital, University of Gothenburg, Bruna Stråket 16, 413 45 Gothenburg, Sweden
e-mail: fredrik.backhed@wlab.gu.se

M. Orešič, A. Vidal-Puig (eds.), *A Systems Biology Approach to Study Metabolic Syndrome,*
DOI 10.1007/978-3-319-01008-3_9,

and the immune system) and external (e.g., mode of birth, impact of environment, diet, hospitalization, and application of antibiotics) factors. Whereas the initial gut microbiota exhibits relatively low diversity the adult microbiota is composed by $> 1{,}000$ bacterial species of which 160 are found in $> 50\,\%$ of the study population (Qin et al. 2010). The adult gut microbiota is estimated to weigh 1.5 kg, which is similar to the liver, and the microbiome contains 150–200 times as many genes as we have in our own human genome (Qin et al. 2010, Xu and Gordon 2003). The gut microbiota should be considered a metabolic organ that can affect our metabolism and physiology function on the intersection between our genetics and our ingested food. Accordingly, obesity is associated with an altered gut microbiota and germ-free mice have reduced adiposity and resistant to diet-induced obesity (Bäckhed et al. 2004, 2007; Ley et al. 2005, 2006). However, the underlying mechanisms are just being clarified and will require a combination of clinical and animal studies.

9.2 Microbial Extraction of Energy and Storage in Adipose Tissue

The gut microbiota contains an impressive suite of glycosidases and lyases that are essential for degrading otherwise non-degradable polysaccharides from the diet and ferment them to short chain fatty acids (SCFA) (Xu et al. 2003). Interestingly, the microbiota of obese mice exhibit increased capacity to degrade complex polysaccharides and store these extra calories in as adipose tissue (Turnbaugh et al. 2006). Similarly, germ-free mice have reduced capacity to degrade polysaccharides and the reduced levels of SCFA reaching the liver from portal circulation in these mice is associated with reduced lipogenesis (Bäckhed et al. 2004). In addition to salvage energy by fermenting carbohydrates to SCFA the gut microbiota promotes glucose absorption from the small intestine by a yet unidentified mechanism (Bäckhed et al. 2004). The increased levels of circulating SCFA and glucose act as substrates for *de novo* lipogenesis in the liver by activating the key lipogenic transcription factors sterol regulatory element binding protein 1c (SREBP-1c) and carbohydrate element binding protein (ChREBP). These transcription factors in turn regulate the rate limiting lipogenic enzymes fatty acid synthase and acetyl-CoA caboxylase (Bäckhed et al. 2004). The increased lipogenesis was confirmed when we applied LC-MS based lipidomics to investigate how the gut microbiota affects the host lipidome. We identified ~ 100 triglyceride species that were increased in livers of conventionally raised mice (Velagapudi et al. 2010). Furthermore, colonization was associated with increased hepatic VLDL production which is transported to the adipose tissue for storage (Bäckhed and Crawford 2010).

The gut microbiota does not only promote lipogenesis and VLDL production it also facilitate storage in adipose tissue by increasing lipoprotein lipase (LPL) activity. In white fat, LPL is regulated posttranscriptionally by nutritional status: fasting reduces and refeeding increases enzyme activity. Angiopoietin-like protein (Angptl4, also known as fasting-induced adipose factor) is an important regulator of LPL activity (Yoshida et al. 2002). Expression analysis revealed that the gut microbiota regulates the expression intestinal Angptl4 expression and analysis of germ-free *Angptl4-/-* mice revealed that the gut microbiota increase LPL activity by suppressing Angptl4 and that this suppression was associated with increased adiposity (Bäckhed et al. 2004, 2007). Accordingly, the gut microbiota has profound effects not only on energy extraction but also on energy partitioning and can thus be considered as an environmental factor that modulates adiposity.

9.3 Microbial Modulation of Host Metabolism

The gut microbiota does not only affect energy extraction but also affect host metabolism by other means. We found that germ-free mice are resistant to diet-induced obesity (Bäckhed et al. 2007), which suggests that the gut microbiota may have large effects on both central and peripheral metabolism. The observation that germ-free mice are protected from developing diet-induced obesity has recently been confirmed by several independent laboratories (Rabot et al. 2010; Fleissner et al. 2010; Ding et al. 2010) and some additional mechanistic insights are currently being made. Since 'Western' diet and other high-fat diets do not contain complex polysaccharides the gut microbiota promotes obesity by other means than carbohydrate fermentation and recent evidence suggested that the gut microbiota modulates several important pathways implicated in energy metabolism (Greiner and Backhed 2011).

The lean phenotype of germ-free mice is associated with increased levels of phosphorylated AMP-activated protein kinase (AMPK), and its downstream molecular targets involved in fatty acid oxidation (acetylCoA carboxylase; carnitinepalmitoyltransferase) in skeletal muscle and liver (Bäckhed et al. 2007). AMPK is an enzyme that functions as a "fuel gauge", which monitors cellular energy status (Kahn et al. 2005) and may thus be responsible for the ability of germ-free mice to increase fatty acid oxidation in skeletal muscle (Bäckhed et al. 2007). As mentioned above the gut microbiota suppresses the intestinal expression of Angptl4, which in addition to its effects on regulating LPL, also is a potent regulator of fatty acid oxidation (Mandard et al. 2006). In support of these findings we observed that germ-free mice lacking *Angptl4* lost their protection from developing diet-induced obesity (Bäckhed et al. 2007). Further phenotypic characterization revealed that genes involved in fatty acid oxidation were controlled by Angptl4 under germ-free conditions. Notably, Angptl4 did not signal via AMPK, which suggests that germ-free mice are protected against diet induced obesity by two complementary but independent mechanisms that result in increased fatty acid metabolism. In addition the gut microbiota may affect expression and function of other hormones that modulate host metabolism.

9.4 The Gut Microbiota and Obesity

The gut microbiota is altered in obesity but is there direct evidence that the gut microbiota may affect obesity? Elegant experiments revealed such link by transplanting the gut microbiota of lean and obese mice into germ-free recipients (Turnbaugh et al. 2006). Surprisingly, the lean and obese phenotypes were transferred to the recipients where mice transplanted with an obese microbiota gained significantly more adipose mass. A recent study identified that Toll-like receptor 5 (Tlr5), a pattern recognition receptor that recognizes the bacterial protein flagellin, as an important modulator of gut microbial ecology (Vijay-Kumar et al. 2010). Strikingly, the *Tlr5*-deficient mice are associated with inflammatory bowel disorder as well as obesity and the Metabolic Syndrome (Vijay-Kumar et al. 2007, 2010). Importantly, the obesity phenotype was transferrable to germ-free mice demonstrating that the phenotype was caused by the microbiota. Further analysis revealed that the altered gut microbiota in Tlr5-deficient mice was associated by increased food intake thus suggesting that the gut microbiota may affect the central nervous system. Furthermore, it is becoming increasingly clear that the gut microbiota modulates behavioral aspects of animal physiology. For example, the germ-free mice have increased locomotors activity and display altered anxiety behavior (Bäckhed et al. 2007; Heijtz et al. 2011). Accordingly, the gut microbiota does not only modulate peripheral metabolism but can also affect eating and behavioral phenotypes.

9.5 The Gut Microbiota as Modulator of Glucose Metabolism

In addition, to microbial effects on host adiposity the gut microbiota also contributes to metabolic abnormalities by promoting low-grade metabolic inflammation (Cani et al. 2007). Endotoxin is taken up from the gut together with chylomicrons or alternatively through increased gut permeability (reviewed in (Caesar et al. 2010)). Activation of TLR4 in macrophages that are recruited to the adipose tissue promotes inflammation that reduces insulin sensitivity (Saberi et al. 2009). Germ-free mice are exposed to very low levels of endotoxin from the diet but the levels are very low compared to those observed in colonized mice. As expected germ-free mice have reduced adipose inflammation and improved insulin sensitivity compared with colonized counterparts (Bäckhed et al. 2004; 2007; Rabot et al. 2010; Reigstad et al. 2009). Interestingly, antibiotic treatment of obese mice with antibiotics reduces plasma endotoxin levels, adipose inflammation, adiposity, liver triglycerides, and improves host glucose metabolism (Cani et al. 2008; Membrez et al. 2008). Taken together these findings suggest that the gut microbiota can contribute directly to host metabolism by affecting energy harvest from the diet and by modulating metabolic and/or inflammatory signaling pathways.

9.6 Systems Biology Approaches to Investigate Host Microbial Interactions

The gut microbiota has an immense propensity to affect the metabolism of ingested compounds such as dietary, drug, and xenobiotic components. Recent metabolomic and lipidomic studies of germ-free and colonized mice have revealed that the gut microbiota affects several important biotransformations that may have large effects on host physiology (Claus et al. 2008; Velagapudi et al. 2010; Wikoff et al. 2009). Untargeted mass spectrometry-based metabolomics of serum from germ-free and conventionally raised mice demonstrated that a vast majority of the serum metabolites were modulated by the gut microbiota (Wikoff et al. 2009). Many of these metabolites corresponded to increased xenobiotic metabolism in colonized mice and included metabolites that were sulfated, glucoronidated, or conjugated to glycine (Wikoff et al. 2009; Claus et al. 2008), which renders hydrophobic compounds water soluble. Tyrosine can be metabolized to *p*-cresol sulfate sulfatation by the gut microbiota and thus germ-free mice have reduced serum levels of *p*-cresol sulfate and corresponding elevated levels of tyrosine (Wikoff et al. 2009). The capacity of the gut microbiota to perform such modifications contributes to an individual's capacity to perform xenobiotic metabolism as individuals with high bacterially mediated p-cresol generation have reduced with O-sulfonation of widely used analgesic paracetamol (acetaminophen; (Clayton et al. 2009)). These findings may provide a mechanism for why pre-dose urinary composition can be used to predict the extent of liver damage sustained after paracetamol administration (Clayton et al. 2006).

Metabolism of other amino acids such as tryptophan is also modulated by the gut microbiota. Serum levels of tryptophan are accordingly reduced in colonized mice compared with germ-free counterparts (Velagapudi et al. 2010; Wikoff et al. 2009), which can be explained by that the gut microbiota expresses tryptphanase which converts tryptophan to indole, pyruvate, and ammonia. Further metabolism of indole to indole-3-propionic acid is exclusively microbially mediated and thus IPA was only detected in the serum of conventionally raised mice (Wikoff et al. 2009). IPA is further metabolized by host enzymes in the liver to indoxyl and indoxyl sulfate. Furthermore, serotonin is produced from tryptophan by enterochromaffin cells in the gut and accordingly germ-free rats have increased volumes of serotonin producing cells (Uribe et al. 1994). Taken together, microbial metabolism of amino acids may have profound effects on host physiology.

9.7 Using Metabolomics to Identify Microbial Metabolites in Disease

The above examples illustrate how the gut microbiota effects may affect host physiology by modulating metabolism of dietary components. Accordingly, metabolomic investigations of urine from mice predisposed or resistant to developing nonalcoholic fatty liver disease revealed that susceptible mice were associated with increased microbial conversion of phosphatidylcholine to trimethylamines (TMA), which are excreted into the urine. Mechanistically, the reduced levels of phosphatidylcholines may promote steatosis and insulin resistance since phosphatidylcholines are required for VLDL secretion and impaired VLDL secretion could lead to triglyceride accumulation in the liver. However, recent findings revealed that metabolism of dietary choline to TMA is associated with cardiovascular disease and plasma levels of choline-derived metabolites are a excellent predictor of cardiovascular disease (Wang et al. 2011). Accordingly, dysbalanced choline metabolism by altered gut microbial structure/function may affect host metabolism and physiology.

9.8 Targeted Approaches to Determine Microbial Regulation of Specific Classes of Metabolites

9.8.1 Bile Acids

Metabolism of bile acids is another example of mammalian-microbial co-metabolism that may have great physiological effects (Claus et al. 2008; Martin et al. 2007; Swann et al. 2010). The gut microbiota is important for deconjugation, dehydrogenation, and dehydroxylation of bile acids, which results in secondary bile acids and increases the chemical diversity of these signaling molecules (Midtvedt 1974). Accordingly, germ-free animals have very simplified bile acids characterized by conjugated cholic and murocholic acids, which are absorbed before they reach the distal gut (Claus et al. 2008; Swann et al. 2010; Wostmann 1973). Bile acids have different capacity to induce the nuclear receptor FXR and the G-coupled receptor TGR5 (Thomas et al. 2008). Furthermore, bile acids have been shown to regulate not only their own synthesis and enterohepatic recirculation, but also triglyceride and cholesterol homeostasis through activation of FXR (Sinal et al. 2000). Interestingly, bile acids reduce diet-induced obesity and prevent hyperglycaemia in rodents, which suggests that they also have effects on energy homeostasis (Ikemoto et al. 1997; Prawitt et al. 2011; Watanabe et al. 2011).

In contrast, bile acid activation of the GPCR TGR5 in brown adipose tissue increases energy expenditure by producing active tri-iodothyronine (T_3), which subsequently increases metabolic rate and energy expenditure (Watanabe et al. 2006). Accordingly, stimulation of TGR5 prevents diet-induced obesity (Thomas et al. 2009). Recent data demonstrated that TGR5 also is expressed by L-cells in the colon and activation of TGR5 in L-cells induce secretion of the incretin GLP-1, which promotes improved pancreas function and glucose metabolism in obese mice (Reimann et al. 2008; Thomas et al. 2009). However, it is currently unclear whether this protection requires TGR5 activation in brown adipose tissue or in L-cells.

The immense capacity of the gut microbiota to modulate bile acid diversity clearly indicates that signaling through FXR and TGR5 are influenced by the gut microbiota.

This was recently illustrated by that the gut microbiota modulated FXR-regulated pathway transcripts in gnotobiotic rats (Swann et al. 2010) and suggests that new pharmacological approaches targeting bile acid signaling networks must take the person's gut microbiota into account.

9.8.2 *Short Chain Fatty Acids*

Fermentation of fibers in the distal gut produces SCFAs, which has important signaling functions in the gut through the GPCRs GPR41 and GPR43. Enteroendocrine cells express the SCFA receptor GPR41 and may thus be regulated through the fermentation capacity of the gut microbiota. There are no apparent differences in the body composition of germ-free wildtype and *Gpr41*-deficient mice (Samuel et al. 2008). However, colonized *Gpr41*-deficient mice were leaner, which was associated with decreased expression of the hormone PYY. GPR43 was first identified as a modulator of inflammatory responses in the gut as a chemoattractant receptor on neutrophiles (Maslowski et al. 2009). Besides its effects on host inflammation GPR43 has profound effects on host physiology. Bjursell et al., recently found that *Gpr43*-deficient mice were resistant to diet-induced obesity (Bjursell et al. 2010). Protection against developing diet-induced obesity was, at least in part, explained by increased energy expenditure in *Gpr43*-deficient mice and the reduced fat mass was accompanied by improved glucose tolerance. Despite no difference in adipocyte size *Gp43-/-* mice contained fewer macrophages in the white adipose tissue, which may explain the improved glucose metabolism.

9.8.3 *Endocannabinoids*

Anandamide (AEA) is an endogenous endocannabinoid (eCB) that is synthesized from membrane bound phosphatidylethanolamine. AEA binds the eCB receptor CB_1, which has profound effects on host metabolism. Genetic and pharmacological interference with the CB_1 receptor protects against the development of hepatic steatosis and diet-induced obesity (Osei-Hyiaman et al. 2005, 2008). Recent results using several animal models demonstrated that the gut microbiota modulates AEA levels in the colon but not in the small intestine and that the elevated levels correlated with reduced expression of fatty acid amide hydrolase (FAAH), which degrades AEA (Muccioli et al. 2010).

Obesity is associated with increased production of adipose derived eCB, increased CB_1 expression, and increased gut permeability (Cani et al. 2007; Di Marzo 2008). Pharmacologic inhibition of CB_1 in obese mice improved tight junction, reduced gut permeability, leading to reduced serum endotoxin levels (Muccioli et al. 2010). Thus it is possible that the gut microbiota controls gut permeability through modulating the endocannabinoid system. As indicated above the gut microbiota has extensive effects on host adiposity, as well as on the eCB system in adipose tissue. Strikingly, prebiotic-mediated decreases in fat mass, reduced CB_1 mRNA expression, and reduced AEA levels were associated with increased expression of markers of adipocyte differentiation and lipogenesis suggesting a role of the gut microbiota. Importantly, these results were phenocopied by blocking CB_1 receptor signaling (Muccioli et al. 2010). Conversely, agonists of CB_1 increased adipogenesis in lean mice. Since CB_1 function is associated with plasma LPS levels Muccioli et al hypothesized that LPS could modulate eCB -induced adipogenesis. Accordingly, LPS reduced the expression of adipogenic and lipogenic markers in adipose tissue. Furthermore, LPS completely abolished the adipogenic effects of eCB receptor activation. Taken together, these data suggest that the eCB system regulates adipogenesis, which is controlled by LPS.

In summary, the gut microbiota affects several important signaling systems that contribute to controlling host metabolism these include, but are likely not limited to the SCFA, bile acid, and eCB system.

9.9 Conclusions

The recent identification of the gut microbiota as an environmental factor that contributes to metabolic diseases has spurred research all over the world. At present several studies have demonstrated an altered gut microbiota that is associated with obesity. However, the results are highly variable between studies, likely in part due to different analytical procedures, heterogeneous populations, and relatively small study groups. Accordingly, well controlled studies are required to parse out the relevant differences between the gut microbiota between lean and obese individuals, furthermore, the gut microbiota should also be studied in other metabolic diseases such as diabetes and cardiovascular disease. Accordingly, systems biology approaches to construct metabolic networks and putative functions encoded in the microbiome, which are associated with specific metabolic phenotypes in the host (including obesity, diabetes, and cardiovascular disease), will be of utmost importance to gain understanding in how the gut microbiota may affect host metabolism. However, one limitation with this approach is the lack of direct mechanistic understanding. An attractive tool to investigate mechanisms is to use 'personalized' gnotobiotics where germ-free mice are colonized with defined microbial communities and subsequently phenotyped. In these models interventions can be tested as well as the impact of specific macro- and micronutrients (Kau et al. 2011). The usage of genetically manipulated mouse models, especially tissue specific, will provide further evidence for host factors involved in specific signaling pathways or biological processes. However, to fully understand the intricate interplay between the gut microbiota, host, and metabolism we will need to integrate data from several 'omics' technologies such as metagenomics, transcriptomics, proteomics, and metabolomics

References

Bäckhed F, Crawford PA (2010) Coordinated regulation of the metabolome and lipidome at the host-microbial interface. Biochim Biophys Acta 1801:240–245

Bäckhed F, Ding H, Wang T, Hooper LV, Koh GY, Nagy A, Semenkovich CF, Gordon JI (2004) The gut microbiota as an environmental factor that regulates fat storage. Proc Natl Acad Sci U S A 101:15718–15723

Bäckhed F, Manchester JK, Semenkovich CF, Gordon JI (2007) Mechanisms underlying the resistance to diet-induced obesity in germ-free mice. Proc Natl Acad Sci U S A 104:979–984

Bjursell M, Admyre T, Goransson M, Marley AE, Smith DM, Oscarsson J, Bohlooly YM (2010) Improved glucose control and reduced body fat mass in free fatty acid receptor 2 (Ffar2) deficient mice fed a high fat diet. Am J Physiol Endocrinol Metab

Caesar R, Fåk F, Bäckhed F (2010) Effects of gut microbiota on obesity and atherosclerosis via modulation of inflammation and lipid metabolism. J Intern Med 268:320–328

Cani PD, Amar J, Iglesias MA, Poggi M, Knauf C, Bastelica D, Neyrinck AM, Fava F, Tuohy KM, Chabo C, Waget A, Delmee E, Cousin B, Sulpice T, Chamontin B, Ferrieres J, Tanti JF, Gibson GR, Casteilla L, Delzenne NM, Alessi MC, Burcelin R (2007) Metabolic endotoxemia initiates obesity and insulin resistance. Diabetes 56:1761–1772

Cani PD, Bibiloni R, Knauf C, Waget A, Neyrinck AM, Delzenne NM, Burcelin R (2008) Changes in gut microbiota control metabolic endotoxemia-induced inflammation in high-fat diet-induced obesity and diabetes in mice. Diabetes 57:1470–1481

Claus SP, Tsang TM, Wang Y, Cloarec O, Skordi E, Martin F-P, Rezzi S, Ross A, Kochhar S, Holmes E, Nicholson JK (2008) Systemic multicompartmental effects of the gut microbiome on mouse metabolic phenotypes. Mol Syst Biol 4:219

Clayton JJ, Baig W, Reynolds GW, Sandoe JA (2006) Endocarditis caused by Propionibacterium species: a report of three cases and a review of clinical features and diagnostic difficulties. J Med Microbiol 55:981–987

Clayton TA, Baker D, Lindon JC, Everett JR, Nicholson JK (2009) Pharmacometabonomic identification of a significant host-microbiome metabolic interaction affecting human drug metabolism. Proc Natl Acad Sci U S A 106:14728–14733

Costello EK, Lauber CL, Hamady M, Fierer N, Gordon JI, Knight R (2009) Bacterial community variation in human body habitats across space and time. Science 326:1694–1697

Di Marzo V (2008) Targeting the endocannabinoid system: to enhance or reduce? Nat Rev Drug Discov 7:438–455

Ding S, Chi MM, Scull BP, Rigby R, Schwerbrock NM, Magness S, Jobin C, Lund PK (2010) High-fat diet: bacteria interactions promote intestinal inflammation which precedes and correlates with obesity and insulin resistance in mouse. PLoS One 5:e12191

Dominguez-Bello MG, Costell EK, Contreras M, Magris M, Hidalgo G, Fierer N, Knight R (2010) Delivery mode shapes the acquisition and structure of the initial microbiota across multiple body habitats in newborns. Proc Natl Acad Sci U S A 107:11971–11975

Fleissner CK, Huebel N, Abd El-BaryMM, Loh G, Klaus S, Blaut M (2010) Absence of intestinal microbiota does not protect mice from diet-induced obesity. Br J Nutr 104:919–929

Greiner T, Backhed F (2011) Effects of the gut microbiota on obesity and glucose homeostasis. Trends Endocrinol Metab 22:117–123

Heijtz RD, Wang S, Anuar F, Qian Y, Bjorkholm B, Samuelsson A, Hibberd ML, Forssberg H, Pettersson S (2011) Normal gut microbiota modulates brain development and behavior. Proc Natl Acad Sci U S A 108:3047–3052

Ikemoto S, Takahashi M, Tsunoda N, Maruyama K, Itakura H, Kawanaka K, Tabata I, Higuchi M, Tange T, Yamamoto TT, Ezaki O (1997) Cholate inhibits high-fat diet-induced hyperglycemia and obesity with acyl-CoA synthetase mRNA decrease. Am J Physiol 273:E37–45

Kahn BB, Alquier T, Carling D, Hardie DG (2005) AMP-activated protein kinase: ancient energy gauge provides clues to modern understanding of metabolism. Cell Metab 1:15–25

Kau AL, Ahern PP, Griffin NW, Goodman AL, Gordon JI (2011) Human nutrition, the gut microbiome and the immune system. Nature 474:327–336

Ley RE, Bäckhed F, Turnbaugh P, Lozupone CA, Knight RD, Gordon JI (2005) Obesity alters gut microbial ecology. Proc Natl Acad Sci U S A 102:11070–11075

Ley RE, Turnbaugh PJ, Klein S, Gordon JI (2006) Microbial ecology: human gut microbes associated with obesity. Nature 444:1022–1023

Mandard S, Zandbergen F, Straten E van, Wahli W, Kuipers F, Muller M, Kersten S (2006) The fasting-induced adipose factor/angiopoietin-like protein 4 is physically associated with lipoproteins and governs plasma lipid levels and adiposity. J Biol Chem 281:934–944

Martin FP, Dumas ME, Wang Y, Legido-Quigley C, Yap IK, Tang H, Zirah S, Murphy GM, Cloarec O, Lindon, JC, Sprenger N, Fay LB, Kochhar S, Bladeren P van, Holmes E, Nicholson JK (2007) A top-down systems biology view of microbiome-mammalian metabolic interactions in a mouse model. Mol Syst Biol 3:112

Maslowski KM, Vieira AT, Ng A, Kranich J, Sierro F, Yu D, Schilter HC, Rolph MS, Mackay F, Artis D, Xavier RJ, Teixeira MM, Mackay CR (2009) Regulation of inflammatory responses by gut microbiota and chemoattractant receptor GPR43. Nature 461:1282–1286

Membrez M, Blancher F, Jaquet M, Bibiloni R, Cani PD, Burcelin RG, Corthesy I, Mace K, Chou CJ (2008) Gut microbiota modulation with norfloxacin and ampicillin enhances glucose tolerance in mice. FASEB J 22:2416–2426

Midtvedt T (1974) Microbial bile acid transformation. Am J Clin Nutr 27:1341–1347

Muccioli GG, Naslain D, Bäckhed F, Reigstad CS, Lambert DM, Delzenne NM, Cani PD (2010) The endocannabinoid system links gut microbiota to adipogenesis. Mol Syst Biol 6:392

Osei-Hyiaman D, Depetrillo M, Pacher P, Liu J, Radaeva S, Batkai S, Harvey-White J, Mackie K, Offertaler L, Wang L, Kunos G (2005) Endocannabinoid activation at hepatic CB_1 receptors stimulates fatty acid synthesis and contributes to diet-induced obesity. J Clin Invest 115:1298–1305

Osei-Hyiaman D, Liu J, Zhou L, Godlewski G, Harvey-White J, Jeong WI, Batkai S, Marsicano G, Lutz B, Buettner C, Kunos G (2008) Hepatic CB_1 receptor is required for development of diet-induced steatosis, dyslipidemia, and insulin and leptin resistance in mice. J Clin Invest 118:3160–3169

Prawitt J, Abdelkarim M, Stroeve JH, Popescu I, Duez H, Velagapudi VR, Dumont J, Bouchaert E, Dijk TH van, Lucas A, Dorchies E, Daoudi M, Lestavel S, Gonzalez FJ, Oresic M, Cariou B, Kuipers F, Caron S, Staels B (2011) Farnesoid X receptor deficiency improves glucose homeostasis in mouse models of obesity. Diabetes

Qin J, Li R, Raes J, Arumugam M, Burgdorf KS, Manichanh C, Nielsen T, Pons N, Levenez F, Yamada T, Mende DR, Li J, Xu J, Li S, Li D, Cao J, Wang B, Liang H, Zheng H, Xie Y, Tap J, Lepage P, Bertalan M, Batto JM, Hansen T, Le Paslier D, Linneberg A, Nielsen HB, Pelletier E, Renault P, Sicheritz-Ponten T, Turner K, Zhu H, Yu C, Jian M, Zhou Y, Li Y, Zhang X, Qin N, Yang H, Wang J, Brunak S, Dore J, Guarner F, Kristiansen K, Pedersen O, Parkhill J, Weissenbach J, Bork P, Ehrlich SD (2010) A human gut microbial gene catalogue established by metagenomic sequencing. Nature 464:59–65

Rabot S, Membrez M, Bruneau A, Gerard P, Harach T, Moser M, Raymond F, Mansourian R, Chou CJ (2010) Germ-free C57BL/6J mice are resistant to high-fat-diet-induced insulin resistance and have altered cholesterol metabolism. FASEB J

Reigstad CS, Lunden GO, Felin J, Bäckhed F (2009) Regulation of serum amyloid A3 (SAA3) in mouse colonic epithelium and adipose tissue by the intestinal microbiota. PLoS One 4:e5842

Reimann F, Habib AM, Tolhurst G, Parker HE, Rogers GJ, Gribble FM (2008) Glucose sensing in L cells: a primary cell study. Cell Metab 8:532–539

Reinhardt C, Reigstad CS, Bäckhed F (2009) Intestinal microbiota during infancy and its implications for obesity. J Pediatr Gastroenterol Nutr 48:249–256

Saberi M, Woods NB, De Luca C, Schenk S, Lu JC, Bandyopadhyay G, Verma IM, Olefsky JM (2009) Hematopoietic cell-specific deletion of toll-like receptor 4 ameliorates hepatic and adipose tissue insulin resistance in high-fat-fed mice. Cell Metab 10:419–429

Samuel BS, Shaito A, Motoike T, Rey FE, Bäckhed F, Manchester JK, Hammer, RE, Williams SC, Crowley J, Yanagisawa M, Gordon JI (2008) Effects of the gut microbiota on host adiposity are modulated by the short-chain fatty-acid binding G protein-coupled receptor, Gpr41. Proc Natl Acad Sci U S A 105:16767–16772

Sinal CJ, Tohkin M, Miyata M, Ward JM, Lambert G, Gonzalez FJ (2000) Targeted disruption of the nuclear receptor FXR/BAR impairs bile acid and lipid homeostasis. Cell 102:731–744

Swann JR, Want EJ, Geier FM, Spagou K, Wilson ID, Sidaway JE, Nicholson JK, Holmes E (2010) Microbes and Health Sackler Colloquium: Systemic gut microbial modulation of bile acid metabolism in host tissue compartments. Proc Natl Acad Sci U S A

Thomas C, Pellicciari R, Pruzanski M, Auwerx J, Schoonjans K (2008) Targeting bile-acid signalling for metabolic diseases. Nat Rev Drug Discov 7:678–693

Thomas C, Gioiello A, Noriega L, Strehle A, Oury J, Rizzo G, Macchiarulo A, Yamamoto H, Mataki C, Pruzanski M, Pellicciari R, Auwerx J, Schoonjans K (2009) TGR5-mediated bile acid sensing controls glucose homeostasis. Cell Metab 10:167–177

Turnbaugh PJ, Ley RE, Mahowald MA, Magrini V, Mardis ER, Gordon JI (2006) An obesity-associated gut microbiome with increased capacity for energy harvest. Nature 444:1027–1031

Turnbaugh PJ, Hamady M, Yatsunenko T, Cantarel BL, Duncan A, Ley RE, Sogin ML, Jones WJ, Roe BA, Affourtit JP, Egholm M, Henrissat B, Heath AC, Knight R, Gordon JI (2009) A core gut microbiome in obese and lean twins. Nature 457:480–484

Uribe A, Alam M, Johansson O, Midtvedt T, Theodorsson E (1994) Microflora modulates endocrine cells in the gastrointestinal mucosa of the rat. Gastroenterology 107:1259–1269

Velagapudi VR, Hezaveh R, Reigstad CS, Gopalacharyulu PV, Yetukuri L, Islam S, Felin J, Perkins R, Boren J, Oresic M, Bäckhed F (2010) The gut microbiota modulates host energy and lipid metabolism in mice. J Lipid Res 51:1101–1112

Vijay-Kumar M, Sanders CJ, Taylor RT, Kumar A, Aitken JD, Sitaraman SV, Neish AS, Uematsu S, Akira S, Williams IR, Gewirtz AT (2007) Deletion of TLR5 results in spontaneous colitis in mice. J Clin Invest 117:3909–3921

Vijay-Kumar M, Aitken JD, Carvalho FA, Cullender TC, Mwangi S, Srinivasan S, Sitaraman SV, Knight R, Ley RE, Gewirtz AT (2010) Metabolic syndrome and altered gut microbiota in mice lacking Toll-like receptor 5. Science 328:228–231

Wang Z, Klipfell E, Bennett BJ, Koeth R, Levison BS, Dugar B, Feldstein AE, Britt EB, Fu X, Chung Y-M, Wu Y, Schauer P, Smith JD, Allayee H, Tang WHW, Didonato JA, Lusis AJ, Hazen SL (2011) Gut flora metabolism of phosphatidylcholine promotes cardiovascular disease. Nature 472:57–63

Watanabe M, Houten SM, Mataki C, Christoffolete MA, Kim BW, Sato H, Messaddeq N, Harney JW, Ezaki O, Kodama T, Schoonjans K, Bianco AC, Auwerx J (2006) Bile acids induce energy expenditure by promoting intracellular thyroid hormone activation. Nature 439:484–489

Watanabe M, Horai Y, Houten SM, Morimoto K, Sugizaki T, Arita E, Mataki C, Sato H, Tanigawara Y, Schoonjans K, Itoh H, Auwerx J (2011) Lowering bile acid pool size with a synthetic FXR agonist induces obesity and diabetes through reduced energy expenditure. J Biol Chem

Wikoff WR, Anfora AT, Liu J, Schultz PG, Lesley SA, Peters EC, Siuzdak G (2009) Metabolomics analysis reveals large effects of gut microflora on mammalian blood metabolites. Proc Natl Acad Sci U S A 106:3698–3703

Wostmann BS (1973) Intestinal bile acids and cholesterol absorption in the germfree rat. J Nutr 103:982–990

Xu J, Gordon JI (2003) Inaugural Article: Honor thy symbionts. Proc Natl Acad Sci U S A 100:10452–10459

Xu J, Bjursell MK, Himrod J, Deng S, Carmichael LK, Chiang HC, Hooper LV, Gordon JI (2003) A genomic view of the human-*Bacteroides thetaiotaomicron* symbiosis. Science 299:2074–2076

Yoshida K, Shimizugawa T, Ono M, Furukawa H (2002) Angiopoietin-like protein 4 is a potent hyperlipidemia-inducing factor in mice and inhibitor of lipoprotein lipase. J Lipid Res 43:1770–1772

Part III
Key Technologies to Study Metabolic Syndrome

This section introduces some of the key technologies used to tackle metabolic syndrome, and review how they have been applied using systems biology approach.

Chapter 10
Proteomics in the Systems-Level Study of the Metabolic Syndrome

Isabel Rubio-Aliaga, Irma Silva-Zolezzi, Michael Affolter, Loïc Dayon, Alexandre Panchaud and Martin Kussmann

Abstract Proteomics has come a long way from the initial qualitative analysis of proteins present in a given sample to the large-scale characterization of proteomes, their interactions and dynamic behaviour in time and space. Originally enabled by breakthroughs in protein separation and visualization (by two-dimensional gels electrophoresis) and protein identification (by mass spectrometry), the discipline now encompasses a large body of protein and peptide separation, labelling, detection and sequencing tools supported by stable-isotope- and label-free techniques and computational data processing for quantitative proteomics. The key functional importance to investigate the protein complement has driven the study of proteomes in numerous physiological and pathological conditions. Proteomics has been mainly applied in discovering novel biomarkers of disease, evidenced by the fact that most clinical tests today measure proteins in blood. Moreover, understanding the proteome helps investigate health and disease states and understand the mechanisms of action of specific molecules, e.g., nutrients or drugs. Here, we briefly recapitulate proteomic technologies and cover their evolution to today's and future cutting-edge platforms. Then we review and discuss proteomic applications to the study of the metabolic syndrome.

Keywords Proteomics · Diabetes · Obesity · Metabolic syndrome · Mass spectrometry · Systems biology

---

M. Kussmann (✉) · L. Dayon
Molecular Biomarkers Core, Nestlé Institute of Health Sciences SA,
EPFL Campus Quartier de l'innovation, bâtiment H, 1015 Lausanne, Switzerland
e-mail: martin.kussmann@rd.nestle.com

L. Dayon
e-mail: loic.dayon@rd.nestle.com

M. Kussmann
Faculty of Life Sciences, Ecole Polytechnique Fédérale Lausanne (EPFL),
Lausanne, Switzerland

Faculty of Science, Aarhus University, Aarhus, Denmark

I. Rubio-Aliaga · I. Silva-Zolezzi · M. Affolter · A. Panchaud
BioAnalytical Science Department, Nestlé Research Centre,
Vers-chez-les-Blanc, 1000, Lausanne, Switzerland

M. Orešič, A. Vidal-Puig (eds.), *A Systems Biology Approach to Study Metabolic Syndrome*, DOI 10.1007/978-3-319-01008-3_10, © Springer International Publishing Switzerland 2014

10.1 Proteomics, a Key Technology in Systems Biology

Proteomics is the large scale measurement of the proteome (the protein complement of a cell, organ or even organism) at a given time and state. In systems biology, proteomics complements genomics, transcriptomics and metabolomics, to holistically investigate biological phenomena at the molecular level. Main challenges in proteomics are: (i) the interrogation of not only the expression of proteins, but also their interactions, functions, modifications and structures; (ii) the large variety of proteins found in biological samples due to post-translational modifications; and (iii) the dynamic range of the proteome, i.e., the wide concentration range of proteins, especially in plasma. This complexity has led to the development of a multitude of technologies for large-scale measurements of proteins. Previously, proteins had been classically studied individually or in small numbers. Advances in two key technologies have contributed to the rapid rise of proteomics : the development of soft ionization techniques for mass spectrometry, namely electrospray ionisation (ESI) and matrix-assisted laser desorption/ionisation (MALDI), enabling sequencing of proteins and peptides; and advanced liquid chromatography (LC) (Mallick and Kuster 2010). In this first part an overview of the status, advances and perspectives of proteomic technologies and approaches in biology is given.

10.1.1 Proteomic Concepts and Challenges

The "proteomic gold rush" in the later 1990's and early 2000's was mainly justified by several findings and assets that make proteomics an essential activity in life science research:

i. The availability of complete genomic sequences is not sufficient to derive biological function.
ii. Proteomics complements genomics by focusing on the gene products active in a cell, and therefore delivers information closer to the observed phenotype.
iii. Proteins are involved in all biological processes, such as trafficking, signalling, catalysis and they have structural functions; therefore, proteins are highly informative to understand biology.
iv. Gene expression and protein abundance do not always correlate linearly and, therefore, the proteins have to be determined directly (Gygi et al. 1999b).
v. Even with modern bioinformatic algorithms prediction of gene products from genomic data remains difficult; hence, verification of a gene product by proteomic methods is important for annotation.
vi. Prediction of protein modification or localization is barely possible at the DNA level; hence proteomics is indispensable for the elucidation of protein isoforms, post-translational modifications and gene product localization.
vii. Protein regulation mechanisms, protein-protein interactions or protein complex structures can be only determined at the protein level.

Unfortunately, during the last two decades, the three hopes set on proteomics, namely (i) a complete map of the human proteome; (ii) identification of many new clinical biomarkers; and (iii) the development of new drug candidates, have only been partly fulfilled. Mainly two factors have tempered these hopes:

i. The complexity of the proteome.
ii. The dynamics of the proteome.

In comparison to the human genome with approximately 25,000 genes, the proteome has been estimated to encompass around 1 million different protein species (Jensen 2004). This divergence in quantity is due to more than 300 post-translational (e.g., protein lysine acetylation) and post-transcriptional modifications (e.g., phosphorylation). Moreover, proteogenomics has emerged, i.e., the identification of protein coding genes by proteomics, suggesting that the amount of coding genes may be higher than assumed (Ahrens et al. 2010). This complexity of the proteome is further augmented by the fact that proteins are rarely functioning as single molecules but rather as complexes with other proteins or smaller molecules. In addition to this complexity, proteins in biological samples display an enormous concentration range. In human plasma, the estimated protein concentration range is 10^{10}–10^{12}, and represents 6–8 orders of magnitude difference in human cells (Lescuyer et al. 2007). Additionally, the hallmark of a proteome is its ability to regulate protein expression dynamically in response to physiological, pathophysiological and environmental conditions. This complexity renders proteomics a challenging task and the fact that protein analysis lacks the equivalent of polymerase chain reaction (PCR) amplification does not make it simpler (Domon and Broder 2004).

Despite these challenges, remarkable progress has been achieved in the last decades, positioning proteomics as a leading technology in life sciences research. Depletion of abundant and/or enrichment of less prominent proteins have been combined and extensive protein and peptide pre-fractionation techniques are in place to achieve deeper proteome coverage. In addition, current mass spectrometric (MS) platforms can cover a wider dynamic range due to improved instrumentation that offers better sensitivity and specificity, the latter based on superior mass accuracy and resolution (e.g., Orbitrap analyzer (Hu et al. 2005)). Finally, excellent software for peptide sequencing and protein identification is in place (e.g., Phenyx (Colinge et al. 2003), OMSSA (Geer et al. 2004), X!Tandem (Craig and Beavis 2004) or VEMS (Matthiesen et al. 2003; Matthiesen et al. 2004; Matthiesen et al. 2005)). The computing tools for proteomic data processing have evolved from scoring mass spectra with regard to their fidelity for peptide sequence read out to assessing the trade-off between false-positives (specificity) and false-negatives (sensitivity) in a data-dependent manner, eventually leading to computational data validation (PeptideProphet (Keller et al. 2002) and ProteinProphet (Nesvizhskii et al. 2003)).

Yet, we are far from looking at the totality of the plasma proteome, to mention the clinically most relevant sample for protein biomarkers. Filling the clinical biomarker and drug discovery pipelines has not become a reality because of the difficulty to translate protein identifications into candidate markers and these into targets that deserve follow-up. Additionally, many of the marker/target validation technologies are still low-throughput and require multi-centre collaboration: *in vivo* validation takes much longer than *in vitro* discovery (Alaiya et al. 2005). Moreover, the pay-off for taking a protein biomarker to the bed-side test has so far been small in economic terms (Alaiya et al. 2005). However, the fact that many if not most clinical tests today measure proteins should encourage further efforts in the use and improvement of proteomics-based biomarker discovery strategies. Probably one of the better known examples is the use of liver transaminases for liver cell destruction (Amacher 1998). A few successful proteomics-derived biomarkers can be cited: (i) The 14-3-3 proteins in cerebrospinal fluid (CSF) as markers for types of brain destruction such as those found in Creutzfeldt-Jakob Disease (CJD) (Hsich et al. 1996); (ii) stroke biomarkers discovered in the spinal fluid that have been clinically validated in serum (Lescuyer et al. 2007). Recently, the Food and Drug Administration (FDA) approved the first proteomics-discovered diagnostic test, OVA1, a panel of proteins that allows the identification of malignant ovarian tumors preoperatively. Further technological advances allow the reliable and consistent measurement of plasma proteins in such low concentrations as nanogram per millilitre in high-throughput matter; together with well-established guidelines for designing clinical studies, standardized procedures and quality assessment should improve the efficiency of clinical biomarker studies using proteomics in the near future (Surinova et al. 2011).

10.1.2 Protein and Peptide Analysis

No general method enables the identification and quantification of all proteins in a given sample by using one single technology, in contrast to DNA microarray (Schena et al. 1995), where the whole gene expression profile can be obtained from the total extracted mRNA with automated systems (Raymond et al. 2006). While genome-wide microarrays are available (MacBeath and Schreiber 2000; Ptacek et al. 2005), proteomics laboratories have developed and tested a wide range of alternative techniques and use an ample biochemical toolbox to achieve their goals.

The typical methodology applied in proteomics involves protein extraction from biological samples, digestion of the proteins typically with trypsin, followed by extensive fractionation of the resulting peptides, and analysis of these in each fraction using a mass spectrometer (Ahrens et al. 2010). There are five basic types of mass analyzers currently used in proteomics (Aebersold and Mann 2003; Ptacek et al. 2005) : ion trap (IT), time-of-flight (TOF), quadrupole (Q), Fourier-transform ion cyclotron resonance (FT-ICR), and Orbitrap (OT) (Hu et al. 2005; Ptacek et al. 2005). Often, they work as stand-alone mass analyzer, but the current trend is to combine the advantages of different analyzers in one hybrid mass spectrometer: triple-Q, Q-IT, Q-TOF, IT-TOF, TOF-TOF, IT-FT-ICR, IT-OT or Q-OT tandem mass spectrometers are all capable of protein or peptide sequencing. IT-FT-ICRs and IT-OTs are especially efficient when combined with fragmentation techniques such as electron-capture dissociation (ECD) or electron-transfer dissociation (ETD) (Ptacek et al. 2005; Syka et al. 2004a).

Apart from the traditional and increasingly phased out 2D gel electrophoresis (GE), two main concepts are currently applied to deal with proteome complexity and dynamic range: extensive fractionation using separation techniques such as multi-dimensional protein identification technology (MudPIT) (Washburn et al. 2001) or enrichment/depletion using affinity-based techniques (Righetti et al. 2005; Righetti et al. 2006; Righetti and Boschetti 2007).

Most of LC tandem mass spectrometry (MS/MS)-based proteomic studies rely on on-line LC-ESI systems. However, to combine features of the MALDI technique, where peptides can be immobilized on a target plate, with the separation power of liquid chromatography, LC-MALDI was developed in an automated off-line mode (Zhen et al. 2004). Comparisons of LC-ESI-MS/MS and LC-MALDI-MS/MS have revealed the complementarities between the two strategies.

10.1.3 Protein and Peptide Identification and Validation

In MS/MS a precursor ion with known mass selected from the survey scan is fragmented and provides sequence data, which enables more confident identification of peptides. This process is described as peptide mass sequencing (PMS) as opposed to PMF (peptide mass fingerprinting, i.e., using sets of peptide masses only). Identification of proteins using MS/MS data is nowadays performed using three different approaches: (i) peptide sequence tag (Mann et al. 1993; Mann and Wilm 1994; Sunyaev et al. 2003; Tabb et al. 2003); (ii) cross-correlation method (Eng et al. 1994; Hansen et al. 2001); and (iii) probability-based matching (Colinge et al. 2003; Craig and Beavis 2004; Matthiesen et al. 2003; Matthiesen et al. 2004; Matthiesen et al. 2005; Perkins et al. 1999).

The simple case scenario in terms of identification would be to use only the mass information of a peptide as a unique signature without the need to spend time for MS/MS. Such approaches have already been described as the "accurate mass tag" (AMT) approach (Conrads et al. 2000; Zubarev et al. 1996). In this technique, instruments providing sub-ppm (0.1 ppm) mass accuracy are needed. However, even with such accuracy, high levels of confidence in protein identifications can only be achieved in small eukaryotic systems (e.g., yeast). Additional information such as isoelectric point or LC elution time are needed to confidently identify a protein without using MS/MS (Cargile and Stephenson 2004; Norbeck et al. 2005; Palmblad et al. 2004).

Liquid chromatography combined with MS/MS is the preferred and most established method to conduct large-scale characterization of proteomes, together with peptide mass sequencing search engines. One of the major problems of such sequence search engines is that they return false positive results. In the case of small datasets, this can be solved by manually verifying and validating the spectrum-to-peptide assignment. In the case of large datasets with tens of thousands of spectra, such a time-consuming approach is not feasible. Two main strategies have been developed to meet this challenge: the first is based on a robust and accurate statistical model exemplified by PeptideProphet and ProteinProphet in the Trans-Proteomic Pipeline (Keller et al. 2002; Nesvizhskii et al. 2003). Each spectrum-to-peptide assignment is compared with machine-learning techniques to all other assignments in the datasets and computes for each spectrum-to-peptide assignment a probability of being correct. The second strategy relies on database search using a target-decoy database (Elias et al. 2005; Elias and Gygi 2007; Peng et al. 2003; Perkins et al. 1999) : first an appropriate "target" protein sequence database is generated and then, a "decoy" database typically generated by reversing or scrambling the target protein sequences, is created. The search is done against both the target and the composite databases. One can estimate the total number of false positives (FPs) assuming that no correct peptides are found in both the target and decoy entities and that incorrect assignments from target or decoy sequences are equally likely. Neither one of the approaches can claim to remove all false positives but both help researchers to share, compare and publish data with a defined false discovery rate (Lisacek et al. 2006; Matthiesen 2007).

10.1.4 Protein and Peptide Quantification

In proteomics we distinguish between relative quantification, defined as the amount of a protein relative to another measurement of the same protein in another state (e.g., protein expression changes after drug treatment or nutritional intervention), and absolute quantification, which delivers the absolute amount of proteins in a given sample.

All current discovery proteomic methods are relative methods. Absolute quantification of peptides as protein surrogates can be achieved with isotopically labelled synthetic peptides and mass spectrometry, a method known as AQUA (Gerber et al. 2003). Yet, the availability of such standard peptides in defined amounts is a limited. A way to partially overcome this bottleneck is the QCAT (Beynon et al. 2005) or QconCAT (Pratt et al. 2006) technique, in which an artificial gene is used for expression, labelling and purification of a corresponding artificial protein, which represents a concatemer of tryptic peptides representing several known proteins. However such method requires *a priori* knowledge of the target proteins and the preparation of isotopically labelled synthetic peptides for each targeted protein. MS peptide detection has a Boolean nature, and therefore, so called "proteotypic" peptides—peptides being a unique identifier and optimally representing the protein in chromatography and MS behaviour—have to be synthesized to ensure absolute quantification at global scale (Kuster et al. 2005; Mallick et al. 2007).

While still being widely deployed, "gel-based" quantitative proteomics has been largely superseded by "gel-free" MS-based quantitative proteomics approaches. As in the gel approach, where the protein staining intensity within a gel is not necessarily proportional to the amount present in the sample, in both MALDI- and ESI-MS the relationship between the amount of protein or peptides present and the measured signal intensity is incompletely understood. Moreover, the reproducibility of a peptide/protein signal between different runs is limited. To improve quantification performance of mass spectrometers different approaches have been developed:

A first solution came with the technique of stable-isotope dilution (SID). Pairs of chemically identical molecules (in this case peptide pairs), but with different stable-isotope composition (^{13}C instead of ^{12}C, ^{2}H instead of ^{1}H, ^{18}O instead of ^{16}O or ^{15}N instead of ^{14}N) can be differentiated in a mass spectrometer owing to their mass difference. Thus, the ratio of signal intensities should be a direct and accurate measure of the abundance ratio between the two peptides/proteins. Three main approaches exist today: (i) metabolic stable-isotope labelling, (ii) isotope tagging by chemical reaction and, (iii) stable-isotope incorporation via enzyme reaction.

A well established metabolic method is stable isotope labelling with amino acids in cell culture (SILAC) (Ong et al. 2002), in which amino acids containing stable isotopes, like arginine with six ^{13}C atoms, are supplied in growth media. Several amino acids have been used such as leucine, which labels 70 % of tryptic peptides (Foster et al. 2003). Taking advantage of trypsin preferentially cleaving proteins at arginine and lysine residues at the C-terminus, the incorporation of heavy lysine and arginine combined with subsequent tryptic digestion results in stable-isotope labelling of all peptides (Ibarrola et al. 2003). The principal advantages of metabolic labelling over chemical labelling are the earliest possible introduction of the label into the living cells, immediate pooling of case and control and the concomitant reduction of parallel sample preparation bias and the absence of chemical side reactions. These methods can be applied to cultured cells like bacteria or yeast, and also to higher organisms by incorporating the labelled amino acids in their food. Examples include *Caeonorhabditis elegans*, *Drosophila melanogaster* (Krijgsveld et al. 2003), plants (Ippel et al. 2004) or even a rat by using ^{15}N-labeled algae (Wu et al. 2004).

All isotopically labelled chemical reagents target reactive sites on a protein or peptide and the two proteomes to be compared are labelled with the light and heavy reagent, respectively. Isotope-coded affinity tagging (ICAT) was the first approach described in 1999 (Gygi et al. 1999a). A recent evolution of this method is called catch-and-release (CAR) (Gartner et al. 2007). Isotope-coded protein labelling (ICPL) (Schmidt et al. 2005), targets all amino groups at the protein level. Isobaric tags for relative and absolute quantification (iTRAQ) (Ross et al. 2004) use the same chemistry as ICPL (labelling at protein or peptide level) but generates a specific reporter ion for quantification in tandem mass spectra (mass 114, 115, 116, 117) and this with isobaric mass at MS level. Multiplexing (currently eight-plex) is an interesting feature of this technique as it allows comparing more than two conditions.

Carboxylic groups have also been labelled using either methyl (Goodlett et al. 2001) or ethyl (Syka et al. 2004b) esterification at peptide level. Yet, both methods use deuterium atoms and therefore bear the risk of chromatographic discrimination; plus the mass offsets of 2 Da (methyl) and 4 Da (ethyl) pose problems of isotopic overlap of the peptide pairs. A technique called AniBAL (Panchaud et al. 2008) labels both amino and carboxylic groups at the protein level using two symmetrical tags with six times ^{13}C, thereby addressing the previously mentioned caveats.

A clear advantage of these chemical approaches is the availability of functional groups in proteins allowing the design of almost any kind of quantitative tag. Enrichment is also an asset, as it reduces sample complexity without losing quantitative information. However, reactions must be specific, proceed to completion, minimize sample handling and avoid side reactions. Despite these constraints, chemical stable-isotope labelling has contributed importantly to quantitative proteome data, mainly due to its chemical versatility and its applicability to any biological sample.

Promising "label-free" approaches have emerged in quantitative proteomics. The peptide match score summation (PMSS) is semi-quantitative and based on the assumption that a protein score is the sum of identification scores of its peptides and that the protein score is correlated with abundance (Allet et al. 2004). Another similar approach relies on counting the spectra that identify a protein, the so-called spectrum sampling (SpS) (Liu et al. 2004). Protein abundance indices (PAIs) are believed to be more reliable as they are based on observable parameters, e.g., the number of peptides identifying a protein that rises with increasing protein amount. As a larger protein will statistically generate more measurable peptides than a smaller one, a simple PAI can be derived by normalizing the number of observed peptides with the number of observable peptides for the protein under consideration (Rappsilber et al. 2002; Sanders et al. 2002). Ishihama and colleagues have described an exponentially modified PAI (emPAI) by observing a logarithmic relationship between the number of peptides observed and the protein amount within a given sample (Ishihama et al. 2005).

Two proteomes can be directly compared label-free when analyzed one after the other and in exactly the same way (Chelius and Bondarenko 2002; Lasonder et al. 2002). Software exists to extract the intensities of the same peptide observed in two separate runs to compare and determine their relative abundance (e.g., MSight (Palagi et al. 2005), SuperHirn (Mueller et al. 2007), MapQuant (Leptos et al. 2006), SpecArray (Li et al. 2005) or VEMS (Matthiesen et al. 2004; Matthiesen et al. 2005; Matthiesen 2007)). A clear advantage of such method is the absence of any label and the applicability to any type of instrument. Clear drawbacks are multiple occasions for quantification errors occurring during parallel sample processing, analysis and the need for very accurate and reproducible replicate LC-MS runs.

10.1.5 Uncoupling Intact Mass Acquisition from Sequencing and Transferring Peptide Separation from the Liquid to the Gas Phase

Figure 10.1 summarizes contemporary techniques to (pre-)separate, mass-measure, sequence and quantify proteomes and peptidomes. Essentially, today's proteomics starts with a tissue, cell or body fluid sample which is subsequently decomplexified, typically at both biological (organelle isolation) and biochemical (protein/peptide fractionation) level. Separation has traditionally consisted of protein-level gel-based techniques, which are increasingly outperformed and replaced by peptide-level LC

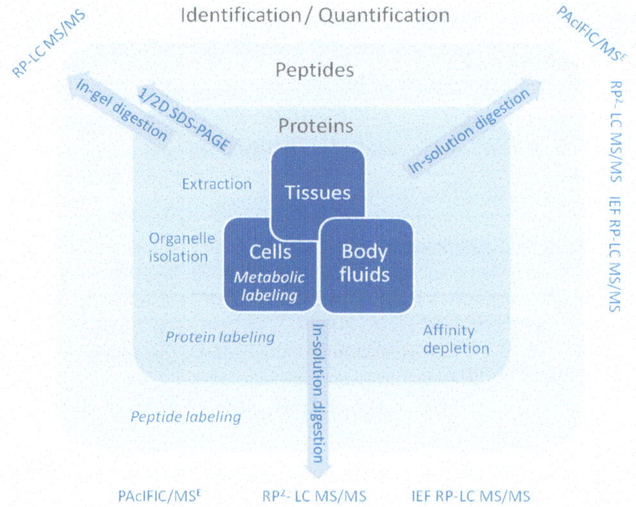

Fig. 10.1 Schematic overview on modern proteomics workflows including sample preparation, protein/peptide fractionation, cell/protein/peptide-level stable-isotope labeling, and mass spectrometric (*MS*) data acquisition. (*IEF*) isoelectric focusing; (*LC*) liquid chromatography; MS^E: precursor-independent acquisition of intact and fragment masses; MS/MS: tandem mass spectrometry; (*PAcIFIC*) precursor acquisition independent from ion count; (*RP*) reversed phase; RP^2: tandem reversed phase

or isoelectric focussing (IEF). Quantification is enabled by either protein staining or protein/peptide mass labelling or without any label at all and, finally, data acquisition has classically been done in a data-dependent acquisition (DDA) manner, i.e., by selecting precursor peptide ions for further fragmentation.

Today's proteomics workflows need to optimize the trade-off between effort and yield, i.e., between throughput and resource-use on the one hand and completeness of proteome coverage on the other. This is increasingly being achieved by: (i) maximizing sample specificity if sample amount allows for it (e.g., pure cell populations or cell organelles); (ii) multiplexed depletion of the most abundant proteins (possibly complemented by some biochemical enrichment); and, very importantly, (iii) transferring peptide fractionation from the liquid to the gas phase (i.e., restricting liquid phase separation to 1D (or on-line 2D applying a pH step) LC plus gas phase fractionation (GPF)); plus, equally relevant, (iv) data-independent acquisition (DIA) of intact and fragment peptide masses, which results in uncoupling mass determination from sequencing (Precursor acquisition independent from ion count (PAcIFIC) (Panchaud et al. 2009); precursor-independent acquisition of intact and fragment masses (MS^E) (Bateman et al. 2002; Geromanos et al. 2009); and the so-called swath-MS technique (Gillet et al. 2012). Approaches (iii), (iv) and (v) result in a split into smaller mass acquisition windows plus acquisition of all intact masses and/or, independently, all fragment masses (with a retrospective software-based reconstruction of the peptide/protein context). Such GPF (DDA) and PAcIFIC (DIA) routines have shown to

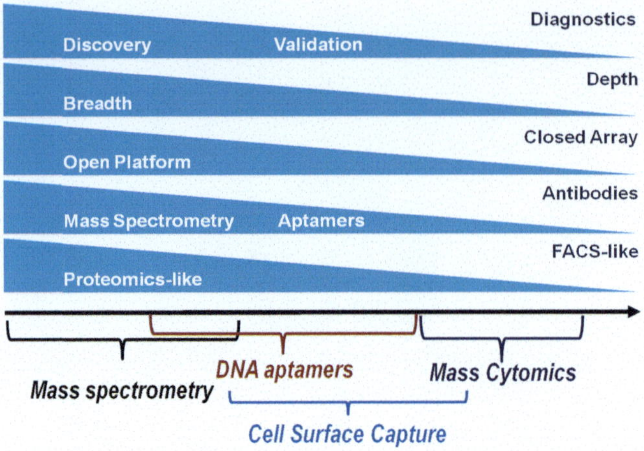

Fig. 10.2 The continuum and trade-off of proteomic platform properties. The range spans from highly versatile, "open" and flexible mass spectrometric approaches; via more sensitive, but also more "closed" and restricted array-like platforms such as DNA aptamers (Brody et al. 2010; Gold et al. 2010); to cell surface proteome-oriented technologies such as cell surface capture (Wollscheid et al. 2009) and mass cytomics (Bendall et al. 2011)

deliver deeper proteome coverage and better reproducibility (Panchaud et al. 2009; Panchaud et al. 2011; Scherl et al. 2008), because they alleviate the highly stochastic nature of classical DDA MS/MS experiments and because fractionation in the mass spectrometer is more reproducible than liquid-phase separation.

10.1.6 Proteomics Beyond Classical "Mass Spectrometry-Only" Platforms

Mass spectrometry has developed into the key technology player in proteomics and will most likely consolidate this role, mainly thanks to the information-richness of mass spectra and the wide compatibility of mass spectrometers in terms of samples, molecules, methods and point-of-analysis. However, the present and future challenges of delivering complete, accurate, time-resolved, and quantitative proteomes of higher organisms—especially in view of the emerging complexity of genome regulation at epigenetic and other levels—cannot be met by classical, exclusively MS-based platforms alone, even considering the continuing impressive improvements in accuracy, reproducibility, robustness, user friendliness and throughput. Rather, depending on the nature of the proteomic task, a set of complementary technologies is already and will increasingly be required.

Figure 10.2 shows the continuum and trade-off of proteomic platform properties: the range spans from highly versatile, flexible MS approaches; via more sensitive, but also more "closed" or restricted array-like platforms such as DNA aptamers (Brody et al. 2010; Gold et al. 2010); to cell surface proteome-oriented technologies such as cell surface capture (Wollscheid et al. 2009) and mass cytomics (Bendall et al. 2011).

DNA aptamers: A new aptamer-based proteomic technology for biomarker discovery has recently been described. The platform can simultaneously measure thousands of proteins from a few millilitres of human serum or plasma. The described assay measured over 800 proteins with a median LOD of 1 pM, covered a dynamic range from 100 fM to 1 mM, and exhibited only 5 % median coefficient of variation. The basis of this technology is a new generation of DNA aptamers, i.e., chemically modified nucleotides with greatly expanded physicochemical diversity. The measurement process transforms protein concentrations values into DNA aptamer concentration signatures, with the latter being quantified on a DNA microarray, thereby enabling a PCR-like amplification within a proteomics assay. In essence, this method combines the structural specificity of proteomics with the detection sensitivity of transcriptomics (Gold et al. 2010).

Cell surface capture: With the objective to phenotype cells without antibodies in an unbiased, hypothesis-free fashion, multiplexed mass-spectrometric identification of several hundreds of N-linked glycosylation sites specifically from cell surface–exposed glycoproteins has recently been demonstrated. This new cell surface–capturing (CSC) technology covalently labels extracellular glycan moieties on live cells, and can thereby detect and compare the cell surface N-glycoproteomes of e.g., T and B cells. The method has also been shown to monitor abundance changes of cell surface N-glycoprotein markers during T-cell activation and the controlled differentiation of embryonic stem cells into the neural lineage. Such snapshots of the cell surface N-glycoproteins are expected to unravel N-glycoprotein panels as potential differentiation markers currently inaccessible by traditional means such as FACS (Wollscheid et al. 2009).

Mass cytomics: Combining flow cytometry with MS, single-cell "mass cytometry" has recently been developed and applied to examine healthy human bone marrow, simultaneously measuring 34 parameters in single cells (binding of 31 antibodies, viability, DNA content, and relative cell size). The resulting data set enabled the assembly of related cell types defined by surface antigen expression, providing a map of cell signalling responses in combination with drug inhibition. The analysis revealed to-date unnoticed signalling responses confined to conventionally defined cell subsets, and also more continuous phosphorylation responses across different cell populations. Such single-cell analyses provided system-wide views of immune signalling in healthy human hematopoiesis (Bendall et al. 2011).

10.1.7 The Human Proteome Project (HPP)

More than 20 years ago a shift from initiating a Human Proteome Project to a Human Genome Project took place, mainly due to political changes and the rapid advances in molecular biology techniques (Banks et al. 2000). Yet, after the success of the Human Genome Project, the need to launch a Human Proteome Project emerged again. Currently, approximately 30 % of the predicted genes lack experimental evidence at the protein level and for many known proteins their localization, distribution, interaction and function is not known (Legrain et al. 2011). Therefore, in 2010 the Human Proteome Organisation (HUPO) officially launched the Human Proteome Project (HPP) designed to map the entire human protein set, elucidate biological/molecular functions and advance disease diagnostics/treatments. The HPP is based on three technological pillars: shotgun and targeted MS, monoclonal and polyclonal antibodies and an integrated knowledge base. The long-term objectives of the project are to provide a complete map of the proteins in human cells, tissues and organs in adults and at different developmental stages, and under various physiological and pathological conditions. Yet, as a proteome map is useful but not sufficient, the focus of the HPP will be not only to generate but also to disseminate knowledge to enable researchers to reliably measure the proteome of interest for their investigation (Hoofnagle et al. 2011). In parallel, in China, the Beijing Proteome Research Center has launched another ambitious project to map the human proteome (Cyranoski 2010). Finally, adjacent to the HPP are biology-driven projects, e.g., to study cancer, diabetes or energy metabolism pathways that will further contribute in the investigation of the metabolic syndrome using proteomic technologies.

10.2 Proteomics in the Study of the Metabolic Syndrome

Like the other *omics* technologies, proteomics has been applied extensively in the study of the metabolic syndrome. A search in Medline (Search engine OvidSP (Ovid Technologies, Inc.), Date: October 2011) using *proteom** and *metabolic syndrome* or *diabetes* or *insulin resistance* or *obesity* as keywords, in the title or abstract delivered 520 references (excluding reviews), published mainly during the last 10 years. Although some references included a combination of various terms, most publications have focused on diabetes (405 references) followed by obesity (152 references). Most studies have been carried out in humans (303 references), in the model organism rat or mouse (255 references) or a combination of both (62 references). Finally, the samples of choice in most studies were plasma or blood, followed by adipose tissue, an organ where biopsies are not rare in obesity or diabetes studies. Main focuses were to identify biomarkers of the metabolic syndrome or related metabolic conditions like diabetes and to elucidate mechanisms of actions or altered pathways. Fewer proteomic approaches have been used in combination with intervention studies or clinical trials. Early proteomic studies focused on detecting the relative protein abundance differences between two phenotypic states. In the last years, investigations aimed also at quantifying these differences and integrating the proteomic data with other *omics* technology-derived data, following a systems level approach. In the second part of this chapter, various examples of the approaches using proteomics in the study of the metabolic syndrome are presented.

10.2.1 Proteome Analysis of Adipose Tissue

With the discovery of leptin in 1995, adipose tissue was recognized as an endocrine organ and not only as a fat and energy depot (Halaas et al. 1995). Adipose tissue is distributed between different depots as subcutaneous or omental fat. The roles of the depots are different. Yet, adipose tissue is mainly responsible for the storage of triglycerides in adipocytes and in signalling information about the amount stored for the regulation of energy intake and expenditure.

Adipose tissue has therefore become an interesting organ for studying proteomic patterns related to the metabolic syndrome. The Swiss 2D-PAGE database provides 2D-GE protein maps of mouse white and brown adipose tissue (Lanne et al. 2001; Sanchez et al. 2001). Protein recovery from adipose tissue is challenging due to the high fat content of the tissue. Addition of hydroxyethyl disulfide in the preparation of adipose-tissue recovered proteins improves the solubilisation and isoelectric separation in 2D-GE proteomic analysis (Corton et al. 2004; Sanchez et al. 2001). Adachi *et al.* further investigated the adipocyte proteome by combining high-accuracy, high-sensitivity protein identification technology with subcellular fractionation of nuclei, mitochondria, membrane and cytosol of 3T3-L1 adipocytes (Adachi et al. 2007). Approximately 3,300 proteins were identified. Comparison with microarray data showed that the mRNA abundance of detected versus non-detected proteins differed by less than 2-fold and proteomics covered a large proportion of the insulin signalling pathway.

Hydrolysis of triacylglycerols and cholesteryl esters by lipases in the adipose tissue is a key event in energy homeostasis. Proteomic studies with fluorescent activity tags mimicking lipid substrates and 2D difference gel electrophoresis (DIGE) combined with LC-MS/MS help understand energy homeostasis as new lipases have been identified (Birner-Gruenberger et al. 2005). To better elucidate insulin action, a catalogue of the tyrosine-phosphoproteome of the insulin signalling pathway was established with 40 identified insulin-induced effectors, using high-resolution MS in combination with phosphotyrosine immunoprecipitation and SILAC in differentiated brown adipocytes (Kruger et al. 2008).

Other proteomic studies have focused on the identification of differentially expressed proteins between the different adipose tissue depots: 43 proteins were found to be differentially expressed between omental and subcutaneous fat using 2D-DIGE and LC-MS/MS (Perez-Perez et al. 2009). These proteins covered processes such as glucose and lipid metabolism, lipid transport, protein synthesis, protein folding and response to stress and inflammation, suggesting a high metabolic activity and increased cell stress in omental compared to subcutaneous fat. Additionally, the interest in adipose tissue development triggered the identification of secreted proteins during differentiation of pre-adipocytes to adipocytes (Kratchmarova et al. 2002), and a comparative shotgun label-free analysis of the secretory proteome of human adipose stromal vascular fraction cells during adipogenesis (Kim et al. 2010): 177 proteins were significantly upregulated during adipogenesis and 60 proteins significantly down-regulated. Members of developmental pathways like Wnt signalling were differentially expressed and expression changes of several proteins could be confirmed by quantitative RT-PCR and immunoblotting.

Since leptin, many other hormones, the so-called adipokines, have been identified to be secreted by the adipose tissue. Most show a positive correlation between their circulation levels and the amount of adipose tissue mass. Moreover, imbalance in some adipokine levels probably due to enlarged adipose tissue, have been correlated with obesity, type 2 diabetes mellitus (T2DM) and cardiovascular problems (Deng and Scherer 2010). A combination of cleavable isotope-coded affinity tags (cICAT) and label-free quantification showed 317 proteins differentially secreted by 3T3-L1 adipocytes with or without insulin treatment; 179 proteins were significantly upregulated and 53 downregulated. Western blot analysis of the reported adipokines, like adiponectin or resistin, confirmed the quantitative results from MS, revealing individualized secreting patterns of these proteins by increasing insulin dose (Zhou et al. 2009). Although the list of adipokines is already long, proteomic profiling of media from differentiated primary human adipocytes revealed a novel adipokine, DPP4 (dipeptidyl- peptidase 4). DDP4 is mainly known in energy metabolism as a transmembrane glycoprotein that cleaves the incretins glucagon-like peptide-1 (GLP-1) and gastric inhibitory polypeptide (GIP). These incretins are released from the intestine after food intake and contribute to the postprandial insulin secretion. Yet, the proteomic approach suggested that DPP4 is also released from the adipose tissue to the circulation and their levels are positively correlated with adipose tissue mass. Interestingly, obese individuals showed a two-fold increased secretion compared to lean subjects (Lamers et al. 2011).

10.2.2 Proteome Analysis of Other Organs and Organelles Relevant to the Metabolic Syndrome

Pancreatic beta cells. The pancreatic beta cell plays a key role in the maintenance of glucose homeostasis primarily by secreting insulin after certain stimuli. As with adipose tissue, proteomic studies using beta cells have focused on cataloguing the proteins for a better understanding of beta cell (dys)function. 2D-LC-MS/MS analysis of human pancreatic islets revealed 3,365 proteins (Metz et al. 2006). The major islet proteins (insulin, glucagon and somatostatin) were detected as well as various beta cell-enriched secretory proteins, ion channels and transcription factors. Moreover, a wide range of metabolic enzymes and cellular pathway proteins was covered, like the integrin signalling cascade, MAP kinase, NFκβ and JAK/STAT pathway.

Pancreatic islets. Other proteomic projects focused on finding new molecular targets associated with islet cell (dys)function and protection. A ground-laying study of glucose-responsive proteins in pancreatic islets from rats separated and visualized by 2D-GE was already performed in 1992 (Collins et al. 1992). More recently, 78 proteins showed changes in their levels after glucose and 3-isobutyl-1-methylxanthine stimulation using 2D-DIGE and MS (Suss et al. 2009). Most of these were proposed as novel targets of rapid regulation in beta cells. Schrimpe-Rutledge et al. updated this with novel glucose-regulated proteins in isolated human pancreatic islets using LC-MS/MS-based proteomics (Schrimpe-Rutledge et al. 2012) : their analysis revealed 256 differentially abundant proteins after 24 h of high-glucose exposure from more than 4'500 identified proteins in total. Several novel glucose-regulated proteins were elevated under high glucose, including regulators of mRNA splicing, processing and function. Many proteins found to be regulated by high-glucose stimulation are still uncharacterized or hypothetical. Another study catalogued the proteins in the insulin secretory granules of pancreatic beta cells by fractionating and characterizing the rat insulin-secreting cell line INS-1E and deploying nano-LC-ESI-MS/MS. This catalogue including 130 proteins, of them 110 newly reported to insulin secretory granules should help understand impaired insulin secretion (Brunner et al. 2007).

Skeletal muscle. In order to elucidate molecular alterations associated with insulin resistance in muscle tissue, human skeletal muscle biopsies from patients with T2DM were compared to healthy controls with 2D-GE and eight proteins were suggested as potential markers of T2DM in fasting state in muscle. The observed changes in protein expression indicated elevated cellular stress, and perturbations in ATP (re)synthesis and mitochondrial metabolism (Hojlund et al. 2003). In a later study, 15 proteins were found differentially expressed between muscle tissues of lean, obese and type 2 diabetes subjects. An overall decreased mitochondrial protein abundance was observed, together with altered cytoskeletal structure, chaperone function and proteasome subunits, indicating changes in muscle structure, protein degradation and folding (Hwang et al. 2010). Quantitative interaction proteomics based on SILAC technology and whole cell lysate from the differentiated murine muscle cell line C2C12 has recently added new insights into the insulin signalling pathway: it revealed the phosphotyrosine interactome of the insulin receptor family and its substrates IRS-1 and IRS-2 (Hanke and Mann 2009).

Kidney. Expanding from evident energy metabolism-related tissues like islets or muscle, renal cortical proteome analysis of db/db mice using 2D-GE combined with MALDI-TOF, MALDI-TOF/TOF, and LC-MS/MS yielded 278 proteins whose expression levels were significantly increased or decreased more than two-fold by diabetes. Pathway analysis indicated that PPAR-α was the most common node of interaction for the numerous enzymes whose expression levels were influenced by diabetes (Tilton et al. 2007).

Mitochondria are membrane-enclosed cell organelles that generate most of the biochemical energy ATP required in eukaryotic cells. Mitochondria are also involved in biological roles like cell growth and death. Energy metabolism-linked pathologies like diabetes or obesity have been associated with mitochondrial dysfunction (Wallace 1999). Currently, the most complete mitochondrial protein catalogue comprises over 1,000 proteins and seems to cover 85 % of the predicted-to-be-complete mitochondrial proteome (Calvo and Mootha 2010). In a further study, 1,091 mitochondrial proteins, 228 phosphoproteins, and 355 hydroxyproteins were identified in the mitochondrial proteome (Deng et al. 2010). In addition, proteins involved in beta-oxidation, the tricarboxylic acid cycle, oxidative phosphorylation, and other bio-energetic processes were up-regulated in T2DM. Increased oxidative stress was evidenced by altered post-translational modifications, like more abundant hydroxylated proteins with progressing diabetes. Finally, the necessity to integrate transcriptomic and proteomic data was confirmed as many alterations could not be identified by either proteomics or genomics approaches alone.

10.2.3 Potential Biomarkers of Metabolic Syndrome and Related Pathologies

Discovering novel biomarkers by proteomics is one of the most challenging but also most promising applications of proteomic technologies and numerous studies focussed on finding new biomarkers for metabolic syndrome and related pathologies. For example, early biomarkers of diabetes should facilitate interventions designed to prevent diabetes, or delay its onset, or slow down its progression. Also, diabetic patients would greatly benefit from biomarkers for disease monitoring and therapeutic success.

Using a label-free quantitative shotgun strategy, 1,377 human serum proteins were detected in a first approach. Then, so-termed "localized statistics of protein abundance distribution" (LSPAD) revealed 68 proteins significantly over-represented in diabetic sera compared to control sera of human volunteers. Pathway analysis showed an over-representation of the complement system pathway. In addition, an upstream activator of the complement pathway, ficolin-3, was confirmed to be over-represented in the serum of an additional cohort of type 2 diabetic patients. Therefore, ficolin-related complement activation is a candidate biomarker in diabetes (Li et al. 2008).

In diabetic patients, the need for a non-invasive diagnostic method is of great interest to circumvent the repeated needle prick. Therefore, several efforts in detecting urinary or salivary biomarkers have been published. Analysis of whole saliva from control and type-2 diabetic individuals by 2D-LC-MS/MS and label-free quantification, identified 65 proteins with greater than 2-fold difference in abundance between control and type-2 diabetes samples. Independent validation of a subset of potential biomarkers by immunodetection confirmed their differential expression. Interestingly, a trend of increase in their abundance with progression from the pre-diabetic to the diabetic state was observed (Rao et al. 2009).

A model for monitoring proteome alterations with T2DM disease progression is to study those changes over time in the urinary proteome of a streptozotocin-induced rat model of diabetes that occur during the initial response to the hyperglycaemic state. 2D-LC-MS/MS and label-free quantification were executed on urine samples obtained at different time points from treated and control rats (Schlatzer et al. 2009) : differences in urinary protein abundance occurred in this animal model before observable pathophysiological changes appeared and include increased pro-alpha collagen, which is regulated by circulating levels of insulin and/or glucose. Collagen fragments have been suggested as earlier biomarkers for diabetes-induced renal damage with respect to the currently used urinary albumin. Subjects with diabetic nephropathy were identified after profiling their urinary proteome by capillary electrophoresis MS

(Alkhalaf et al. 2010). In a previous study, urinary protein abundance was investigated in patients with selected glomerular diseases: focal segmental glomerulosclerosis, lupus nephritis, membranous nephropathy, or diabetic nephropathy (Varghese et al. 2007). Twenty-one 2D-GE protein spots were found significant for the differentiation of the diseases; 11 proteins were identified by MS indicating that diseases causing nephrotic syndrome change glomerular protein permeability. These proteins are candidate biomarkers but need validation in further studies.

10.2.4 Proteomics in the Elucidation of Mechanisms in Metabolic Syndrome

Most mechanistic proteomic studies on the metabolic syndrome have been performed in rat or mouse models, such as the leptin-deficient mouse (*Lep*$^{-/-}$). This choice of model also applies to many other pathological and physiological conditions, because of tissue availability and the possibility to further genetically modify the organism.

Angiopoietin-like protein 4 (ANGPTL4) has been suggested as an adipokine in energy metabolism, through unknown mechanisms though. Therefore, livers of db/db mice were treated with or without ANGPTL4 and analyzed by 2D-DIGE followed by MALDI-TOF and LC-MS/MS (Wang et al. 2007). This study suggested that the adipokine ANGPTL4 might elicit its metabolic effects through modulating mitochondrial function and methionine metabolic cycles in the liver tissue. In another study, 2D-GE followed by MALDI-TOF of an immortalized rat mesangial cell line suggested that the Akt/GSK-3beta pathway and the adapter protein 14-3-3zeta may play an important role in IGF-1 signalling and survival of mesangial cells in diabetic nephropathy (Singh et al. 2007). Combining 2D-GE with MALDI-TOF/TOF-MS, ubiquitin and ubiquitin carboxyl-terminal esterase L1 (UCHL-1) in the hypothalamus were suggested to influence the genesis of diet-resistant or diet-induced obesity in rats by interfering with the integrated signalling network that control energy balance and feeding (Wang et al. 2011b).

Proteomics has recently given mechanistic insights into endoplasmic reticulum (ER) stress in obesity. Fu *et al.* performed 1D nano-LC MS/MS proteomics in the livers of *Lep*$^{-/-}$ mice and their wild-type littermates (Fu et al. 2011) : from a total of 2,021 proteins identified, 120 were significantly regulated in obese hepatic ER samples. The up-regulated proteins were mainly involved in lipid metabolism where as the down-regulated were implicated in protein synthesis and transport, indicating a hepatic ER function shift from protein to lipid synthesis and metabolism. Further lipidomic analysis, biochemical and genetic experiments identified and confirmed a lipid-driven calcium transport, which—together with ER stress— provided a framework to understand liver pathogenesis in obesity.

Proteomics has been deployed not only to reveal mechanisms of action in metabolic syndrome and its pathogenesis, but also to investigate pathways in the consequences of these pathogeneses, like diabetes-induced renal injury. Starkey *et al.* labelled the protein fractions isolated from the renal cortex of control and *db/db* mice with^{18}O- and^{16}O-water (Starkey et al. 2010). Tryptic peptides were first separated by 2D-LC, further by nano-LC, and then identified and quantified by MS. This analysis revealed that retinoic acid metabolism in the diabetic kidney may be dysregulated. A further proteomic study showed that increased 3-hydroxy-3-methylglutaryl-CoA (HMG-CoA) synthase 2 (HMGCS2) abundance leads to excess ketogenic activity in diabetic kidneys (Zhang et al. 2011).

10.2.5 Proteomics in Human Intervention Studies Within the Scope of Metabolic Syndrome

Proteomics has also been applied to understand the mechanisms underlying nutritional or pharmacological interventions in metabolic syndrome and related pathologies.

One widely used anti-diabetic drug, rosiglitazone, is an insulin sensitizer metabolite known to act on peroxisome proliferator-activated receptor alpha (PPAR-α). This drug has anti-diabetic benefits but it has also been observed to augment the risk of cardiovascular disease in some patients. In the last decade, several intervention studies have investigated the underlying mechanisms of action to better understand both desired and undesired effects of the drug and to find alternative targets. Some of these studies have applied proteomics : in one of these, 122 proteins spots separated by 2D-GE were found differentially expressed in obese healthy subjects before and two weeks after rosiglitazone treatment (Ahmed et al. 2010). Major changes were observed in cytoskeletal rearrangement, insulin and calcium signaling, and inflammatory and redox signals that up-regulate GLUT4 granule trafficking in human adipose tissue. In another study, control and *Lep*$^{-/-}$ mice were treated with rosiglitazone and proteomic changes were monitored in liver plus white and brown adipose tissue by 2D-GE-MS/MS (Sanchez et al. 2003) : here, 34 proteins were found differently expressed between lean control and obese leptin deficient mice. Eleven proteins were significantly modulated after treatment in the obese, but not in the lean mice. These proteins may reveal new drug targets in insulin resistance. Another PPAR-α agonist (Wy 14,643) has earlier been shown to up-regulate 16 liver proteins in leptin-deficient mice as elucidated by a 2D-GE-MS/MS approach (Edvardsson et al. 1999).

By using 2D-LC, MALDI-TOF/TOF and ESI-TOF analysis, the levels of transthyretin, AMBP, haptoglobin precursor were found to be down-regulated, while albumin, zinc alpha 2 glycoprotein, RBP4 and E cadherin were observed up-regulated in the urine of diabetic patients as compared to controls (Riaz et al. 2011). Treatment with thiamine decreased albumin by 34 % in urine samples of diabetic patients compared to controls and placebo-treated diabetic patients, while the other urinary protein markers did not show any significant change after the therapy.

Proteome changes were also monitored when applying a low-calorie diet in a clinical obesity study: the expression of 6 adipocyte proteins was found significantly changed in eight overweight/obese subjects who underwent an intervention of

5 weeks of a very low-calorie diet followed by 3 weeks of a normal diet; these changes suggested a reduced intracellular scaffolding of GLUT4, an increased uptake of fatty acids, an improved inflammatory profile of the adipose tissue and a change in fat droplet organization (Bouwman et al. 2009).

10.2.6 (Prote) omics-*Driven Obesity Research Networks*

During the European 5th Framework Programme (started in 2001) a project applied *omics* technology in the study of obesity, *NuGenOb: Nutrition, Genes and Obesity*. The main objectives of this consortium were to (i) identify and characterise novel nutrient-sensitive candidate genes for obesity; (ii) analyse differential gene expression in adipose tissue in relation to the acute intake of a high-fat meal and of a long term hypocaloric diet with either a high or a low fat content; (iii) assess the effects of functional variants of the candidate genes on physiological responses in obese subjects to a high-fat test meal; and (iv) identify predictors of changes in body weight and composition during dietary intervention, including changes in fat intake. Single nucleotide polymorphisms and transcriptomics analysis in parallel to classical anthropometric and biochemical parameters were measured. Targeted proteomic analysis showed that moderate caloric restriction but not macronutrient composition influences the production and secretion of adipose tissue-derived proteins during a hypocaloric diet (Arvidsson et al. 2004).

Not only targeted but also hypothesis-free discovery proteomics were applied in a project within the 6th European Framework Programme, namely *DiOGenes: Diet, Obesity and Genes*. This European, multicentre, randomized, controlled dietary intervention study showed that a diet rich in protein and with a low glycaemic index is slightly more efficient in maintaining weight (after intended weight loss) than a high-protein/high glycaemic index diet or a low-protein diet with either low or high glycaemic index (Larsen et al. 2010). Classical biochemical, anthropometrical and *omics* measurements were applied to investigate the mechanisms behind, identify biomarkers and/or asses the physiological/biological consequences of the intervention. Targeted proteomic analysis was applied to validate and identify biomarkers of successful weight maintenance (Wang et al. 2011a; Wang et al. 2012), whereas proteomic analysis searched for the mechanisms accompanying weight maintenance (Rubio-Aliaga et al. 2011).

The EU FP7 “beta-JuDO” project (http://ec.europa.eu/research/health/medical-research/diabetes-and-obesity/projects/beta-judo_en.html) aims at studying hyperlipidemia in young obese individuals and is based on the study of lipotoxicity on human islets in culture, but also on a cohort of 3,000 patients. Also within the 7th European Framework Programme the obesity-related *omics* technology project FLORINASH, (http://ec.europa.eu/research/health/medical-research/severe-chronic-diseases/projects/florinash_en.html) has started in 2010, studying the role of the intestinal microflora in non-alcoholic fatty liver disease (NAFLD). In accordance with the evolution of systems biology in the last decade, advanced systems-level *omics* metrics (leveraging proteomic, metabolomic, trancriptomic and metagenomic platforms) combined with bioinformatic and chemometric modelling of the phenotypes will be applied.

Important for the present scope and context, the Human Proteome Organisation (HUPO, see 10.1.7) has recently announced several biology/disease-driven human proteome projects (HPPs), among which the one on diabetes has started with a recently established consortium (http://cms.unige.ch/medecine/hdpp.info/node/1). The plan is to leverage this international constellation of expertise to generate systems-level insights into diabetes-associated cellular changes by gathering multivariate data sets over time from specialised cells and organs of healthy and diabetes-affected individuals.

10.3 Future Perspectives

Novel approaches in study design, technological advances, standardization of applied methodologies, comprehensive cataloguing of the human proteome and the integration with other *omics* technologies (genomics, transcriptomics, metabolomics and epigenomics), will facilitate proteomic progress in the delivery of novel biomarkers, the understanding of mechanisms of action and/or the delivery of new drug targets in the context of metabolic syndrome and other common complex traits and diseases.

These expectations are nourished by the following recent developments or emerging trends:

i. Ultra-high resolution and -mass accuracy mass spectrometers.
ii. Workflows combining efficient reduction of sample complexity and deep and broad proteome coverage with reasonable effort and high throughput.
iii. Robust bioinformatics for efficient data processing, assessment, validation, visualization and interpretation.
iv. Expanding beyond classical discovery/targeted mass spectrometry and embracing new and complementary proteomic technologies such as DNA aptamer assays and cell surface capture.

All the above relates to technology improvements. Moreover and at least equally important, the design of systems biology studies may need to be improved considering the following aspects:

i. A key to healthy energy metabolism and to to avoiding/delaying the onset of metabolic syndrome is maintenance or (re-)establishment or of metabolic flexibility and elasticity. Therefore, repeated observations of the very same study subject over time and under (repeated) exposure to a safe metabolic challenge, should enable to capture the moment of "derailing from healthy trajectory", i.e., the early biomarker signature.
ii. These repeated observations need to be systems-based, technology-rooted but not technology-driven: truly integrated measurements from all possible angles

[(epi-)genomics, transcriptomics, proteomics, lipidomics, metabonomics] are required.

iii. Each study subject should serve, when possible, as its own case and control, to better address the influence of genomic individuality.

iv. Interesting or extreme phenotypes observed within the range of responses to challenges can be mechanistically investigated and hypothesis should be tested in human clinical studies.

v. Ideally, such metabolic studies should not only sample the host, but also its gut residing microbiome, because of its impact on host health and metabolism.

Taken together, these improvements will in our opinion facilitate a time-resolved systems view of a complex chronic disease such as the metabolic syndrome.

References

Adachi J, Kumar C, Zhang Y, Mann M (2007) In-depth analysis of the adipocyte proteome by mass spectrometry and bioinformatics. Mol Cell Proteomics 6(7):1257–12739

Aebersold R, Mann M (2003) Mass spectrometry-based proteomics. Nature 422(6928):198–207

Ahmed M, Neville MJ, Edelmann MJ, Kessler BM, Karpe F (2010) Proteomic analysis of human adipose tissue after rosiglitazone treatment shows coordinated changes to promote glucose uptake. Obesity 18(1):27–34

Ahrens CH, Brunner E, Qeli E, Basler K, Aebersold R (2010) Generating and navigating proteome maps using mass spectrometry. Nat Rev Mol Cell Bio 11(11):789–801

Alaiya A, Al-Mohanna M, Linder S (2005) Clinical cancer proteomics: promises and pitfalls. J Proteome Res 4(4):1213–1222

Alkhalaf A, Zurbig P, Bakker SJ, Bilo HJ, Cerna M, Fischer C, Fuchs S, Janssen B, Medek K, Mischak H, Roob JM, Rossing K, Rossing P, Rychlik I, Sourij H, Tiran B, Winklhofer-Roob BM, Navis GJ, Group PREDICTIONS (2010) Multicentric validation of proteomic biomarkers in urine specific for diabetic nephropathy. PLoS ONE (Electronic Resource) 5(10):e13421

Allet N, Barrillat N, Baussant T, Boiteau C, Botti P, Bougueleret L, Budin N, Canet D, Carraud S, Chiappe D, Christmann N, Colinge J, Cusin I, Dafflon N, Depresle B, Fasso I, Frauchiger P, Gaertner H, Gleizes A, Gonzalez-Couto E, Jeandenans C, Karmime A, Kowall T, Lagache S, Mahe E, Masselot A, Mattou H, Moniatte M, Niknejad A, Paolini M, Perret F, Pinaud N, Ranno F, Raimondi S, Reffas S, Regamey PO, Rey PA, Rodriguez-Tome P, Rose K, Rossellat G, Saudrais C, Schmidt C, Villain M, Zwahlen C (2004) In vitro and in silico processes to identify differentially expressed proteins. Proteomics 4(8):2333–2351

Amacher DE (1998) Serum transaminase elevations as indicators of hepatic injury following the administration of drugs. Regul Toxicol Pharmacol 27(2):119–130

Arvidsson E, Viguerie N, Andersson I, Verdich C, Langin D, Arner P (2004) Effects of different hypocaloric diets on protein secretion from adipose tissue of obese women. Diabetes 53(8):1966–1971

Banks RE, Dunn MJ, Hochstrasser DF, Sanchez JC, Blackstock W, Pappin DJ, Selby PJ (2000) Proteomics: new perspectives, new biomedical opportunities. Lancet 356(9243):1749–1756

Bateman RH, Carruthers R, Hoyes JB, Jones C, Langridge JI, Millar A, Vissers JP (2002) A novel precursor ion discovery method on a hybrid quadrupole orthogonal acceleration time-of-flight (Q-TOF) mass spectrometer for studying protein phosphorylation. J Am Soc Mass Spectrom 13(7):792–803

Bendall SC, Simonds EF, Qiu P, Amir el-AD, Krutzik PO, Finck R, Bruggner RV, Melamed R, Trejo A, Ornatsky OI, Balderas RS, Plevritis SK, Sachs K, Pe'er D, Tanner SD, Nolan GP

(2011) Single-cell mass cytometry of differential immune and drug responses across a human hematopoietic continuum. Science 332(6030):687–696

Beynon RJ, Doherty MK, Pratt JM, Gaskell SJ (2005) Multiplexed absolute quantification in proteomics using artificial QCAT proteins of concatenated signature peptides. Nat Methods 2(8):587–589

Birner-Gruenberger R, Susani-Etzerodt H, Waldhuber M, Riesenhuber G, Schmidinger H, Rechberger G, Kollroser M, Strauss JG, Lass A, Zimmermann R, Haemmerle G, Zechner R, Hermetter A (2005) The lipolytic proteome of mouse adipose tissue. Mol Cell Proteomics 4(11):1710–1717

Bouwman FG, Claessens M, Baak MA van, Noben JP, Wang P, Saris WH, Mariman EC (2009) The physiologic effects of caloric restriction are reflected in the in vivo adipocyte-enriched proteome of overweight/obese subjects. J Proteome Res 8(12):5532–5540

Brody EN, Gold L, Lawn RM, Walker JJ, Zichi D (2010) High-content affinity-based proteomics: unlocking protein biomarker discovery. Expert Rev Mol Diagn 10(8):1013–1022

Brunner Y, Coute Y, Iezzi M, Foti M, Fukuda M, Hochstrasser DF, Wollheim CB, Sanchez JC (2007) Proteomics analysis of insulin secretory granules. Mol Cell Proteomics 6(6):1007–1017

Calvo SE, Mootha VK (2010) The mitochondrial proteome and human disease. Annu Rev Genomics Hum Genet 11:25–44

Cargile BJ, Stephenson JL Jr (2004) An alternative to tandem mass spectrometry: isoelectric point and accurate mass for the identification of peptides. Anal Chem 76(2):267–275

Chelius D, Bondarenko PV (2002) Quantitative profiling of proteins in complex mixtures using liquid chromatography and mass spectrometry. J Proteome Res 1(4):317–323

Colinge J, Masselot A, Giron M, Dessingy T, Magnin J (2003) OLAV: towards high-throughput tandem mass spectrometry data identification. Proteomics 3(8):1454–1463

Collins H, Najafi H, Buettger C, Rombeau J, Settle RG, Matschinsky FM (1992) Identification of glucose response proteins in two biological models of beta-cell adaptation to chronic high glucose exposure. J Biol Chem 267(2):1357–1366

Conrads TP, Anderson GA, Veenstra TD, Pasa Tolic L, Smith RD (2000) Utility of accurate mass tags for proteome-wide protein identification. Anal Chem 72:3349–3354 (15 Jul 2000)

Corton M, Villuendas G, Botella JI, San Millan JL, Escobar-Morreale HF, Peral B (2004) Improved resolution of the human adipose tissue proteome at alkaline and wide range pH by the addition of hydroxyethyl disulfide. Proteomics 4(2):438–441

Craig R, Beavis RC (2004) TANDEM: matching proteins with tandem mass spectra. Bioinformatics 20(9):1466–1467

Cyranoski D (2010) China pushes for the proteome. Nature 467:380

Deng Y, Scherer PE (2010) Adipokines as novel biomarkers and regulators of the metabolic syndrome. Ann N Y Acad Sci 1212:E1–E199

Deng WJ, Nie S, Dai J, Wu JR, Zeng R (2010) Proteome, phosphoproteome, and hydroxyproteome of liver mitochondria in diabetic rats at early pathogenic stages. Mol Cell Proteomics 9(1):100–116

Domon B, Broder S (2004) Implications of new proteomics strategies for biology and medicine. J Proteome Res 3(2):253–260

Edvardsson U, Alexandersson M, Brockenhuus LH von, Nystrom AC, Ljung B, Nilsson F, Dahllof B (1999) A proteome analysis of livers from obese (ob/ob) mice treated with the peroxisome proliferator WY14,643. Electrophoresis 20(4–5):935–942

Elias JE, Gygi SP (2007) Target-decoy search strategy for increased confidence in large-scale protein identifications by mass spectrometry. Nat Methods 4(3):207–214

Elias JE, Haas W, Faherty BK, Gygi SP (2005) Comparative evaluation of mass spectrometry platforms used in large-scale proteomics investigations. Nat Methods 2(9):667–675

Eng JK, McCormack AL, Yates JR (1994) An approach to correlate tandem mass spectral data of peptides with amino acid sequences in a protein database. J Am Soc Mass Spectrom 5:976–989

Foster LJ, De Hoog CL, Mann M (2003) Unbiased quantitative proteomics of lipid rafts reveals high specificity for signaling factors. Proc Natl Acad Sci U. S. A. 100(10):5813–5818

Fu S, Yang L, Li P, Hofmann O, Dicker L, Hide W, Lin X, Watkins SM, Ivanov AR, Hotamisligil GS (2011) Aberrant lipid metabolism disrupts calcium homeostasis causing liver endoplasmic reticulum stress in obesity. Nature 473(7348):528–531

Gartner CA, Elias JE, Bakalarski CE, Gygi SP (2007) Catch-and-release reagents for broadscale quantitative proteomics analyses. J Proteome Res 6(4):1482–1491

Geer LY, Markey SP, Kowalak JA, Wagner L, Xu M, Maynard DM, Yang X, Shi W, Bryant SH (2004) Open mass spectrometry search algorithm. J Proteome Res 3(5):958–964

Gerber SA, Rush J, Stemman O, Kirschner MW, Gygi SP (2003) Absolute quantification of proteins and phosphoproteins from cell lysates by tandem MS. Proc Natl Acad Sci U. S. A. 100(12):6940–6945

Geromanos SJ, Vissers JP, Silva JC, Dorschel CA, Li GZ, Gorenstein MV, Bateman RH, Langridge JI (2009) The detection, correlation, and comparison of peptide precursor and product ions from data independent LC-MS with data dependant LC-MS/MS. Proteomics 9(6):1683–1695

Gillet LC, Navarro P, Tate S, Röst H, Selevsek N, Reiter L, Bonner R, Aebersold R (2012) Targeted data extraction of the MS/MS spectra generated by data independent acquisition: a new concept for consistent and accurate proteome analysis. Mol Cell Proteomics 11(6) doi: 10.1074/mcp.O111.016717:1–17

Gold L, Ayers D, Bertino J, Bock C, Bock A, Brody EN, Carter J, Dalby AB, Eaton BE, Fitzwater T, Flather D, Forbes A, Foreman T, Fowler C, Gawande B, Goss M, Gunn M, Gupta S, Halladay D, Heil J, Heilig J, Hicke B, Husar G, Janjic N, Jarvis T, Jennings S, Katilius E, Keeney TR, Kim N, Koch TH, Kraemer S, Kroiss L, Le N, Levine D, Lindsey W, Lollo B, Mayfield W, Mehan M, Mehler R, Nelson SK, Nelson M, Nieuwlandt D, Nikrad M, Ochsner U, Ostroff RM, Otis M, Parker T, Pietrasiewicz S, Resnicow DI, Rohloff J, Sanders G, Sattin S, Schneider D, Singer B, Stanton M, Sterkel A, Stewart A, Stratford S, Vaught JD, Vrkljan M, Walker JJ, Watrobka M, Waugh S, Weiss A, Wilcox SK, Wolfson A, Wolk SK, Zhang C, Zichi D (2010) Aptamer-based multiplexed proteomic technology for biomarker discovery. PLoS.One 5(12):e15004

Goodlett DR, Keller A, Watts JD, Newitt R, Yi EC, Purvine S, Eng JK, Haller P von, Aebersold R, Kolker E (2001) Differential stable isotope labeling of peptides for quantitation and de novo sequence derivation. Rapid Commun Mass Spectrom 15(14):1214–1221

Gygi S, Rist B, Gerber S, Turecek F, Gelb M, Aebersold R (1999a) Quantitative analysis of complex protein mixtures using isotope-coded affinity tags. Nat Biotechnol 17:994–999

Gygi SP, Rochon Y, Franza BR, Aebersold R (1999b) Correlation between protein and mRNA abundance in yeast. Mol Cell Biol 19(3):1720–1730

Halaas JL, Gajiwala KS, Maffei M, Cohen SL, Chait BT, Rabinowitz D, Lallone RL, Burley SK, Friedman JM (1995) Weight-reducing effects of the plasma protein encoded by the obese gene. Science 269(5223):543–546

Hanke S, Mann M (2009) The phosphotyrosine interactome of the insulin receptor family and its substrates IRS-1 and IRS-2. Mol Cell Proteomics 8(3):519–534

Hansen BT, Jones JA, Mason DE, Liebler DC (2001) SALSA: a pattern recognition algorithm to detect electrophile-adducted peptides by automated evaluation of CID spectra in LC-MS-MS analyses. Anal Chem 73(8):1676–1683

Hojlund K, Wrzesinski K, Larsen PM, Fey SJ, Roepstorff P, Handberg A, Dela F, Vinten J, McCormack JG, Reynet C, Beck-Nielsen H (2003) Proteome analysis reveals phosphorylation of ATP synthase beta -subunit in human skeletal muscle and proteins with potential roles in type 2 diabetes. J Biol Chem 278(12):10436–10442

Hoofnagle AN, Aebersold R, Anderson NL, Felsenfeld A, Liebler DC (2011) Painting a moving picture: large-scale proteomics efforts and their potential for changing patient care. Clin Chem 57(10):1357–1360

Hsich G, Kenney K, Gibbs CJ, Lee KH, Harrington MG (1996) The 14-3-3 brain protein in cerebrospinal fluid as a marker for transmissible spongiform encephalopathies. N Engl J Med 335(13):924–930

Hu Q, Noll RJ, Li H, Makarov A, Hardman M, Graham CR (2005) The Orbitrap: a new mass spectrometer. J Mass Spectrom 40(4):430–443

Hwang H, Bowen BP, Lefort N, Flynn CR, De Filippis EA, Roberts C, Smoke CC, Meyer C, Hojlund K, Yi Z, Mandarino LJ (2010) Proteomics analysis of human skeletal muscle reveals novel abnormalities in obesity and type 2 diabetes. Diabetes 59(1):33–42

Ibarrola N, Kalume DE, Gronborg M, Iwahori A, Pandey A (2003) A proteomic approach for quantitation of phosphorylation using stable isotope labeling in cell culture. Anal Chem 75(22):6043–6049

Ippel JH, Pouvreau L, Kroef T, Gruppen H, Versteeg G, Putten P van den, Struik PC, Mierlo CPM van (2004) In vivo uniform 15N-isotope labelling of plants: using the greenhouse for structural proteomics. Proteomics 4(1):226–234

Ishihama Y, Oda Y, Tabata T, Sato T, Nagasu T, Rappsilber J, Mann M (2005) Exponentially modified protein abundance index (emPAI) for estimation of absolute protein amount in proteomics by the number of sequenced peptides per protein. Mol Cell Proteomics 4(9):1265–1272

Jensen ON (2004) Modification-specific proteomics: characterization of post-translational modifications by mass spectrometry. Curr Opin Chem Biol 8(1):33–41

Keller A, Nesvizhskii AI, Kolker E, Aebersold R (2002) Empirical statistical model to estimate the accuracy of peptide identifications made by MS/MS and database search. Anal Chem 74(20):5383–5392

Kim J, Choi YS, Lim S, Yea K, Yoon JH, Jun DJ, Ha SH, Kim JW, Kim JH, Suh PG, Ryu SH, Lee TG (2010) Comparative analysis of the secretory proteome of human adipose stromal vascular fraction cells during adipogenesis. Proteomics 10(3):394–405

Kratchmarova I, Kalume DE, Blagoev B, Scherer PE, Podtelejnikov AV, Molina H, Bickel PE, Andersen JS, Fernandez MM, Bunkenborg J, Roepstorff P, Kristiansen K, Lodish HF, Mann M, Pandey A (2002) A proteomic approach for identification of secreted proteins during the differentiation of 3T3-L1 preadipocytes to adipocytes. Mol Cell Proteomics 1(3):213–222

Krijgsveld J, Ketting, RF Mahmoudi T, Johansen J, Artal-Sanz M, Verrijzer CP, Plasterk RH, Heck AJ (2003) Metabolic labeling of C. elegans and D. melanogaster for quantitative proteomics. Nat Biotechnol 21(8):927–931

Kruger M, Kratchmarova I, Blagoev B, Tseng YH, Kahn CR, Mann M (2008) Dissection of the insulin signaling pathway via quantitative phosphoproteomics. Proc Natl Acad Sci U. S. A. 105(7):2451–2456

Kuster B, Schirle M, Mallick P, Aebersold R (2005) Scoring proteomes with proteotypic peptide probes. Nat Rev. Mol Cell Biol 6(7):577–583

Lamers D, Famulla S, Wronkowitz N, Hartwig S, Lehr S, Ouwens DM, Eckardt K, Kaufman JM, Ryden M, Muller S, Hanisch FG, Ruige J, Arner P, Sell H, Eckel J (2011) Dipeptidyl peptidase 4 is a novel adipokine potentially linking obesity to the metabolic syndrome. Diabetes 60(7):1917–1925

Lanne B, Potthast F, Hoglund A, Brockenhuus LH von, Nystrom AC, Nilsson F, Dahllof B (2001) Thiourea enhances mapping of the proteome from murine white adipose tissue. Proteomics 1(7):819–828

Larsen TM, Dalskov SM, Baak M van, Jebb SA, Papadaki A, Pfeiffer AF, Martinez JA, Handjieva-Darlenska T, Kunesova M, Pihlsgard M, Stender S, Holst C, Saris WH, Astrup A (2010) Diets with high or low protein content and glycemic index for weight-loss maintenance. N Engl J Med 363(22):2102–2113

Lasonder E, Ishihama Y, Andersen JS, Vermunt AM, Pain A, Sauerwein RW, Eling WM, Hall N, Waters AP, Stunnenberg HG, Mann M (2002) Analysis of the Plasmodium falciparum proteome by high-accuracy mass spectrometry. Nature 419(6906):537–542

Legrain P, Aebersold R, Archakov A, Bairoch A, Bala K, Beretta L, Bergeron J, Borchers CH, Corthals GL, Costello CE, Deutsch EW, Domon B, Hancock W, He F, Hochstrasser D, Marko-Varga G, Salekdeh GH, Sechi S, Snyder M, Srivastava S, Uhlen M, Wu CH, Yamamoto T, Paik YK, Omenn GS (2011) The human proteome project: current state and future direction. Mol Cell Proteomics 10(7):M111

Leptos KC, Sarracino DA, Jaffe JD, Krastins B, Church GM (2006) MapQuant: open-source software for large-scale protein quantification. Proteomics 6(6):1770–1782

Lescuyer P, Hochstrasser D, Rabilloud T (2007) How shall we use the proteomics toolbox for biomarker discovery? J Proteome Res 6(9):3371–3376

Li XJ, Yi EC, Kemp CJ, Zhang H, Aebersold R (2005) A software suite for the generation and comparison of peptide arrays from sets of data collected by liquid chromatography-mass spectrometry. Mol Cell Proteom 4:1328–1340 (Sep 2005)

Li RX, Chen HB, Tu K, Zhao SL, Zhou H, Li SJ, Dai J, Li QR, Nie S, Li YX, Jia WP, Zeng R, Wu JR (2008) Localized-statistical quantification of human serum proteome associated with type 2 diabetes. PLoS ONE (Electronic Resource) 3(9):e3224

Lisacek F, Cohen Boulakia S, Appel RD (2006) Proteome informatics II: Bioinformatics for comparative proteomics. Proteomics 6:5445–5466 (Oct 2006)

Liu HB, Sadygov RG, Yates JR III (2004) A model for random sampling and estimation of relative protein abundance in shotgun proteomics. Anal Chem 76(14):4193–4201

MacBeath G, Schreiber SL (2000) Printing proteins as microarrays for high-throughput function determination. Science 289(5485):1760–1763

Mallick P, Kuster B (2010) Proteomics: a pragmatic perspective. Nat Biotechnol 28(7):695–709

Mallick P, Schirle M, Chen SS, Flory MR, Lee H, Martin D, Ranish J, Raught B, Schmitt R, Werner T, Kuster B, Aebersold R (2007) Computational prediction of proteotypic peptides for quantitative proteomics. Nat Biotechnol 25(1):125–131

Mann M, Hojrup P, Roepstorff P (1993) Use of mass spectrometric molecular weight information to identify proteins in sequence databases. Biol Mass Spectrom 22(6):338–345

Mann M, Wilm M (1994) Error-tolerant identification of peptides in sequence databases by peptide sequence tags. Anal Chem 66(24):4390–4399

Matthiesen R (2007) Methods, algorithms and tools in computational proteomics: a practical point of view. Proteomics 7(16):2815–2832

Matthiesen R, Lundsgaard M, Welinder KG, Bauw G (2003) Interpreting peptide mass spectra by VEMS. Bioinf 19(6):792–793

Matthiesen R, Bunkenborg J, Stensballe A, Jensen ON, Welinder KG, Bauw G (2004) Database-independent, database-dependent, and extended interpretation of peptide mass spectra in VEMS V2.0. Proteomics 4(9):2583–2593

Matthiesen R, Trelle MB, Hojrup P, Bunkenborg J, Jensen ON (2005) VEMS 3.0: algorithms and computational tools for tandem mass spectrometry based identification of post-translational modifications in proteins. J Proteome Res 4(6):2338–2347

Metz TO, Jacobs JM et al (2006) Characterization of the human pancreatic islet proteome by two-dimensional LC/MS/MS. J Proteome Res 5(12):3345–3354

Mueller LN, Rinner O, Schmidt A, Letarte S, Bodenmiller B, Brusniak MY, Vitek O, Aebersold R, Muller M (2007) SuperHirn—a novel tool for high resolution LC-MS-based peptide/protein profiling. Proteomics 7(19):3470–3480

Nesvizhskii AI, Keller A, Kolker E, Aebersold R (2003) A statistical model for identifying proteins by tandem mass spectrometry. Anal Chem 75(17):4646–4658

Norbeck AD, Monroe ME, Adkins JN, Anderson KK, Daly DS, Smith RD (2005) The utility of accurate mass and LC elution time information in the analysis of complex proteomes. J Am Soc Mass Spectrom 16(8):1239–1249

Ong SE, Blagoev B, Kratchmarova I, Kristensen DB, Steen H, Pandey A, Mann M (2002) Stable isotope labeling by amino acids in cell culture, SILAC, as a simple and accurate approach to expression proteomics. Mol Cell Proteomics 1(5):376–386

Palagi PM, Walther D, Quadroni M, Catherinet S, Burgess J, Zimmermann Ivol CG, Sanchez JC, Binz PA, Hochstrasser DF, Appel RD (2005) MSight: an image analysis software for liquid chromatography-mass spectrometry. Proteomics 5:2381–2384 (Jun 2005)

Palmblad M, Ramstrom M, Bailey CG, McCutchen-Maloney SL, Bergquist J, Zeller LC (2004) Protein identification by liquid chromatography-mass spectrometry using retention time prediction. J Chromatogr B Analyt Technol Biomed Life Sci 803(1):131–135

Panchaud A, Hansson J, Affolter M, Rhlid RB, Piu S, Moreillon P, Kussmann M (2008) ANIBAL—stable-isotope-based quantitative proteomics by ANIline and Benzoic acid labeling of amino and carboxylic groups. Mol Cell Proteomics 7(4):800–812

Panchaud A, Scherl A, Shaffer SA, Haller PD von, Kulasekara HD, Miller SI, Goodlett DR (2009) Precursor acquisition independent from ion count: how to dive deeper into the proteomics ocean. Anal Chem 81(15):6481–6488

Panchaud A, Jung S, Shaffer SA, Aitchison JD, Goodlett DR (2011) Faster, quantitative, and accurate precursor acquisition independent from ion count. Anal Chem 83(6):2250–2257

Peng J, Elias JE, Thoreen CC, Licklider LJ, Gygi SP (2003) Evaluation of multidimensional chromatography coupled with tandem mass spectrometry (LC/LC-MS/MS) for large-scale protein analysis: the yeast proteome. J Proteome Res 2(1):43–50

Perez-Perez R, Ortega-Delgado FJ, Garcia-Santos E, Lopez JA, Camafeita E, Ricart W, Fernandez-Real JM, Peral B (2009) Differential proteomics of omental and subcutaneous adipose tissue reflects their unalike biochemical and metabolic properties. J Proteome Res 8(4):1682–1693

Perkins DN, Pappin DJ, Creasy DM, Cottrell JS (1999) Probability-based protein identification by searching sequence databases using mass spectrometry data. Electrophoresis 20(18):3551–3567

Pratt JM, Simpson DM, Doherty MK, Rivers J, Gaskell SJ, Beynon RJ (2006) Multiplexed absolute quantification for proteomics using concatenated signature peptides encoded by QconCAT genes. Nat Protocols 1(2):1029–1043

Ptacek J, Devgan G, Michaud G, Zhu H, Zhu X, Fasolo J, Guo H, Jona G, Breitkreutz A, Sopko R, McCartney RR, Schmidt MC, Rachidi N, Lee SJ, Mah AS, Meng L, Stark MJ, Stern DF, De VC, Tyers M, Andrews B, Gerstein M, Schweitzer B, Predki PF, Snyder M (2005) Global analysis of protein phosphorylation in yeast. Nature 438(7068):679–684

Rao PV, Reddy AP, Lu X, Dasari S, Krishnaprasad A, Biggs E, Roberts CT, Nagalla SR (2009) Proteomic identification of salivary biomarkers of type-2 diabetes. J Proteome Res 8(1):239–245

Rappsilber J, Ryder U, Lamond AI, Mann M (2002) Large-scale proteomic analysis of the human spliceosome. Genome Res 12(8):1231–1245

Raymond F, Metairon S, Borner R, Hofmann M, Kussmann M (2006) Automated target preparation for microarray-based gene expression analysis. Anal Chem 78(18):6299–6305

Riaz S, Skinner V, Srai SK (2011) Effect of high dose thiamine on the levels of urinary protein biomarkers in diabetes mellitus type 2. J Pharm Biomed Anal 54(4):817–825

Righetti PG, Boschetti E (2007) Sherlock Holmes and the proteome—a detective story. FEBS J 274(4):897–905

Righetti PG, Boschetti E, Lomas L, Citterio A (2006) Protein equalizer technology: the quest for a "democratic proteome". Proteomics 6(14):3980–3992

Righetti PG, Castagna A, Antonioli P, Boschetti E (2005) Prefractionation techniques in proteome analysis: the mining tools of the third millennium. Electrophoresis 26(2):297–319

Ross PL, Huang YN, Marchese JN, Williamson B, Parker K, Hattan S, Khainovski N, Pillai S, Dey S, Daniels S, Purkayastha S, Juhasz P, Martin S, Bartlet-Jones M, He F, Jacobson A, Pappin DJ (2004) Multiplexed protein quantitation in saccharomyces cerevisiae using amine-reactive isobaric tagging reagents. Mol Cell Proteomics 3(12):1154–1169

Rubio-Aliaga I, Marvin-Guy LF, Wang P, Wagniere S, Mansourian R, Fuerholz A, Saris WH, Astrup A, Mariman EC, Kussmann M (2011) Mechanisms of weight maintenance under high- and low-protein, low-glycaemic index diets. Mol Nutr Food Res 55(11):1603–1612

Sanchez JC, Chiappe D, Converset V, Hoogland C, Binz PA, Paesano S, Appel RD, Wang S, Sennitt M, Nolan A, Cawthorne MA, Hochstrasser DF (2001) The mouse SWISS-2D PAGE database: a tool for proteomics study of diabetes and obesity. Proteomics 1(1):136–163

Sanchez JC, Converset V, Nolan A, Schmid G, Wang S, Heller M, Sennitt MV, Hochstrasser DF, Cawthorne MA (2003) Effect of rosiglitazone on the differential expression of obesity and insulin resistance associated proteins in lep/lep mice. Proteomics 3(8):1500–1520

Sanders SL, Jennings J, Canutescu A, Link AJ, Weil PA (2002) Proteomics of the eukaryotic transcription machinery: identification of proteins associated with components of yeast TFIID by multidimensional mass spectrometry. Mol Cell Biol 22(13):4723–4738

Schena M, Shalon D, Davis RW, Brown PO (1995) Quantitative monitoring of gene expression patterns with a complementary DNA microarray. Science 270(5235):467–470

Scherl A, Shaffer SA, Taylor GK, Kulasekara HD, Miller SI, Goodlett DR (2008) Genome-specific gas-phase fractionation strategy for improved shotgun proteomic profiling of proteotypic peptides. Anal Chem 80(4):1182–1191

Schlatzer DM, Dazard JE, Dharsee M, Ewing RM, Ilchenko S, Stewart I, Christ G, Chance MR (2009) Urinary protein profiles in a rat model for diabetic complications. Mol Cell Proteomics 8(9):2145–2158

Schmidt A, Kellermann J, Lottspeich F (2005) A novel strategy for quantitative proteomics using isotope-coded protein labels. Proteomics 5(1):4–15

Schrimpe-Rutledge AC, Fontes G, Gritsenko MA, Norbeck AD, Anderson DJ, Waters KM, Adkins JN, Smith RD, Poitout V (2012) Discovery of novel glucose-regulated proteins in isolated human pancreatic islets using LC-MS/MS-based proteomics. J Proteome Res 11(7):3520–3532

Singh LP, Jiang Y, Cheng DW (2007) Proteomic identification of 14-3-3zeta as an adapter for IGF-1 and Akt/GSK-3beta signaling and survival of renal mesangial cells. Int J Biol Sci 3(1):27–39

Starkey JM, Zhao Y, Sadygov RG, Haidacher SJ, Lejeune WS, Dey N, Luxon BA, Kane MA, Napoli JL, Denner L, Tilton RG (2010) Altered retinoic acid metabolism in diabetic mouse kidney identified by O isotopic labeling and 2D mass spectrometry. PLoS ONE (Electronic Resource) 5(6):e11095

Sunyaev S, Liska AJ, GolodA, Shevchenko A, Shevchenko A (2003) MultiTag: multiple error-tolerant sequence tag search for the sequence-similarity identification of proteins by mass spectrometry. Anal Chem 75(6):1307–1315

Surinova S, Schiess R, Huttenhain R, Cerciello F, Wollscheid B, Aebersold R (2011) On the development of plasma protein biomarkers. J Proteome Res 10(1):5–16

Suss C, Czupalla C, Winter C, Pursche T, Knoch KP, Schroeder M, Hoflack B, Solimena M (2009) Rapid changes of mRNA-binding protein levels following glucose and 3-isobutyl-1-methylxanthine stimulation of insulinoma INS-1 cells. Mol Cell Proteomics 8(3):393–408

Syka JE, Coon JJ, Schroeder MJ, Shabanowitz J, Hunt DF (2004a) Peptide and protein sequence analysis by electron transfer dissociation mass spectrometry. Proc Natl Acad Sci U. S. A. 101(26):9528–9533

Syka JE, Marto JA, Bai DL, Horning S, Senko MW, Schwartz JC, Ueberheide B, Garcia B, Busby S, Muratore T, Shabanowitz J, Hunt DF (2004b) Novel linear quadrupole ion trap/FT mass spectrometer: performance characterization and use in the comparative analysis of histone H3 post-translational modifications. J Proteome Res 3(3):621–626

Tabb DL, Saraf A, Yates JR III (2003) GutenTag: high-throughput sequence tagging via an empirically derived fragmentation model. Anal Chem 75(23):6415–6421

Tilton RG, Haidacher SJ, Lejeune WS, Zhang X, Zhao Y, Kurosky A, Brasier AR, Denner L (2007) Diabetes-induced changes in the renal cortical proteome assessed with two-dimensional gel electrophoresis and mass spectrometry. Proteomics 7(10):1729–1742

Varghese SA, Powell TB, Budisavljevic MN, Oates JC, Raymond JR, Almeida JS, Arthur JM (2007) Urine biomarkers predict the cause of glomerular disease. J Am Soc Nephrol 18(3):913–922

Wallace DC (1999) Mitochondrial diseases in man and mouse. Science 283(5407):1482–1488

Wang Y, Lam KS, Lam JB, Lam MC, Leung PT, Zhou M, Xu A (2007) Overexpression of angiopoietin-like protein 4 alters mitochondria activities and modulates methionine metabolic cycle in the liver tissues of db/db diabetic mice. Mol Endocrinol 21(4):972–986

Wang P, Holst C, Andersen MR, Astrup A, Bouwman FG, Otterdijk S van, Wodzig WK, Baak MA van, Larsen TM, Jebb SA, Kafatos A, Pfeiffer AF, Martinez JA, Handjieva-Darlenska T, Kunesova M, Saris WH, Mariman EC (2011a) Blood profile of proteins and steroid hormones predicts weight change after weight loss with interactions of dietary protein level and glycemic index. PLoS. One 6(2):e16773

Wang QM, Yang H, Tian DR, Cai Y, Wei ZN, Wang F, Yu AC, Han JS (2011b) Proteomic analysis of rat hypothalamus revealed the role of ubiquitin-proteasome system in the genesis of DR or DIO. Neurochem Res 36(6):939–946

Wang P, Holst C, Astrup A, Bouwman FG, Otterdijk S van, Wodzig WK, Andersen MR, Baak MA van, Rasmussen LG, Alfredo MJ, Jebb SA, Pfeiffer AF, Kafatos A, Handjieva-Darlenska T, Hlavaty P, Saris WH, Mariman EC (2012) Blood profiling of proteins and steroids during weight maintenance with manipulation of dietary protein level and glycaemic index. Br J Nutr 107:106–119

Washburn MP, Wolters D, Yates JR III (2001) Large-scale analysis of the yeast proteome by multidimensional protein identification technology. Nat Biotechnol 19(3):242–247

Wollscheid B, Bausch-Fluck D, Henderson C, O'Brien R, Bibel M, Schiess R, Aebersold R, Watts JD (2009) Mass-spectrometric identification and relative quantification of N-linked cell surface glycoproteins. Nat Biotechnol 27(4):378–386

Wu CC, MacCoss MJ, Howell KE, Matthews DE, Yates JR III (2004) Metabolic labeling of mammalian organisms with stable isotopes for quantitative proteomic analysis. Anal Chem 76(17):4951–4959

Zhang D, Yang H, Kong X, Wang K, Mao X, Yan X, Wang Y, Liu S, Zhang X, Li J, Chen L, Wu J, Wei M, Yang J, Guan Y (2011) Proteomics analysis reveals diabetic kidney as a ketogenic organ in type 2 diabetes. Am J Physiol Endocrinol Metab 300(2):E287–E295

Zhen Y, Xu N, Richardson B, Becklin R, Savage JR, Blake K, Peltier JM (2004) Development of an LC-MALDI method for the analysis of protein complexes. J Am Soc Mass Spectrom 15(6):803–822

Zhou H, Xiao Y, Li R, Hong S, Li S, Wang L, Zeng R, Liao K (2009) Quantitative analysis of secretome from adipocytes regulated by insulin. Acta Biochim Biophys Sin 41(11):910–921

Zubarev RA, Hakansson P, Sundqvist B (1996) Accuracy requirements for peptide characterization by monoisotopic molecular mass measurements. Anal Chem 68(22):4060–4063

Chapter 11
Metabolomics in the Systems-Level Study of the Metabolic Syndrome

Tuulia Hyötyläinen

Abstract Metabolic syndrome (MetS) is a combination of medical disorders, including abnormalities in insulin, glucose and lipid metabolism, hypertension and abdominal obesity, causing several metabolic complications. Metabolomics has been increasingly employed for investigating obesity and obesity-related diseases in recent years. Because metabolome is dependent on the physiological, environmental, and genetic status of an organism, metabolomics can help to understand the biochemical networks that underlie metabolic homeostasis in obesity. The ongoing development of mass spectrometry (MS) and chromatography has enabled more comprehensive coverage of the metabolome. Many obesity-related metabolites have already been identified by metabolomics and demonstrated to be disturbed significantly in both animal models and in humans. This chapter describes the methodology used in metabolomics and summarizes the current knowledge about the effect of obesity and MetS on metabolic pathways.

Keywords Metabolomics · Chromatography · Mass spectrometry · NMR · Metabolic syndrome

11.1 Introduction

Metabolomics involves the comprehensive characterization of the small molecule metabolites (MW < 1500 amu) found in an organism. The metabolites include a large variety of compounds, including nucleotides, amino acids, bile acids, steroids, fatty acids, carbohydrates, hormones and lipids, among others. The metabolome is closely tied to the genotype of an organism, and is affected by the environmental factors such as lifestyle, diet and gut microbiota (Holmes et al 2011; Vrieze et al 2010; Wikoffa et al. 2009). Therefore, metabolomics offers a unique opportunity to look at genotype-phenotype as well as genotype-envirotype relationships. The levels of metabolite indicate the status of the organism's physiology, for example, metabolites

T. Hyötyläinen (✉)
VTT Technical Research Centre of Finland,
Tietotie 2, P. O. Box 1000, 02044 VTT, Espoo, Finland
e-mail: tuulia.hyotylainen@vtt.fi

M. Orešič, A. Vidal-Puig (eds.), *A Systems Biology Approach to Study Metabolic Syndrome*, 213
DOI 10.1007/978-3-319-01008-3_11,

such as citrate, lactate, and glucose can reflect situations such as apoptotic alterations, hypoxia and oxidative stress.

The advances in metabolomics have been driven by the development of analytical technologies, in combination with advances in mathematical and statistical modeling for the data mining and data interpretation. In spite of the novel developments, currently there is no single metabolomics platform that allows for the comprehensive analysis of full metabolome. The metabolic analyzes are challenging because the high number of metabolites, the large concentration range, over eight orders of magnitude, and chemical diversity of metabolites. No single analytical method can cover analysis of all the metabolites. It should also be noted that not only is the chemical analysis difficult enough, but the mathematical modeling needed to make useful deductions is often even more challenging, owing to the colossal amount of data produced by the novel analytical methodologies. In metabolomics the appropriate design of the work protocol is essential, including all the steps from study design to sampling, selection of both the analytical techniques, data preprocessing and statistical tools, to identification of candidate biomarkers, and finally, to the elucidation of the biological meaning of the findings.

In recent years, metabolomics has been increasingly employed for investigating obesity and obesity-related diseases. Metabolomics can help to understand the biochemical networks that underlie metabolic homeostasis in obesity. Indeed, many obesity-related metabolites have already been identified by metabolomics and demonstrated to be disturbed significantly in both animal models and humans.

11.2 Selection of the Analytical Tools

The analytical workflow covers all steps from sampling, storage, sample preparation, analysis and data preprocessing. Two types of approaches are used in the metabolomics, namely targeted selective analysis and more comprehensive, non-targeted profiling methods (Fig. 11.1). In the targeted analysis, only a limited number of preselected metabolites are analyzed with an optimized, selective and typically quantitative analytical protocol. This approach allows very sensitive and robust determination of the metabolites, and is suitable for the hypothesis driven approach where the key metabolic pathways are already known. However, the targeted approach gives relatively limited information of the total metabolic system. In the non-targeted approaches, for hypothesis generating studies, the endogenous metabolites of potential interest are not selected a priori, and the aim is to cover as many metabolites as possible. The non-targeted, global methods are typically only semi-quantitative, however, and it is not possible to optimize the methodology for all individual compounds.

The quality control is an essential step in metabolomics particularly when large sample sets are analyzed, and the methodological variation must be corrected. Typically, a suitable set of internal standards are used for normalization of the methodological variation. Moreover, the performance of the method must be followed by using sufficient amount of standard samples, blank samples and control samples. The best control sample is to take an aliquot of each sample of the batch and

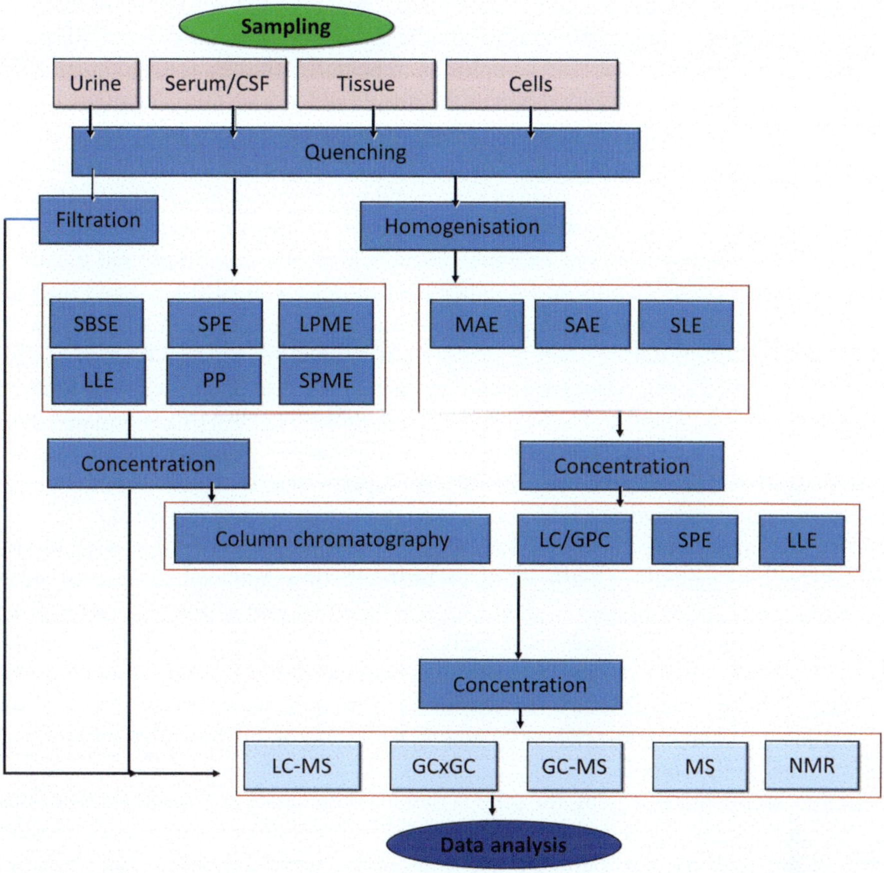

Fig. 11.1 The work flow for sample preparation for metabolomics. *GPC* gel permeation chromatography, *LLE* liquid-liquid extraction, *LPME* liquid phase micro extraction, *MAE* microwave assisted extraction, *PP* protein precipitation, *SAE* sonication assisted extraction, *SBSE* stir-bar sorptive extraction, *SLE* solid-liquid extraction, *SPE* solid phase extraction, *SPME* solid-phase micro extraction

pool them together, when possible. In addition, a batch sample for controlling of the long-time variation should be analyzed within each sample set. In routine analyzes the number of quality control samples should be ca. 5 % of the total sample amount (1:20 samples) but in more complex analyzes, quality control samples should be analyzed even more frequently (1:5 to 1:10 samples). The randomization of both sample preparation and analysis order is also important.

11.2.1 Sampling and Sample Preparation

In metabolomics, the sample type may be gaseous, such as breath samples, liquid such as urine, saliva, blood, serum, plasma, CSF, or (semi)soli such as cells, and tissue biopsies. With biological samples particularly, the sampling, storage and sample preparation are in crucial role because unsuitable protocols can lead to biased results due to conversion or degradation of metabolites (Álvarez-Sánchez et al. 2010a).

The most common types of samples in metabolic studies are blood-derived serum and plasma because blood metabolic profile reflects systemic changes in the metabolome. In addition, blood samples require relatively non-invasive sampling. It should be noted, however, that the metabolic profiles of serum and plasma are not identical because of the preparation of serum and plasma are quite different. Also, in plasma samples, the anticoagulant used for the plasma collection has also an effect to the metabolic composition. For example, citric acid which is an important metabolite of the TCA cycle, is present in higher concentrations in EDTA plasma than in serum, most probably due to complex formation between citric acid and Ca^{2+} and Mg^{2+} ions in serum during sample preparation, leading to precipitation in serum (Gronwald et al. 2008). Also, the overall protein content in serum is lower than in plasma which is a benefit in the analysis of metabolites because of the reduced matrix effects in e.g. LC-MS determinations. For the analysis of lipids, the serum and EDTA or citrate plasma give very similar results, however, for the determination of carboxylic acids and amino acids, it is best to use serum instead of plasma.

Also the temperature during the sampling and sample preprocessing may be crucial for specific metabolites, particularly for unstable compounds such as eicosanoids. Deactivation of metabolism, i.e. quenching, is essential part of the procedure. It is particularly critical in the analysis of cells and tissues when the goal is to elucidate the metabolic profile inside (endometabolome) and outside (exometabolome) the cell (Álvarez-Sánchez et al. 2010b). Deactivation can be done by very rapid freezing of the sample, by changing the pH, or by adding a mixture of enzyme inhibitors and/or antioxidants to the sample.

In the sample pretreatment, the goal is to isolate the analytes of interest from the matrix, and to improve the selectivity, detectability, reliability, accuracy and repeatability of the analysis. The sample preparation includes several steps, depending on the sample type (Fig. 11.1). The sample preparation is in most cases necessary due to the complexity of the samples. It should be emphasized that sample preparation is the major cause of errors in the analytical results because typically, multistep and often manual procedures are required in this stage. The lack of homogeneous standard protocols for sample preparation also introduces a barrier to comparing results among laboratories and to reproducing metabolomics experiments. In most cases, extraction is needed before the analysis, except in cases where analysis is to be performed directly on the tissue, as is the case with matrix-assisted laser desorption/ionization (MALDI) imaging (Trim et al. 2008; Pól et al. 2011) or desorption electrospray ionization (DESI) (Wu et al. 2009). For tissue samples, homogenization is required before the extraction. Homogenization can be done using cryogenic homogenizator or more conventionally, using ball mills. The advantage of the cryogenic homogenizator is that the samples are deep frozen during the homogenization, minimizing alteration of the sample during this step.

It should also be noted that in the sample preparation stage, different approaches are needed for targeted analysis and for global profiling. For targeted analyzes, highly selective sample pretreatment is beneficial while non-selective methodology that is not biased towards any certain groups of metabolites is needed for global profiling (Hyötyläinen 2012; Hyötyläinen 2010; Álvarez-Sánchez et al. 2010b). For extraction, the most common methods are liquid extraction with suitable organic solvents, or solid-phase extraction for liquid samples. In many cases, simple protein precipitation with methanol or acetonitrile is sufficient. For targeted methods, the extraction conditions can be carefully adjusted, and often the crude sample extract, e.g. after liquid extraction, is purified using solid-phase extraction to selectively remove disturbing matrix compounds such as phospholipids. For global profiling, simple solvent extraction is typically used, such as Folch extraction for lipids and other hydrophobic metabolites while methanol or acetonitrile is often used for more polar metabolites. In the profiling, it is difficult to find optimal conditions for all types of analytes, so some compromised has to be done in the selection of conditions (Álvarez-Sánchez et al. 2010b; Hyötyläinen 2012; Theodoridis et al. 2012).

Novel approaches in this field utilize techniques such as cryogenic homogenization of tissues, solid-phase extraction with novel formats and automated at-line sample pretreatment systems can help to minimize the challenges in the sample preparation step.

11.2.2 *Analytical Methodologies*

At present, the methodologies used in metabolomics for final separation and identification are based mainly on mass spectrometry (MS), often combined with chromatographic methods, such as liquid chromatography (LC) and gas chromatography (GC). Also NMR is used to some extent. Capillary electrophoretic techniques (CE) have also been utilized in some metabolic studies, however, they have not gained wide popularity, mainly because of the less repeatable migration times hinders the use of CE in any large scale analyses.

Although MS detection alone can be used for metabolomics, using high-resolution chromatographic separation prior to MS detection, which can be achieved with HPLC or GC, or even more preferable, UHPLC or GCxGC, is better suited to global profiling of the metabolome. Optimized chromatographic separation improves peak shape and sensitivity of the detection, thus dramatically decreasing signal variability in the MS detection. Moreover, the separation of components by the chromatography prior to their detection minimizes matrix effects, such as ion suppression, charge competition, and other possible interactions between matrix components. Moreover, isobaric components can be analyzed, as long as they can be separated by the chromatographic step. With MS alone, it is not possible to analyze isobaric compounds they have identical elemental compositions and therefore identical *m/z* ratios.

11.2.2.1 MS Analyzers

Several types of MS systems are used in metabolomics, and the selection of the instrument depends on the application (Dettmer et al. 2007). If the goal is to determine

Table 11.1 Comparison of the MS instruments in terms of accuracy, resolution and scanning speed. (modified from http://www.chromedia.org)

Analyzer	Mass range (m/z, $*10^{-3}$)	Resolution (Δm/z)	Resolving power, FWHM ($*10^{-3}$)	Mass accuracy (ppm)	Scan rate (Hz)
Single quadrupole	3	0.1–0.7	–	< 100	0.5–20
QqQ	3	0.1–0.7	–	< 100	0.5–4
Ion trap	4–20	0.1–0.7	–	50–200	5–10
Linear ion trap	4–20	0.1–0.7	–	50–200	1–4
Orbitrap	6	–	50–100	0.5–1	0.5–2
TOF	20–40	–	15–20	2–5	20
QTOF	10–40	–	15–25	3–5	20
IM-TOF	20–100	–	20–40	0.5–1	30
Double focusing magnetic sector	5–15	–	30–80	1–5	0.1–0.5
FTICR	5–15	–	100–1500	0.1–1	0.3–1

elemental composition of specific metabolites, high-resolution, accurate mass determination (HRAM), typically together with tandem mass measurements for structural characterization, is required. For targeted, qualitative analyzes, low-resolution mass spectrometers such as quadrupole MS are typically used; as triple quadrupole systems for LC-MS, and single quadruple systems for GC-MS.

In LC-MS combinations, as well as in shotgun MS, the use of hybrid quadrupole mass spectrometer-time-of-flight MS (Q-TOF) and Fourier-trap-MS (FTMS) instruments, both the cyclotron and the Orbitrap type, allow for accurate mass measurements and, subsequently, the identification of metabolites. Desired detector characteristics include high resolution (greater than 10,000) and accurate mass (less than 5 ppm) in full-scan mode. Table 11.1 compares the critical features of the different MS systems. TOF technologies typically offer higher scanning speed (more than 20 Hz) and achieve the required mass accuracy with internal calibration standards. Orbitrap and FT-ICR mass spectrometers offer excellent measurement accuracy with external calibration and a wide resolution range. Particularly the FT-Orbitrap-MS has been gaining popularity in combination with LC separation because it a high-speed, high ion-transmission instrument, due to shorter accumulation times. The maximum resolving power and mass accuracy are lower than those for FT-ICR-MS but clearly higher than for QTOFMS. The novel tandem MS system combining ion mobility with TOFMS enable even the separation of isobaric compounds.

Targeted analyses, using LC for the separation are typically done using triple-quadrupole (QqQ) mass spectrometer, owing to its high duty cycle and a large dynamic range in MS/MS mode. The QqQ system utilizes collision-induced dissociation while monitoring unique precursor to product ion transitions, affording superior sensitivity, precision, and accuracy, especially for simultaneous analysis of multiple analytes (selected reaction monitoring, multiple reaction monitoring). It should be noted, however, that novel HRAM MS systems are now achieving sensitivity comparable to that of QqQ instruments while providing full-scan MS data on all analytes, which will certainly increase their utility for quantitative assays. An especially attractive feature of the HRAM approach is that, unlike multiple reaction monitoring experiments, data are collected over a preselected wide mass range, and the ability to measure unexpected analytes is preserved. In GC-MS, the quadrupole MS is still the most common instrument, but also TOFMS and other accurate mass instruments are gaining popularity.

Shotgun MS. The shotgun methods uses direct infusion of sample extracts into the mass spectrometer (Schwudke et al. 2007) and in principle, any type of (ionizable) compound can be analyzed. Typically, high-resolution mass spectrometers are used in the shot-gun approach. The shotgun approach is a relatively simple and rapid way to profile crude extracts of the biological samples. The major complication in this methodology is the ion suppression and compounds present in trace amount are often not detected due to the suppression phenomena (Moco et al. 2007). Moreover, because the composition of the samples can vary substantially, the ion suppression can change from sample to sample, and thus the quantitative results for global profiling is not very reliable. With labeled standards, quantitative results can be obtained for those, rather limited number of compounds that the standards are available. To avoid ion suppression, more careful sample pre-treatment, using *e.g.* fractionation can be applied, however, then the main advantage of the shotgun approach, i.e. simplicity and speed are then lost. Thus, the applicability of shot-gun MS in the search of novel, previously unknown biomarkers is relatively restricted. However, the shot-gun MS is a viable tool for clinical practices for a group of identified biomarkers, particularly when combined with suitable, very selective sample preparation method.

LC–MS. In metabolomics, LC-MS based methodologies have several advantages and this technique has been widely used for both for targeted and non-targeted analyses, using various types of mass spectrometers, from a simple single quadrupole to hybrid instruments and to high-resolution Orbitrap instruments (Nygren et al. 2011; Theodoridis et al. 2012). The sensitivity in LC-MS is typically high, and identification of novel compounds in possible. The novel fast LC methodologies, utilizing very high pressures, elevated temperatures and novel column materials allow high-through-put analyses.

GC–MS. GC is suitable for the targeted analysis of sufficiently volatile compounds, such as fatty acids, small polar compounds, such as carboxylic acids, sugars, amino acids and sterols. The latter require derivatization before the analysis, complicating the sample preparation. This has been considered as the main bottleneck in the metabolomic analyses and reason why many of the GC-based methodologies are currently done by LC-MS. However, the GC-based methods have several advantages over the LC-MS based approaches. The main advantage is the availability of huge commercial spectral libraries for the identification of unknown metabolites. An additional advantage of all GC based methods is the repeatable characteristics of retention, particularly when nonpolar stationary phases are used. This makes it possible to use universal retention indexes which are highly useful in the identification of unknown compounds, in combination with mass spectral detection. Moreover, the quantitation is more reliable, because the matrix effects in GC-MS are not an issue.

The most commonly used derivatization methods are trimethylsilylation and tert-butyldimethylsilylation. In addition, keto-(oxo-) groups are usually oximated in order to improve their GC properties and prevent enolization reactions which can introduce multiple reaction products. The derivatization can be done in automatic fashion even in the injector systems of the modern GC instruments, thus simplifying the procedure.

Recently, two-dimensional gas chromatography combined with time-of-flight MS (GCxGC-TOFMS) has been gaining popularity in the metabolomics as it is one of the most powerful analytical tools for the separation of organic compounds in complex samples, and particularly suitable for the non-targeted profiling, allowing the separation of several hundreds of compounds in a single analysis (Castillo et al. 2011; Hyötyläinene et al. 2012; Dettmer et al. 2007).

11.2.2.2 NMR

NMR (^{1}H-NMR is a fast and robust method, it is non-invasive and non-destructive and the analytical costs per-sample are low (although the instruments are very expensive) (Malet-Martinoa et al. 2011). NMR has the theoretical possibility to detect and structurally elucidate any molecule containing hydrogen and also being very quantitative in its nature. However, the main disadvantage of NMR is the inherent lack of sensitivity, typical detection limits being in the range milli- to high micromolar. In addition, due to the large number of metabolites, the one-dimensional spectra gets very dense with spectral information resulting in more abundant metabolites masking signal of less abundant ones. The complexity of the data, together with the limited sensitivity, limit quantitative profiling to less than100 metabolites in most biological samples by current methods. For NMR, minimal sample preparation is required for urine and other low-molecular-weight metabolite-containing fluids, whereas blood, plasma, and serum require extraction or NMR-weighted techniques to separate polar and lipophilic metabolites (Aranibar et al. 2006).

11.2.2.3 Identification of Metabolites

The identification of unknown metabolites is a very challenging task. If authentic standards are available, a metabolite can be identified unambiguously by directly comparing its mass, retention time and fragmentation spectrum with those of the standards. However, this is not always possible, and in any case, the metabolite has to be first tentatively identified before standards can be applied.

In GC(xGC)-EI-MS, comprehensive commercial libraries are available (NIST/EPA/NIH (Babushok et al. 2007), Fiehnlib (Kind et al. 2009) and Golm (Kopka et al. 2005) based on electron ionization using high internal energy (70 eV). The NIST library, for example, have a mass spectra for 192 108 compounds. In GC-based applications it is also possible to utilize retention indexes in the identification, together with the spectra, making the identification much more reliable. It should also be noted that a large amount of compounds still cannot be identified as they are not found in the spectral libraries. However, even for unknown compounds, due to the reliable fragmentation patterns in GC(xGC)-EI-MS, group level identification can typically be made, making the classification of the functional group(s) of the metabolites possible (Hummel et al. 2010). The elemental composition of a compound can be determined by GC equipped with chemical ionization (CI) and accurate mass detector (Kumari et al. 2011). With electron impact ionization, the molecular ion often is not detected due to the extensive fragmentation, making the identification of unknown compounds more challenging. With GC-CI-MS, the molecular ion can usually be detected; however, the mass spectral libraries for GC-CI-MS are not as extensive as for GC-EI-MS. The best approach to identify unknown compounds is to analyse the samples with both GC-EI-MS and GC-CI-MS, thus combine the fragment information and accurate mass determination of the molecular ion.

In LC-MS and shotgun-MS, the lack of comprehensive mass-spectral libraries is a major limitation in the identification of unknown compounds. Some spectral libraries are available, including the Human Metabolome Database (HMDB) (http://www.hmdb.ca), the METLIN Metabolite Database (http://metlin.scripps.edu) (Smith et al. 2005) and the MassBank (http://www.massbank.jp) (Horai et al. 2010). Unfortunately, construction of universal spectral databases for API-MS is challenging due to the poor reproducibility and high interinstrument variability of fragmentation patterns. In the interpretation of the mass spectra from both shotgun MS and LC-MS studies, first step is to ensure that the signal of interest really corresponds to a monoisotopic ion and not to a natural isotopologue ion or an adduct ion (Xu et al. 2010; Ralf et al. 2011). Elemental compositions can be deduced from accurate mass measurements and used for further database queries. To aid the identification, a number of chemical databases containing huge amount of chemical structures to aid in structural elucidation after fragmentation of the metabolite are available (e.g. ChemSpider, http://www.chemspider.com, http://msbi.ipb-halle.de/MetFrag) (Wolf et al. 2010). Also several in-silico software packages for the prediction of *in-silico* spectra for compounds are available (Neumann et al. 2010; Schymanski et al. 2009).

11.2.3 Data Analysis Tools

The correct choice of the mathematical tools is essential for efficient data mining and correct information of the data. It is challenging to link the huge amount of metabolic information obtained with the current methodologies with available clinical and genetic data. The work flow in data analysis compromises of signal processing, data normalization, transformation, and assessment followed by application of statistical methods for comparison of groups and the construction of predictive models (Katajamaa et al. 2007; Sumner et al. 2008).

Often, data on hundreds of metabolites is obtained in the modern metabolic analyzes. However, in practise the models using information of hundreds of metabolites are not realistic. For robust models, the number of significant variables should be less than ca. 25, derived from a limited number of metabolic markers (3-10 compounds). With a larger number of variables/metabolites, there is a high risk of over fitting of the data. Therefore, the first step of the statistical analysis is typically data reduction, using e.g. pattern recognition processes; either in unsupervised or supervised manner. In unsupervised data analysis, the data analysis is conducted without any preconceptions or preselection, i.e. without biasing the results by the introduction of prior information of the samples. The unsupervised methods include hierarchical cluster analysis and principal component analysis, and they are good in identification of patterns of the data. In the supervised approach, such as principal component regression and neural networks, each sample or metabolite is first associated to already known class, and this prior information is then used in generation of the clusters of patterns (Lee et al. 2008). For the discovery of predefined metabolic pathways or altered biological networks several methods are available (Aggio et al. 2010; Xia and Wishart 2010; Kankainen et al. 2011). In addition, new computational approaches have been recently developed for analysing high-dimensional datasets and elucidate the genotype to phenotype relationship using metabolic constraint-based modelling (CBM) approach (Duarte et al. 2007; Jerby et al. 2010; Trawick and Schilling 2006; Shlomi et al. 2009; Lanpher et al. 2006). In these methods, thermodynamic and stoichiometric constraints are given to enable the genome-scale analysis and prediction of metabolic phenotypes under diverse physiological and genetic conditions (Duarte et al. 2007; Jerby et al. 2010).

11.2.4 Comparison of Different Methodologies for Metabolomics

The key features of the analytical techniques utilized in metabolomics are shown in Table 11.2. In Human serum Metabolome study, five different metabolic profiling methods were used to experimentally characterize as much of the known serum metabolome as possible using 1) NMR; 2) GC-MS; 3) lipid mediators by LC-ESI-MS/MS; 4) lipidomics profiling via TLC/GC-FID-MS; and 5) direct infusion MS/MS (N. Psychogios et al. 2011). Using these five platforms, a total of 3564 distinct metabolites including several exogenous compounds could be identified from the serum. The study showed clear differences in the ability of the platforms to identify and quantify metabolites: NMR spectroscopy was able to identify and quantify 49 compounds, GC-MS was able to identify 90 and quantify 33 compounds, targeted ESI-MS/MS identified and quantified 96 compounds, TLC/GC-FID-MS identified and quantified 3381 compounds while direct infusion MS/MS identified and quantified 139 compounds. Some metabolites could be detected with several platforms, and for example, comparison of amino acid concentrations as measured by NMR, GC-MS and DFI MS/MS showed that the quantitative results are in relatively good agreement. The study showed that of the known, quantifiable serum metabolome (4229 known and probable metabolites; 665 literature derived metabolites and

Table 11.2 A comparison of the key features of the metabolomics methods. (modified from Hyötyläinen 2012)

	NMR	shot-gun-MS	UPLC-MS	UPLC-MS	GC-MS	GCxGC-TOFMS
Approach	*Profiling + targeted*	*Targeted*	*Profiling*	*Targeted*	*Targeted*	*Profiling + targeted*
Sample preparation	simple	Simple-moderate	Moderate	Tedious	Tedious	Tedious
Sample amount[a]	60–500 μl	5–100 μl	5–100 μl	5–100 μl	5–100 μl	5–100 μl
Sample throughput	ca 10 min/sample	50–100/day	40–80/day	20–80/day	20–80/day	20–30/day
Reproducability	good	Moderate	Moderate	Good	Good	Good
Sensitivity	Millimolar—high micromolar	Nanomolar	Nanomolar-picomolar	Nanomolar-picomolar	Nanomolar-picomolar	Nanomolar-picomolar
Cost/sample	Low	Low	Medium	Medium	Low-medium	Medium
Instrument cost	High-very high	High	Moderate-high	Moderate-high	Moderate-high	Moderate-high
Comments	Good spectral libraries, may require high sample volumes	Isobaric compounds cannot be identified, ion suppression is a major problem	Suitable for wide range of metabolites, identification of unknowns requires tandem MS	Suitable for wide range of metabolites	Only suitable for small metabolites (MW < 700), polar compounds require derivatization, huge spectral libraries available, retention indexes can be used for verification of identification	Only suitable for small metabolites (MW < 700) polar compounds require derivatization, huge spectral libraries available, clean spectra's due to very high separation efficiency, retention indexes can be used for verification of identification

[a]For biofluids, such as urine, serum/plasma

3564 experimentally derived or predicted), NMR was able to measure ~1.2 % (49/4229) of the human serum metabolome, GC-MS was able to measure 2.13 % (90/4229), ESI-MS/MS was able to measure 2.3 % (96/4229), TLC/GC-FID-MS (general lipidomics) was able to measure 79.9 % (3381/4229) while direct infusion MS/MS was able to measure 3.3 % (139/4229) of the serum metabolome. It should be noted that the study included only known metabolites. With more advance methodologies, such as global profiling with GCxGC-TOFMS, typically 1200 metabolites can be detected from human serum samples, of these however, only 10–30 % can be identified. Also, global profiling of lipids with UHPLC-QTOFMS typically allows detection of ca. 600 lipids in human serum, and of these, ca. 150 lipids can be identified.

11.3 Metabolic Profile in Metabolic Syndrome

Metabolic Syndrome (MetS) is a combination of medical disorders, including abnormalities in insulin, glucose and lipid metabolism, hypertension and abdominal obesity, causing several metabolic complications, such as type 2 diabetes, non-alcoholic fatty liver disease (NALFD) and coronary heart diseases (CHD). Abdominal adiposity and insulin resistance appear to be at the core of the pathophysiology of the MetS and its individual components. Insulin, on the other hand, has an effect on several key functions, such as amino acid uptake, protein synthesis, proteolysis, adipose tissue triglyceride lipolysis, lipoprotein lipase activity, very low density lipoprotein (VLDL) triglyceride secretion, muscle and adipose tissue glucose uptake, muscle and liver glycogen synthesis, and endogenous glucose production. Adipose tissue and liver are the two key organs associated with obesity, metabolic syndrome and diabetes risk (Fig. 11.2) (Cohen et al. 2011; Kotronen and Yki-Jarvinen 2008).

Current research indicates that the capacity of adipose tissue to safely store fat may be more vital than the actual amount of fat an individual has. In the case of excess calory intake, the extra energy is stored as fat, preferably in the adipocytes (Pietiläinen et al. 2011). However, the adipocytes do not have an unlimited capacity to store extra fat, and when their capacity is exceeded, the fat starts to accumulate in skeletal muscles, in liver and in other organs, leading to potentially toxic effects in peripheral tissues via the excessive accumulation of reactive lipid species. Particularly, the fat accumulation in skeletal muscle and in liver is associated with insulin resistance which in turn causes major changes in glucose and lipid metabolism. Insulin affects on amino acid uptake, protein synthesis, proteolysis, adipose tissue triglyceride lipolysis, lipoprotein lipase activity, very low density lipoprotein (VLDL) triglyceride secretion, muscle and adipose tissue glucose uptake, muscle and liver glycogen synthesis, and endogenous glucose production. Thus, metabolic analyses covering intermediates of these pathways can be utilized for in depth studies of the obesity associated diseases.

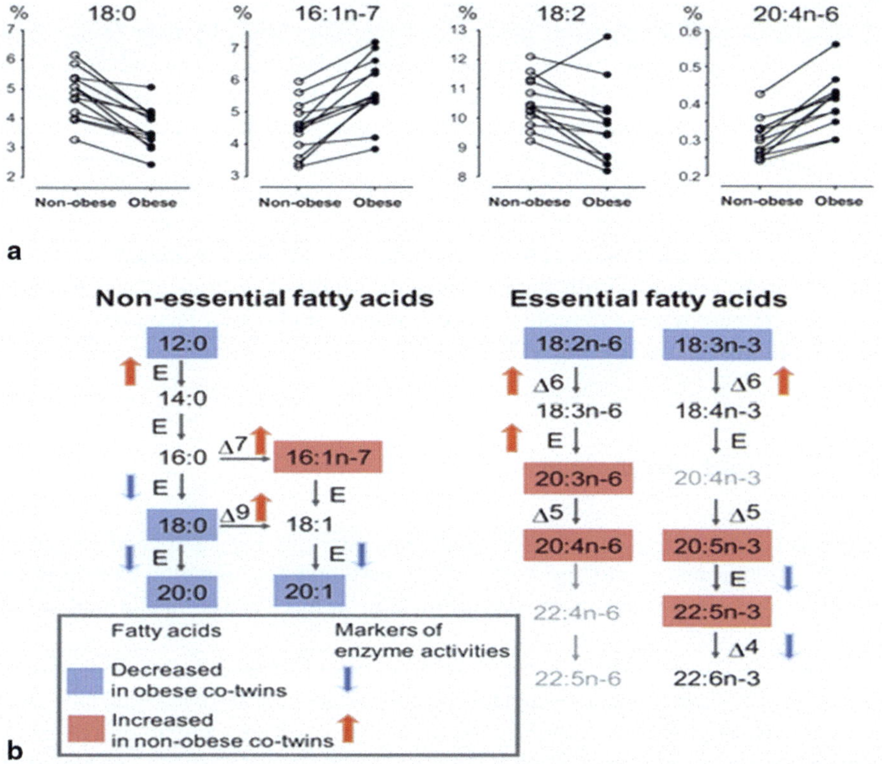

Fig. 11.2 Fatty acid composition of adipose tissue in acquired obesity. **a** Selected fatty acid relative amounts in twin pairs discordant for BMI. Lines connect the pairs of twins. **b** Schematic representation of fatty acid compositional changes when comparing heavy and lean obesity-discordant twins. Significant changes ($p<0.05$; pairwise t test) are color-coded. The activities of specific fatty acid elongation or desaturation steps are estimated by appropriate fatty acid concentration ratios. From Pietiläinen et al. 2011

11.3.1 Glucose Metabolism and Tricarboxylic Acid Cycle

Glucose metabolism and energy metabolism has been found to be severely altered in obesity in both animal and human studies. Elevated levels of lactate, a glucose precursor, has been observed both in blood of obese mice and Zucker rats (Serkova et al. 2006; Duggan et al. 2011) and in several human studies (Newgard et al. 2009; Felig et al. 1974). The concentration of lactate in blood is predominately determined by the balance of rate of lactate production in skeletal muscle, the erythrocytes, brain and adipose tissue, relative to the removal of lactate by other tissues and urinary excretion. It has been shown that subcutaneous fat is a significant source of lactate, and this could explain the higher blood lactate levels of obese subjects, due to increased release of lactate by the excess adipose tissue (Jansson et al. 1994).

However, obese subjects have also shown to have significantly higher net uptake of lactate, as well as glycerol, alanine and free fatty acids by liver (Felig et al. 1974). It has been suggested that elevated levels of lactate in obese subjects is due to upregulated anaerobic glycolysis, in comparison with lean subjects. The current results thus indicate perturbation of hepatic glucose production and lipid synthesis. Also pyruvate, the endpoint of glycolysis, has been found out to be upregulated in obese subjects (Felig et al. 1974). Under aerobic conditions, pyruvate enters the TCA cycle via acetyl coenzyme A to produce energy via formation of citrate, and other TCA cycle metabolites. The concentration of citrate is regulated by insulin, glucose levels, fatty acid utilization, cholesterol synthesis, liver clearance and renal excretion. Higher levels of citrate has been observed in high fat feed obese animals with insulin resistance (Li et al. 2008).

11.3.2 Lipid Metabolism

Lipids are a diverse group of compounds with multiple key biological functions: they function as energy storage sources, they participate in signaling pathways and constitute the cellular structural building blocks in both cell and organelle membranes (Orešič et al. 2008; Fahy et al. 2011).

Obesity is linked with increased basal lipolysis in adipose tissue, and elevated circulating FFAs. In blood, elevated levels of FFAs or cholesterol in blood have been used already for a long time as indicators of disease risk, as their alteration is related to some cardiovascular diseases, such as coronary heart disease, diabetes, and hypertension (Orvall et al. 1996; Laaksonen et al. 2002; Green et al. 2002). Traditionally, FAs secreted from adipocytes have been considered to function entirely as energy sources for other tissues of the body. However, it has been shown that FAs function also as endocrine factors that regulate metabolic function in target tissues and the obesity-associated insulin resistance may be explained by competition between increased FFAs and glucose for oxidative metabolism in insulin-responsive cells (Randle et al. 1963). Later studies have implied that glucose uptake rather than intracellular glucose metabolism is the rate-limiting step for FA-induced insulin resistance (Shulman 2000). Also the serum FFA composition, especially the proportion of saturated FAs has been observed to correlate positively with the development of obesity and diabetes (Wang et al. 2003). Diabetic subjects has been shown to have elevated levels of long-chain free fatty acids, such as oleic, palmitoleic and palmitic acids, and decreased levels of arachnidonic acid than non-diabetic subjects (Fiehn et al. 2010). Also the concentrations of blood TGs is affected in obesity due to tissue resistance to insulin mediated glucose uptake, which in turn accelerates the very low density lipoprotein (VLDL), TG production rate and leads to endogenous hypertriglyceridemia (Reaven et al. 1967; Barter and Nestel 1973; Kissebah et al. 1976).

Applying global lipid profiling, combined with targeted FA analysis to plasma samples from obese individuals has shown that fatty liver overproduces triglycerides with lower carbon number and double bond content (Westerbacka et al. 2010), indicative of increased de novo hepatic lipogenesis (Kotronen et al. 2009a). Further analysis of lipid profiles in lipoprotein particles showed that these lipids are also enriched in VLDL particles of insulin resistant subjects (Kotronen et al. 2009b). Recent research also implicates increased mitochondrial pathways and gluconeogenesis in NAFLD (Sunny et al. 2011). Induction of lipid oxidation is required for the endergonic steps of gluconeogenesis and ureagenesis, pathways that are partially localized in liver mitochondria and constitutively upregulated during insulin resistance (Sunny et al. 2011). Thus, unlike skeletal muscle, the insulin-resistant and fatty liver may activate oxidative metabolism (Iozzo et al. 2010; Sanyal et al. 2001). A prevalent model for development of diet-induced insulin resistance confirms these findings, stating that when mitochondrial fatty acid oxidation is inadequate to deal with the large load of dietary fat, the result is the accumulation of lipid-derived metabolites such as diacylglycerols (DAGs) and ceramides that can activate stress kinases to interfere with insulin action (Savage et al. 2007; Holland et al. 2008).

Recently, identification of numerous lipid and protein signals secreted from white adipose tissue has led to its recognition as a major endocrine organ (Rondinone 2006; Trayhurn and Bing 2006). In an animal study, male transgenic apolipoprotein E3 Leiden (ApoE3Leiden) mice were studied during high-fat feeding (Caesar et al. 2010). This animal study combined time-resolved microarray analyzes of mesenteric-, subcutaneous-, and epididymal adipose tissue (EWAT) with histology and targeted lipidomics. The results showed upregulated levels of linoleic acid and alpha-linolenic acid in EWAT in comparison with other depots. The authors hypothesized that this could be due to the androgen receptor, which expression was higher in EWAT than in other tissues, causing depot-dependent differences in de novo lipogenesis rate. The authors therefore suggested that dietary essential FAs are accumulated in EWAT as a result of sex-steroid mediated suppression of lipogenesis, providing an adaptive strategy to provide precursors for epididymal PUFA synthesis (Caesar et al. 2010). The same mouse model has been used also in another study, reporting that specific plasma free fatty acids (C16:1, C16:0, C18:1, C18:0 and C18:2, C22:6) as well as their ratio can be used to predict future glucose intolerance in these ApoELeiden mice (Wopereis et al. 2012).

In a recent human study, young and healthy obesity-discordant MZ twins were studied, allowing to focus on the effects of obesity independent of genetic factors (Pietiläinen et al. 2011). In this study, the lipid composition of adipose tissue was studied. The study subjects were monozygotic twin pairs, the other twin being obese, and the other lean. The results showed multiple changes in membrane phospholipids composition in obese, but still healthy co-twins. The changes in phospholipids were highly selective, showing both functional group as well as fatty acid specificity (Fig. 11.3). Despite that the obese co-twins had a lower dietary polyunsaturated fatty acid intake, they had increased levels of palmitoleic and arachidonic acids in their adipose tissue, including upregulated levels of ethanolamine plasmalogens containing arachidonic acid. In comparison to non-obese co-twins, the obese co-twins had increased levels of lysophosphatidylcholines, which are lipids found in proinflammatory (Yang et al. 2005) and proatherogenic conditions, as well as decreased levels of ether phospholipids, which are known to exert antioxidative properties (Engelmann 2004). These lipid changes were associated with insulin resistance. The results of the next part of the study, using molecular dynamics simulations of lipid bilayers, suggested that the observed lipid remodeling maintains the biophysical properties of lipid membranes, at the price, however, of increasing their vulnerability to inflammation (Pietiläinen et al. 2011). Next, a separate group of morbidly obese subjects was studied. Interestingly, in morbidly obese subjects, the levels of these plasmalogens containing arachidonic acid in the adipose tissue was distinctly decreased, on contrary to the results obtained for obese, but still metabolically healthy subjects. These results were further supported by *in vitro* adipocyte confirmatory model. The study also showed that that the lipid network regulating the observed remodeling may be responsive to genetic modulation. The authors therefore suggested that by a new generation of therapies directed at several targets in the lipid metabolism pathways, it may be possible to correct these abnormalities as proper management of obesity, and favorably modify the risk, course and outcome of diabetes and cardiovascular diseases. This study allowed to identify adaptive mechanisms that may lay behind the characteristic remodeling of the AT lipidome in response to positive-energy-balance-induced adipose tissue expansion during the evolution of obesity (Pietiläinen et al. 2011).

11.3.3 Amino Acid Metabolism

Amino acids have been in several studies to display a significant obesity associated variations. It has been reported already decades ago several amino acids, including branched chain amino acids (BCAAs), are elevated in the blood of obese, insulin-resistant, or type 2 diabetic subjects relative to healthy controls (Felig and Wahren 1971). BCAAs have an important role in protein synthesis as well as in glucose metabolism and oxidation and they can also regulate leptin secretion from fat and food intake. A more recent study suggests that obesity induces increased catabolism of BCAA and correlated with insulin resistance even in obese subject that are healthy, although more insulin resistant than the lean controls. Five amino acids, namely three BCAAs (isoleucine, leucine, valine) and two aromatic amino acids (tyrosine and phenylalanine) have been shown to have highly significant associations with future diabetes in a two large, longitudinal studies (Wang et al. 2011). Moreover, a combination of three amino acids predicted future diabetes. The change in concentration of these five amino acids during oral glucose tolerance test was not associated with incident diabetes, suggesting that concentrations after OGTT did not add predictive information to the baseline concentrations. Results of amino acid supplementation, both in a high-fat diet animal study and as an infusion of a AA mix in a human study (Tremblay et al. 2005) showed that the AA supplementation decreased the insulin sensitivity, thus indicating that these metabolites are not merely by-products of the disease process, but contribute to development of metabolic dysfunction.

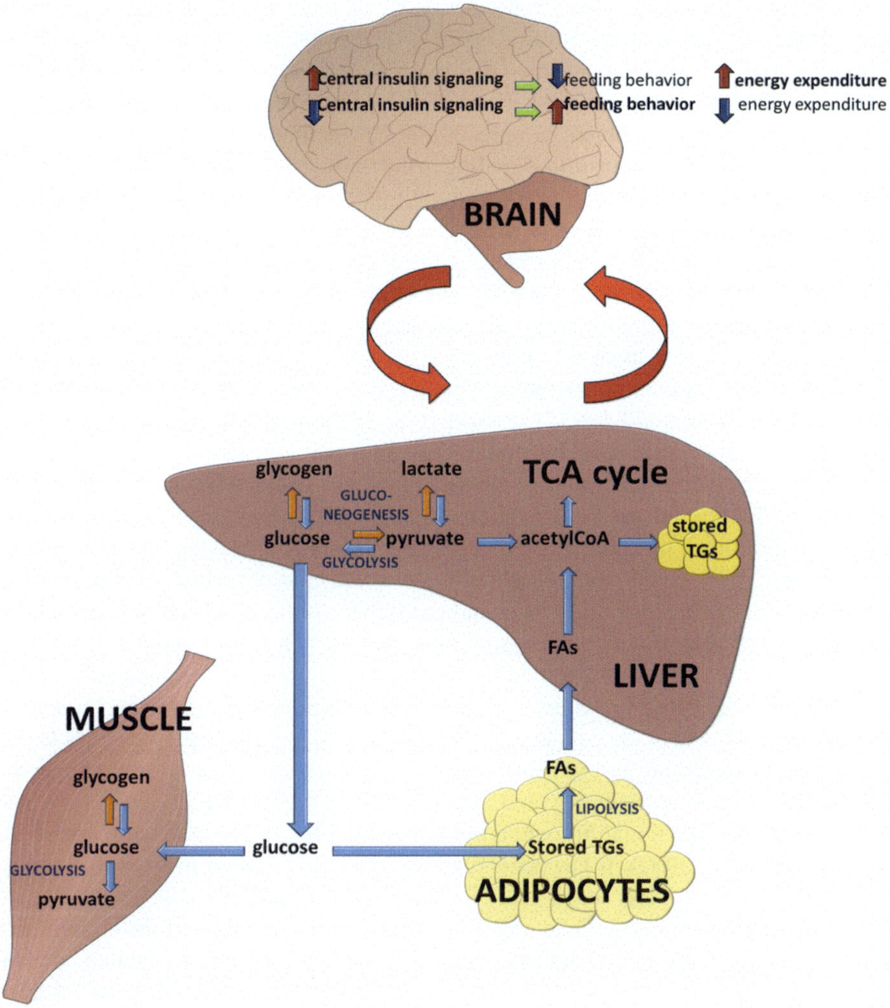

Fig. 11.3 The obesity associated metabolic pathway

11.3.4 Metabolites Related to Mitochondrial Dysfunction and Oxidative Stress

Mitochondrial functions include regulation of several key pathways related to energy homeostasis, including biogenesis, thermogenesis, and glucose and fatty acid metabolism, and obesity has shown to alter these functions which in turn may lead to obesity-related pathologies. Excess dietary intake can lead to significant mitochondrial dysfunction, in part due to the harmful effects of highly reactive molecular species (ROS) which can cause oxidative stress to the system. The oxidative stress is defined as a persistent imbalance between the production of ROS and antioxidant defenses of the organism.

Particularly, defects in mitochondrial oxidative metabolism of fatty acids have been linked to diet-induced obesity and the development of insulin resistance in adipose tissue and skeletal muscle (Kusunoki and Kanatani 2006; Maassen et al. 2007). Insulin resistance may arise from defects in these mitochondrial functions, which in turn lead to increases in intracellular FA metabolites (fatty acyl-CoA and diacylglyerol) that disrupt insulin signaling in the muscle as well as the liver (Qatanani and Lazar 2007).

Obesity has shown to cause systemic oxidative stress to the organism. The oxidative stress has been shown to correlate with fat accumulation in humans and mice (Halliwell 1995; Rosen et al. 2001; Evans et al. 2002). Studies have shown that ROS are upregulated already in prediabetic stage, which is suggested to be due to obesity-related elevations of FAs that cause oxidative stress due to increased mitochondrial uncoupling and β oxidation, leading to the increased production of ROS (Wojtczak and Schonfeld 1993; Carlsson et al. 1999; Rao and Reddy 2001). This hypothesis has been supported by studies that show that infusion of FAs into healthy subjects causes increased oxidative stress and insulin resistance that is reversed by infusion with antioxidants such as glutathione (Paolisso et al. 1992). In addition, there is a significant inverse correlation between the fasting plasma FFA concentration and ratio of reduced/oxidized glutathione in T2D patients (Palisso et al. 1996). In vitro studies have shown that ROS and oxidative stress lead also to the activation of multiple serine/threonine kinase signaling cascades (Kyriakis and Avruch 1996; Evans et al. 2003). These activated kinases can act on a number of potential targets in the insulin signaling pathway, including the insulin receptor and the family of IRS proteins. However, the molecular mechanisms by which oxidative stress leads to insulin resistance are not yet fully clarified. Also the decrease in mitochondrial function associated with obesity and insulin resistance is not totally clear, as it has been shown that functional mitochondria are needed for an FA-induced increase in ROS (Evans et al. 2002). It has been suggested that an increase in ROS due to FA oxidation may occur early during the development of insulin resistance and prior to mitochondrial dysfunction (Quatatni et al. 2007). At a later stage of developing insulin resistance, ROS might lead to a decrease in mitochondrial function that on turn can lead to the accumulation of fat in the muscle and liver, exacerbating the insulin resistance phenotype. Thus, the recent studies support the hypothesis that oxidative stress is a causative factor in the development of insulin resistance (Fridlyand and Philipson 2006; Houstis and Rosen 2006).

11.3.5 Metabolic Biomarkers of MetS

Early biomarkers for MetS and related diseases have been searched intensively. Current studies indicate several different metabolites characteristic for T2DM. Glucose, 2-hydroxyisobutyric acid (α-HB), linoleic acid, palmitic acid and phosphate have been indicated to be potential biomarkers in some studies (Li et al. 2009). Particularly, serum α-HB has been considered as an early marker for both insulin resistance and impaired glucose regulation (Gall et al. 2010). Association between α-HB levels and HOMA-IR has also been observed in specific lipoprotein fraction, namely in VLDL fraction (Hyötyläinen et al. 2012). Upregulation of several intermediates of pentose-phosphate pathway, such as xylose, has also been observed in other recent studies in diabetic subjects (Fiehn et al. 2010).

11.4 Conclusions

Metabolomics is a valuable tool for a detailed study of the pathophysiological mechanisms in obesity and the mechanisms in obesity-related diseases. The novel analytical tools allow detailed characterization of the metabolic profiles in biofluids and tissues. Moreover, integration of the metabolic data with clinical, genomic and transcriptomics data facilitates the development of metabolic models to understand the alterations of metabolism during obesity and related diseases. Several of the core metabolic pathways have been already been defined, however, a detailed knowledge of the metabolic pathways is still missing. Evaluation of the metabolic pathways across multiple tissues is crucial for understanding the compensatory mechanisms that operate at organismal level to maintain energy homeostasis. Detailed understanding of metabolic pathways may help in developing new generation of therapies for proper management of obesity.

References

Aggio RB, Ruggiero K, Granato Villas-Bôas S (2010) Pathway activity profiling (PAPi): from the metabolite profile to the metabolic pathway activity. Bioinformatics 26:2969–2976

Álvarez-Sánchez B, Priego-Capote F, Luque de Castro MD (2010a) Metabolomics analysis I. Selection of biological samples and practical aspects preceding sample preparation. Trends Anal Chem 29:111–119

Álvarez-Sánchez B, Priego-Capote F, Luque de Castro MD (2010b) Metabolomics analysis II. Preparation of biological samples prior to detection. Trends Anal Chem 29:120–127

Aranibar N, Ott KH, Roongta V et al (2006) Metabolomic analysis using optimized NMR and statistical methods. Anal Biochem 355:62–70

Babushok VI, Linstrom PJ, Reed JJ, Zenkevich IG, Brown RL, Mallard WG, Stein SE (2007) Development of a database of gas chromatographic retention properties of organic compounds. J Chromatogr A 1157:414–421

Barter PJ, Nestel PJ (1973) Precursors of plasma triglyceride fatty acids in obesity. Metabolism 22:779–783

Caesar R, Manieri M, Kelder T, Boekschoten M, Evelo C, Muller M, Kooistra T, Cinti S, Kleemann R, Drevon CA (2010) A combined transcriptomics and lipidomics analysis of subcutaneous, epididymal and mesenteric adipose tissue reveals marked functional differences. PloS ONE 5:e11525

Carlsson C, Borg LA, Welsh N (1999) Sodium palmitate induces partial mitochondrial uncoupling and reactive oxygen species in rat pancreatic islets in vitro. Endocrinology 140:3422–3428

Castillo S, Mattila I, Miettinen J, Orešič M, Hyötyläinen T (2011) Data analysis tool for comprehensive two-dimensional gas chromatography-time of flight mass spectrometry, Anal Chem 83:3058–3067

Cohen JC, Horton JD, Hobbs HH (2011) Human fatty liver disease: old questions and new insights. Science 332:1519–1523

Dettmer K, Aronov PA, Hammock BD (2007) Mass spectrometry-based metabolomics. Mass Spectrom Rev 26:51–78

Duarte NC, Becker SA, Jamshidi N et al (2007) Global reconstruction of the human metabolic network based on genomic and bibliomic data. Proc Natl Acad Sci USA 104:1777–1782

Duggan GE, Hittel DS, Hughey CC, Weljie A, Vogel HJ, Shearer J (2011) "Differentiating short- and long-term effects of diet in the obese mouse using (1) H-nuclear magnetic resonance metabolomics". Diabetes Obes Metab 13:859–862

Engelmann B (2004) Plasmalogens: targets for oxidants and major lipophilic antioxidants. Biochem Soc Trans 32:147–150

Evans JL, Goldfine ID, Maddux BA, Grodsky GM (2002) Oxidative stress and stress-activated signaling pathways: a unifying hypothesis of type 2 diabetes. Endocr Rev 23:599–622

Evans JL, Goldfine ID, Maddux, BA, Grodsky GM (2003) Are oxidative stress-activated signaling pathways mediators of insulin resistance and ß-cell dysfunction? Diabetes 52:1–8

Fahy E, Cotter D, Sud M, Subramaniam S (2011) Lipid classification, structures and tools. Biochim Biophys Acta 1811:637–47

Felig P, Wahren J (1971) Influence of endogenous insulin secretion on splanchnic glucose and amino acid metabolism in man. The J Clin Inv 50:1702–1711

Felig P, Wahren J, Hendler R, Brundin T (1974) Splanchnic glucose and amino acid metabolism in obesity. J Clin Invest 53:582–590

Fiehn O, Garvey WT, Newman JW, Lok KH, Hoppel CL, Adams SH (2010) Plasma metabolomic profiles reflective of glucose homeostasis in non-diabetic and type 2 diabetic obese African-American women. PloSOne 5:e15234

Fridlyand LE, Philipson LH (2006) Reactive species and early manifestation of insulin resistance in type 2 diabetes. Diabetes Obes Metab 8:136–145

Gall WE, Beebe K, Lawton KA, Adam K-P, Mitchell MW, Nakhle PJ, Ryals JA, Milburn MV, Nannipieri M, Camastra S, Natali A, Ferrannini E (2010) Alpha-hydroxybutyrate is an early biomarker of insulin resistance and glucose intolerance in a nondiabetic population. PLoS ONE 5:e10883

Green R, Kwok S, Durrington PN (2002) Preventing cardiovascular disease in hypertension: effects of lowering blood pressure and cholesterol. QJM 95:821–826

Gronwald W, Klein MS, Kaspar H et al (2008) Urinary metabolite quantification employing 2D NMR spectroscopy. Anal Chem 80:9288–9297

Halliwell B (1995) Antioxidant characterization: methodology and mechanism. Biochem Pharmacol 49:1341–1348

Holland WL, Summers SA (2008) Sphingolipids, insulin resistance, and metabolic disease: new insights from in vivo manipulation of sphingolipid metabolism. Endocr Rev 29:381–402

Holmes E, Li JV, Athanasiou T, Ashrafian H, Nicholson JK (2011) Understanding the role of gut microbiome-host metabolic signal disruption in health and disease. Tr Microbiol 19:349–359. doi:10.1016/j.tim.2011.05.006

Horai H, Arita M, Kanaya S et al (2010) T. metabolomic profiling for identification of novel potential biomarkers in cardiovascular diseases. J Mass Spectrom 45:703–714

Houstis N, Rosen ED (2006) Lander, ES reactive oxygen species have a causal role in multiple forms of insulin resistance. Nature 440:944–948

Hummel J, Strehmel N, Selbig J, Walther D, Kopka J (2010) Metabolomics 6:322–333

Hyötyläinen T (2010) Analytical methodologies utilized in the search of biomarkers for chronic diseases. Bioanalysis 2/5:919–923

Hyötyläinen T (2012) Novel methodologies in metabolic profiling with a focus on molecular diagnostic applications. Expert Rev Mol Diagn 12:527–538

Hyötyläinen T, Mattila I, Wiedmer S, Koivuniemi A, Taskinen M-R, Yki-Järvinen H, Orešič M (2012) Metabolomic analysis of polar metabolites in lipoprotein fractions identifies lipoprotein-specific metabolic profiles and their association with insulin resistance. Mol Biosyst 8:2559–2565

Iozzo P, Bucci M, Roivainen A, Någren K, Järvisalo MJ, Kiss J, Guiducci L, Fielding B, Naum AG, Borra R, Virtanen K, Savunen T, Salvadori PA, Ferrannini E, Knuuti J, Nuutila P (2010) Fatty acid metabolism in the liver, measured by positron emission tomography, is increased in obese individuals. Gastroenterology 139:846–56

Jansson A, Larsson A, Smith U, Lonnroth P (1994) Lactate release from the subcutaneous tissue in lean and obese men. J Clin Invest 93:240–246

Jerby L, Shlomi T, Ruppin E (2010) Computational reconstruction of tissue-specific metabolic models: application to human liver metabolism. Mol Syst Biol 6:401

Kankainen M, Gopalacharyulu P, Holm L, Orešič M (2011) MPEA—metabolite pathway enrichment analysis. Bioinformatics 27:1878–1879

Katajamaa M, Oresic M (2007) Data processing for mass spectrometry- based metabolomics. J Chromatogr A 1158:318–328

Kind T, Wohlgemuth G, Lee do Y, Lu Y, Palazoglu M, Shahbaz S, Fiehn O (2009) FiehnLib: mass spectral and retention index libraries for metabolomics based on quadrupole and time-of-flight gas chromatography/mass spectrometry. Anal Chem 81:10038–10048

Kissebah AH, Alfarsi S, Adams PW, Wynn V (1976) The metabolic fate of plasma lipoproteins in normal subjects and in patients with insulin resistance and endogenous hypertriglyceridaemia. Diabetologia 12:501–509

Kopka J, Schauer N, Krueger S, Birkemeyer C, Usadel B, Bergmuller E, Dormann P, Weckwerth W, Gibon Y, Stitt M, Willmitzer L, Fernie AR, Steinhauser D (2005) GMD@CSB.DB: the golm metabolome database. Bioinformatics 21:1635–1638. doi:10.1093/bioinformatics/bti236

Kotronen A, Seppanen-Laakso T, Westerbacka J, Kiviluoto T, Arola J, Ruskeepaa AL, Oresic M, Yki-Jarvinen H (2009) Hepatic stearoyl-CoA desaturase (SCD)-1 activity and diacylglycerol but not ceramide concentrations are increased in the nonalcoholic human fatty liver. Diabetes 58:203–8

Kotronen A, Velagapudi VR, Yetukuri L, Westerbacka J, Bergholm R, Ekroos K, Makkonen J, Taskinen MR, Oresic M, Yki-Jarvinen H (2009) Serum saturated fatty acids containing triacylglycerols are better markers of insulin resistance than total serum triacylglycerol concentrations. Diabetologia 52:684–90

Kotronen A, Yki-Järvinen H (2008) Fatty liver: a novel component of the metabolic syndrome. Arterioscler Thromb Vasc Biol 28:27–38

Kumari S, Stevens D, Kind T, Denkert C, Fiehn O (2011) Applying in-silico retention index and mass spectra matching for identification of unknown metabolites in accurate mass GC-TOF mass spectrometry, applying in-silico retention index and mass spectra matching for identification of unknown metabolites in accurate mass GC-TOF mass spectrometry. Anal Chem 83(15):5895–5902

Kusunoki J, Kanatani A (2006) Moller DE modulation of fatty acid metabolism as a potential approach to the treatment of obesity and the metabolic syndrome. Endocrine 29:91–100

Kyriakis JM, Avruch J (1996) Sounding the alarm: protein kinase cascades activated by stress and inflammation. J Biol Chem 271:24313–24316

Laaksonen DE, Lakka TA, Lakka HM et al (2002) Serum fatty acid composition predicts development of impaired fasting glycaemia and diabetes in middle-aged men. Diabetic Medicine 19:456–464

Lanpher B, Brunetti-Pierri N, Lee B (2006) Inborn errors of metabolism: the flux from mendelian to complex diseases. Nat Rev Genet 7:449–460

Lee JK, Williams PD, Cheon S (2008) Data mining in genomics. Clin Lab Med 28:145–66

Li H, Xie Z, Lin J et al (2008) Transcriptomic and metabonomic profiling of obesity-prone and obesity-resistant rats under high fat diet. J Prot Res 7:4775–4783

Li X, Xu Z, Lu X, Yang X, Yin P, Kong H, Yu Y, Xu G (2009) Comprehensive two-dimensional gas chromatography/time-of-flight mass spectrometry for metabonomics: biomarker discovery for diabetes mellitus. Anal Chim Acta 633:257–62

Maassen JA, Romijn JA, Heine RJ (2007) Fatty acid-induced mitochondrial uncoupling in adipocytes as a key protective factor against insulin resistance and beta cell dysfunction: a new concept in the pathogenesis of obesity-associated type 2 diabetes mellitus. Diabetologia 50:2036–2041

Malet-Martinoa M, Holzgrabeb U (2011) NMR techniques in biomedical and pharmaceutical analysis. J Pharm Biomed Anal 55:1–15

Moco S, Bino RJ, De Vos RCH, Vervoort J (2007) Metabolomics technologies and metabolite identification. Trends Anal Chem 26:855–866

Neumann S, Böcker S (2010) Computational mass spectrometry for metabolomics: identification of metabolites and small molecules. Anal Bioanal Chem 398:2779–2788

Newgard CB, An J, Bain JR et al (2009) "A branched-chain amino acid-related metabolic signature that differentiates obese and lean humans and contributes to insulin resistance." Cell Metabolism 9:311–326

Nygren H, Seppänen-Laakso T, Castillo S, Hyötyläinen T, Orešič M (2011) LC/MS-based lipidomics for studies of body fluids and tissues. Methods Mol Biol 708:247–257

Orešič M, Hänninen V, Vidal-Puig A (2008) Lipidomics: a new window to biomedical frontiers. Trends Biotechnol 26:647–652

Orvall M, Berglund L, Salminen I, Lithell H, Aro A, Vessby B (1996) The serum cholesterol ester fatty acid composition but not the serum concentration of alpha tocopherol predicts the development of myocardial infarction in 50-year-old men: 19 years follow-up. Atherosclerosis 127:65–71

Paolisso G, Giugliano D (1996) Oxidative stress and insulin action: is there a relationship? Diabetologia 3(9):357–363

Paolisso G, Giugliano D, Pizza G, Gambardella A, Teasauro P, Varricchio M, D´Onofrio F (1992) Glutathione infusion potentiates glucose-induced insulin secretion in aged patients with impaired glucose tolerance. Diabetes Care 15:1–7

Paolisso G, Gambardella A, Tagliamonte MR, Saccomanno F, Salvatore T, Gualdiero P, D'Onofrio MV, Howard BV (1996) Does free fatty acid infusion impair insulin action also through an increase in oxidative stress? J Clin Endocrinol Metab 81:4244–4248

Pietiläinen K, Róg T, Seppänen-Laakso T, Virtue S, Gopalacharyulu P, Tang J, Rodriguez-Cuenca S, Maciejewski A, Naukkarinen J, Rissanen A, Ruskeepää A-L, Niemelä P, Yetukuri L, Yew Tan C, Velagapudi V, Castillo S, Nygren H, Hyötyläinen T, Kaprio J, Yki-Järvinen H, Vattulainen I, Vidal-Puig A, Orešič M (2011) Remodeling of adipose tissue lipidome as adaptation to acquired obesity: benefits and costs. PLoS Biol 9:e1000623

Pól J, Vidová V, Hyötyläinen T, Volný M, Novák P, Strohalm M, Kostiainen R, Havlíček V, Wiedmer SK, Holopainen JM (2011) Spatial distribution of glycerophospholipids in R. the ocular lens. PLoS ONE 6:e19441

Psychogios N, Hau DD, Peng J, Chi Guo A, Mandal R, Bouatra S, Sinelnikov I, Krishnamurthy R, Eisner R, Gautam B, Young N, Xia J, Knox C, Dong E, Huan P, Hollander Z, Pedersen TL, Smith SR, Bamforth F, Greiner R, McManus B, Newman JW, Goodfriend T, Wishart DS (2011) Plos One 6:e16957

Qatanani M, Lazar MA (2007) Mechanisms of obesity-associated insulin resistance: many choices on the menu. Genes & Dev 21:1443–1455

Randle PJ, Garland PB, Hales CN, Newsholme EA (1963) The glucose fatty-acid cycle: its role in insulin sensitivity and the metabolic disturbances of diabetes mellitus. Lancet 1:785–789

Rao MS, Reddy JK (2001) Peroxisomal β-oxidation and steatohepatitis. Semin Liver Dis 21:43–55

Reaven GM, Lerner RL, Stern MP, Farquhar JW (1967) Role of insulin in endogenous hypertriglyceridemia. J Clin Invest 46:1756–1767

Rondinone CM (2006) Adipocyte-derived hormones, cytokines, and mediators. Endocrine 29:81–90

Rosen P, Nawroth PP, King G, Moller W, Tritschler HJ, Packer L (2001) The role of oxidative stress in the onset and progression of diabetes and its complications: a summary of a congress series sponsored by UNESCO-MCBN, the American Diabetes Association and the German Diabetes Society. Diabetes Metab Res Rev 17:189–212

Sanyal AJ, Campbell-Sargent C, Mirshahi F, Rizzo WB, Contos MJ, Sterling RK, Luketic VA, Shiffman ML, Clore JN (2001) Nonalcoholic steatohepatitis: association of insulin resistance and mitochondrial abnormalities. Gastroenterology 120:1183–92

Savage DB, Petersen KF, Shulman GI (2007) Disordered lipid metabolism and the pathogenesis of insulin resistance. Physiol Rev 87:507–520

Schwudke D, Liebisch G, Herzog R, Schmitz G, Shevchenko A (2007) Shotgun lipidomics by tandem mass spectrometry under data-dependent acquisition control. Methods Enzymol 433:175–91

Schymanski EL, Meringer M, Brack W (2009) Matching structures to mass spectra using fragmentation patterns: are the results as good as they look? Anal Chem 81:3608–3617

Serkova NJ, Jackman M, Brown JL et al (2006) Metabolic profiling of livers and blood from obese Zucker rats. J Hepatol 44:956–962

Shlomi T, Cabili MN, Ruppin E (2009) Predicting metabolic biomarkers of human inborn errors of metabolism. Mol Syst Biol 5:263

Shulman GI (2000) Cellular mechanisms of insulin resistance. J Clin Invest 106:171–176

Smith CA, O´Maille G, Want EJ et al (2005) METLIN: a metabolite mass spectral database. Ther Drug Monit 27:747–751

Sumner LW, Urbanczyk-Wochniak E, Broecklin CD (2008) Metabolomics data analysis, visualization, and integration. Methods Mol Biol 406:409–436

Sunny NE, Parks EJ, Browning JD, Burgess SC (2011) Excessive hepatic mitochondrial TCA cycle and gluconeogenesis in humans with nonalcoholic fatty liver disease. Cell Metab 14:804–810

Theodoridis GA, Gika HG, Want EJ, Wilson ID (2012) Liquid chromatography-mass spectrometry based global metabolite profiling: a review. Anal Chim Acta 711:7–16

Trawick JD, Schilling CH (2006) Use of constraint-based modeling for the prediction and validation of antimicrobial targets. Biochem Pharmacol 71:1026–1035

Trayhurn P, Bing C (2006) Appetite and energy balance signals from adipocytes. Philos Trans R Soc Lond B Biol Sc 361:1237–1249

Tremblay F, Krebs M, Dombrowski L, Brehm A, Bernroider E, Roth E, Nowotny P, Waldhäusl W, Marette A, Roden M (2005) Overactivation of S6 kinase 1 as a cause of human insulin resistance during increased amino acid availability. Diabetes 54:2674–2684

Trim PJ, Atkinson SJ, Princivalle AP, Marshall PS, West A, Clench MR (2008) Matrix-assisted laser desorption/ionisation mass spectrometry imaging of lipids in rat brain tissue with integrated unsupervised and supervised multivariant statistical analysis. Rapid Commun Mass Spectrom 22:1503–1509

Vrieze A, Holleman F, Zoetendal EG, de Vos WM, Hoekstra JBL, Nieuwdorp M (2010) The environment within: how gut microbiota may influence metabolism and body composition. Diabetologia 53:606–613

Weber RJ, Southam AD, Sommer U, Viant MR (2011) Characterization of Isotopic Abundance Measurements in high resolution FT-ICR and orbitrap mass spectra for improved confidence of metabolite identification. Anal Chem 83:3737–3743

Wang L, Folsom AR, Zheng ZJ, Pankow JS, Eckfeldt JH (2003) Plasma fatty acid composition and incidence of diabetes in middle-aged adults: the Atherosclerosis Risk in Communities (ARIC) Study. Am J Clin Nutr 78:91–98

Wang TJ, Larson MG, Vasan RS, Cheng S, Rhee EP, McCabe E, Lewis GD, Fox CS, Jacques PF, Fernandez C, O'Donnell CJ, Carr SA, Mootha VK, Florez JC, Souza A, Melander O, Clish CB, Gerszten RE (2011) Metabolite profiles and the risk of developing diabetes. Nat Med 17:448–453

Westerbacka J, Kotronen A, Fielding BA, Wahren J, Hodson L, Perttila J, Seppanen-Laakso T, Suortti T, Arola J, Hultcrantz R, Castillo S, Olkkonen VM, Frayn KN, Oresic M, Yki-Jarvinen

H (2010) Splanchnic balance of free fatty acids, endocannabinoids and lipids in subjects with NAFLD. Gastroenterology doi:10.1053/j.gastro.2010.06.064

Wikoff WR, Anfora AT, Liu J, Schultz PG, Lesley SA, Peters EC, Siuzdak G (2009) Metabolomics analysis reveals large effects of gut microflora on mammalian blood metabolites. PNAS 106:3698–3703

Wojtczak L, Schonfeld P (1993) Effect of fatty acids on energy coupling processes in mitochondria. Biochim Biophys Acta 1183:41–57

Wolf S, Schmidt S, Muller-Hannemann M, Neumann S (2010) In silico fragmentation for computer assisted identification of metabolite mass spectra. BMC Bioinf 11:148

Wopereis S, Radonjic M, Rubingh C, Erk Mv, Smilde A, Duyvenvoorde Wv, Cnubben N, Kooistra T, Ommen Bv, Kleemann R (2012) Identification of prognostic and diagnostic biomarkers of glucose intolerance in ApoE3Leiden mice. Physiol Genomics 1(44):293–304

Wu C, Ifa DR, Manicke NE, Cooks RG (2009) Rapid, direct analysis of cholesterol by charge labeling in reactive desorption electrospray Ionization. Anal Chem 81:7618–7624

Xia J, Wishart DS (2010) MSEA: a web-based tool to identify biologically meaningful patterns in quantitative metabolomic data. Nucleic Acids Res 38:W71–W77

Xie B, Waters MJ, Schirra H (2012) Investigating potential mechanisms of obesity by metabolomics. J Biomed Biotechn 2012:805683

Xu Y, Heilier JF, Madalinski G, Genin E, Ezan E, Tabet JC, Junot C (2010) Evaluation of accurate mass and relative isotopic abundance measurements in the LTQ-Orbitrap mass spectrometer for further metabolomics database building. Anal Chem 82:5490–5501

Yang LV, Radu CG, Wang L, Riedinger M, Witte ON (2005) Gi-independent macrophage chemotaxis to lysophosphatidylcholine via the immunoregulatory GPCR G2A. Blood 105:1127–1134

Chapter 12
Fluxomics

Marta Cascante, Adrián Benito, Igor Marín de Mas, Josep J. Centelles, Anibal Miranda and Pedro de Atauri

Abstract Metabolic processes are dynamic, finely regulated and interconnected. In order to characterize metabolic networks and their functional operation, quantitative knowledge of intracellular fluxes is required. Whereas metabolite concentrations can be directly estimated, the set of molecular fluxes through each reaction within a metabolic network can only be estimated indirectly. Isotope labelling experiments with ^{13}C-labelled tracers, using nuclear magnetic resonance or mass spectrometry, are emerging as powerful strategies used to measure fluxes in complex interconnected metabolic networks. In this chapter, we review these methods together with the computational resources for flux analysis. Current challenges and limitations in fluxomics applied to Metabolic Syndrome are discussed.

Keywords Fluxomics · Fluxome · ^{13}C labelling · Isotopomer · Isotopologue · Isotopic steady state

12.1 Methods in Fluxomics: an Overview

As stated in the previous chapter, the term *metabolome* refers to the collection of all metabolites in a biological entity (cell, tissue, organ or organism), these metabolites being the end products of gene expression, mRNA synthesis and protein activity (Kuchel 2010; Dudley et al. 2010; Oresic 2009). In brief, the metabolome of an organism is the result of the *in vivo* activity of gene products and is closely tied to its physiology and environment (what the organism eats or breathes). Accordingly, *metabolomics* has been defined as the measurement of the composition of low-molecular-weight molecules in biological samples. However, knowledge of the complete set of metabolites is not enough to predict the phenotype of a cell as it gives only an instantaneous 'snapshot' of the physiology of that cell (Castro-Perez et al. 2011; Lorkowski 2011).

M. Cascante (✉) · A. Benito · I. Marín de Mas · Josep J. Centelles · A. Miranda · Pedro de Atauri
Department of Biochemistry and Molecular Biology, Faculty of Biology, University of Barcelona, Av Diagonal 643, Edifici Nou, planta -2, 08028 Barcelona, Spain
e-mail: martacascante@ub.edu

M. Orešič, A. Vidal-Puig (eds.), *A Systems Biology Approach to Study Metabolic Syndrome*, DOI 10.1007/978-3-319-01008-3_12,

The distinct metabolic processes involved in metabolite production and degradation are dynamic, finely regulated and interconnected. Therefore, in order to comprehensively characterize metabolic networks and their functional operation (metabolite turnover), quantitative knowledge of intracellular metabolic fluxes is required. *Fluxomics* is the field of 'omics' research dealing with the dynamic changes in metabolites over time, i.e. the quantitative analysis of fluxes through metabolic pathways (Cascante and Marin 2008; Paul Lee et al. 2010). Fluxomics involves the quantitative monitoring of the operation of the whole metabolic network by methods of flux analysis, providing a global perspective on integrated regulation at the transcriptional, translational and metabolic levels (Cascante and Marin 2008; Bothwell and Griffin 2011). Molecular fluxes through metabolic networks (the *fluxome*) reflect the integration of genetic and metabolic regulation and determine the cellular phenotype (Sauer 2004). Thus, whereas transcriptomics, proteomics and metabolomics provide a measure of metabolic network capacities and thermodynamic driving forces, fluxomics provides a measure of the flux through each reaction in the network, which can be considered a direct measure of the metabolic phenotype (Krömer et al. 2009).

Whereas metabolite concentrations can be directly estimated using the methods described in Chap. 11, the set of molecular fluxes through each reaction within a metabolic network can only be estimated indirectly. Thus, measurement over time of metabolites in the extracellular media that enter or exit the cell permits estimation of the fluxes in uptake and secretion. The balance of these input and output fluxes, within a stoichiometric model of the reaction network, gives the overall quantitative flux distribution. However, quantifying fluxes by simple flux balancing is only possible for simple metabolic pathways, e.g. biosynthetic linear pathways (Zamboni and Sauer 2005), due to futile cycles and a high degree of interconnection between metabolic pathways.

The quantification of fluxes through complex interconnected networks has only been possible with the use of labelled metabolites that can enter and be metabolized inside the cell, e.g. ^{13}C-labelled metabolites. Nowadays, isotope labelling experiments are emerging as very powerful strategies used to measure fluxes in metabolic networks (for a review see (Cascante and Marin 2008; Paul Lee et al. 2010; Metallo et al. 2009; Krömer et al. 2009; Matsuoka and Shimizu 2010; Selivanov et al. 2010)).

The isotopes used in such experiments can be radioactive (^{3}H, ^{32}P and ^{14}C) or nonradioactive stable isotopes (^{2}H, ^{15}N and most commonly ^{13}C). Experiments using either radioactive, or preferably stable isotopes, have been extensively performed at the level of cell culture as well as at the level of tissues, organs and whole organisms (Lane et al. 2011). The specific conversion of substrates into products through each metabolic pathway imprints different characteristic label patterns in intermediate metabolites and products. This increases the capacity of stoichiometric flux balancing to trace the flux in more complex non-linear pathways.

Metabolomics and fluxomics can accurately and comprehensively depict both the steady physiological state of a cell or organism and their dynamic responses to genetic and environmental modulation. Thus, fluxomics offers a unique opportunity to look at relationships between genotype and phenotype and how these are influenced by the environment and the overall state of the organism, including the influence of nutrition and disease.

An increasing number of papers are providing a detailed description of the fluxes throughout metabolic pathways at the cellular level. These descriptions are based on the combination of experimental techniques, using ^{13}C substrates, and computational approaches, providing a detailed picture of the metabolic pathways. Below, a description is provided of the principles on which the analysis of metabolic pathways in mammalian cells is based. An overview of the computational approaches is given, beginning with methods based on biochemical and biophysical constraints, followed by an explanation of the coupling of computational methods and stable-isotope assisted experiments.

12.2 Flux Analysis Using Biochemical and Biophysical Constraints

Various computational approaches have been used to estimate the flux distribution in metabolic pathways. Early attempts to model metabolism in cells began with rate equations of individual enzymes in each metabolic pathway, which constrain the metabolic behaviour. This type of model (kinetic model) has been constructed not only for unicellular organisms such as *Escherichia coli* and yeast, but also for highly complex cells such as hepatocytes (e.g. (Rodriguez-Prados et al. 2009; Curto et al. 1995; Teusink et al. 2000; Raftos et al. 2010; Curien et al. 2009; Nikerel et al. 2009; Maher et al. 2003)).

The regulation of enzyme activity by metabolites that can act as inhibitors or activators of the enzyme is taken into account in these kinetic models. The principal difficulty when constructing this type of quantitative flux model of metabolic networks in a particular cell, tissue or organism is the availability of rate equations and reliable values for the required kinetic parameters (e.g. K_m, K_i, V_{max} etc.). Another critical step is to decide the boundaries of the models, e.g. metabolites assumed to be constant (de Atauri et al. 2000).

In modern systems biology it is widely accepted that detailed kinetic models are not always necessary, as many specific questions such as which enzymes should be modified to successfully alter the flux, or metabolite concentration, in pathways of clinical relevance, can be tackled by techniques such as *Metabolic Control Analysis* (Heinrich and Schuster 1996; Stephanopoulos et al. 1998; Cascante et al. 2002; Moreno-Sanchez et al. 2008; Bruggeman et al. 2008).

Moreover, the fact that biological information is typically incomplete led to the development of constraint-based modelling approaches for analysing flux distribution in metabolic networks (for a review see (Llaneras and Pico 2008; Ruppin et al. 2010; Orth et al. 2010)), which do not include rate equations for enzymes. In practice, these approaches enable the steady state distribution of metabolic fluxes to be estimated based on the addition of successive constraints. In this context, flux balances around metabolites are constraints describing the network topology and the associated stoichiometry for each reaction step. Additional constraints can include experimental data, e.g. measurements of substrate uptakes, production rates, thermodynamics of reactions, etc. All these constraints are limiting the solution space of flux distributions, where a solution of flux distribution is such a complete assignation of flux values of the reaction steps that all the constraints are satisfied. The set of all possible solutions is the solution space of flux distributions. The solution is provided by solving problems of maximization (or minimization), e.g. maximizing fluxes of essential cell components, such as biomass or ATP, which should have a selective advantage (Schuster et al. 2008; Feist and Palsson 2010; Orth et al. 2010). These maximizations are applied to compute optimal flux distributions, which can be non-unique. The minimum and maximum flux for each reaction in the network can be analysed while satisfying some state of the network, e.g. 90 % of maximal biomass production (Mahadevan and Schilling 2003; Gudmundsson and Thiele 2010). The idea underlying these approaches is to describe a metabolic network by a set of constraints that characterize its possible metabolic behaviour and permit calculation of the metabolic flux states that would be exhibited under different assumptions.

12.3 Flux Analysis Using Labelled Substrates

12.3.1 Experimental Approaches for Stable Isotope-Assisted Fluxomics

Uncertainty regarding intracellular fluxes comes in large part from the degrees of freedom associated with complex metabolic networks that involve ramifications, bifurcations and cycles. The use of stable isotope-labelled substrate provides a direct means of estimating flux distributions, which can be combined to satisfy the above discussed constraints. The distribution of atoms from a labelled substrate (*tracer*) allows the quantitative determination of the substrate–product relationship (Cascante and Marin 2008; Paul Lee et al. 2010; Metallo et al. 2009; Krömer et al. 2009; Matsuoka and Shimizu 2010; Selivanov et al. 2010). The propagation of the label from labelled substrates to different metabolic products is analysed in order to estimate the relative contribution of the internal fluxes involved in the complex metabolic networks.

Figure 12.1 illustrates the use of label propagation. The propagation after one turn of the Krebs cycle is represented, with ^{13}C coming from [1,2–$^{13}C_2$]D-glucose. Here, a hypothetically labelled molecule of pyruvate and its transformation to acetyl-CoA for lipid production is represented. In this example, carbons from lipids synthesized *de novo* are labelled with ^{13}C, enabling estimation of glucose contribution for lipid synthesis (Boren et al. 2003). After more turns of the cycle the picture regarding labelled and non-labelled carbons becomes more complex. In this example all reactions affecting label propagation also affect the concentration of intermediaries. However, some reactions produce changes in label distribution but not in chemical

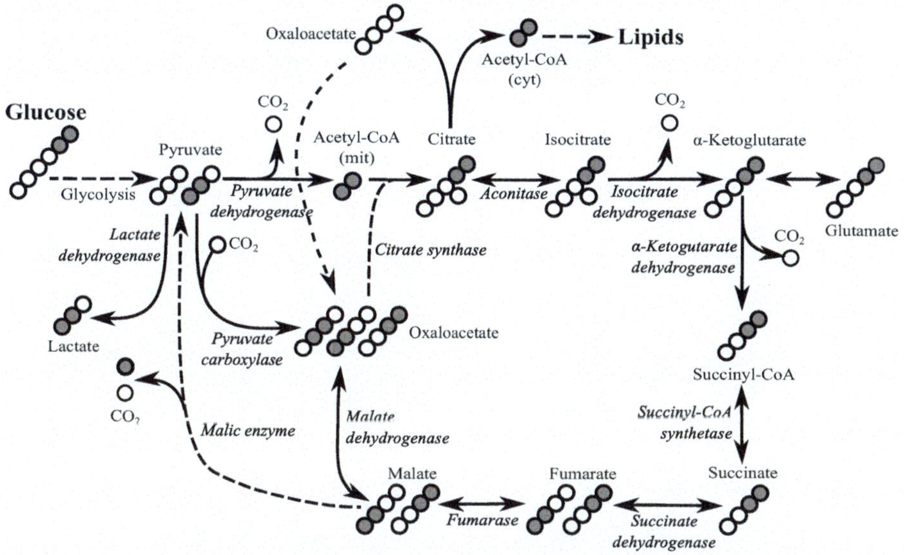

Fig. 12.1 Label propagation. Scheme of the carbon skeleton of Krebs cycle intermediates, where each carbon is represented by a circle. The non-labelled (*white circles*) and labelled (*gray circles*) carbons come from a labelled substrate (in this case [1,2–$^{13}C_2$]D-glucose) after only one turn of the cycle. Reactions with discontinuous arrows were not considered

balance (Selivanov et al. 2005). These 'invisible reactions' can be followed using labelled metabolites. Figure 12.2 illustrates invisible reactions involved in the mechanism of transketolase. The example focuses on the case of ribose-5-phosphate and sedoheptulose-7-phosphate (discontinuous circle), where there is an 'invisible' cycle between them that does not affect concentrations but does affect label distribution (discontinuous arrows).

Selection of the appropriate labelled substrates and products is a major challenge that is determined by the intracellular fluxes to be quantified, and strongly influences the quality of the estimated fluxes (Metallo et al. 2009). Depending on the fluxes under study, a wide range of tracers exists. Among them, stable-isotope labelled substrates such as ^{13}C-glucose or ^{13}C-glutamine are the most widely used. Traditionally, ^{13}C tracers have been employed in the study of microbial metabolism, where microorganisms are generally fed with single carbon source. The issue of tracer choice is more complex in mammalian cell systems, which are cultured with multiple carbon sources and are grown in complex media containing all amino acids and often additional compounds such as nucleotides or lipids. Researchers have used a variety of tracers in order to study particular metabolic pathways, including labelled glucose, glutamine, acetate and atypical substrates such as propionate (Rodriguez-Prados et al. 2010; Gaglio et al. 2011; Collins et al. 2011; Burgess et al. 2007). For instance, [1,2–$^{13}C_2$]-glucose provides information on glycolysis, pentose phosphate pathway, pyruvate oxidation versus carboxylation (Marin et al. 2003; Vizan

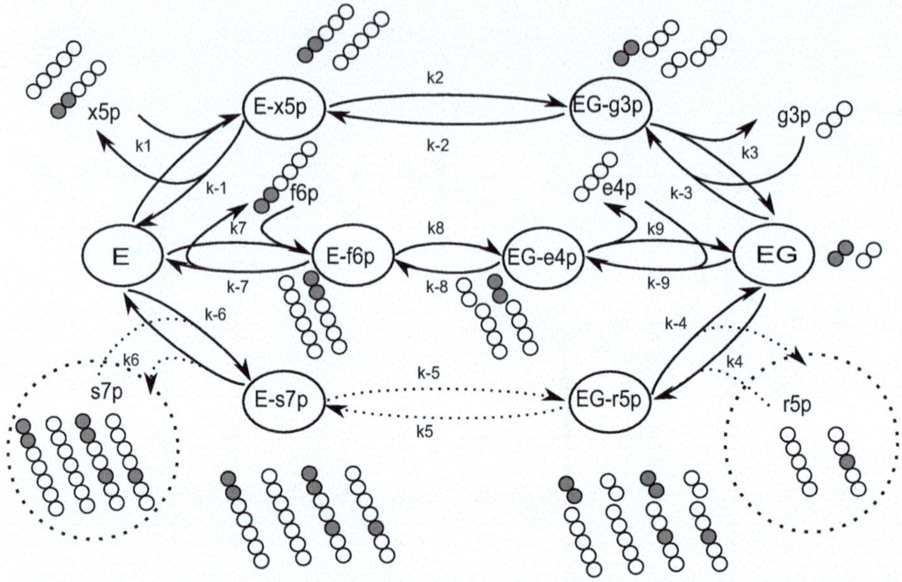

Fig. 12.2 Invisible reactions involving transketolase. Transketolase reversibly catalyzes the transfer of a glycoaldehyde from ketoses to aldoses by a ping-pong mechanism: transketolase (E) reacts with a ketose (x5p, s7p or f6p) to form an enzyme-glycoaldehyde form (EG) and liberates an aldose (g3p, r5p or e4p, respectively). Then, the reaction cycle is completed when EG covalently links the glycoladehyde with another aldose, thus producing a new ketose and E. Binding the same type of aldose as the previously released, results in returning to the initial state: no concentrations are changed. However, if this bound aldose differs in the isotopomer composition, such a reaction cycle can change the isotopomer distribution of the media without changing the total metabolite concentrations. Such a reaction will be called 'invisible reaction', e.g. s7p + r5p ↔ r5p + s7p. Metabolites: e4p, erythrose-4-phosphate; f6p, fructose-6-phosphate; g3p, glyceraldehyde-3-phosphate; r5p, ribose-5-phosphate; s7p, sedoheptulose-7-phosphate; x5p, xylulose-5-phosphate

et al. 2009) and *de novo* synthesis of lipids (Fig. 12.1) (Boren et al. 2003). Specific products can be selected to determine the ratios of specific intracellular fluxes. Thus, ^{13}C propagation from [1,2–$^{13}C_2$]-glucose to lactate is analysed to determine the approximate ratio of fluxes through the pentose-phosphate pathway with respect to the carbon flow descending through glycolysis, whereas ^{13}C propagation from [1,2–$^{13}C_2$]-glucose to RNA ribose is analysed to determine the approximate ratio of fluxes through the oxidative and non-oxidative pentose-phosphate pathway. In addition, ^{13}C propagation from [1,2–$^{13}C_2$]-glucose to glutamate is analysed to estimate the approximate relative contributions of pyruvate carboxylase and pyruvate dehydrogenase to the Krebs cycle (Marin et al. 2003). However, labelled glucose often fails when assessing fluxes in the Krebs cycle, since most ^{13}C atoms can be excreted in the form of lactate and alanine. In this case, labelled glutamine can be used to better resolve the activity of the Krebs cycle.

In the context of ^{13}C labelling, *isotopologue* refers to isomers with a specific number of ^{13}C substitutions and *isotopomer* ('isotope isomers') to isomers with

^{13}C substitution in a specific carbon position[1], which can be determined via various analytical platforms. Alternatively, such isomers can be referred to as *mass isotopomer* and *positional isotopomer*, respectively (Paul Lee et al. 2010). Thus, isotopologues account for those reaction products with zero, one, two, etc. ^{13}C substitutions, which are denoted as *m0, m1, m2*, etc. respectively. For a specific metabolite, the number of possible isotopomers, 2^n, depends on the number, n, of carbons for each metabolite. For instance, for lactate, with three carbons, there are $2^3 = 8$ isotopomers. A lactate compound denoted as *m2* contains three possible isotopomers: $^{13}C_1$–$^{13}C_2$–$^{12}C_3$, $^{13}C_1$–$^{12}C_2$–$^{13}C_3$ and $^{12}C_1$–$^{13}C_2$–$^{13}C_3$.

The detection and quantification of isotopomers and isotopologues is a challenging issue. There are currently two main methods of detecting and measuring isotopomer/isotopologue distributions: *nuclear magnetic resonance* (NMR) and *mass spectrometry* (MS) (Lane et al. 2009; Wiechert 2001; Li et al. 2011). The NMR technique is able to distinguish isotopomers by the magnetic properties of the atom nucleus, and provides information of positional labelling. However, it requires a large amount of sample, long analysis times and expensive equipment. On the other hand MS is widely accessible, more affordable and provides high sensitivity in the detection of isotopologues ('mass' isotopomers).

MS distinguishes between isotopologues by differences of mass, thus allowing ^{13}C-labelled compounds to be distinguished from unlabelled compounds containing ^{12}C, as the former are heavier than the latter. However, a major limitation of this method compared with the NMR method is that only isotopologues can be distinguished unless tandem MS analysis is used, which potentially increases the amount of labelling data (Choi and Antoniewicz 2011). Coupled to MS, various chromatographic separations can be used, such as liquid (LC) (Henry et al. 2011) or gas chromatography (GC) (Selivanov et al. 2010; Hiller et al. 2010). A high resolution is achieved by GC-MS, although samples must be chemically derivatized to form volatile compounds before fractionation by GC and ionization by electron impact or chemical ionization. Afterwards, the resulting ions are detected using their *mass-to-charge ratio* (m/z), and as the ions usually have a single charge, the m/z value is equivalent to the molecular weight of the fragment in mass units. For LC-MS no derivatization is required. Nevertheless, isotopologue analysis by LC-MS is infrequent.

12.3.2 Flux Analysis Under Isotopic and Non-Isotopic Steady State

The relative abundance of labelled product (isotopomers and isotopologues) depends on the labelled status of the substrates (e.g. labelled glucose) and the flux distribution

[1] For a more general definition of *isotopologue* and *isotopomer* we refer the reader to (*IUPAC Compendium of Chemical Terminology—the Gold Book 2011*)

throughout the metabolic network. Thus, by proposing an appropriate metabolic network, at fixed relative abundances of the isotopic forms of labelled substrates, one or more flux distributions throughout the assumed networks are expected that will fit the measured distribution of labelled products, also satisfying the biochemical and biophysical constraints associated with the metabolic network. Other experimental measures, such as measured inputs or outputs, can also be fitted or fixed as constraints to be satisfied. However, additional factors must be considered (Wiechert 2001). Among them, we highlight the natural abundance of ^{13}C, and in particular, the fact that both the steady state of concentration and fluxes and the isotopic enrichments of all metabolite pools undergo a transient phase before the steady state of isotopomer distribution is finally reached (*isotopic steady state*). Thus, the duration of the experiment and the size of the metabolic pools are fundamental factors when determining the strategy to be applied to solve fluxes based on labelled measurements (Deshpande et al. 2009; Selivanov et al. 2010; Wiechert 2001). Larger pools experience a longer transient phase, which may be even longer than the duration of the experiment, especially when the metabolites belong to pools corresponding to macromolecular biomass constituents, such as proteins or RNA, large pools of internal metabolites, such as glycogen, or pools of intermediates of central carbohydrate metabolism that are secreted to or exchanged with the culture medium, such as lactate and other amino acids or lipids. Although intracellular metabolites may reach the isotopic steady state within minutes, the existence of these stores delays the time required to establish the isotopic steady state (Selivanov et al. 2010).

When metabolic and isotopic steady state is required, cells need to be incubated with the tracer for a sufficiently long period to obtain complete isotope incorporation, which can take a long time in mammalian systems. Thus, it is crucial to study the metabolic and isotopic behaviour prior to deducing any physiological conclusions from a tracer-assisted experiment (Deshpande et al. 2009). Under isotopic steady state, isotopomer and isotopologue abundances can be predicted from the balances around isotopomers, which are analogous to the flux balances around metabolites, but also taking into account label transitions (Antoniewicz et al. 2006; Antoniewicz et al. 2007). These balances describe the dependency of isotopomer abundances on fluxes and isotopomer abundances of other metabolites. The goal is to find the set of fluxes that minimizes the difference between observed and simulated isotopomer measurements (Antoniewicz et al. 2006).

Alternatively, under (isotopic) non steady state, time-series of isotopic labelling can be measured to compute fluxes during dynamic phases. This kind of analysis is useful when internal stores such as glycogen, amino acids, lipids etc. delay the (isotopic) steady state of the internal and external metabolites. Estimation of fluxes during dynamic phases requires alternative mathematical approaches (Noh et al. 2006; Baxter et al. 2007; Selivanov et al. 2010). In simulations using detailed kinetic models, comparison with experimental time courses of such 'slow' variables provides additional restrictions that help to evaluate the fluxes. Taking into account the thermodynamic relationships between kinetic and equilibrium constants restricts the domain of estimated fluxes, as illustrated in the detailed modelling of the kinetic mechanisms of transaldolase and transketolase (Selivanov et al. 2005).

In summary, the tools for the theoretical analysis of both isotopic steady and non-steady state have been developed in conjunction with the experimental techniques of stable isotope tracing. There are many examples of application to the study of metabolism in different organisms, such as fungi (Jouhten et al. 2009) and yeast (Jouhten et al. 2008; Aboka et al. 2009), as well as mammalian cells in normal and disease states, as shown in man (Fan et al. 2009), in primary cultures (Amaral et al. 2010), and in cultured cancer cells (Selivanov et al. 2010).

12.4 Computational Resources for Flux Analysis

Several strategies for metabolic flux estimation have been described in this chapter. In order to obtain such estimates in a complex and dynamic system such as the cell metabolism, the use of specialized software is required. To this end a large number of computational resources are available for analysis associated with fluxes and ^{13}C labelling. Examples of the algorithms and software used for flux analysis from carbon-labelled experiments are '13C-FLUX' (Wiechert et al. 2001), 'Fiatflux' (Zamboni et al. 2005), 'OpenFLUX' (Quek et al. 2009), FIA (Srour et al. 2011) and 'Isodyn' (Selivanov et al. 2004, Selivanov et al. 2005; Selivanov et al. 2006).

In order to illustrate how modelling is used to analyse flux based on ^{13}C tracer data, the software Isodyn will be discussed in more depth. Isodyn (from 'isotopomer dynamics'), which was developed in C++, represents a simulation environment for the dynamics of metabolite labelling by ^{13}C isotopes in metabolic reactions of living cells. Isodyn simulates ^{13}C isotopomer distribution in the same way as classical kinetic models simulate the time-course of metabolite concentrations.

Figure 12.3 provides a detailed flow chart description of Isodyn. The program is organized into separated modules for simulations of kinetic of metabolite concentrations, isotopomer concentrations and statistical analysis. A first module is a classical kinetic model of metabolic pathways represented by a system of *ordinary differential equations* (ODE). Numerical solution of this ODE system gives the time course of evolution of metabolite concentrations and fluxes. If metabolic steady state is simulated, the basic kinetic simulation lasts until the metabolic steady state is reached. Kinetic model is linked, through total metabolite concentration and fluxes, with a module that computes the distribution of ^{13}C isotopomers. The program automatically constructs hundreds of ODEs describing the evolution of all isotopomers in the system. The simulation of isotopomer distribution starts after simulating the evolution of metabolic fluxes and concentrations in accordance with the scheme of metabolic pathways and a set of parameters and initial concentrations, implemented in the kinetic model. The isotopomer distributions are fitted by modifying the values of model parameters using a modified *Simulated Annealing algorithm* (Kirkpatrick et al. 1983). The goodness of fit is evaluated by χ^2 (square of difference between the measured data and the predicted values, normalized by standard deviation σ of the measurement) and numbers of degrees of freedom, as described in (Press et al. 2002). The χ^2 values obtained for the best fit identify the acceptability of the model and the reliability of the evaluated fluxes.

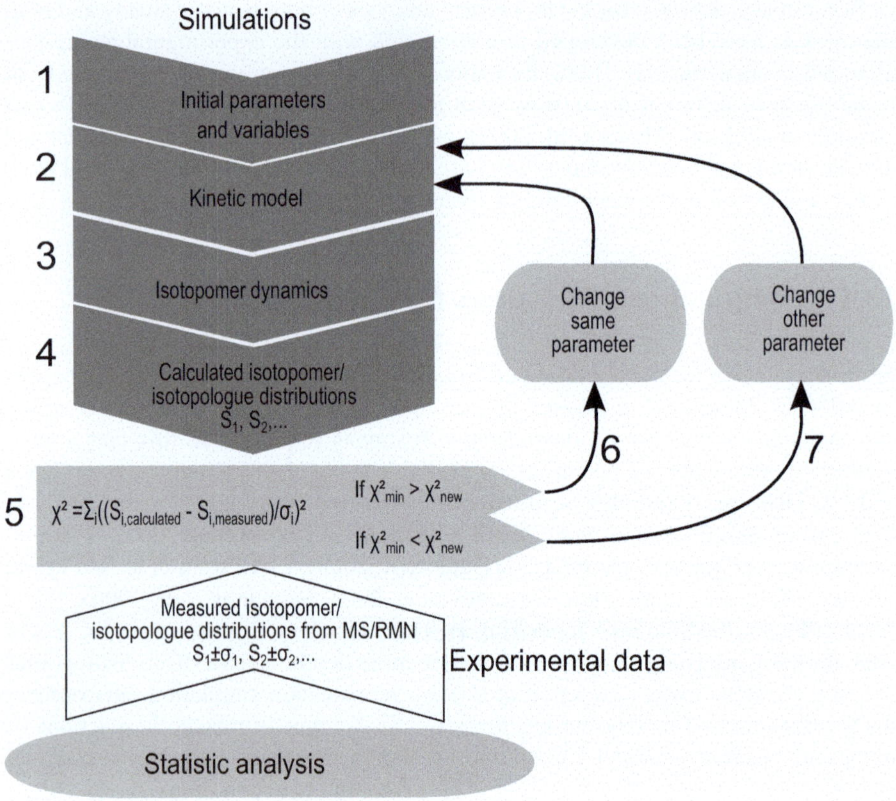

Fig. 12.3 Isodyn flow chart. Steps: *1*) Starts with an initial set of model parameters; *2*) Metabolite concentrations and fluxes are determined by dynamic simulation; *3*) In order to simulate isotopomer dynamics, metabolite concentrations are used as initial values for unlabeled isotopomers (labeled internal isotopomers initially have zero concentrations) and fluxes are used to define the sum of reaction rates for all isotopomers; *4*) Simulation of isotopomer dynamics provides isotopomer distributions (S_i being a particular isotopomer/isotopologue); *5*) The χ^2 (see main text) is determined; *6*) The value of a randomly chosen parameter is modified to decrease χ^2, the process from step 2 to 5 is repeated, and the new χ^2 is compared with the previous one (the process is repeated by changing the value of the same parameter while the new χ^2 remains lower than the minimum, previously obtained); and *7*) A new model parameter is randomly selected and step 6 is repeated for this parameter

12.5 Current Challenges and Limitations in Fluxomics Applied to Metabolic Syndrome

Metabolic Syndrome (MetS), being a multifactorial disorder, brings about an array of alterations in metabolite concentrations that occur in tissues and cells that are distant from those containing the primary lesions. One of the main challenges for fluxomics is to be able to map all human metabolic network fluxes, including compartmentalization and crosstalk between organs. The achievement of this objective will require:

1) Collecting sufficient information from biofluids (blood, urine, breath, saliva. . . .) from tissue biopsies and/or from cell lines established from patient tissues.
2) A view of metabolism and metabolic fluxes with the focus not only on maps of the metabolic stoichiometric reaction networks catalyzed by enzymes, but also on the supramolecular organization of the metabolism. This overall view of the metabolism needs to include interactions between different cell components and new modes of metabolic regulation (i.e. interactions with structural elements, interfering RNAs, etc. (Kuchel 2010)).

It is expected that recent advances in transcriptomics, proteomics, metabolomics and fluxomics with reference to MetS will, in future, provide sufficient information from a patient to enable early diagnosis, prognosis, prediction of treatment efficiency and follow-up of response to treatment.

Since MetS requires a complete and simultaneous understanding of several organs at the same time, *in vivo* approaches are required to fully understand this disease. Current approaches to *in vivo* fluxomics rely on the study of the systemic circulation of metabolites, and usually involve estimation of the rates of hepatic glucose production and clearance (Vaitheesvaran et al. 2010a), rates of lipolysis through glycerol production (Vaitheesvaran et al. 2010b), and general pathway utilization after glucose administration (Xu et al. 2002). More accurate results have been obtained by combining techniques such as GC-MS and NMR after infusion with ^{13}C-labelled metabolites of perfused rat hearts (Li et al. 2011), or applied to the study of the metabolism and clearance of xenobiotics (Zhang et al. 2009; Gu et al. 2010).

Acknowledgements This work was supported by the European Commission Seventh Framework Programme FP7 (Etherpaths KBBE-grant n°222639); the Spanish Government and the European Union FEDER funds (SAF2011-25726); ISCIII-RTICC & European Regional Development Fund (RD06/0020/0046); Generalitat de Catalunya (2009SGR1308 and ICREA Academia price to MC). AB was granted by CSIC (Programa JAE Predoc).

References

Aboka FO, Heijnen JJ, Van Winden WA (2009) Dynamic 13C-tracer study of storage carbohydrate pools in aerobic glucose-limited saccharomyces cerevisiae confirms a rapid steady-state turnover and fast mobilization during a modest stepup in the glucose uptake rate. FEMS Yeast Res 9:191–201

Amaral AI, Teixeira AP, Martens S, Bernal V, Sousa MF, Alves PM (2010) Metabolic alterations induced by ischemia in primary cultures of astrocytes: merging 13C NMR spectroscopy and metabolic flux analysis. J Neurochem 113:735–748

Antoniewicz MR, Kelleher JK, Stephanopoulos G (2006) Determination of confidence intervals of metabolic fluxes estimated from stable isotope measurements. Metab Eng 8:324–37

Antoniewicz MR, Kelleher JK, Stephanopoulos G (2007) Elementary metabolite units (EMU): a novel framework for modeling isotopic distributions. Metab Eng 9:68–86

Baxter CJ, Liu JL, Fernie AR, Sweetlove LJ (2007) Determination of metabolic fluxes in a non-steady-state system. Phytochemistry 68:2313–2319

Boren J, Lee WN, Bassilian S, Centelles JJ, Lim S, Ahmed S, Boros LG, Cascante M (2003) The stable isotope-based dynamic metabolic profile of butyrate-induced HT29 cell differentiation. J Biol Chem 278:28395–28402

Bothwell JH, Griffin JL (2011) An introduction to biological nuclear magnetic resonance spectroscopy. Biol Rev Camb Philos Soc 86:493–510
Bruggeman FJ, Snoep JL, Westerhoff HV (2008) Control, responses and modularity of cellular regulatory networks: a control analysis perspective. IET Syst Biol 2:397–410
Burgess SC, He T, Yan Z, Lindner J, Sherry AD, Malloy CR, Browning JD, Magnuson MA (2007) Cytosolic phosphoenolpyruvate carboxykinase does not solely control the rate of hepatic gluconeogenesis in the intact mouse liver. Cell Metab 5:313–320
Cascante M, Boros LG, Comin-Anduix B, De Atauri P, Centelles JJ, Lee PW (2002) Metabolic control analysis in drug discovery and disease. Nat Biotechnol 20:243–249
Cascante M, Marin S (2008) Metabolomics and fluxomics approaches. Essays Biochem 45:67–81
Castro-Perez JM, Roddy TP, Shah V, Mclaren DG, Wang SP, Jensen K, Vreeken RJ, Hankemeier T, Johns DG, Previs SF, Hubbard BK (2011) Identifying static and kinetic lipid phenotypes by high resolution UPLC-MS: unraveling diet-induced changes in lipid homeostasis by coupling metabolomics and fluxomics. J Proteome Res 10:4281–4290
Collins JM, Neville MJ, Pinnick KE, Hodson L, Ruyter B, Van Dijk TH, Reijngoud DJ, Fielding MD, Frayn KN (2011) De novo lipogenesis in the differentiating human adipocyte can provide all fatty acids necessary for maturation. J Lipid Res 52:1683–1692
Curien G, Bastien O, Robert-Genthon M, Cornish-Bowden A, Cardenas ML, Dumas R (2009) Understanding the regulation of aspartate metabolism using a model based on measured kinetic parameters. Mol Syst Biol 5:271
Curto R, Sorribas A, Cascante M (1995) Comparative characterization of the fermentation pathway of Saccharomyces cerevisiae using biochemical systems theory and metabolic control analysis: model definition and nomenclature. Math Biosci 130:25–50
Choi J, Antoniewicz MR (2011) Tandem mass spectrometry: a novel approach for metabolic flux analysis. Metab Eng 13:225–233
De Atauri P, Sorribas A, Cascante M (2000) Analysis and prediction of the effect of uncertain boundary values in modeling a metabolic pathway. Biotechnol Bioeng 68:18–30
Deshpande R, Yang TH, Heinzle E (2009) Towards a metabolic and isotopic steady state in CHO batch cultures for reliable isotope-based metabolic profiling. Biotechnol J 4:247–263
Dudley E, Yousef M, Wang Y, Griffiths WJ (2010) Targeted metabolomics and mass spectrometry. Adv Protein Chem Struct Biol 80:45–83
Fan TW, Lane AN, Higashi RM, Farag MA, Gao H, Bousamra M, Miller DM (2009) Altered regulation of metabolic pathways in human lung cancer discerned by (13)C stable isotope-resolved metabolomics (SIRM). Mol Cancer 8:41
Feist AM, Palsson BO (2010) The biomass objective function. Curr Opin Microbiol 13:344–349
Gaglio D, Metallo CM, Gameiro PA, Hiller K, Danna LS, Balestrieri C, Alberghina L, Stephanopoulos G, Chiaradonna F (2011) Oncogenic K-Ras decouples glucose and glutamine metabolism to support cancer cell growth. Mol Syst Biol 7:523
Gu L, Zhang GF, Kombu RS, Allen F, Kutz G, Brewer WU, Roe CR, Brunengraber H (2010) Parenteral and enteral metabolism of anaplerotic triheptanoin in normal rats. II. Effects on lipolysis, glucose production, and liver acyl-CoA profile. Am J Physiol Endocrinol Metab 298:E362–371
Gudmundsson S, Thiele I (2010) Computationally efficient flux variability analysis. BMC Bioinformatics 11:489
Heinrich R, Schuster S (1996) The regulation of cellular systems. Chapman and Hall, New York
Henry O, Jolicoeur M, Kamen A (2011) Unraveling the metabolism of HEK-293 cells using lactate isotopomer analysis. Bioprocess Biosyst Eng 34:263–273
Hiller K, Metallo CM, Kelleher JK, Stephanopoulos G (2010) Nontargeted elucidation of metabolic pathways using stable-isotope tracers and mass spectrometry. Anal Chem 82:6621–6628
Jouhten P, Rintala E, Huuskonen A, Tamminen A, Toivari M, Wiebe M, Ruohonen L, Penttila M, Maaheimo H (2008) Oxygen dependence of metabolic fluxes and energy generation of Saccharomyces cerevisiae CEN.PK113-1A. BMC Syst Biol 2:60
Jouhten P, Pitkanen E, Pakula T, Saloheimo M, Penttila M, Maaheimo H (2009) 13C-metabolic flux ratio and novel carbon path analyses confirmed that Trichoderma reesei uses primarily the respirative pathway also on the preferred carbon source glucose. BMC Syst Biol 3:104

Kirkpatrick S, Gelatt CD Jr, Vecchi MP (1983) Optimization by simulated annealing. Science 220:671–680

Krömer J, Quek LE, Nielsen LK (2009) 13C-Fluxomics: a tool for measuring metabolic phenotypes. Australian Biochemist 40:17–20

Kuchel PW (2010) Models of the human metabolic network: aiming to reconcile metabolomics and genomics. Genome Med 2:46

Lane AN, Fan TW, Higashi RM, Tan J, Bousamra M, Miller DM (2009) Prospects for clinical cancer metabolomics using stable isotope tracers. Exp Mol Pathol 86:165–173

Lane AN, Fan TW, Bousamra M II, Higashi RM, Yan J, Miller DM (2011) Stable isotope-resolved metabolomics (SIRM) in cancer research with clinical application to nonsmall cell lung cancer. Omics 15:173–182

Li W, Bian F, Chaudhuri P, Mao X, Brunengraber H, Yu X (2011) Delineation of substrate selection and anaplerosis in tricarboxylic acid cycle of the heart by 13C NMR spectroscopy and mass spectrometry. NMR Biomed 24:176–187

Llaneras F, Pico J (2008) Stoichiometric modelling of cell metabolism. J Biosci Bioeng 105:1–11

Lorkowski S (2011) Chemistry meets nutrition: toward a system biological description of human metabolism. Pure Appl Chem 83:151–165

Mahadevan R, Schilling CH (2003) The effects of alternate optimal solutions in constraint-based genome-scale metabolic models. Metab Eng 5:264–276

Maher AD, Kuchel PW, Ortega F, De Atauri P, Centelles J, Cascante M (2003) Mathematical modelling of the urea cycle. A numerical investigation into substrate channelling. Eur J Biochem 270:3953–3961

Marin S, Chiang K, Bassilian S, Lee WN, Boros LG, Fernandez-Novell JM, Centelles JJ, Medrano A, Rodriguez-Gil JE, Cascante M (2003) Metabolic strategy of boar spermatozoa revealed by a metabolomic characterization. FEBS Lett 554:342–346

Matsuoka Y, Shimizu K (2010) Current status of 13C-metabolic flux analysis and future perspectives. Process Biochemistry 45:1873–1881

Metallo CM, Walther JL, Stephanopoulos G (2009) Evaluation of 13C isotopic tracers for metabolic flux analysis in mammalian cells. J Biotechnol 144:167–174

Moreno-Sanchez R, Saavedra E, Rodriguez-Enriquez S, Olin-Sandoval V (2008) Metabolic control analysis: a tool for designing strategies to manipulate metabolic pathways. J Biomed Biotechnol 2008:597913

Nikerel IE, Van Winden WA, Verheijen PJ, Heijnen JJ (2009) Model reduction and a priori kinetic parameter identifiability analysis using metabolome time series for metabolic reaction networks with linlog kinetics. Metab Eng 11:20–30

Noh K, Wahl A, Wiechert W (2006) Computational tools for isotopically instationary 13C labeling experiments under metabolic steady state conditions. Metab Eng 8:554–577

Oresic M (2009) Metabolomics, a novel tool for studies of nutrition, metabolism and lipid dysfunction. Nutr Metab Cardiovasc Dis 19:816–824

Orth JD, Thiele I, Palsson BO (2010) What is flux balance analysis? Nat Biotechnol 28:245–248

Paul Lee WN, Wahjudi PN, Xu J, Go VL (2010) Tracer-based metabolomics: concepts and practices. Clin Biochem 43:1269–1277

Press W, Flannery B, Teukolsky S, Vetterling W (2002) Numerical recipes in C: the art of scientific computing. Cambridge University Press, New York

Quek LE, Wittmann C, Nielsen LK, Kromer JO (2009) OpenFLUX: efficient modelling software for 13C-based metabolic flux analysis. Microb Cell Fact 8:25

Raftos JE, Whillier S, Kuchel PW (2010) Glutathione synthesis and turnover in the human erythrocyte: alignment of a model based on detailed enzyme kinetics with experimental data. J Biol Chem 285:23557–23567

Rodriguez-Prados JC, De Atauri P, Maury J, Ortega F, Portais JC, Chassagnole C, Acerenza L, Lindley ND, Cascante M (2009) In silico strategy to rationally engineer metabolite production: a case study for threonine in Escherichia coli. Biotechnol Bioeng 103:609–620

Rodriguez-Prados JC, Traves PG, Cuenca J, Rico D, Aragones J, Martin-Sanz P, Cascante M, Bosca L (2010) Substrate fate in activated macrophages: a comparison between innate, classic, and alternative activation. J Immunol 185:605–614

Ruppin E, Papin JA, De Figueiredo LF, Schuster S (2010) Metabolic reconstruction, constraint-based analysis and game theory to probe genome-scale metabolic networks. Curr Opin Biotechnol 21:502–510

Sauer U (2004) High-throughput phenomics: experimental methods for mapping fluxomes. Curr Opin Biotechnol 15:58–63

Schuster S, Pfeiffer T, Fell DA (2008) Is maximization of molar yield in metabolic networks favoured by evolution? J Theor Biol 252:497–504

Selivanov VA, Puigjaner J, Sillero A, Centelles JJ, Ramos-Montoya A, Lee PW, Cascante M (2004) An optimized algorithm for flux estimation from isotopomer distribution in glucose metabolites. Bioinformatics 20:3387–3397

Selivanov VA, Meshalkina LE, Solovjeva ON, Kuchel PW, Ramos-Montoya A, Kochetov GA, Lee PW, Cascante M (2005) Rapid simulation and analysis of isotopomer distributions using constraints based on enzyme mechanisms: an example from HT29 cancer cells. Bioinformatics 21:3558–3564

Selivanov VA, Marin S, Lee PW, Cascante M (2006) Software for dynamic analysis of tracer-based metabolomic data: estimation of metabolic fluxes and their statistical analysis. Bioinformatics 22:2806–2812

Selivanov VA, Vizan P, Mollinedo F, Fan TW, Lee PW, Cascante M (2010) Edelfosine-induced metabolic changes in cancer cells that precede the overproduction of reactive oxygen species and apoptosis. BMC Syst Biol 4:135

Srour O, Young JD, Eldar YC (2011) Fluxomers: a new approach for 13C metabolic flux analysis. BMC Syst Biol 5:129

Stephanopoulos G, Aristidou A, Nielsen J (1998) Metabolic Engineering. Principles and Methodologies. San Diego Academic Press

Teusink B, Passarge J, Reijenga CA, Esgalhado E, Van Der Weijden CC, Schepper M, Walsh MC, Bakker BM, Van Dam K, Westerhoff HV, Snoep JL (2000) Can yeast glycolysis be understood in terms of in vitro kinetics of the constituent enzymes? Testing biochemistry. Eur J Biochem 267:5313–5529

Vaitheesvaran B, Chueh FY, Xu J, Trujillo C, Saad MF, Lee WN, Mcguinness OP, Kurland IJ (2010a) Advantages of dynamic "closed loop" stable isotope flux phenotyping over static "open loop" clamps in detecting silent genetic and dietary phenotypes. Metabolomics 6:180–190

Vaitheesvaran B, Leroith D, Kurland IJ (2010b) MKR mice have increased dynamic glucose disposal despite metabolic inflexibility, and hepatic and peripheral insulin insensitivity. Diabetologia 53:2224–2232

Vizan P, Alcarraz-Vizan G, Diaz-Moralli S, Solovjeva ON, Frederiks WM, Cascante M (2009) Modulation of pentose phosphate pathway during cell cycle progression in human colon adenocarcinoma cell line HT29. Int J Cancer 124:2789–2796

Wiechert W (2001) 13C metabolic flux analysis. Metab Eng 3:195–206

Wiechert W, Mollney M, Petersen S, De Graaf AA (2001) A universal framework for 13C metabolic flux analysis. Metab Eng 3:265–283

Xu J, Xiao G, Trujillo C, Chang V, Blanco L, Joseph SB, Bassilian S, Saad MF, Tontonoz P, Lee WN, Kurland IJ (2002) Peroxisome proliferator-activated receptor alpha (PPARalpha) influences substrate utilization for hepatic glucose production. J Biol Chem 277:50237–50244

Zamboni N, Fischer E, Sauer U (2005) FiatFlux—a software for metabolic flux analysis from 13C-glucose experiments. BMC Bioinformatics 6:209

Zamboni N, Sauer U (2005) Fluxome profiling in microbes. In: Vaidyanathan S, Harrigan GG, Goodacre R (eds) Metabolome analyses: strategies for systems biology. Springer, New York

Zhang GF, Kombu RS, Kasumov T, Han Y, Sadhukhan S, Zhang J, Sayre LM, Ray D, Gibson KM, Anderson VA, Tochtrop GP, Brunengraber H (2009) Catabolism of 4-hydroxyacids and 4-hydroxynonenal via 4-hydroxy-4-phosphoacyl-CoAs. J Biol Chem 284:33521–33534

Chapter 13
In vitro Colon Model to Study Metabolic Syndrome

Anna-Marja Aura

Abstract Metabolic syndrome (MetS) is a metabolic disorder, tightly linked to obesity, and predisposing to cardiovascular disease and type-2 *diabetes mellitus* (T2D). There is already some knowledge related to intake of diet rich in phytochemicals and improved biological effects *e.g.* adipose tissue function or biomarkers of improved insulin sensitivity. This chapter aims to describe the connection between plant food and beverage intake and circulating metabolites by using the *in vitro* colon model. The *in vitro* colon model coupled with metabolomic systems biology is applied to connect food components, over-expressed *in vitro* metabolite profiles and corresponding *in vivo* metabolite profiles. When the diet relation has been made, the same metabolites are identified from the human body fluids. The traditional end-points related to health and disease can be connected with the biomarkers of intake of the same volunteers. Finally, novel biomarkers of health status can be sought using novel lipidomic and proteomic techniques, which has potential to answer to the urgent need for early biomarkers of disease and improved health.

Keywords Metabolic syndrome · Colonic metabolism and *in vitro* digestion models · Flavanols and proanthocyanins · Hydroxycinnamic acids · Anthocyanins

13.1 Introduction

Metabolic Syndrome (MetS) is a metabolic disorder, tightly linked to obesity, and predisposing to cardiovascular disease and type-2 *diabetes mellitus* (T2D). MetS is characterized by central obesity, insulin resistance, dysglycemia, dyslipidemia and hypertension (Eckel et al. 2005; Möller and Kaufman 2005). MetS is also connected with adipose tissue metabolism, skeletal muscles, liver steatotosis and systemic oxidative stress. MetS associated inflammation is indicated as elevated levels of C-reactive protein and levels of pro-inflammatory cytokines in obese people and genetically modified obese (ob/ob) mice (Visser et al. 1999; Zahorska-Markiewicz

A.-M. Aura (✉)
VTT Technical Research Centre of Finland, Tietotie 2, Espoo,
P.O. Box 1000, 02044 VTT, Finland
e-mail: anna-marja.aura@vtt.fi

M. Orešič, A. Vidal-Puig (eds.), *A Systems Biology Approach to Study Metabolic Syndrome*,
DOI 10.1007/978-3-319-01008-3_13,

et al. 2000; Dube et al. 2008). MetS patients exhibit activation of biochemical pathways leading to systemic oxidative stress, which regulates expression of genes governing lipid and glucose metabolism (Grattagliano et al. 2008).

MetS patients are a good target group for linking plant food intake and improvement of health status, because they have already biomarkers related to unbalanced health status, however, those biomarkers are still reversible, if the factors affecting regulation of the balance are corrected. There is already some knowledge related to dietary fibre, plant food or beverage intake and improved biological effects in MetS, *e.g.* adipose tissue function as biomarker of improved insulin sensitivity. This chapter aims to describe the connection between plant food and beverage intake and metabolism, the *in vitro* colon model as a tool for filling in the gap between the intake and its circulating metabolites, and finally to describe research on colonic metabolism of those compounds from the diet which have shown a relation to MetS or its end-points. The chapter will conclude with prospects and considerations related to this work.

13.2 Dietary Fibre Components and Health

Different types of dietary fibre (DF) have been reported to take part in the control of body weight, glucose and lipid homeostasis, insulin sensitivity and in the regulation of many inflammation markers involved in the pathogenesis of MetS (Galisteo et al. 2008; Ma et al. 2008; Priebe et al. 2010). American Dietetic Association (ADA) has taken a position that the intake of DF, coming from variety of plant foods, has beneficial effect on risk factors for developing several chronic diseases, including MetS and T2D (ADA 2008). The novel definition of DF in the European Commission directive 29th Oct 2008 (2008/100/EC) includes non-digestible and non-absorbable carbohydrate polymers, oligomers and monomers and phytochemicals such as phenolic compounds and sterols.

Meta-analysis (14 cohorts with 48 000 individuals in European descent) was performed on interaction of the dietary whole grain intake with fasting glucose- and insulin levels: greater whole grain intake was associated with lower fasting glucose and insulin levels and it was related to relevant genetic logi independent of demographic, other dietary or life style factors (Nettleton et al. 2010). Cereal fibre intake and effects in improved lipid, glucose and energy metabolism has been reviewed recently, indicating also qualitative factors related to physicochemical properties (e.g. solubility, viscosity, fermentability), fibre extract molecular weight, particle size and microstructural features (Smith and Tucker 2011; de Mello et al. 2011). A stronger protective role association was shown for insoluble fibre than for soluble fibre, suggesting that other characteristics than fermentability of carbohydrates to short-chain fatty acids is associated with DF intake and reduced risk of diabetes (Weickert and Pfeiffer 2008). A diet rich in fatty fish, bilberries and wholegrain products was associated with improved markers of endothelial function and inflammation in overweight and obese individuals at high risk of developing diabetes. The study was conducted as a randomised controlled diet (de Mello et al. 2011).

Table 13.1 shows that intake of whole grain and DF associated polyphenolic components has exerted systemic effects *in vivo* on biomarkers of MetS in humans and animals, including adipose tissue function and glucose and insulin metabolism. Table 13.1 also displays some *in vitro* mechanistic studies related to the effects of DF associated phenolic compounds on the glucose and insulin metabolism.

Furthermore, ellagitannins exhibited strong activities in both stimulating insulin-like glucose uptake and inhibition of 3T3-L1-adipocyte differentiation *in vitro* (Bai et al. 2008). It should be noted that not only such solid phenolic sources as DF matrices, but also tea consumption resulted in improvement of biomarkers related to insulin sensitivity (Stote and Baer 2008). Green tea epigallocatechin gallate was shown to direct the lipid accumulation to the cytosol of hepatocyte cell model (McA-RH7777), decreasing apoB-100 very low density lipoprotein secretion thus affecting its assembly (Li et al. 2006). Also citrus flavonoids affected lipid metabolism in HepG2 cells by suppressing ApoB secretion (Lin et al. 2011). This evidence suggests that diet rich in DF and or phenolic compounds can be associated with reduction of MetS-related biomarkers. Verzelloni et al. (2011) demonstrated that ellagitannin colonic metabolites (urolithins and pyrogallol) were effective *in vitro* antiglycative agents, whereas chlorogenic acid-derived metabolites (dihydrocaffeic or –ferulic acids or feruloylglycine) were neuroprotective *in vitro*, suggesting to be counteractive in diabetic complications i.e. protein glycation and neurodegeneration. The question remains to be answered how components which are enclosed in the DF matrix or large in size can regulate pathways active in reduction of MetS. The studies exposing the cell lines to dietary phenolics as is do not take into account the situation where polyphenols are converted to their intestinal metabolites. From this perspective it is more likely that the tissues in the body are rather subjected to the intestinal metabolites of food components than to those originally present in the food.

DF is by definition non-digestible and is thus subjected to colonic microbial interaction. Colonic microbiota degrades the DF complex, forms short-chain fatty acids (SCFA) from carbohydrates, releases phytochemicals from the plant matrix and transforms them to chemically different forms such as phenolic acids and lactones (Selma et al. 2009; Aura 2008). Microbial metabolites can form a link to food exposure, because they are abundant in circulation after exposure (Lafay and Gil-Izquierdo 2008). The studies have verified the connection between plant foods and beverages and pure phenolic compounds as precursors of colonic metabolites, which furthermore have also been identified in the plasma and urine of humans after consumption of the precursors, having a long residence time in the body (Kuijsten et al. 2005; Sawai et al. 1987; Seeram et al. 2006). This has led to a hypothesis that the health promotion can be mediated by colon derived metabolites as signal molecules alone or in synergy with the plant food components. Therefore identification of the plant food- and beverage-derived microbial metabolites is needed by studying colonic metabolism alone and excluding the host metabolism in the liver or intestinal epithelia.

Table 13.1 Effects of dietary fibre and its components on metabolic syndrome (MetS) biomarkers

Component	Study design	Effect	References
Whole-meal rye bread and bread with whole kernels	Human intervention with healthy Finnish subjects	Lower insulin response, but not a lower glucose response	(Juntunen et al. 2003)
Rye-pasta diet vs. white wheat bread—oat-potato diet	Human intervention with persons with mild metabolic syndrome	Early insulin secretion enhanced, gene expression modulated in subcutaneous abdominal adipose tissue; modulated inflammatory status	(Laaksonen et al. 2005; Kallio et al. 2007; Kallio et al. 2008)
High-polyphenol dark chocolate	Human intervention with hypertensive patients	Blood pressure reduced and insulin sensitivity increased in glucose intolerant, hypertensive animals after 15 days of consumption	(Grassi et al. 2008)
Tea (-)-Epigallo-catechin-3-gallate (EGCG)	Long-term animal trial on C57BL/6J mice	ECGC inhibits obesity, metabolic syndrome markers and fatty-liver disease in high-fat-fed mice during 16 week dietary regime	(Bose et al. 2008)
Cacao liquor proanthocyanidins	Diabetic obese mouse model	Prevents elevation of blood glucose levels	(Tomaru et al. 2007)
Proanthocyanidin polymer or oligomer preparations from persimmon peel	Diabetic mouse model	Especially oligomers affected inflammatory reactions and ameliorated hyper-lipidaemia and hyper-glycaemia	(Lee et al. 2008)
Short-chain fructo-oligosaccharides	Obese and lean dogs	Adipose tissue transcriptome showed modulated genes in fatty acid and glucose metabolism	(Respondek et al. 2008)
Proanthocyanidins	Streptozotozin-induced diabetic rats and tissue cell-lines *in vitro*	Postulated mechanism: direct effect on glucose uptake in adipose tissue and muscle: insulin—signalling pathways involving PI3K and p38 MAPK activation and GLUT-4 translocation	(Pinent et al. 2006)
Hydroxycinnamic acids *p*-methoxycinnamic acid and ferulic acid	*In vivo* rat pancreas and *in vitro* rat pancreatic cells	Regulated blood glucose level by stimulating insulin secretion from pancreatic β-cells.	(Adisakwattana et al. 2008)
Ellagitannin rich raspberry and strawberry extracts	*In vitro* inhibition of α-amylase and α-glucosidase tests	Synergistic inhibition of starch digestion	(McDougall et al. 2005)
Anthocyanins and anthocyanidins	*In vitro* rodent pancreatic β-cells for testing insulin secretion	Delphinidin, cyanidin and pelargonidin glycosides increased insulin secretion	(Jayaprakasam et al. 2005)
Anthocyanins from muscadine grapes	*In vitro* inhibition of α-amylase and lipase activities	Lower digestibility of starch and fat	(You et al. 2011)
Polyphenols from leaves of molokheia (*Corchorus olitorius* L.)	LDL-deficient mice	Reduced body weight, liver weight and hepatic triglycerides by reduction of oxidative stress and increased β-oxidation in the liver	(Wang et al. 2011)

13.3 Colon Models

In vitro models have traditionally been used to study fermentation characteristics of DF components both carbohydrates as precursors of SCFAs, especially propionic and butyric acids (Barry et al. 1995; Rumney and Rowland 1992; Tabernero et al. 2011). *In vitro* colon models include continuous, semi-continuous or batch culture systems, which maintain colonic microbiota, usually obtained from human faeces, under strictly anaerobic conditions. The methods differ from each other in the structure of the fermentor, substrate-to-inoculum ratios, media, operating conditions and sampling (Rumney and Rowland 1992; Edwards et al. 1996). Media have varied from a simple mineral salt solution to a complex medium containing vitamins, hemin, SCFA, yeast extract and trypticase, a reducing agent and different buffers. The cultures have been static or several mixing methods have been used: regular swirling, a shaking water bath and periodic mixing (Edwards et al. 1996).

Different types of colon models can be used for different purposes. Batch models due to its small scale are suitable for the studies of microbial metabolism of DF, resistant starch and isolated phenolic compounds, as examples (Barry et al. 1995; Edwards et al. 1996; Deprez et al. 2000; Aura et al. 2002; Aura et al. 2005b). A European interlaboratory study was performed for testing a simple reproducible batch system for DF fermentation (Barry et al. 1995). The inoculum was composed of fresh human faeces mixed with a carbonate-phosphate-buffer complex supplemented with trace elements and urea. Five different carbohydrate sources were compared in five laboratories on three occasions to determine pH, residual non-starch polysaccharides and SCFA production during fermentation. Interlaboratory differences could be reduced either by adding less substrate during incubations or using a dense inoculum. The optimal amount of carbohydrates was 10 g/l in a faecal suspension of 167 g/l. There was also a close correspondence between the *in vitro* data and an *in vivo* rat experiment using supplemented diets with the same sources of DF (Barry et al. 1995). This batch model has been further applied to various isolated phenolic compounds and foods containing them in search for the food-related metabolites.

In the batch models nutrients are consumed and products are accumulated causing altered cultural conditions during the incubation, which may result in a distorted interpretation of microbial metabolism in batch cultures at later time points. This can be avoided by detecting the metabolites profiles at several time points to elucidate the metabolite dynamics and excluding artefacts occurring late during the incubation. However, continuous and semi-continuous systems have additional benefits because they do not accumulate the metabolites and their more dilute microbiota adapts to the substrates with more nutritive media, than applied in batch models (Barry et al. 1995). Thus continuous systems are also suitable for the study of the ecosystem and bacteriology (Miller and Wolin 1981; Allison et al. 1989; Bearne et al. 1990; Campbell et al. 1992).

A single-stage continuous system cannot reproduce the heterogeneity of physiological conditions and nutrient availabilities that occur at different parts of colon, which is achieved in multi-stage chemostats (Allison et al. 1989). To correct this problem, a two-stage continuous system was first developed by Bearne et al. (1990).

Furthermore, a three-stage continuous culture system was investigated in terms of carbohydrate metabolism and amino acid metabolism. Carbohydrate breakdown and SCFA production occurred mainly in the first reactor, whereas amino acid fermentation producing branch-chain fatty acids occurred mainly in the reactors 2 and 3. Thus the reactions in the three reactors corresponded to reactions occurring in the ascending, transverse and descending colon *in vivo*, respectively (Macfarlane et al. 1998). Molly and co-workers took this development even further when they presented a 5-step multi-chamber reactor as a simulation of the human intestinal microbial ecosystem. This so called SHIME model comprises of two vessels simulating duodenum and jejunum, and ileum, respectively, and the 3rd, 4th and 5th corresponding caecum and ascending colon, transverse colon and descending colon, respectively (Molly et al. 1993).

The latest development of a continuous, anaerobic colon model is a validated computer controlled *in vitro* system, comprising of TNO Intestinal Models 1 and 2 (TIM-1 and TIM-2), simulating the upper intestine and colon, respectively (Tabernero et al. 2011; Venema et al. 2004). When the substrate is passed from one compartment to another through the model, temperature, pH, peristaltic movements, absorption of water and fermentation products are continuously monitored and controlled. The model has been applied with high density complex microbiota. The model has been used for fermentation of DF and phenolic compounds, formation of ammonia, investigation of probiotic bacteria and bioconversion of glucosinolates (Venema et al. 2004).

Identification of food-specific intestinal metabolites is impossible without ruling out the metabolism from the tissues in the human body. An anaerobic *in vitro* colon model by Barry et al. (1995) was adopted and modified for the purpose of release and conversion of dietary phytochemicals. The time course and extent of microbial metabolite formation can be measured from complex food mixtures after enzymatic digestion *in vitro* (Aura et al. 1999) or from isolated compounds with tailor-made approaches without enzymatic digestion. The model is performed in strictly anaerobic conditions at human body temperature and by using pooled human faeces from several healthy donors to provide diverse microbiota and reproducible data. The model has already been applied to study metabolism of the DF complex of cereals (Aura et al. 2005a; Aura et al. 2006), fruits (Bazzocco et al. 2008; Aura et al. 2012) and various isolated phenolic compounds (Aura et al. 2002; Aura et al. 2005b; Pealvo et al. 2005) and it has been coupled with metabolomics platform based on data-processing from mass-spectrometry (Aura et al. 2012; Aura et al. 2008) for search of phenolic compound-related metabolites. Thus *in vitro* colon model coupled with metabolomic systems biological tools is applied to connect food components, over-expressed *in vitro* metabolite profiles and corresponding *in vivo* metabolite profiles. When the diet relation is made by using the *in vitro* model and the same metabolites are identified from the human body fluids, the traditional end-points related to health and disease can be connected with these biomarkers of intake and compliance. There is an urgent need for early biomarkers of disease and for biomarkers designating the improvement of health status. The role of *in vitro* colon model in the research related to MetS is displayed in Fig. 13.1.

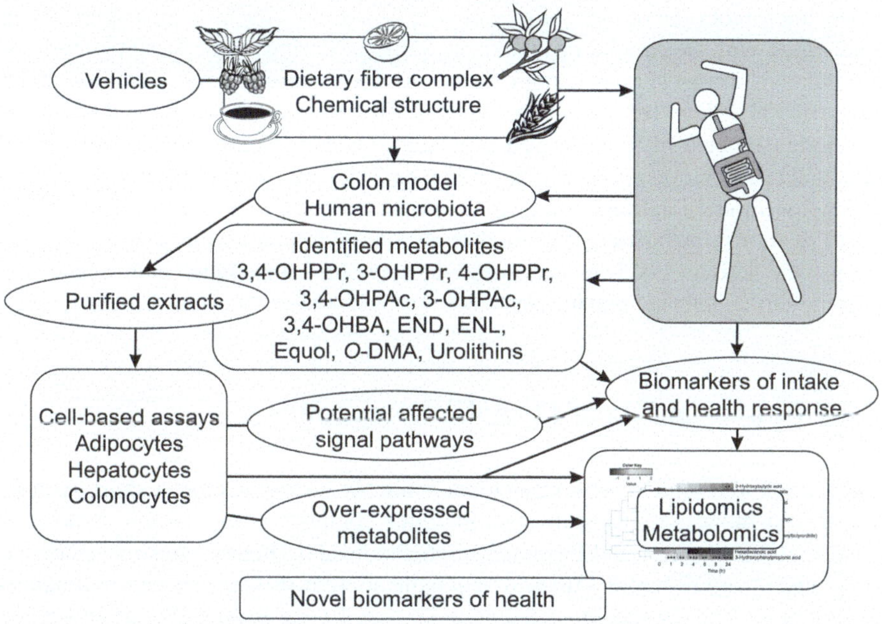

Fig. 13.1 Colon model in studies of Metabolic Syndrome (MetS). Colon model helps the identification of food-related metabolites by indicating compliance biomarkers of the food intake. Biomarkers of MetS include biomarkers involved in glucose and insulin, liver fatty acid, adipose tissue and skeletal muscle metabolism. Mechanistic studies are performed for finding the regulated pathways in the cell assays under influence of the metabolite extracts from the same foods as has been included in the diet of the clinical trial. Mechanistic studies reveal the pathway regulation patterns and explain potential novel biomarkers, which can be investigated from human body fluids and tissues with metabolomic and lipidomic approaches. Finally, biomarkers of compliance, traditional and novel biomarkers of MetS are combined to understand the causal relationships between diet and health

13.4 Impact of Microbiota in Health

Colonic microbiota consists of 1000 phylogenetic types of microorganisms constituting an ecologically dynamic community, which is closely related to human physiology (Rajilic-Stojanovic et al. 2007). Microbiota is involved in maturation and proliferation of intestinal cells, help in maintenance of the homeostasis with pro-inflammatory and anti-inflammatory responses, relating to MetS (Hattori and Taylor 2009). Changes in microbiota is also associated with induced inflammation by lipopolysaccharides from the gut in connection with high-fat diet, which induces MetS and insulin resistance (Cani et al. 2007; Cani et al. 2008). Furthermore, modulation of the microbiota by prebiotics has been linked to improvement of glucose homeostasis, leptin sensitivity and enteroendocrine cell activity in obese and diabetic mice (Everard et al. 2011). It has also been reported that colonic microbial activity influences the fatty acid composition in liver and adipose tissue and reduces proinflammatory cytokines (Wall et al. 2009).

The microbiota can be associated with the formation of the food related microbial metabolites, with the circulating human metabolome during the dietary interventions, and with the biological responses in the human body (Turnbaugh and Gordon 2008; Wishart 2008). However, there is a challenge from the large individual differences which are expected in food-derived microbial metabolite profiles, because human microbial composition and their conversion capacity vary between subjects. The *in vitro* colon model facilitates finding biomarkers of exposure and establishing the causal relationships between the food intakes and circulating metabolites *in vivo,* because the activities of liver and intestinal epithelia are excluded and reduces the diversity of the detected metabolites.

13.5 Bioavailability of Phenolic Compounds

In order to understand how dietary phytochemicals enter the colon, a short description of their bioavailability is needed. Phenolic compounds serve as biomarkers of exposure to plant food and beverage phytochemicals. Dietary intake of phenolic compounds from different fruit and vegetable sources differs between individuals depending on their habitual diets but the average has been estimated at about 1 g/day (Scalbert and Williamson 2000). Mediterranean diet, rich in fruits and vegetables, and Nordic diet rich in rye and berries, contain complex set of monomeric and polymeric phenolic compounds which are converted to specific microbial metabolites (Aura 2008). Bioaccessibility *in vitro* and bioavailability *in vivo* of phenolic compounds are dependent on the resilience and nature of the food and beverage matrices to digestion in the upper intestine, the degradation and release properties of the matrix in the colon and structures of the phenolic compounds. The degradation and release properties of the food and beverage matrices also determines in which stage and to what extent components are accessible for the colonic microbiota and its conversion activities, and finally, what is the composition of metabolite pool in the blood circulation.

When flavonoids and related phenolic compounds are consumed in the diet, they are released from the matrix after mastication. The stomach reduces food particle size, which further enhances the release of phenolic compounds (Scalbert et al. 2002). Phenolic acids can be absorbed in the free form in the stomach (Zhao et al. 2004; Lafay et al. 2006). Flavonoid glycosides can be deglycosylated in the mouth by microbiota or by oral epithelial cells (Walle et al. 2005). Aglycones can also be formed in the small intestine by the action of membrane–bound lactase phloridzin hydrolase and they are absorbed passively through the epithelium or flavonoids can be transported through the epithelium as glycosides by sugar transporters. In the epithelial cells, cytosolic β-glucosidase hydrolyses the glycosides and aglycones are formed after absorption (Day et al. 2000; Nemeth et al. 2003). Generally, absorption is affected by the structure of phenolic compounds (glycosylation, molecular weight and esterification). Once absorbed, aglycones and phenolic acids undergo conjugation in the ileal epithelium or in the liver (Scalbert et al. 2002). A part of the hepatic metabolites (methylated, sulfated or glucuronidated conjugates) (Donovan et al. 1999; Felgines et al. 2003; Natsume et al. 2003) is returned to the lumen via bile (Roberts et al. 2002) and re-absorbed after deconjugation by the luminal bacteria from the colon (entero-hepatic re-circulation). Subsequently, luminal non-absorbed or re-cycled phenolic compounds and those still bound into the food matrix enter the colon.

In the colon, cleavage of the glycosyl or glucuronosyl moiety from the phenolic backbone is catalysed by microbial enzymes resulting in the transient appearance of aglycones prior to further catabolism by the microbiota (Aura et al. 2002; Rechner et al. 2004; Hein et al. 2008). Deconjugation is catalysed by fecal microbial enzymes (α-rhamnosidase, β-glucosidase and β-glucuronidase), the specific activities of which reflect the *in vitro* deconjugation rates of phenolic compounds. Many flavonoid aglycones undergo C-ring-fission, as a result of which hydroxylated aromatic compounds are formed from the A-ring and phenolic acids from the B-ring. (Aura et al. 2002; Aura et al. 2005a; Rechner et al. 2004). In contrast to deconjugation, sulfate conjugation of phenolic hydroxyl groups is catalysed by arylsulfotransferases originating from human intestinal bacteria. This enzyme activity has been detected in human and rat faeces (Kim and Kobashi 1986a; Kim et al. 1986b). Furthermore, lactones can be formed from ellagic acid and plant lignans (Heinonen et al. 2001; Cerda et al. 2004; Cerda et al. 2005).

Pharmacokinetic studies show that microbial metabolites have up to a 24–48 h residence time in the bloodstream after a single dose of their parent compounds (Kuijsten et al. 2005; Sawai et al. 1987; Gross et al. 1996; Nesbitt et al. 1999; Juntunen et al. 2000; Kilkkinen et al. 2003). After formation, complex microbial metabolites such as lactones are absorbed from the colon and are again substrates for liver metabolism, resulting in the production of glucuronidated, methylated, glycinated and sulfated derivatives (Adlercreutz et al. 1995; Lampe 2003). These microbial and hepatic metabolites undergo entero-hepatic re-circulation, which contributes to the low diurnal variation of the metabolite concentrations in the blood (Bach Knudsen et al. 2003). Finally, lactone forms are excreted via urine as hepatic conjugates (Seeram et al. 2006; Adlercreutz et al. 1995), and microbial phenolic acid metabolites mainly in a free form (Sawai et al. 1987). This may also indicate that the free phenolic acids are better accessible to peripheral tissues such as skeletal muscle and adipose tissue, when their molecular size is not increased by conjugation.

The daily flux of water through the colonic epithelium is estimated to be over one litre calculated from the daily volumes of the chyme entering the caecum ($\sim$ 1500 mL) and that of the fecal excretion ($\sim$ 200 mL) (Guyton and Hall 1996). As the metabolites described above and their hepatic conjugates are found in plasma and urine, therefore they circulate through the body and may exhibit both local and systemic effects. Phenolic metabolite levels in urine are at the *micro*-molar range (Gross et al. 1996; Knust et al. 2006), whereas plasma levels range from low to high *nano*-molar concentrations (Sawai et al. 1987; Kilkkinen et al. 2001; Kern et al. 2003; Johnsen et al. 2004; Kuijsten et al. 2006). In peripheral tissues the concentrations of anthocyanidin glucuronides have been shown to be at range of 0.01–0.50 nano-moles per kilogram fresh weight of brain tissues (Milbury and Kalt 2010).

Personal differences in the enzymatic conversion activities, intestinal diseases or medication cause qualitative and quantitative variation of the metabolite profiles *in vivo*, which is applicable to colon model, as well. High individual variation in the extent of metabolism is a reflection of the inter- and intra-individually variable colonic microbiota (Cerda et al. 2005; Lampe 2003; Kilkkinen et al. 2002). When a pooled inoculum from several donors is applied to a colon model, the repeatability of the responses can be enhanced. As a consequence, the results from different experiments have appeared to be qualitatively similar and having smaller standard deviations than those observed from corresponding animal trials or human experiments. Furthermore, the control is from the same origin and thus comparison between the control (no food added) and the food (food with microbiota) shows the over-expression of the metabolites, which is useful especially when a non-targeted analytical approach is applied (Aura et al. 2012; Aura et al. 2008). Use of proton nuclear magnetic resonance 1 H-NMR, gas chromatography coupled to mass spectrometry (GC-MS), two-dimensional GC coupled to time-of-flight MS (GC × GCTOFMS) and liquid chromatography coupled to MS (LC-MS) methods enable a metabolomic approach to profiling of the colonic conversions. Furthermore, these techniques can be applied to samples derived from colon models and urine and faecal water from human intervention studies (Aura et al. 2008; Grün et al. 2008; Jacobs et al. 2008; Pettersson et al. 2008; Hanhineva et al. 2012; Aura et al. 2012).

The complex microbial community provides a high metabolic potential and interaction with the host via the colonic epithelium. The individual metabolite profiles show considerable inter- and intra-individual differences. The quantitative metabolite concentrations are more variable than the qualitative metabolite compositions among subjects, suggesting that different colonic communities share general metabolic activities, which convert food components to specific metabolite profiles (Jacobs et al. 2009). The identification of the phenolic microbial metabolites reveals that a rather large number of different dietary phenolic precursors (flavonoids, phenolic acids and proanthocyanidins) are converted to the same microbial phenolic acid metabolites which comprise the predominant metabolites as far as we know today. The next chapter will elucidate the formation of microbial metabolism more in detail.

13.6 Identification of the Metabolites from the Key Foods Associated with MetS

As mentioned in the Table 13.1 key phenolic groups concerned in terms of possible interaction with biochemical pathways associated with MetS are proantocyanidins, monomeric flavanols and hydroxycinnamic acids, all sharing phenylpropionic and phenylvaleric acids as primary microbial metabolites. Anthocyanins have also been considered as possible affective phenolic compound.

13.6.1 Monomeric Flavanols and Proanthocyanidins

Flavan-3-ols can be divided into monomers such as (+)-catechin and (–)-epicatechin and their galloylated derivatives which are in high amounts in green tea, and dimeric, oligomeric and polymeric proanthocyanidins, which are abundant in cocoa, apples, grapes and, as a consequence, beverages thereof such as cider and wine. Metabolites from monomeric, dimeric and polymeric flavan-3-ols are collated in Table 13.2.

Hydroxyphenylvaleric acids were shown to be formed from (+)-catechin by the *in vitro* action of rabbit intestinal microbiota (Scheline 1970). When two different batches of human microbiota was used in the *in vitro* incubations, stereoisomerism did not have a marked effect on the conversion of (+)-catechin and (–)-epicatechin, but the site of ring-fission was dependent on the microbiota. (+)-Catechin and (–)-epicatechin were transformed either to hydroxyphenylpropionic or hydroxyphenylvaleric acid derivatives in two separate experiments using pools of microbiota obtained from different donors of the faecal inocula (Aura et al. 2008; Meselhy et al. 1997). (–)-Epicatechin, (–)-epigallocatechin and their 3-*O*-gallates were catabolised extensively by human faecal suspensions, whereas gallates resisted degradation by a rat caecal suspension, suggesting species differences in catabolic activity of the two microbiota (Meselhy et al. 1997). In a recent study (–)-epicatechin-3-*O*-gallate incubated *in vitro* with rat caecal microbiota was converted to trihydroxy- or dihydroxyphenyl-trihydroxypropanol, and dihydroxyphenylhydroxyvaleric acid (Table 13.2) (Takagaki and Nanjo 2010). A series of catechin isomers and their gallates were incubated in pig caecal microbiota and all had a similar qualitative metabolite profile, including phenylpropionic acids, phenylacetic acids, benzoic acids and phloroglucinol (vant't Slot and Humpf 2009).

The first *in vivo* study identified 11 metabolites in human urine after (+)-catechin intake. The major components were hydroxyphenylpropionic acid and hydroxyphenyl- γ-valerolactone (Table 13.2) (Das 1971). In a pharmacokinetic study of tea catechins, ring-fission catabolites appeared in significant amounts after 3 h and peaked in plasma at 8–15 h after consumption (Lee et al. 2002). Ring-fission metabolites from tea polyphenols were excreted in urine as di- and trihydroxylated derivatives of phenyl-γ-valerolactone and they accounted for 1.5–16 % of the ingested catechins (Meng et al. 2002).

A substantial proportion of ingested green tea flavan-3-ols passes through the small intestine to the large intestine (Stalmach et al. 2009). Roowi et al. (2010) showed that a minimum of 40 % of the intake of flavan-3-ols from green tea is converted to phenolic acid metabolites in the colon, whereas about 8 % is excreted as flavan-3-ol methyl, glucuronide and sulfate metabolites, the metabolites which reflect the small intestine absorption. Furthermore, colonic metabolites undergo hepatic metabolism, since urinary metabolites comprise of hydroxybenzoic acid, hippuric acid and methoxylated-hydroxyphenylacetic acid, which do not accumulate in the faecal suspensions. In contrast, dihydroxyphenyl-γ-valerolactone and dihydroxyphenyl-γ-valeric acid, which were major (–)-epicatechin metabolites in faecal suspensions, were not excreted in urine (Table 13.2) (Roowi et al. 2010).

Table 13.2 Identified metabolites from flavan-3-ols, hydroxycinnamic acids and anthocyanins

Component	Study design	Metabolites	References
Flavan-3-ol monomers			
(+)-Catechin (−)-Epicatechin	*In vitro* human colon model with two different faecal pools as a source of microbiota	3-(3′,4′-dihydroxyphenyl)propionic acid 3-(3′-hydroxyphenyl)propionic acid 5-(3′,4′-dihydroxyphenyl)-γ-valeric acid 3-(3′,4′-dihydroxyphenyl)propionic acid 3-(3′-hydroxyphenyl)propionic acid 5-(3′,4′-dihydroxyphenyl)-γ-valeric acid 5-(3′-hydroxyphenyl)-γ-valeric acid	(Aura et al. 2008)
(−) Epicatechin-3-O-gallate	*In vitro* model rat caecal microbiota	1-(3′,4′,5′-trihydroxyphenyl)-3-(2″,4″,6″-trihydroxy)propan-2-ol 1-(dihydroxyphenyl-3-(2″,4″,6″-trihydroxy)propan-2-ol 5-(3′,5′-dihydroxyphenyl)-4-hydroxyvaleric acid	(Takagaki and Nanjo 2010)
(+)-Catechin	*In vivo* study on human urine	3-(3′-hydroxyphenyl)propionic acid 5-(3′-hydroxyphenyl)-γ-valerolactone	(Das 1971)
Green tea flavan-3-ols (−)-Epicatechin	Human faecal suspension from an *in vivo* study Urinary hepatic microbial metabolites	5-(3′,4′-dihydroxyphenyl)-γ-valerolactone 5-(3′,4′-dihydroxyphenyl)-γ-valeric acid 4-hydroxybenzoic acid, hippuric acid, 3′-methoxy-4′-hydroxyphenylacetic acid	(Roowi et al. 2010)
Grape seed proanthocyanidin dimers B1-B4 and two unidentified	*In vitro* human colon model	3′,4′-dihydroxyphenylacetic acid, 5-(3′,4′-dihydroxyphenyl)-γ-valerolactone	(Appeldoorn et al. 2009)
[14C]-Proanthocyanidin polymer	*In vitro* human colon model	2-(3′-hydroxyphenyl)acetic acid, 2-(4′-hydroxyphnyl)acetic acid, 5-(3′-hydroxyphenyl)valeric acid, 3-phenylpropionic acid, 3-(3′-hydroxyphenyl)propionic acid, 3-(4′-hydroxyphenyl)propionic acid	(Deprez et al. 2000)

Table 13.2 (continued)

Component	Study design	Metabolites	References
Hydroxycinnamic acids			
Phenolic acids from whole grain wheat: a matrix bound ferulic acid	*In vitro* human colon model	3-(3′-hydroxyphenyl)propionic acid, 3-phenylpropionic acid	(Mateo Anson et al. 2009; Mateo Anson et al. 2011)
Phenolic acids from whole grain wheat bread: ferulic acid, sinapic acid, coumaric acid	Human intervention with a single dose and collection of plasma and urine	Ferulic acid, sinapic acid, coumaric acid 3,4-dimethoxybenzoic acid 3-(3′-hydroxyphenyl)propionic acid 3-phenylpropionic acid	
Caffeic acid and its esters, chlorogenic acid and caftaric acid	*In vitro* human colon model	3-(3′-hydroxyphenyl)propionic acid benzoic acid	(Gonthier et al. 2006)
Caffeic acid	*In vitro* human colon model	4-ethylcatechol	(Peppercorn and Goldman 1971)
Ferulic acid dimers with a 8-O-4–linkage	*In vitro* human colon model	monomeric ferulic acid 3-(3′,4′-dihydroxyphenyl)propionic acid	
Ferulic acid dimers with a 5–5-linkage	*In vitro* human colon model	2-(3′,4′-dihydroxyphenyl)acetic acid 3-phenylpropionic acid 3-(4′-hydroxy-3′-methoxyphenyl)pyruvic acid, 3-(3′,4′-dihydroxyphenyl)lactic acid, benzoic acid dimeric products demethylated to caffeic acid, dimeric products with reduced double bonds to dihydroderivatives (propionic acid side chain) and a combination of the two conversions	(Braune et al. 2009)
Chlorogenic acid	*In vivo* rat plasma and urine *In vivo* human urine	Hippuric acid	(Gonthier et al. 2003b)
Cranberry phenolic acids	*In vivo* rat urine	Methylated, sulfated and glucuronidated, conjugates of hydroxyphenylacetic acid and hydroxyphenylpropionic acid	(Prior et al. 2010)

Table 13.2 (continued)

Component	Study design	Metabolites	References
Anthocyanins			
Cyanidin B ring Malvidin B-ring Peonidin B-ring Pelargonidin B-ring Anthocyanin A-ring	*In vitro* human colon model	3,4-dihydroxybenzoic acid (protocatechuic acid) 3,5-dimethoxy-4-hydroxybenzoic acid (syringic acid) 3-methoxy-4-hydroxybenzoic acid (vanillic acid) 4-hydroxybenzoic acid 2′,4′, 6′-trihydroxyphenylacetaldehyde, benzaldehyde, 3-methoxy-4-hydroxybenzoic acid, 2,4,6-trihydroxybenzaldehyde, 2,4,6-trihydroxybenzoic acid, gallic acid	(Aura et al. 2005b; Keppler and Humpf 2005; Fleschhut et al. 2006)
Cyanidin-O-glucosides	*In vivo* human urine and faeces	3,4-dihydroxybenzoic acid (protocatechuic acid)	(Vitaglione et al. 2007)
Delphinidin-, petunidin-, peonidin- and malvidin-3-O-glucosides from Cabernet Sauvignon grapes	*In vitro* model with pig microbiota	3-O-methylgallic acid, 3,4-dimethoxybenzoic acid, 2,4,6-trihydroxybenzaldehyde	(Forrester et al. 2008)

Van't Slot and co-workers found that flavan-3-ols dimers, but not trimers, were degraded by pig caecal microbiota *in vitro* (Slot van't et al. 2010). When a mixture of proanthocyanidin dimers was incubated with human colonic microbiota, the main metabolites were dihydroxylated phenylacetic acid and corresponding γ-valerolactone (Table 13.2) (Appeldoorn et al. 2009). In the reproducible application of the *in vitro* model the metabolism of dimers was slower than that of monomers (Aura et al. 2008).

Stoupi et al. (2010a) also showed that orally administered ^{14}C-labelled dimer B2 appeared to blood of male rats only to a low extent (8–11 %). Because only 63 % of the label was excreted via urine within 4 days after administration, a long residence time of metabolites in the body was shown prior to the final excretion. Despite the similarities of microbial metabolite profiles of (–)-epicatechin and proanthocyanidin B2 (ten common metabolites), five unique metabolites of the dimer B2 were produced by human faecal microbiota (Stoupi et al. 2010b). These were dimeric intermediates, where the C-ring was cleaved either in the upper or in the lower unit (Stoupi et al. 2010c).

Deprez et al. (2000) found that only 9–22 % of the label from [^{14}C]-proanthocyanidin was incorporated into the metabolite pool after *in vitro* fermentation with human colonic microbiota. The metabolites that were produced in this study included several derivatives of phenylvaleric, phenylpropionic, phenylacetic and benzoic acids with different patterns of hydroxylation (Table 13.2). The total yields of metabolites were shown to decrease significantly with an increased polymerization (Bazzocco et al. 2008; Gonthier et al. 2003a; Rios et al. 2003). Phenylcarboxylic acid metabolites derived from monomeric or dimeric catechins are also formed from their analogues in apples or cider, cocoa, almonds and grapes or red wine, (Bazzocco et al. 2008; Aura et al. 2012; Urpi-Sarda et al. 2009a; Urpi-Sarda et al. 2009b). Llorach et al. (2009) detected 27 metabolites using metabolomic profiling of urine samples after cocoa intake. These metabolites included alkaloid derivatives and polyphenol metabolites, as well as processing-derived products such as diketopiperazines.

Further evidence linking a high degree of flavan-3-ol polymerization with reduced rates of degradation was obtained by Bazzocco and Aura with colleagues (Bazzocco et al. 2008; Aura et al. 2012). Bazzocco and colleagues showed that Marie Ménard apples and cider proanthocyanidins, with a respective average degree of polymerisation 8.2 and 2.2, were broken down more extensively and yielded higher metabolite concentrations than Averolles apples and cider with an average degree of polymerisation 71.2 and 7.4, respectively. In grape products, the highest amounts of metabolites were obtained with red wine with the lowest degree of polymerization. Isolated proanthocyanidin fractions from Marie Ménard and Averolles apples, and Syrah grapes surprisingly showed significantly lower short-chain fatty acid formation from the carbohydrates in the samples than faecal control (Aura et al. 2012). The suppression occurred only in the absence of fruit pericarp or beverage matrix and was dose-dependent, which was tested using Syrah proanthocyanidin fraction (Bazzocco et al. 2008; Aura et al. 2012). Polymeric flavan-3-ols have a tendency to bind to proteins, causing astringency and inhibition of enzymes. The binding increases with the degree of polymerisation due to high number of interacting hydroxyl groups (Scalbert 1991). This may well contribute to the lower number of metabolites and to the reduced metabolism of the polymeric proanthocyanidins.

13.6.2 Hydroxycinnamic Acids

Phenolic acids are present in a number of foods with especially high amounts occurring in whole grains, berries and wine. The most abundant phenolic acid in cereals is ferulic acid. Since a high proportion of the ferulic acid in cereals is bound to polysaccharides by ester bonds, its bioavailability requires the presence of degradative esterases from intestinal tissues and microbiota (Kroon et al. 1997; Plumb et al. 1999; Andreasen et al. 2001a; Andreasen et al. 2001b; Rondini et al. 2004). In addition, Andreasen and co-workers (Andreasen et al. 2001a) have demonstrated the release of sinapic acid and *p*-coumaric acid from rye and wheat bran by human colonic esterases (Andreasen et al. 2001b). Hydrolysis by intestinal esterases is probably the major route for release of soluble, conjugated hydroxycinnamic acids *in vivo*. Such free phenolic acids can then be absorbed across the gastrointestinal barrier and enter the peripherial circulation of mammals (Andreasen et al. 2001a). Metabolites derived from hydroxycinnamic acids are collated in Table 13.2.

When eight post-menopausal women consumed rye bread, daily urinary excretion of ferulic acid was 47 % of daily intake (Harder et al. 2004). Only a half of the analysable ferulic acid was apparent, which can be explained partly by its matrix bound nature, by a slow release but also by further metabolism to structurally different compounds. An experiment was performed using wholegrain wheat breads fortified with either bioprocessed (by enzymatic and yeast fermentation) or native wheat bran. Bioaccessibility in the upper intestine and colonic conversions were studied in human *in vitro* gastrointestinal models TIM-1 and TIM-2 described in section 13.3. The bioaccessibility of ferulic acid, *p*-coumaric acid and sinapic acid was increased 5-fold, if the bran in the bread had undergone bioprocessing. Since the contents of *p*-coumaric acid and sinapic acid were low and their release was mainly in the TIM-1 model simulating the upper intestine, the microbial conversion products were better connected to matrix bound ferulic acid (Mateo Anson et al. 2009). The major colonic metabolites in the TIM-2 model from the wheat bran fortified breads were 3-(3′-hydroxyphenyl)propionic acid and 3-phenylpropionic acid (Table 13.2) independent of the bioprocessing, however, the release and conversion were enhanced by enzymatic and yeast fermentation processes of the wheat bran (Mateo Anson et al. 2011). A similar pattern was observed when 300 g of the same breads were digested by eight healthy volunteers, after which plasma and urine samples were collected for 24 h after intake. In addition to ferulic, sinapic and coumaric acids, 3,4-dimethoxybenzoic acid was detected as an early metabolite from the whole grain breads. Bioavailability of the phenolic acids was increased 2–3-fold by bioprocessing and the same phenylpropionic acid metabolites (Table 13.2) were detected as major metabolites presumably from the matrix-bound phenolics entering the colon (Mateo Anson et al. 2011).

Dry fractionation using micronization and electrostatic separation of rye and wheat brans resulted in aleurone fractions of the cereal grains. The formation of 3-(3′-hydroxyphenyl)propionic acid was enhanced compared with the native brans in an *in vitro* colon model using human faecal inoculum as a source of microbiota (Nordlund et al. 2012). The impact of dry fractionation on bioaccessibility of phenolic acids had been shown earlier (Hemery et al. 2010). Reduced particle size after dry fractionation and enzymatic digestion was concluded to have contributed to the release and conversion of ferulic acid from rye and wheat aleurone fractions (Nordlund et al. 2012).

Additional conversions of hydroxycinnamic acids can be observed *in vitro*. Formation of hydroxyphenylpropionic acid and small amounts of benzoic acid (Table 13.2) occurred when caffeic acid and its esters, chlorogenic acid and caftaric acid, were used as substrates in an *in vitro* human colon model (Gonthier et al. 2006). The side-chain shortening of phenylpropionic acid occurs via β-oxidation. Caffeic acid can also be decarboxylated and reduced to 4-ethylcatechol by human fecal microbiota (Table 13.2) (Peppercorn and Goldman 1971).

Ferulic acid dimers are another form of released phenolic acids from cereal matrices (Andreasen et al. 2000). Two ferulic acid moieties can be bound to each other via either an 8-*O*-4– or a 5-5–linkage. Dehydrodiferulic acid 8-*O*-4- and 5-5-derivatives, were incubated with human fecal microbiota. The 8-*O*-4-derivative was shown to degrade transiently to monomeric ferulic acid. In contrast, the 5-5-diferulate derivative was subjected only to demethylation and/or side-chain reduction. The metabolites from ferulic acid dimers are shown in Table 13.2 (Braune et al. 2009).

In vivo studies support the *in vitro* findings. Feeding studies with chlorogenic acid (5-*O*-caffeoylquinic acid) revealed the appearance of hydroxylated cinnamic acids, benzoic acid and hippuric acid in plasma and urine of rats. Hippuric acid is formed by glycination of benzoic acid in the liver (Gonthier et al. 2003b). In humans half of the ingested chlorogenic acid was excreted in urine as hippuric acid (Olthof et al. 2003), while 57 % of the ingested chlorogenic acid was detected in rat plasma and urine as microbial metabolites (Gonthier et al. 2003a). Prior and colleagues studied excretion of metabolites in rats after ingestion of concentrated cranberry powder taking into consideration the colonic conversions and further conjugations of these metabolites. Hydroxyphenylacetic and hydroxyphenylpropionic acids were excreted in urine as methylated, sulfated and glucuronidated conjugates with a degree of conjugation ranging from 65 to 100 % depending on the phenolic acid. Only the degree of conjugation of 3-(4′-hydroxyphenyl)propionic acid was enhanced with the increased dose of the cranberry powder (Prior et al. 2010).

13.6.3 Anthocyanins

Anthocyanins have a labile flavylium cation structure and their microbial metabolism has been investigated only recently. Due to their instability, the detection of anthocyanins from human samples has been challenging. The total amount of anthocyanins in 24 h urine is very low, $< 0.2\,\%$ of the ingested dose. Usually the parent anthocyanins predominate in urine, but methylated and glucuronidated derivatives have also been detected (Vitaglione et al. 2007).

The microbial metabolism of anthocyanin glycosides includes deconjugation and ring-fission as described above for flavonols (Aura et al. 2005b; Keppler and Humpf 2005; Fleschhut et al. 2006). The metabolites rising from different anthocyanin structures reflect the substitution pattern of the B-ring of their precursors, whereas the metabolites suggested to be formed from the A ring are more common to all anthocyanins (Table 13.2) (Keppler and Humpf 2005; Fleschhut et al. 2006). Furthermore, an unidentified derivative of the anthocyanin aglycone with a molecular weight 85 mass units higher than that of the corresponding aglycone was detected for anthocyanidins originating from malvidin-, peonidin-, petunidin-, delphinidin- and cyanidin-3-*O*-glucosides (Aura et al. 2005b). The major *in vivo* metabolite of cyanidin-*O*-glucosides, 3,4-dihydroxybenzoic acid, is excreted in 24-h urine and faeces in amounts corresponding to 73 % of the ingested anthocyanin (Vitaglione et al. 2007). Thus, results *in vitro* and *in vivo* are in agreement.

Compounds similar to colonic anthocyanin metabolites can be formed at neutral pH. Glycosidic anthocyanins exist in equilibrium between the flavylium cation (red coloured and abundant at low pH), blue quinoidal bases and the carbinol and chalcone pseudobases (non-coloured), the three latter compounds being more abundant at pH 7 (Clifford 2000). Anthocyanidins can form a reactive α-diketone, which can readily decompose to phenolic acids and the corresponding aldehydes (Keppler and Humpf 2005). However, under the mildly acidic conditions predominating in the *in vitro* colon model, phenolic acids were not detected, unless active microbiota were present (Aura et al. 2005b). This suggests that both microbial and spontaneous decomposition of anthocyanins can occur in the intestinal environment.

Nurmi and co-workers (Nurmi et al. 2009) followed urinary excretion of phenolic acids after ingestion of a bilberry-lingonberry purée for a 48-h-period. Early excretion of 3,4-dihydroxybenzoic acid was observed, together with caffeic acid, *p*-coumaric acid and ferulic acid. Low level excretion of isoferulic acid and dehydroisoferulic acid was also detected. 3′-Methoxy-4′-hydroxyphenylacetic acid (homovanillic acid) and 3-methoxy-4-hydroxybenzoic acid (vanillic acid) were the most abundant metabolites, peaking after 6 h. 3-(3′,4′-Dihydroxyphenyl) propionic and the corresponding acetic acid appeared 12 h after intake and at the same time their monohydroxylated 3′-derivatives maintained steady profiles (Nurmi et al. 2009). The metabolic pattern of bilberry-lingonberry purée appears to reflect the presence of bilberry anthocyanins and lignonberry phenolic acids and flavonols.

A berry mixture containing bilberries, lingonberries, blackcurrants and chokeberries, was ingested in a placebo controlled human intervention. Phenol intake was 837 mg/day, and quercetin, *p*-coumaric acid, 3′-hydroxyphenylacetic acid, caffeic acid, 3,4-dihydroxybenzoic acid, 3-methoxy-4-hydroxybenzoic acid, 3′-methoxy-4′-hydroxyphenylacetic acid and 3-(3′-hydroxyphenyl)propionic acid increased significantly in the plasma of the berry group compared to that of the control group. The urinary excretion of quercetin, *p*-coumaric acid and 3′-hydroxyphenylacetic acid was increased likewise in the berry group (Koli et al. 2010).

Forrester and Waterhouse (Forrester and Waterhouse 2008) carried out an *in vitro* study incubating an extract of Cabernet Sauvignon grapes with pig microbiota. Delphinidin-, petunidin-, peonidin- and malvidin-3-*O*-glucosides disappeared within 6 h and three major metabolites (3-methoxygallic acid, 3,4-dimethoxybenzoic acid and 2,4,6-trihydroxybenzaldehyde accumulated (Table 13.2). The microbial metabolites from pigs, therefore, resemble those of humans.

Anthocyanin bioavailability has also been investigated in Sprague-Dawley rats after ingestion of wild bilberries. Benzoic acid was the main metabolite in the liver and brain tissues after 4- and 8- week- supplementation (Bò et al. 2010). In addition, anthocyanins and their glucuronides were detected in pig brain tissues after administration of a 2 % blueberry diet. The levels were low (ranging 0.01–0.50 nmol\kg fresh weight) but indicative of passage through the blood-brain barrier (Milbury and Kalt 2010). These studies suggest that phenolic compounds and their metabolites may enter the peripheral tissues and potentially be bioactive at these sites.

13.7 Prospects and Considerations of Colon Model in Relation to MetS

In vitro colon models have been used for dietary fibre research for years. Choice of a suitable model for different applications should be considered carefully. Multi-compartmentalized colon models are excellent tools for studying effect of substrates on changes of the composition of microbiota, whereas simple batch models are not applicable in microbial ecological research due to the stability and high density of the microbial inoculation with faeces. However, batch models are suitable for studying metabolic profiles, because the microbiota is not changing. If the strict anaerobiosis is maintained throughout the processing of the faecal suspension, the interaction between food and the microbiota is genuine. Metabolic comparisons between substrate incubations with the microbiota and the faecal control without substrates give relevant responses: the over-expressed metabolites from the food describing the food-related intestinal metabolome, appearing in the circulation and excreted in urine, when the same foods are eaten.

Metabolomics is an explorative technology, which is still under development. In its untargeted forms, it enables complex metabolic patterns and pathways to be profiled and explored in a reproducible manner. The identification of the metabolite profile from the colon model includes data alignment, statistics and visualization as heat maps displaying fold-changes of metabolites against the control without food or the food without microbiota. The biggest challenge is in the identification of the over-expressed components. Using combination of targeted approach, group specific identification and several databases, the identification of profiles of food metabolites can be achieved. If the compounds cannot be given specific names, the mass spectra can still be used for finding the same biomarkers of intake from human body fluids as biomarkers of intake and compliance. The role of the colon model is in building the connection between specific metabolites and the authentic food.

Metabolomic analysis has revealed the diversity of components in foods as a discriminative or informative fingerprint (Cevallos-Cevallos et al. 2009), but when this diversity is objected to microbial metabolism, the metabolites are shared with many plant foods and the number of designated metabolites is reduced and points out rather a polyphenol-rich diet than a single food item. Furthermore, when the number of metabolites can be reduced by selecting only over-expressed metabolites for finding them from the human body it is much easier to find correlations between the diet and the possible affected biomarkers or pathways related to MetS.

Correlation between diet and a health response without causality is not adequate. Causality requires mechanistic approach using *in vitro* cell and *in vivo* animal models. In order to take into account the circulating metabolites the purified microbial metabolite extracts from a food or a diet can be used in the cell based assays using isolated food extracts without metabolism as a control. Precise purification techniques have to be applied to ensure that the extracts contain only food-related metabolites. The use of controls ensures that the effects in the cell lines are not caused by media, microbial non-food components or other artefacts. The effect could also be verified in the target tissue in animal models, possessing the same metabolites from the same diet. In addition, the metabolite transport across the body and reaching the site of action could be performed by analysis of plasma and urine in humans and if possible, by tissue biopsies. Finally, improvement of the clinical outcome in subjects with MetS is the most important biomarker of the successful positive effect related to the diet. Only then the effect of a diet on MetS is established.

Phenolic metabolites have been shown to correlate with particular food intake, which has been proven to affect MetS biomarkers. Heterogenic symptoms in MetS give a challenge for finding correlations using traditional statistical methods. Novel data processing techniques may offer personalized approaches and new angles to characterize the responses. Nevertheless, the great challenges are apparent in all aspects of systems biological nutritional research. For this reason the combination of different techniques are needed. As a consequence only with genuine multidisciplinary collaboration between scientists these challenges can be overcome.

References

ADA (2008) Position of the American dietetic association: Health implications of dietary fibre. J Am Diet Assoc 108:716–1731

Adisakwattana S, Moonsan P, Yibchok-Anun S (2008) Insulin-Releasing properties of a series of Cinnamic Acid Derivatives *in vitro* and *in vivo*. J Agric Food Chem. 56:7838–7844

Adlercreutz H, Wildt J van der, Kinzel J, Attalla H, Wähälä K, Mäkelä T, Hase T, Fotsis T (1995) Lignan and isoflavonoid conjugates in human urine. J Steroid Biochem Mol Biol 52:97–103

Allison C, Macfarlane C, Macfarlane GT (1989) Studies on mixed populations of human intestinal bacteria grown in single-stage and multistage continuous culture systems. Appl Environ Microbiol 55:672–678

Andreasen MF, Christensen LP, Meyer AS, Hansen A (2000) Content of phenolic acids and ferulic acid dehydrodimers in 17 Rye (*Secale cereale* L.) Varieties. J Agric Food Chem 48:2837–2842

Andreasen MF, Kroon PA, Williamson G, Garcia-Conesa M-T (2001a) Esterase activity able to hydrolyze dietary antoxidant hydroxycinnamates is distributed along the intestine of mammals. J Agric Food Chem 49:5679–5684

Andreasen MF, Kroon PA, Williamson G, Garcia-Conesa M-T (2001b) Intestinal release and uptake of phenolic antioxidant diferulic acids. Free Rad Biol Med 31:304–314

Appeldoorn MM, Vincken J-P, Aura A-M, Hollman PCH, Gruppen H (2009) Procyanidin dimers are metabolised by human microbiota with 2-(3,4-dihydroxyphenyl)-acetic acid and 5-(3,4-dihydroxyphenyl)-gamma-valerolactone as the Major Metabolites. J Agric Food Chem 57:1084–1092

Aura A-M, Mattila I, Hyötyläinen T, Goplacharyulu P, Cheynier V, Souquet J-M, Bes M, Le Bourvellec C, Guyot S, Orešič M (2012) Characterization of microbial metabolism of Syrah grape products in an *in vitro* colon model using targeted and non-targeted analytical approaches Eur J Nutr. doi:10.1007/s00394-012-0391-8

Aura A-M (2008) Microbial metabolism of dietary phenolic compounds in the colon. Phytochem Rev 7:407–429

Aura A-M, Härkönen H, Fabritius M, Poutanen K (1999) Development of an *in vitro* enzymic digestion method for removal of starch and protein and assessment of its performance using rye and wheat breads. J Cereal Sci 29:139–152

Aura A-M, Karppinen S, Virtanen H, Forssell P, Heinonen S-M, Nurmi T, Adlercreutz H, Poutanen K (2005a) Processing of Rye Bran Influences both the fermentation of dietary fibre and the bioconversion of lignans by human faecal flora *in vitro*. J Sci Food Agric 85:2085–2093

Aura A-M, Martin-Lopez P, O'Leary KA, Williamsson G, Oksman-Caldentey K-M, Poutanen K, Santos-Buelga C (2005b) *In vitro* metabolism of anthocyanins by human gut microflora. Eur J Nutr 44:133–142

Aura A-M, Mattila I, Seppänen-Laakso T, Miettinen J, Oksman-Caldentey K-M, Orešič M (2008) Microbial metabolism of catechin stereoisomers by human faecal microbiota: comparison of targeted analysis and a non-targeted metabolomics method. Phytochem Lett 1:18–22

Aura A-M, O'Leary KA, Williamson G, Ojala M, Bailey M, Puupponen-Pimiä R, Nuutila AM, Oksman-Caldentey K-M, Poutanen K (2002) Quercetin derivatives are deconjugated and converted to hydroxyphenylacetic acids but not methylated by human fecal flora *in vitro*. J Agric Food Chem 50:1725–1730

Aura A-M, Oikarinen S, Mutanen M, Heinonen S-M, Adlercreutz HCT, Virtanen H, Poutanen K (2006) Suitability of a batch *in vitro* fermentation model using human faecal microbiota for prediction of conversion of flaxseed lignans to enterolactone with reference to an *in vivo* rat model. Eur J Nutr 45:45–51

Bach Knudsen KE, Serena A, Bjornbak Kjaer AK, Tetens I, Heinonen S-M, Nurmi T, Adlercreutz H (2003) Rye bread in the pigs enhances the formation of enterolactone and increases its levels in plasma, urine and feces. J Nutr 133:1368–1375

Bai N, He K, Roller M, Zheng B, Chen X, Shao Z, Peng T, Zheng Q (2008) Active compounds from lagerstroemia speciosa, insulin-like glucose uptake-stimulatory/inhibitory and adipocyte differentiation-inhibitory activities in 3T3-L1 Cells. J Agric Food Chem 56:11668–11674

Barry JL, Hoebler C, Macfarlane GT, Macfarlane S, Methers JC, Reed KA, Mortensen PB, Norgards I, Rowland IR, Rumney CJ (1995) Estimation of the fermentability of dietary fibre *in vitro*: a European interlaboratory study. Brit J Nutr 74:303–322

Bazzocco S, Mattila I, Guyot S, Renard CMGC, Aura A-M (2008) Factors affecting the conversion of apple polyphenols to phenolic acids and fruit matrix to short-chain fatty acids by human faecal microbiota *in vitro*. Eur J Nutr 47:442–452

Bearne CA, Mallett AK, Rowland IR, Brennan-Craddock WE (1990) Continuous culture of human faecal bacteria as an *in vitro* model for the colonic microflora. Toxic *in vitro* 4:522–525

Bò CD, Ciappellano S, Klimis-Zacas D, Gardana C, Riso P, Porrini M, Martini D (2010) Anthocyanin absorption, metabolism, and distribution from a wild bilberry-enriched diet (Vaccinum angustifolium) is affected by diet duration in the Sprague-Dawley rat. J Agric Food Chem 58:2491–2497

Bose M, Lambert JD, Ju J, Reuhl KR, Shapses SA, Yang CS (2008) The major green tea polyphenol, (-)-epigallocatechin-3-gallate, inhibits obesity, metabolic syndrome, and fatty liver disease in high-fat-fed mice. J Nutr 138:1677–1683

Braune A, Bunzel M, Yonekura R, Blaut M (2009) Conversion of dehydrodiferulic acids by human intestinal microbiota. J Agric Food Chem 57:3356–3362

Braune A, Engst W, Blaut M (2005) Degradation of neohesperidin dihydrochalcone by human intestinal bacteria. J Agric Food Chem 53:1782–1790

Campbell WL, Franklin W, Cerniglia CE (1992) Validation studies on an *in vitro* semicontinuous culture system designed to simulate a bacterial ecosystem of the human intestine. J Microbiol Methods 16:239–252

Cani PD, Amar J, Iglesias MA, Poggi M, Knauf C, Bastelica D, Neyrinck AM, Fava F, Tuohy KM, Chabo C, Waget A, Delmée E, Cousin B, Sulpice T, Chamontin B, Serrières J, Tanti JF, Gibson GR, Casteilla L, Delzenne NM, Alessi MC, Burcelin R (2007) Metabolic endotoxemia initiates obesity and insulin resistance. Diabetes 56:1761–1772

Cani PD, Bibiloni R, Knauf C, Waget A, Neyrinck AM, Delzenne NM, Burcelin R (2008) Changes in gut microbiota control metabolic endotoxemia-induced inflammation in high-fat diet-induced obesity and diabetes in mice. Diabetes 57:1470–1481

Cerda B, Espin JC, Parra S, Martinez P, Tomas-Barberan FA (2004) The potent *in vitro* antioxidant ellagitannins from pomegranate juice are metabolised into bioavailable but poor antioxidant hydroxy-6 H-dibenzopyran-6-one derivatives by the colonic microflora of healthy humans. Eur J Nutr 43:205–220

Cerda B, Tomás-Barberán FA, Espin JC (2005) Metabolism of antioxidant and chemopreventive ellagitannins from strawberries, raspberries, walnuts, and oak-aged wine in humans: identification of biomarkers and individual variability. J Agric Food Chem 53:227–235

Cevallos-Cevallos JM, Reyes-De-Corcuera JI, Etxeberria E, Danyluk MD, Rodrick GE (2009) Metabolomic analysis in food science: a review. Trends Food Sci Technol 20:557–566

Clifford MN (2000) Anthocyanins—nature, occurrence and dietary burden. J Sci Food Agric 80:1063–1072

Das NP (1971) Studies on flavonoid metabolism. Absorption and metabolism of (+)-catechin in man. Biochem Pharmacol 20:3435–3445

Day AJ, Cañada FJ, Diaz JC, Kroon PA, Mclauchlan R, Faulds CB, Plumb GW, Morgan MRA, Williamson G (2000) Dietary flavonoid and isoflavone glycosides are hydrolysed by the lactase site of lactase phlorizin hydrolase. FEBS Lett 468:166–170

De Mello VDE, Schwab U, Kolehmainen M, Koenig W, Siloaho M, Poutanen K, Mykkänen H, Uusitupa M (2011) A diet high in fatty fish, bilberries and wholegrain products improves markers of endothelial function and inflammation in individuals with impaired glucose metabolism in a randomised controlled trial: the sysdimet study. Diabetologia 54:2755–2767

Depréz S, Brezillon C, Rabot S, Philippe C, Mila I, Lapierre C, Scalbert A (2000) Polymeric proanthocyanidins are catabolized by human colonic microflora into low-molecular-weight phenolic acids. J Nutr 130:2733–2738

Donovan JL, Bell JR, Kasim-Karakas S, German JB, Walzem RL, Hansen RJ, Waterhouse AL (1999) Catechin is present as metabolites in human plasma after consumption of red wine. J Nutr 129:1662–1668

Dube MG, Torto R, Kalra SP (2008) Increased leptin expression selectively in the hypothalamus suppresses inflammatory markers CRP and IL-6 in leptin-deficient diabetic obese mice. Peptides 29:593–598

Eckel RH, Grundy SM, Zimmet PZ (2005) The metabolic syndrome. Lancet 365:1415–1428

Edwards CA, Gibson G, Champ M, Jensen B-B, Mathers JC, Nagengast F, Rumney C, Quelh A (1996) *In vitro* method for quantification of the fermentation of starch by human faecal bacteria. J Sci Food Agric 71:209–217

Everard A, Lazarevic V, Derrien M, Girard M, Giumuccioli GM, Neyrinck AM, Possemiers S, Van Holle A, Francois P, de Vos VM, Delzenne NM, Schrenzel J, Cani Pd (2011) Responses of gut microbiota and glucose and lipid metabolism to prebiotics in genetic obese and diet-induced leptin-resistant mice. Diabetes 60:2775–2786

Felgines C, Talavéra S, Gonthier M-P, Texier O, Scalbert A, Lamaison JL, Rémésy C (2003) Strawberry anthocyanins are recovered in urine as glucuro- and sulfoconjugates in humans. J Nutr 133:1296–1301

Fleschhut J, Kratzer F, Rechkemmer G, Kulling SE (2006) Stability and biotransformation of various dietary anthocyanins *in vitro*. Eur J Nutr 45:7–15

Forrester SC, Waterhouse AL (2008) Identification of cabernet sauvignon anthocyanin gut microflora metabolites. J Agric Food Chem 56:9299–9304

Galisteo M, Duarte J, Zarzuelo A (2008) Effects of dietary fibers on disturbances clustered in the metabolic syndrome. J Nutr Biochem 19:71–84

Gonthier M-P, Donovan JL, Texier O, Felgines C, Rémésy C, Scalbert A (2003a) Metabolism of dietary procyanidins in rats. Free Rad Biol Med 35:837–844

Gonthier M-P, Rémésy C, Scalbert A, Cheynier V, Souquet J-M, Poutanen K, Aura A-M (2006) Microbial metabolism of caffeic acid and its esters chlorogenic and caftaric acids by human faecal microbiota *in vitro*. Biomed Pharmacother 60:536–540

Gonthier M-P, Verny M-A, Besson C, Rémésy C, Scalbert A (2003b) Chlorogenic acid bioavailability largely depends on its metabolism by the gut microflora in rats. J Nutr 133:1853–1859

Grassi D, Desideri G, Necozione S, Lippi C, Casale R, Properzi G, Blumberg JB, Ferri C (2008) Blood pressure reduced and insulin sensitivity increased in glucose intolerant, hypertensive subjects after 15 days of consuming of high-polyphenol dark chocolate. J Nutr 138:1671–1676

Grattagliano I, Palmieri VO, Portincasa P, Moschetta A, Palasciano G (2008) Oxidative stress-induced risk factors associated with the metabolic syndrome: a unifying hypothesis. J Nutr Biochem 19:491–504

Gross M, Pfeiffer M, Martini M, Campbell D, Slavin J, Potter J (1996) The quantitation of metabolites of quercetin flavonols in human urine. Cancer Epidemiol Biomark Prev 5:711–720

Grün CH, van Dortsen FA, Jacobs DM, Le Belleguic M, Van Vezen EJJ, Bingham MO, Jansen H-G, Van Duynhoven JPM (2008) GC-MS methods for metabolite profiling of microbial fermentation products of dietary polyphenols in human and *in vitro* intervention studies. J Chromogr B 871:212–219

Guyton AC, Hall JE (1996) Gastrointestinal physiology. In: GUYTON AC, HALL JE (eds.) Textbook of medicinal physiology, 9th ed. W.B. Saunders Company, Philadelphia, pp 793–813

Hanhineva K, Aura A-M, Rogachev I, Matero S, Skov T, Aharoni A, Poutanen K, Mykkänen H (2012) *In vitro* microbiotic fermentation causes an extensive metabolite turnover of rye bran phytochemicals. PLoS ONE, 06/20 7(6):e39322

Harder H, Tetens I, Let MB, Meyer AS (2004) Rye bread intake elevates urinary excretion of ferulic acid in humans, but does not affect the susceptibility of LDL to oxidation. Eur J Nutr 43:230–236

Hattori M, Taylor TD (2009) The human intestinal microbiome: a new frontier of human biology. DNA Res 16:1–12

Hein E-M, Rose K, Van'T Slot G, Friedrich AW, Humpf H-U (2008) Deconjugation and degradation of flavonol glycosides by pig cecal microbiota chracterized by fluorescence in situ hybridization (FISH). J Agric Food Chem 56:2281–2290

Heinonen S, Nurmi T, Liukkonen K, Poutanen K, Wähälä K, Deyama T, Nishibe S, Adlercreutz H (2001) *In vitro* metabolism of plant lignans: new precursors of mammalian lignans enterolactone and enterodiol. J Agric Food Chem 49:3178–3186

Hemery Y, Mateo Anson N, Havenaar R, Haenen GRMM, Noort MWJ, Rouau X (2010) Dry-fractionation of wheat bran increases the bioaccessibility of phenolic acids in breads made from processed bran fractions. Food Res Int 43:1429–1438

Jacobs DM, Deltimple N, van Velzen E, van Dorsten FA, Bingham M, Vaughan EE, van Dyunhoven J (2008) ^{1}NMR metabolite profiling of feces as a tool to assess the impact of nutrition on the human microbiome. NMR Biomed 21:615–626

Jacobs DM, Gaudier E, van Duynhoven J, Vaughan EE (2009) Non-digestible food ingredients, colonic microbiota and the impact on gut health and immunity: a role for metabolomics. Curr Drug Metabol 10:41–54

Jayaprakasam B, Vareed SK, Olson SK, Nair MG (2005) Insulin secretion by bioactive anthocyanins and anthocyanidins present in Fruits. J Agric Food Chem 53:28–31

Johnsen NF, Hausner H, Olsen A, Tetens I, Christensen J, Bach Knudsen KE, Overvald K, Tjonneland A (2004) Intake of whole grains and vegetables determines the plasma enterolactone concentration of danish women. J Nutr 134:2691–2697

Juntunen KS et al (2003) Structural differences between rye and wheat breads but not total fiber content may explain the lower postprandial insulin response to rye bread. Am J Clin Nutr 78:957–964

Juntunen KS, Mazur WM, Liukkonen KH, Uehara M, Poutanen KS, Adlercreutz HCT, Mykkänen HM (2000) Consumption of wholemeal rye bread increases serum consentrations and urinary excretion of enterolactone compared with consumption of white wheat bread in healthy finnish men and women. Brit J Nutr 84:839–846

Kallio P, Kolehmainen M, Laaksonen DE, Kekäläinen J, Salopuro T, Sivenius K, Pulkkinen L, Mykkänen H, Niskanen L, Uusitupa MI, Poutanen KS (2007) Dietary carbohydrate modification induces marked alterations in gene expression of white adipose tissue in persons with metabolic syndrome—the FUNGENUT study. Am J Clin Nutr 85:1417–1427

Kallio P, Kolehmainen M, Laaksonen D, Pulkkinen L, Atalay M, Mykkänen H, Uusitupa M, Poutanen K, Niskanen L (2008) Inflammation markers are modulated by response to diets differing in postprandial insulin responses in individuals with metabolic syndrome. Am J Clin Nutr 87:1497–1503

Keppler K, Humpf HU (2005) Metabolism of anthocyanins and their phenolic degradation products by the intestinal microflora. Bioorg Med Chem 13:5195–5205

Kern SM, Bennett RN, Mellon FA, Kroon PA, Garcia-Conesa MT (2003) Absorption of hydroxycinnamates in humans after high-bran cereal consumption. J Agric Food Chem 51:6050–6055

Kilkkinen A, Pietinen P, Klaukka T, Virtamo J, Korhonen P, Adlercreutz H (2002) Use of oral antimicrobials decreases serum enterolactone concentration. Am J Epidemiol 155:472–477

Kilkkinen A, Stumpf K, Pietinen P, Valsta LM, Tapanainen H, Adlercreutz H (2001) Determinants of serum enterolactone concentration. Am J Clin Nutr 73:1094–1100

Kilkkinen A, Valsta LM, Virtamo J, Stumpf K, Adlercreutz H, Pietinen P (2003) Intake of lignans is associated with serum enterolactone concentration in finnish men and women. J Nutr 133:1830–1833

Kim D-H, Kobashi K (1986) The role of intestinal flora in metabolism of phenolic sulfate esters. Biochem Pharmacol 35:3507–3510

Kim D-H, Konishi L, Kobashi K (1986) Purification, characterization and reaction mechanism of novel arylsulfotransferase obtained from an anaerobic bacterium of human intestine. Biochim Biophys Acta 872:33–41

Knust U, Spiegelhalder B, Strowitzki T, Owen RW (2006) Contribution of lignan intake to urine and serum enterolignan levels in german females: a randomised controlled intervention trial. Food Chem Toxicol 44:1057–1064

Koli R, Erlund I, Jula A, Marniemi J, Mattila P, Alfthan G (2010) Bioavailability of various polyphenols from a diet containing moderate amounts of berries. J Agric Food Chem 58:3927–3932

Kroon PA et al (1997) Release of covalently bound ferulic acid from fiber in the human colon. J Agric Food Chem 45:661–667

Kuijsten A, Arts ICW, Hollman PCH, van't Veer P, Kampman E (2006) Plasma enterolignans are associated with lower colorectal adenoma risk. Cancer Epidemiol Biomark Prev 15:1132–1136

Kuijsten A, Arts ICW, Vree TB, Hollman PCH (2005) Pharmacokinetics of enterolignans in healthy men and women consuming a single dose of secoisolariciresinol diglucoside. J Nutr 135:795–801

Laaksonen DE, Toppinen LK, Juntunen KS, Autio K, Liukkonen KH, Poutanen KS, Niskanen L, Mykkänen HM (2005) Dietary carbohydrate modification enhances insulin secretion in persons with the metabolic syndrome. Am J Clin Nutr 82:1218–1227

Lafa., S, Gil-Izquierdo A (2008) Bioavailability of phenolic acids. Phytochem Rev 7:301–311

Lafay S, Gil-Izquierdo A, Manach C, Morand C, Besson C, Scalbert A (2006) Chlorogenic acid is absorbed in its intact form in the stomach of rats. J Nutr 136:1–6

Lampe JW (2003) Isoflavonoid and lignan phytoestrogens as dietary biomarkers. J Nutr 133:956–964S

Lee M-J, Maliakal P, Chen L, Meng X, Bondoc FY, Prabhu S, Lambert G, Mohr S, Yang CS (2002) Pharmacokinetics of tea catechins after ingestion of green tea and (–)-epicatechin-3-gallate by humans: formation of different metabolites and individual variability. Cancer Epidemiol.Biomark Prev 11:1025–1032

Lee YA, Cho EJ, Yokozawa T (2008) Effects of proanthocyanidin preparations on hyperlipidemia and other biomarkers in mouse model of type 2 diabetes. J Agric Food Chem 56:17

Li L, Stillermark-Billton P, Beck C, Boström P, Andersson L, Rutberg M, Eriksson J, Magnusson B, Marchesan D, Ljungberg A, Boren J, Olofson S-O (2006) Epigallocatechin gallate increases the formation of cytosolic lipid droplets and decreases the secretion of apoB-100 VLDL. J Lipid Res 47:67–77

Lin Y, Vermeer MA, Bos W, Van Buren L, Schuurbiers E, Miret-Catalan S, Trautwein EA (2011) Molecular structures of citrus flavonoids determine their effects on lipid metabolism in hepG2 cells by primarily suppressing apoB secretion. J Agric Food Chem 59(9):4496–4503

Llorach R, Urpi-Sarda M, Jauregui O, Monagas M, Andres-Lacueva C (2009) An LC-MS-based metabolomics approach for exploring urinary metabolome modifications after cocoa consumption. J Proteome Res 8:5060–5068

Ma Y, Hébert JR, Li W, Bertone-Johnson ER, Oledzki B, Pagoto SL, Tinker L, Rosal MC, Ockene IS, Ockene JK, Griffith JA, Liu S (2008) Association between dietary fiber and markers of systemic inflammation in the women's health initiative observational study. Nutrition 24:941–949

Macfarlane GT, Macfarlane S, Gibson GR (1998) Validation of a three-stage compound continuous culture system for investigating the effect of retention time on the ecology and metabolism of bacteria in the human colon. Microb Ecol 35:180–187

Mateo Anson N, Aura A-M, Selinheimo E, Mattila I, Poutanen K, van den Berg R, Havenaar R, Haenen GRMM (2011) Bioprocessing of wheat bran in whole wheat bread increases the bioavailability of phenolic acids in men and exerts anti-inflammatory effects *ex vivo*. J Nutr 141:137–143

Mateo Anson N, Selinheimo E, Havenaar R, Aura A-M, Mattila I, Lehtinen P, Bast A, Poutanen K, Haenen GRMM (2009) Bioprocessing of wheat bran improves *in vitro* bioaccessibility and colonic metabolism of phenolic compounds. J Agric Food Chem 57:6148–6155

McDougall GJ, Shpiro F, Dobson P, Smith P, Blake A, Stewart D (2005) Different polyphenolic components of soft fruits inhibit *alpha*-amylase and *alpha*-glucosidase. J Agric Food Chem 53:2760–2766

Meng X, Sang S, Zhu N, Lu H, Sheng S, Lee M-J, Ho C-T, Yang CS (2002) Identification and characterization of methylated and ring-fission metabolites of tea catechins formed in humans, mice, and rats. Chem Res Toxicol 15:1041–1050

Meselhy MR, Nakamura N, Hattori M (1997) Biotransformation of (–)-epicatechin-3-O-gallate by human intestinal bacteria. Chem Pharm Bull 45:888–893

Milbury PE, Kalt W (2010) Xenobiotic metabolism and berry flavonoid transport across the blood-brain barrier. J Agric Food Chem 58:3950–3956

Miller TL, Wolin MJ (1981) Fermentation by the human large intestine microbial community in an *in vitro* semicontinuous culture system. Appl Environ Microbiol 42:400–407

Molly K, Vande Woestyne M, Verstraete W (1993) Development of a 5-step multi-chamber reactor as a simulation of the human intestinal microbial ecosystem. Appl Microbiol Biotechnol 39:254–258

Möller DE, Kaufman KD (2005) Metabolic syndrome: a clinical and molecular perspective. Annu Rev Med 56:45–62
Natsume M, Osakabe n, Oyama M, Sasaki M, Baba S, Nakamura Y, Osawa T, Terao J (2003) Structures of (–)-epicatechin glucuronide identified from plasma and urine after oral ingestion of (–)-epicatechin: differences between human and rat. Free Rad Biol Med 34:840–849
Nemeth K, Plumb GW, Berrin JG, Juge R, Naim HY, Williamson G, Swallow DM, Kroon PA (2003) Deglycosylation by human intestinal epithelial cell *beta*-glucosidases is a critical step in the absorption and metabolism of dietary flavonoid glycosides in humans. Eur J Nutr 42:29–42
Nesbitt PD, Lam Y, Thompson LU (1999) Human metabolism of mammalian lignan precursors in raw and processed flaxseed. Am J Clin Nutr 69:549–555
Nettleton JA, Mckeon NM, Kanoni S, Lemaitre RN, Hivert M-F, Ngwa J, van Rooij FJA, Sonenstedt E, Wojczynski MK, Ye Z, Tanaka T (2010) Interactions of dietary whole-grain intake with fasting glucose- and insulin-related genetic loci in individuals of European descent. Diabetes Care 33:2684–2691
Nordlund E, Aura A-M, Mattila I, Kössö T, Rouau X, Poutanen K (2012) Formation of phenolic microbial metabolites and short-chain fatty acids from rye, wheat, and oat bran and their fractions in the metabolical *in vitro* colon model. J Agric Food Chem 60:8134–8145
Nurmi T, Mursu J, Heinonen M, Nurmi A, Hiltunen R, Voutilainen S (2009) Metabolism of berry anthocyanins to phenolic acids in humans. J Agric Food Chem 57:2274–2281
Olthof MR, Hollman PCH, Buijsman MNCP, van Amelsvoort JMM, Katan MB (2003) Chlorogenic acid, Quercetin-3-Rutinoside and black tea phenols are extensively metabolized in humans. J Nutr 133:1806–1814
Peñalvo J, Heinonen S-M, Aura A-M, Adlercreutz H (2005) Dietary sesamin is converted to enterolactone in humans. J Nutr 135:1056–1062
Peppercorn MA, Goldman P (1971) Caffeic acid metabolism by bacteria of the human gastrointestinal tract. J Bacteriol 108:996–1000
Pettersson J, Karlsson PC, Choi YH, Verpoorte R, Rafter JJ, Bohlin L (2008) NMR metabolomic analysis of fecal water from subjects on a vegetarian diet. Biol Pharma Bull 31:1192–1198
Pinent M, Bladé C, Salvado MJ, Blay M, Pujadas G, Fernandez-Larrea J, Arola L, Ardévol A (2006) Procyanidin effects on adipocyte-related pathologies. Crit Rev Food Sci Nutr 46:543–550
Plumb GW, Garcia-Conesa MT, Kroon PA, Rhodes M, Ridley S, Williamson G (1999) Metabolism of chlorogenic acid by human plasma, liver, intestine and gut microflora. J Sci Food Agric 79:390–392
Priebe MG, Wang H, Weening D, Schepers M, Preston T, Vonk RJ (2010) Factors related to colonic fermentation of nondigestible carbohydrates of a previous evening meal increase tissue glucose uptake and moderate glucose-associated inflammation. Am J Clin Nutr 91:90–97
Prior RL, Rogers T, Khanal RC, Wilkes SE, Wu X, Howard LR (2010) Urinary excretion of phenolic acids in rats fed cranberry. J. Sci Food Agric 58:3940–3949
Rajilic-Stojanovic M, Smidt H, de Vos WM (2007) Diversity of the human gastrointestinal tract microbiota revisited. Environment Microbiol 9:2125–2136
Rechner AR, Smith MS, Kuhnle G, Gibson GR, Debham ES, Srai SKS, Moore KP, Rice-Evans CA (2004) Colonic metabolism of dietary polyphenols: influence of structure on microbial fermentation products. Free Rad Biol Med 36:212–225
Respondek F, Swanson KS, Belsito KR, Vetser BM, Wagner A, Istasse L, Diez M (2008) Short-chain fructo-oligosaccharides influence insulin sensitivity and gene expression of fat tissue in obese dogs. J Nutr 138:1712–1718
Rios LY, Gonthier M-P, Rémesy C, Mila I, Lapierre C, Lazarus SA, Williamson G, Scalbert A (2003) Chocolate intake increases urinary excretion of polyphenol-derived phenolic acids in healthy human subjects. Am J Clin Nutr 77:912–918
Roberts MS, Magnusson BM, Burczynski FJ, Weiss M (2002) Enterohepatic circulation physiological, pharmacokinetic and clinical implications. Clin Pharmacokinet 41(10):751–790

Rondini L, Peyrat-Maillard M-N, Marsset-Baglieri A, Fromentin G, Durand P, Tomé D, Prost M, Berset C (2004) Bound ferulic acid from bran is more bioavailable than the free compound in rat. J Agric Food Chem 52:4338–4343

Roowi S, Stalmach A, Mullen W, Lean MEJ, Edwards CA, Crozier A (2010) Green tea flavan-3-ols: colonic degradation and urinary excretion of catabolites by humans. J Agric Food Chem 58:1296–1304

Rumney CJ, Rowland IR (1992) *In vivo* and *in vitro* models of the human colonic flora. Crit Rev Food Sci Nutr 31:299–331

Sawai Y, Kohsaka K, Nishiyama Y, Ando K (1987) Serum concentrations of rutoside metabolites after oral administration of a rutoside formulation to humans. Drug Res 37:729–732

Scalbert A (1991) Antimicrobial properties of tannins. Phytochemistry 30:3875–3883

Scalbert A, Morand C, Manach C, Rémésy C (2002) Absorption and metabolism of polyphenols in the gut and impact on health. Biomed Pharmacother 56:276–282

Scalbert A, Williamson G (2000) Dietary intake and bioavailability of polyphenols. J Nutr 130:2073–2085

Scheline RR (1970) The metabolism of (+)-catechin to hydroxyphenylvaleric acids by the intestinal microflora. Biochim Biophys Acta 222:228–230

Seeram NP, Henning SM, Zhang Y, Suchard M, Li Z, Heber D (2006) Pomegranate juice ellagitannin metabolites are present in human plasma and some persist in urine for up to 48 h. J Nutr 136:2481–2485

Selma MV, Espín JC, Tomás-Barberán FA (2009) Interaction between phenolics and gut microbiota: role in human health. J Agric Food Chem 57:6485–6501

Smith CE, Tucker KL (2011) Health benefits of cereal fibre: a review of clinical trials. Nutr Res Rev 15:1–14

Stalmach A, Troufflard S, Serafini M, Crozier A (2009) Absorption, metabolism and excretion of choladi green tea flavan-3-ols by humans. Mol Nutr Food Res 53:44–53

Stote KS, Baer DJ (2008) Tea consumption may improve biomarkers of insulin sensitivity and risk factors for diabetes. J Nutr 138:1584S–1588S

Stoupi S, Williamson G, Viton F, Barron D, King LJ, Brown JE, Clifford MN (2010a) *In vivo* bioavailability, absorption, excretion and pharmacokinetics of [14C] procyanidin B2 in male rats. Drug Metab Disp 38:287–291

Stoupi S, Williamson G, Drynan JW, Barron D, Clifford MN (2010b) A Comparison of the *in vitro* Transformation of (–)-epicatechin and proanthocyanidin B2 by human faecal microbiota. Mol Nutr Food Res 54:747–759

Stoupi S, Williamson G, Drynan JW, Barron D, Clifford MN (2010c) Procyanidin B2 catabolism by human fecal microflora: partial characterization of 'dimeric' intermediates. Arch Biochem Biophys 501:73–76

Tabernero M, Venema K, Maathuis AJH, Saura-Calixto F (2011) Metabolite production during *in vitro* colonic fermentation of dietary fiber: analysis and comparison of two European diets. J Agric Food Chem 59:8968–8975

Takagaki A, Nanjo F (2010) Metabolism of (–)-epigallocatechin gallate by rat intestinal flora. J Agric Food Chem 58:1313–1321

Tomaru M, Takano H, Osakabe N, Yasuda A, Inoue K-I, Yanagisawa R, Ohwatari T, Uematsu H (2007) Dietary supplementation with cacao liquor proanthocyanidins prevents elevation of blood glucose levels in diabetic obese mice. Nutrition 23:351–355

Turnbaugh PJ, Gordon JI (2008) An invitation to the marriage of metagenomics and metabolomics. Cell 134:708–713

Urpi-Sarda M, Garrido I, Monagas M, Gómez-Cordovés C, Medina-Remón A, Andres-Lacueva C, Bartolomé B (2009a) Profile of plasma and urine metabolites after the intake of almonds [*Prunus dulcis* (Mill.) D.A. Webb] polyphenols in humans. J Agric Food Chem 57:10134–10142

Urpi-Sarda M, Monagas M, Khan n, Lamuela-Ravantos RM, Santos-Buelga C, Sacanella E, Castell M, Permanyer J, Andres-Lacueva C (2009b) Epicatechin, proanthocyanidins, and phenolic microbial metabolites after cocoa intake in humans and rats. Anal Bioanal Chem 394:1545–1556 van't Slot G, Humpf H-U (2009) Degradation and metabolism of catechin, epicatechin-3-gallate (EGCG), and related compounds by the intestinal microbiota in the pig cecum model. J Nutr 57:8041–8048

van't Slot G, Mattern W, Rzeppa S, Grewe D, Humpf H-U (2010) Complex flavonoids in cocoa synthesis and degradation by intestinal microbiota. J Agric Food Chem 58:8879–8886

Venema K, Minekus M, Havenaar R (2004) Advanced *in vitro* models of the gastrointestinal tract—novel tools to study functionality of dietary fibres. In: van der Kamp JW, ASP N-G, Miller Jones J, Schaafsma G (eds.) Dietary fibre: bioactive carbohydrates for food and feed. Wageningen. Wageningen Academic Publishers, pp 99–112

Verzelloni E, Pellacani C, Tagliazucchi D, Tagliaferri S, Calani L, Costa LG, Brighenti F, Borges G, Crozier A, Del Rio D (2011) Antiglycative and neuroprotective activity of colon-derived polyphenol catabolites. Mol Nutr Food Res 55(1):35–43

Visser M, Bouter LM, Mcquillan GM, Wener MH, Harris TB (1999) Elevated C-reactive protein levels in overweight and obese adults. The J Am Med Assoc 282(22):2131–2135

Vitaglione P, Donnarumma G, Napolitano A, Galvano F, Gallo A, Scalfi L, Fogliano V (2007) Protocatechuic acid is the major human metabolite of cyanidin-glucosides. J Nutr 137:2043–2048

Wall R, Ross RP, Shanahan F, O'Mahony L, O.Mahony M, Coakley M, Hart O, Lawlor P, Quigley EM, Kiely B, Fitzgerald GF, Stanton C (2009) Metabolic activity of the enteric microbiota influences the fatty acid composition of murine and porcine liver abd adipose tissues. Am J Clin Nutr 89:1393–1401

Walle T, Browning AM, Steed LL, Reed SG, Walle UK (2005) Flavonoid glycosides are hydrolyzed and thus activated on the oral cavity in humans. J Nutr 135:48–52

Wang L, Yamasaki M, Katsube T, Sun X, Yamasaki Y, Shiwaku K (2011) Antiobesity effect of polyphenolic compounds from molokheiya (*Corchorus olitorius* L.) leaves in LDL receptor-deficient mice. Eur J Nutr 50:127–133

Weickert MO, Pfeiffer AFH (2008) Metabolic effects of dietary fibre consumption and prevention of diabetes. J Nutr 138:439–442

Wishart DS (2008) Metabolomics: application to food acience and nutrition research. Trends Food Sci Tech 19:482–493

You Q, Chen F, Wang X, Luo PG, Jiang Y (2011) Inhibitory effects of muscadine anthocyanins on *alpha*-glucosidase and pancreatic lipase activities. J Agric Food Chem 59:9506–9511

Zahorska-Markiewicz B, Janowska J, Olszanecka-Glinianowicz M, Zurakowski A (2000) Serum concentrations of TNF-*alpha* and soluble TNF-*alpha* receptors in obesity. Int J Obes Relat Metab Disord 24:1392–1395

Zhao Z, Egashira Y, Sanada H (2004) Ferulic acid is quickly absorbed from rat stomach as the free form and then conjugated mainly in liver. J Nutr 134:3083–3088

Part IV
Modeling Metabolic Syndrome

This section introduces some of the key modeling strategies and models needed to study metabolic syndrome, with examples of applications.

Chapter 14
Genome-Scale Modeling of Tissue and Cancer Metabolism

Livnat Jerby and Eytan Ruppin

Abstract The emergence of 'omics' technologies has enabled global reconstruction of cellular metabolic networks. While detailed kinetic modeling of these networks remain challenging due to a lack of kinetic, network analysis based on appropriate constraints such as the steady-state condition (constraint-based analysis) is already feasible. Here the methods for the reconstruction of metabolic networks are reviewed along with recent developments in the field, including the modeling techniques such as flux balance analysis and elementary flux modes.

Keywords Constraint based modeling · Human metabolic model · Metabolic networks

14.1 Introduction

Classically, metabolism, as other fields in biology, has been studied in a reductionist manner, by focusing on a small set of reactions often referred to as a metabolic pathway. Although much of our knowledge of metabolism has been discovered this way, this paradigm cannot account for the numerous interactions and emerging propertied characterizing metabolism. Furthermore, technological developments have produced a flood of genomic, transcriptomic and other high-throughput data that cannot be interpreted and integrated via the classical methods. The insufficiency of the reductionist approach, on the one hand, and the growing availability of large-scale data, on the other hand, created the increasingly pressing need for computational, mathematical methods to study metabolism in a systematic fashion.

By and large, the *in-silico* models of metabolism base upon a representation of metabolism as a network, in which an edge denotes a reaction and a node denotes a metabolite. Some properties can be inferred based on the topology

---

E. Ruppin (✉) · L. Jerby
The Blavatnik School of Computer Science, Tel Aviv University, Tel Aviv, Israel

E. Ruppin
The Sackler School of Computer Sciences & School of Medicine,
Tel Aviv University, 69978 Tel Aviv, Israel
e-mail: ruppin@post.tau.ac.il

M. Orešič, A. Vidal-Puig (eds.), *A Systems Biology Approach to Study Metabolic Syndrome*,
DOI 10.1007/978-3-319-01008-3_14,

of the network per se. Mathematical modeling of cellular metabolism has been traditionally performed via kinetic modeling techniques that are composed by a set of differential equations, describing the change in the metabolites concentration over time (Garfinkel and Hess 1964). These models can provide an informative dynamic description of metabolism, though as they require detailed information on kinetic constants and on enzyme and metabolite levels, the lack of accurate cellular information currently limits the scope of the methodology to small-scale systems, as the human red blood cell (Jamshidi et al. 2001; Joshi and Palsson 1989; Lee and Palsson 1991; Mulquiney and Kuchel 1999). An alternative computational approach, constraint-based modeling (CBM), bypasses these hurdles as it does not depend on such detailed information. CBM assumes a metabolic steady-state under which feasible flux distributions satisfy a stoichiometric mass-balance requirement, thermodynamic constraints, and constraints on enzymes' capacities that are based on experimental observations of flux rates. By imposing the set of these governing cellular constraints the requirement for kinetic information is relaxed.

The modeling paradigm of CBM has been extensively applied with considerable success to study microbial physiology (Edwards et al. 2001; Durot et al. 2009; Shlomi et al. 2005; Feist and Palsson 2008; Oberhardt et al. 2009; AbuOun et al. 2009; Kumar and Maranas 2009). Though still less developed, genome-scale modeling of human metabolism is constantly progressing (Goodacre et al. 2004); earlier work has focused on characterizing distinct human metabolic pathways (Kanehisa and Goto 2000; Romero et al. 2004), and modeling specific cell types and organelles (Wiback and Palsson 2002; Vo et al. 2004; Chatziioannou et al. 2003). In 2007, two generic genome-scale human metabolic models were presented based on an extensive evaluation of genomic and bibliomic data (Ma et al. 2007; Duarte et al. 2007).

The generic human metabolic models (Ma et al. 2007; Duarte et al. 2007) consist of a collection of biochemical reactions that take place in different tissues and cell-types, depending on physiological conditions. The potential clinical utility of the generic model has been previously demonstrated by its ability to identify functionally related sets of reactions that are causally related to hemolytic anemia, and potential drug targets for treating hypercholesterolemia (Duarte et al. 2007). The utility of this generic human metabolic model was further demonstrated in predicting metabolic biomarkers whose concentration is altered due to genomic mutations in IEMs (Shlomi et al. 2009).

Nonetheless, the different tissues and cell-types of the human body vary in their metabolic phenotypes (Mathews et al. 2000; Berg et al. 2002). Therefore, a generic model is not sufficient to recapitulate human metabolism, as it should be modeled in a tissue-specific manner. Here we present two complementary computational methods that address this challenge: the first, the integrative Metabolic Analysis Tool (iMAT) (Shlomi et al. 2008), describes a tissue-specific metabolic phenotype, and the latter, the Model Building Algorithm (MBA), reconstructs tissue-specific models. To describe a tissue-specific metabolic phenotype iMAT integrates a generic model with tissue-specific gene and protein expression data (Shlomi et al. 2008). It was applied to predict a variety of metabolic behaviors of different human tissues, including the brain, liver, and kidney. The predicted metabolic behavior characterizes, for each tissue, a single, normal physiological condition under which the expression data that is used as input was measured. MBA reconstructs tissue-specific metabolic models by utilizing a variety of different tissue-specific molecular data sources: literature-based knowledge, transcriptomic, proteomic, metabolomic, and phenotypic data. MBA was applied to generate the first genome-scale metabolic model of the liver, a compartmentalized model of the plant *Arabidopsis thaliana*, as well as the first genome-scale model of cancer metabolism. Lastly we describe two studies of cancer metabolism, where genome-scale metabolic modeling was applied to guide experimental work, leading to the discovery of metabolic phenomena and to the identification of selective drug targets.

14.2 The Underlying Framework: Constraint-Based Modeling (CBM)

The metabolic network is described by the stoichiometric matrix *S*, formed from the stoichiometric coefficients of the reactions that comprise the network. Each column in the matrix denotes a metabolic reaction, and each row denotes a metabolite. The entries in the matrix are stoichiometric coefficients. Meaning, every row describes the reactions in which the metabolite participates, and thus embodies the interconnection between the different reactions. The dynamic behavior of the network can be described by the flux rate of the metabolic reactions. The network flux vector consists of the flux rates of each reaction in the network. By utilizing the stoichiometric matrix one can transform the network flux vector into a vector of the time derivatives of all the metabolites' concentrations (Fig. 14.1).

In other words, the stoichiometric matrix **S** is a linear transformation from a flux vector

$$v = (v_1, v_2, \ldots, v_n) \tag{14.1}$$

to a vector of time derivatives of the concentration vector

$$x = (x_1, x_2, \ldots, x_m) \tag{14.2}$$

as

$$Sv = \frac{d}{dt}x \tag{14.3}$$

The flux vector is composed of a dynamic component and a steady-state component:

$$v = v_{dyn} + v_{ss} \tag{14.4}$$

By definition at steady-state the concentration of the different intracellular metabolites remains constant, meaning, there is no accumulation or depletion of metabolites. Hence, the rate of each metabolite production is equal to its consumption rate. The steady state component satisfies

$$Sv_{ss} = \frac{d}{dt}x = 0 \tag{14.5}$$

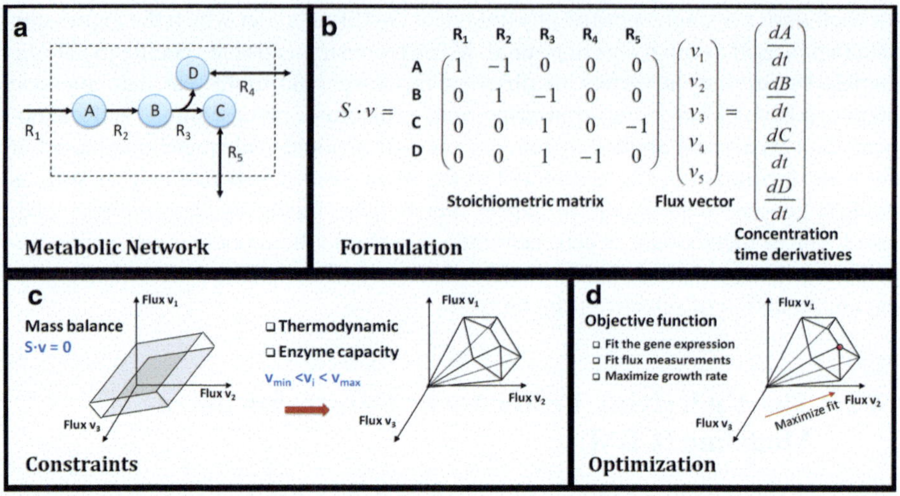

Fig. 14.1 The formulation and conceptual basis of CBM. **a** A schematic illustration of a (toy example) metabolic network. Nodes represent metabolites and edges represent reactions. **b** The network is represented by its stoichiometric matrix S; Columns represents reactions and rows represent metabolites. Matrix S transforms the reactions' flux vector v into a vector that contains the time derivatives of all the metabolites' concentrations. **c** Without constraints, the flux distribution of the network may lie at any point in the space. When mass-balance constraints (imposed by the stoichiometric matrix) and capacity constraints (imposed by the lower and upper bounds) are applied, they define a convex solution space. Only flux distributions within this subspace are valid, while points that lay outside of it are excluded by the constraints. **d** By maximizing an objective function, a narrow-range optimal flux distribution that lies on the edge of the solution space can be identified

Thus, v_{SS} is in the null space of S.

In order to infer the metabolic phenotype at steady state one needs to explore the null space of the stoichiometric matrix, that is, the feasible solutions that satisfy the steady-state constraints. However, the null space, also referred to as the solution space, consists of feasible solutions that are irrelevant in the biological context. By imposing additional constraints one can filter out these inappropriate solutions, and to reduce the space of solutions (Fig. 14.1). Such constraints include linear constraints that limit the upper and lower rate of reactions in the network, in the form of:

$$v_{min} \leq v \leq v_{max} \tag{14.6}$$

Thermodynamic constraints determine the reversibility of reactions by setting the lower bound of irreversible reactions to zero. On the other hand, enzymatic rate constraints impose a limit on the maximal bound $v_{\max}$ according to prior knowledge on the enzymatic capacity. The metabolic network is an open system in which certain metabolites can be secreted or absorbed from the extracellular environment by a set of reactions entitled exchange reaction. By imposing constraints on the upper and lower bounds of these reactions CBM accounts for the availability of metabolites in the extra-cellular surrounding (the medium) and for the prior knowledge regarding secreted metabolites. As the rate of exchange reactions is relatively easier to measure experimentally, it can be integrated by employing Quadratic Programming (QP) to find a flux distribution with a minimum Euclidian distance from a set of the experimentally measured fluxes.

CBM analysis is intimately tied to the mathematical field of optimization. Therefore, in addition to the constraints that govern cellular metabolism, CBM can also account for the optimization criteria that determine the metabolic phenotype (Fig. 14.1d). This is achieved by including a mathematical objective quantified by a scalar function of the unknown flux rates $Z = F(v_1, v_2, \ldots, v_n)$. The function F, termed the *objective function*, further limits the solution space, as only solutions that optimize it are considered. Although the objective function plays a crucial role in CBM, it is not as theoretically sound as physiochemical constraints. It is mostly considered as a working hypothesis, serving as a powerful tool for quantitative predictions, which should be validated experimentally. Surprisingly, certain simple objective functions have been shown to successfully predict diverse biological phenotypic characteristics.

In the realm of microorganisms, the most widely used CBM method is Flux Balance Analysis (FBA) (Kauffman et al. 2003; Varma and Palsson 1994) that assumes the metabolism of a microorganism is evolutionary adopted to maximize growth. To incorporate this assumption in the formulation of the problem a reaction termed biomass reaction is added to the model. This reaction consumes biosynthetic precursors in an appropriate ratio to produce the biomass of the *in-silico* growing organism (Varma and Palsson 1993). Remarkably, FBA has been shown to accurately predict an impressive array of phenotypes observed in microorganisms. These include growth rates (Edwards et al. 2001), uptake rates, by-product secretion (Varma and Palsson 1993), the outcomes of adaptive evolution (Fong and Palsson 2004; Ibarra et al. 2002), gene expression levels (Famili et al. 2003), metabolic flux rates (Segre et al. 2002; Shlomi et al. 2005; Wiback et al. 2004), and knockout lethality (Edwards and Palsson 2000).

When going beyond the realm of microorganisms to that of multi-cellular organisms the definition of the objective function becomes somewhat vague. The biomass maximization hypothesis as utilized in FBA can be considered relevant in certain circumstances, for example, when modeling metabolism during embryonic development, during maturation and growth or in some pathological phenotypes as tissue regeneration, wound healing, and cancer. However, this assumption is not apt to capture the metabolic behavior of most multi-cellular organisms. In multi-tissue organisms additional hurdles are encountered as different tissues apparently have different objective functions that are dynamically changing in an interactive manner according to physiological condition and needs of the entire body. Meaning, although the organism as a whole strives to maintain homeostasis, the role of each one of its tissues varies and is subject to diverse regulatory signals, as hormonal signals for instance.

Nonetheless, the objective function can be utilized as a tool to quantify the capacity of the *biological hardware*, let it be a microorganism or a multi-cellular tissue, to carry out a metabolic function at different conditions. By applying the CBM method

Flux Variability Analysis (FVA) (Mahadevan and Schilling 2003) one can determine the feasible range of each reaction independently within the solution space. FVA is performed by formulating a LP problem that minimizes or maximizes the flux through the reaction of interest. Alternatively, the convex polytope of the solution space can be extensively explored using random Monte Carlo sampling (Schellenberger and Palsson 2009) to obtain distributions of the steady-state flux levels for each reaction in the metabolic network. Despite the computational complexity of this approach it is sometimes preferred over FVA as it gives additional insight into the shape of the high-dimensional solution space.

14.3 Network-based Prediction of Human Tissue-Specific Metabolism

iMAT is a tool for the prediction of human tissue-specific metabolic behavior via the integration of a genome-scale metabolic network with tissue-specific gene expression and protein abundance data (Shlomi et al. 2008). The metabolic phenotype is highly affected by the gene expression and protein abundance levels (Levine et al. 2006; Yanai et al. 2005). It has been reported that there is a strong correlation between gene expression and measured (Daran-Lapujade et al. 2004; Fong and Palsson 2004) as well as predicted (Akesson et al. 2004; Bilu et al. 2006; Famili et al. 2003; Schuster et al. 2002) metabolic fluxes in micro-organisms. Nonetheless, due to various cellular regulatory mechanisms and enzyme kinetics, the correlation is not complete. To account for these mechanisms, iMAT treats the expression levels, not as the final determiners of metabolic activity, but as cues for the likelihood of the corresponding reactions to carry a metabolic flux. Network integration is then used to accumulate these cues into a global, consistent metabolic behavior, which reflects the outcome of putative post-transcriptional regulatory effects (Fig. 14.2). Relying on enzyme expression data to infer tissue-specific metabolic flux activity using iMAT eliminates the need for *a priori* knowledge of tissue-specific objective functions and metabolites exchanged between the tissue and biofluids. Rather, the method provides predictions regarding tissue-specific metabolite uptake and secretion.

To examine iMAT's ability to correctly predict metabolic behavior based on gene expression data, it was first applied to predict the metabolic state of the yeast *S. cerevisiae* under conditions for which reliable data is readily available for validation. Then, the method was applied to the genome-scale human metabolic network model of (Duarte et al. 2007), integrated with tissue-specific enzyme expression data to predict tissue-specific metabolic behavior of ten human tissues. Specifically, for each tissue, a unique view of metabolic activity was obtained. The tissue-specific activity predictions were validated in two ways. First, a comprehensive comparison to known large-scale information on tissue-specificity of genes, reactions and metabolites obtained from various databases was performed. Second, the tissue-specificity of metabolic disease causing genes was studied, demonstrating the method can capture the metabolic behavior of different tissues under normal and disrupted physiological conditions.

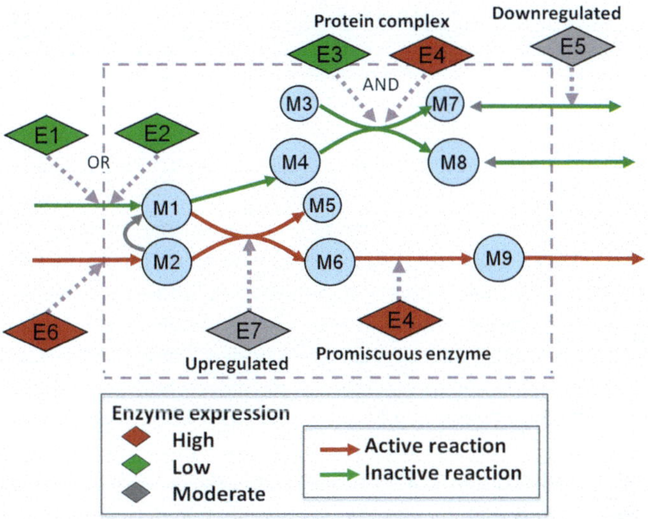

Fig. 14.2 Predicting flux-activity states based on a metabolic model and gene-expression measurements. Circular nodes represent metabolites, whereas diamond nodes represent enzymes. White, red and green represent normal, significantly low and significantly high expression of the enzyme-encoding genes, respectively. Solid edges represent metabolic reactions. Broken edges associate enzymes with the reactions they catalyze. Purple and black arrows denote active and inactive reactions, respectively, according to the predicted steady-state flux distribution. Enzyme E7 is predicted to be post-transcriptionally upregulated and E5 is predicted to be post-transcriptionally downregulated. The figure is adapted from Shlomi et al. (Shlomi et al. 2008)

14.3.1 Integrative Metabolic Analysis Tool (iMAT)

Given a generic metabolic model and two sets of genes, that consist of highly and lowly expressed genes, iMAT initially infers the expression state of each metabolic reaction according to Gene-Protein-Reaction (GRP) associations. By doing so, two sets of reactions are assembled: R_H and R_L that consist of highly and lowly expressed reactions, respectively.

The next step involves formulating a mixed integer linear programming (MILP) problem. The solution to this problem is a steady-state flux distribution satisfying stoichiometric and thermodynamic constraints, while maximizing the number of reactions whose activity is consistent with their expression state (Eq. (14.7)):

$$\max \sum\nolimits_{i\in R_H} \left(y_i^+ + y_i^-\right) + \sum\nolimits_{i\in R_L} (x_i) \tag{14.7}$$

s.t

$$S \cdot v = 0 \tag{14.8}$$

$$v_{min} \leq v \leq v_{max} \tag{14.9}$$

$$\nu_i + \nu_i^+(\nu_{min,i} - \varepsilon) \geq \nu_{min,i} : i \in R_H \quad (14.10)$$

$$\nu_i + \nu_i^-(\nu_{max,i} + \varepsilon) \leq \nu_{max,i} : i \in R_H \quad (14.11)$$

$$(1 - x_i)\nu_{min,i} \leq \nu_i \leq (1 - x_i)\nu_{max,i} : i \in R_L \quad (14.12)$$

Where ν is the flux vector and S is a $m \times n$ stoichiometric matrix, in which m is the number of metabolites and n is the number of reactions. The mass balance constraint is enforced in Eq. (14.8). Thermodynamic constraints that restrict flow direction are imposed in Eq. (14.9), by setting ν_{min} and ν_{max} as lower and upper bounds on flux values, respectively. For each highly expressed reaction, the Boolean variables y^+ and y^- represent whether the reaction is active (in either direction, thus either y^+ or y^- is assigned the value 1) or inactive (when both y^+ and $y^-$are 0). For each lowly expressed reaction, the Boolean variable x represents whether the reaction is inactive (when x is 1) or active (when x is 0). Specifically, a highly expressed reaction is considered to be active if it carries a flux with an absolute value greater than a positive threshold ε(Eqs. (14.10) and (14.11)). Alternatively, a lowly expressed reaction is considered to be inactive if it does not carry a flux that is greater than 0 in either direction (Eq. (14.12)). The optimization maximizes the number of reactions whose activity is similar to their expression state.

A solution found by the MILP solver to the problem formulated above is guaranteed to be optimal, meaning, to maximize the similarity to expression. On the other hand, the solution may not be unique as a space of alternative optimal solutions may exist. In this case, the space of optimal solutions represents alternative steady-state flux distributions attaining the same similarity with the expression data. To account for these alternative solutions, the optimal similarity to the expression is calculated. Then, the model is constrained to attain this optimal similarity, which reduces the solution space significantly. Then the optimal solution space is traversed by employing Flux Variability Analysis (FVA) (Mahadevan and Schilling 2003). FVA obtains for each reaction its minimal and maximal attainable flux, subject to the aforementioned constraints. A reaction is considered to be confidently active if its absolute flux is above some active threshold. Alternatively a reaction is considered to be confidently inactive if its minimal and maximal flux values are 0. Due to solver inaccuracies a small ε around 0 is permitted. The activity state of a reaction whose minimal and maximal flux predications as calculated by FVA do not sustain either condition is considered to be undetermined (unconfident). In summary, for some of the reactions the flux activity state can be uniquely determined to be active or inactive, while for others, the activity state cannot be uniquely determined because of potential alternative flux distributions with the same overall similarity to the expression data, mostly due to isozymes or alternative pathways. For each metabolic reaction iMAT determines its activity state as described above. Notably, because expression levels are not enforced as exclusive determinants of metabolic flux, the flux activity states of reactions may deviate from their expression states. Hence, genes are considered to be post-transcriptionally up-or-down regulated based on a difference between their measured expression level and the predicted flux activity state of their associated reactions.

14.4 Computational Reconstruction of Specific Metabolic Models

MBA is a computational method for reconstructing functional metabolic network models of human tissues based on a variety of different tissue-specific molecular data sources, rather than predicting a single metabolic state of a tissue as done by iMAT (Jerby et al. 2010). The obtained models can be used to explore the metabolic state of a tissue under various genetic and physiological conditions through standard CBM methods, without requiring additional context-specific molecular data. More specifically, by exploiting the metabolic model one could simulate and study the effects of genetic perturbations or drug applications in a straightforward manner. MBA is based on heuristically pruning the generic human metabolic model to derive a sub-model that is as consistent as possible with the pertaining tissue-specific molecular data sources. Applying MBA to generate the first genome-scale metabolic model of the liver, the resulting model is shown to be consistent with a variety of hepatic data sources, and to successfully permit the simulation of various metabolic states under different physiological conditions, in a manner superior to that obtained using the generic model.

14.4.1 The Model Building Algorithm (MBA)

MBA derives a tissue-specific metabolic model from a generic model based on network integration with various molecular data sources. First, based on tissue-specific data *core reactions* are identified. These reactions are included in the generic model, and should be included in the tissue-specific model. Core reactions that have a high *vs.* moderate probability to be carried out in the tissue are differentiated. The core is therefore divided into two sets, C_H and C_M, for high and moderate likelihood reactions, respectively. By and large the C_H set includes human curated tissue specific pathways and the C_M set includes reactions testified by molecular data. Both the C_H and the C_M are subsets of the generic model. The input to the algorithm is a generic model that is composed of a union of reactions that may exist in different, individual tissues and the core, tissue-specific reactions (Fig. 14.3). The core by itself is not a *consistent model*, as it includes many reactions that cannot carry flux. These reactions are termed dead-end reactions. The goal is then to derive the most parsimonious tissue-specific *consistent model*, which includes all the tissue-specific high probability reactions (C_H), a maximal number of moderate probability reactions (C_M), and a limited set of additional reactions from the generic model that is required for gap filling. The latter enable the activation of all core reactions, resulting in a fully viable and consistent tissue-specific model. This is done by searching for a minimal set of reactions that should be added to obtain a consistent model, where for each reaction there is a feasible flux distribution in which it is active. Aiming to find the most parsimonious model on one hand, and maximizing the number of moderate probability reactions included in the model on the other hand, induces a tradeoff. It

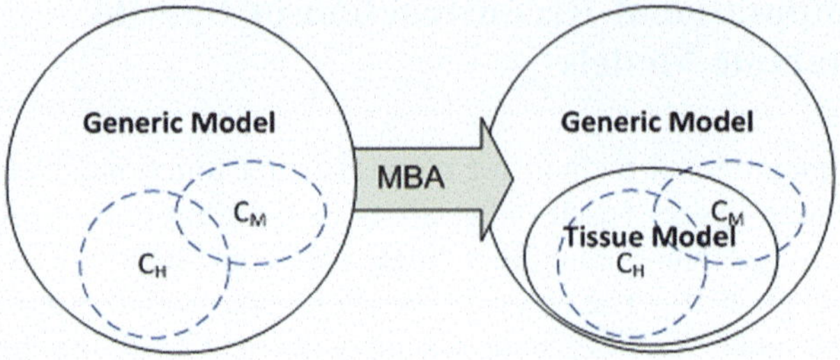

Fig. 14.3 MBA model reconstruction. MBA is given a generic model and tissue-specific core reactions sets (C_H and C_M) as input. It reconstructs a tissue model, containing all the C_H reactions, as many C_M reactions, and a minimal set of other generic model reactions that are required for overall model consistency

is tuned via a parameter that weighs both optimization criteria to obtain a score that evaluates the quality of a model. Sensitivity analysis was preformed to examine the reliance of the resulting model on this optimization parameter, showing that MBA's performance is fairly robust and does not hinge upon a choice of a narrow range of optimization parameter values.

To find an optimal model with a maximal score, a greedy heuristic search is employed. Starting form a generic consistent model non-core reactions are iteratively pruned from it, in a random order, while maintaining its consistency. In each pruning step, a reaction is removed only if its removal does not prevent the activation of reactions in C_H, and increases the model's score. Since the reactions' scanning order may affect the resulting model, the algorithm is executed with different, random pruning orders (1000 in our implementation) to construct multiple candidate models. Each reaction is then assigned a confidence score that is computed as the sum of candidate models in which it is included. To construct the final tissue model, reactions, ordered by their scores, are iteratively added to C_H, until a final, minimal and consistent model is obtained.

14.4.2 Generating Tissue and Cancer Metabolic Models

14.4.2.1 Genome-scale Metabolic Model of the Human Liver

MBA can serve for the rapid development of an array of metabolic models of a variety of human tissues. It provides a computational platform to probe the metabolism of tissues as the kidney, heart and brain on a genomic scale. MBA was applied to automatically generate the first genome-scale metabolic model of the liver (Jerby et al. 2010). The starting point was the generic human metabolic model of (Duarte et al. 2007), accounting for 2766 metabolites, 3742 reactions, 1905 genes, and 100 metabolic pathways. The essential core, C_H, was a collection of metabolic pathways that are known to be active in the liver based on the literature (Bock et al. 1991;

Baynes and Dominiczak 2004; Gropper and Smith 2008; Chisari and Fausto 2001). It consists of 37 intact metabolic pathways that are involved in central metabolism, carbohydrate, lipid, and amino acids metabolism as well as specific hepatic metabolism as drug metabolism, and bile acid biosynthesis. These pathways involve in sum 779 reactions and 873 metabolites. The more permissive core, C_M, consists of a set of 304 reactions, and 484 metabolites. It was assembled from tissue-specific data sources, including metabolomics (Wishart et al. 2007) transcriptomic (Yanai et al. 2005; Shmueli et al. 2003), proteomic (He 2005; Yan and Sadee 2000; Saier et al. 2006), and phenotypic data (McKusick 2007) of the liver. Each reaction was included in C_M only if it was supported by at least two of the data sources listed above or if it was necessary for the inclusion of a liver metabolite that appeared in the metabolomics data (Wishart et al. 2007). The total core hence comprises of 1083 reactions and 1187 metabolites.

The resulting liver model obtained via MBA consists of 1827 reactions and 1360 metabolites. Out of the reactions in the derived liver model that are not included in either reaction cores, 50 % are transport reactions that transfer metabolites across compartments. Most of these reactions are not associated with genes in the model and hence could not be included in the core reaction sets. Notably, despite the heuristic nature of the MBA algorithm, the participation of many of the reactions in the resulting liver model is consistently predicted across the different random reaction elimination orders. For 70.65 % of the non C_H reactions the algorithm provides completely unanimous predictions, that is, the reaction is either present in all models or absent in all of them.

The model was verified using standard cross-validation procedures, and through its ability to carry out hepatic metabolic functions. The model's flux predictions correlate with flux measurements across a variety of hormonal and dietary conditions, and improve upon the predictive performance obtained using the original, generic human model (prediction accuracy of 0.67 versus 0.46). Finally, the model better predicts biomarker changes in genetic metabolic disorders than the generic human model (accuracy of 0.67 versus 0.59).

14.4.2.2 Plant Metabolism: Multi-Tissue Metabolic Model of Arabidopsis

MBA is not restricted to the modeling of human tissues. It can be used to generate tissue models for any organism for which a generic model exists. Indeed, MBA has been applied to construct an array of tissue-specific metabolic models of the plant *Arabidopsis thaliana*. In this work a generic compartmentalized model of *Arabidopsis* was integrated with tissue-specific protein expression data (Schmid et al. 2005) from various tissues, including leaves, flowers, roots, siliques, seeds, and cell cultures. The reconstructed tissue models are fully functional and are amenable for CBM analysis, and significantly extend upon previous model reconstruction attempts. The models were validated by the prediction of measured fluxes in metabolically engineered seed strains. They were then utilized for the computational design of genetic manipulations that could potentially increase vitamin E production, a significant nutrient for human health.

14.4.2.3 Cancer Metabolism: From a General to a Type-Specific Model

Aberrant metabolism is one of the main driving forces in the initiation and development of cancer (Hanahan and Weinberg 2011; Ward and Thompson 2012). During carcinogenesis, selective pressures lead to diverse metabolic alterations, imposed by multiple molecular mechanisms (Cairns et al. 2011; Tamada et al. 2012; Miller et al. 2012). These metabolic adaptations enable cancer cells not only to proliferate and cope with high energetic demands, but also to avoid apoptosis, evade the immune system and control the rate of mutagenesis (Prendergast 2011; Sotgia et al. 2011; Cairns et al. 2011). Characterizing the unique metabolic dependencies of different cancer cells can potentially pave the way towards the development of selective treatments and diagnostic tools (Vander Heiden 2011; Meijer et al. 2012). To this end MBA was applied to construct the first genome scale metabolic model of cancer (Folger et al. 2011). The utility of the model was demonstrated by its ability to predict cytostatic drug targets, of which 40 % are targeted by known, approved or experimental anticancer drugs. Furthermore, the model was applied to identify combinations of synthetic lethal drug targets, whose synergy was validated using available drug efficacy and gene expression measurements across the NCI-60 cancer cell line collection. Finally, potential selective treatments for specific cancers that depend on cancer type-specific downregulation of gene expression and somatic mutations were compiled.

A type-specific model generates more accurate drug-target predictions for a specific type of cancer. Accordingly, MBA was applied to construct a specific metabolic model of *Hereditary Leiomyomatosis and Renal-Cell Cancer* (*HLRCC*) (Frezza et al. 2011). HLRCC is caused by a germline mutation in the gene encoding *Fumarate Hydratase* (*FH*), followed by a somatic mutation in its second allele. By analyzing the specific metabolic model of the FH-deficient cells, the survival mechanism that enables the cells to operate the mitochondrial electron transport-chain despite the mutation was unraveled. According to the computational predictions the FH-deficient cells produce NADH, the driving force of the electron transport-chain, by activating a linear metabolic pathway beginning with glutamine uptake and ending with bilirubin excretion (Fig. 14.4). This pathway, through the biosynthesis and degradation of heme, permits FH-deficient cells a partial mitochondrial NADH production, as it prevents the lethal accumulation of TCA cycle metabolites. In agreement, according to the model, numerous synthetic lethal pairs of FH are located along the heme biosynthesis pathway (Fig. 14.4). These synthetic lethal predictions have been confirmed experimentally *in-vitro*, showing that targeting a key enzyme on this pathway (HMOX) renders only the FH-deficient cells non-viable, selectively sparing wild-type cells. This provides a new potential target for treating HLRCC patients by a drug that is potentially selective, with minimal side-effects on healthy renal cells. However, these studies, as many others, were based on data from cancer cell lines, which might fail to depict the metabolism of the cancer *in-vivo*. To provide a system level view of cancer metabolism *in-vivo* a new method, termed *Metabolic Phenotypic Analysis* (*MPA*) was developed and applied to study breast cancer metabolism based on data of clinical samples.

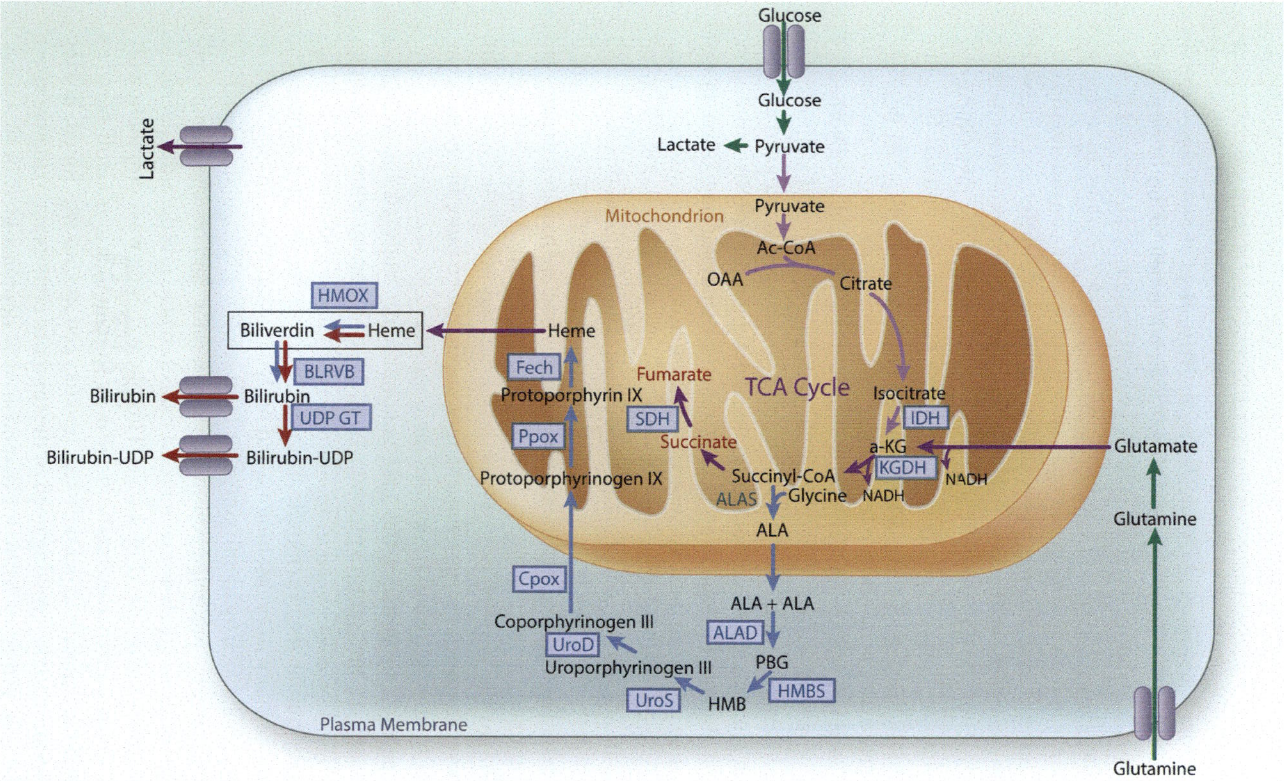

Fig. 14.4 Study of renal cancer metabolism via an MBA-reconstructed model. The scheme depicts the metabolic flux rearrangement observed in FH-deficient cells, based on model-driven predictions and experimental validations (Frezza et al. 2011). Blue arrows indicate FH synthetic lethal metabolic reactions predicted by the metabolic model; red arrows indicate genes and reactions found to be upregulated in FH-deficient cells. The scheme also shows the truncation of the TCA cycle observed in FH-deficient cells. Fumarate and succinate are significantly accumulated (in *red*). The flux through the first part of the TCA cycle is reduced in FH-deficient cells due to decreased pyruvate entry and absence of recycling of metabolites through the TCA cycle. Glutamine uptake and glycolytic production of lactate (in *green*) are induced in FH-deficient cells. Figure 14.4 adopted from Jerby et al. (Jerby and Ruppin 2012)

14.5 Metabolic Phenotypic Analysis (MPA)

Previous methods for incorporating contextual gene expression or protein abundance measurements within a generic CBM metabolic model have focused on describing the metabolic state by restricting the model to obtain an optimal fit to the data (Shlomi et al. 2008; Akesson et al. 2004; Zur et al. 2010; Becker and Palsson 2008; Jensen and Papin 2011). These approaches have shown much value in providing context-dependent metabolic descriptions. However, by requiring an optimal fit they have ignored the ability of cells to adaptively reinstate lost functions by inducing even small changes in their overall gene expression. This, in turn, can potentially lead to false predictions of reaction-inactivity, and mask observed differences between metabolic states. In contradistinction, given a gene expression or protein abundance signature of a sample, MPA provides a genome-scale view of its metabolism by considering solutions that may deviate to some extent from the optimal fit – this yields an estimation of the adaptive potential of the sample to carry out an array of metabolic processes in a given context. In the model, a metabolic process is defined by its medium (that is, which metabolites can be secreted or absorbed by the cell), and its end-reaction. Based on a curated, literature-based definition of metabolic processes (Gille et al. 2010), each sample-process pair is assigned an MPA score: First, the consistency of the sample molecular signature (mRNA expression, proteomics, as defined in (Shlomi et al. 2008)) with the metabolic state of the model is computed, when requiring the activation of the given process in its medium. Then, this consistency score is divided by the optimal consistency that can be obtained between the signature and the model under the same medium, without this additional activation constraint. The result is the final MPA score for this sample-process pair. It quantifies the extent of adaptive significant flux changes that are required to carry out the process examined, given the observed molecular signature. A high MPA score (close to 1) denotes that a given process can be carried out in a given context with minimal adaptive flux or transcriptional changes and is hence more likely to occur, while a low MPA score (close to 0) denotes the opposite.

MPA was applied to conduct the first genome-scale study of breast cancer metabolism based on the gene expression of a large cohort of clinical samples (Jerby et al. 2012). It was validated by examining its ability to capture known differences across three human tissues (muscle, liver, and adipose tissue), approximate the lipid production capacity of breast tumors, and bridge the gap between gene expression and protein abundance by identifying post-transcriptionally regulated metabolic reactions. MPA correctly predicted cell lines' growth rates, and revealed a subtype-independent "go or grow" dichotomy in breast cancer, where proliferation decreases as the tumor becomes metastatic (Fig. 14.5a). This phenomenon was computationally predicted and validated experimentally *in-vitro*. A stoichiometric tradeoff between proliferation and the capacity to detoxify reaction oxygen species was found, linking the slow proliferation to the growing need of the tumor to counteract oxidative stress (Fig. 14.5b, c). The utility of MPA for diagnosis was demonstrated in two ways. First it was shown to improve breast cancer prognosis prediction. Then its potency

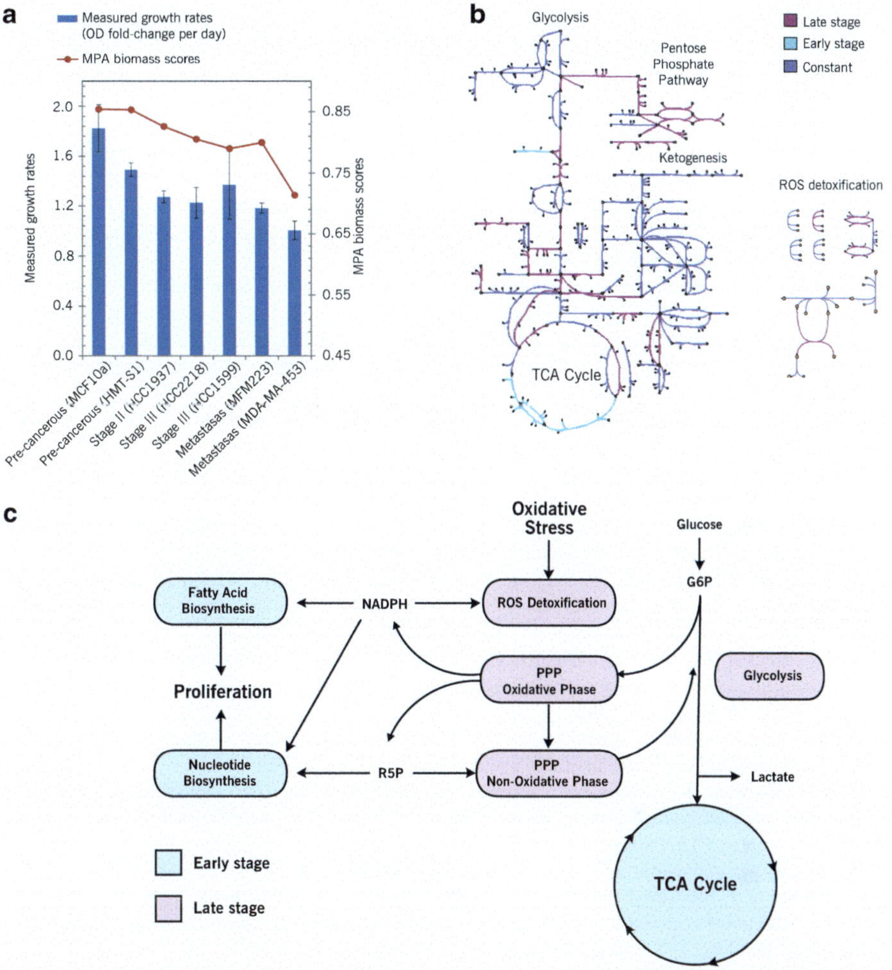

Fig. 14.5 Breast cancer metabolic progression. **a** The average measured (*blue* bars) and MPA-predicted (*red* plot) growth rates of the different breast cancer cell lines, ordered according to their stage. **b** The differences between the central metabolism of early- and late-stage tumors according to MPA. The MPA scores of the reactions colored in azure (*pink*) are significantly higher in early (late) stage tumors. **c** A pathway-level description of the differences between the metabolic states of early- and late-stage tumors, as suggested by MPA. In azure (*pink*) are processes that are more active in early (late) stage tumors. Figure 14.5 is adopted from Jerby et al. (Jerby et al. 2012)

for biomarker identification was validated by comparing its prediction to experimental data; employing it a set of novel biomarkers for metastatic breast cancer was compiled. This study demonstrates how metabolic modeling can, with complementing experimental work, pave the way for a system-level understanding of cancer metabolism that is vital for its diagnosis and treatment.

References

Abuoun M, Suthers PF, Jones GI, Carter BR, Saunders MP, Maranas CD, Woodward MJ, Anjum MF (2009) Genome scale reconstruction of a Salmonella metabolic model: comparison of similarity and differences with a commensal Escherichia coli strain. J Biol Chem 284:29480–29488

Akesson M, Forster J, Nielsen J (2004) Integration of gene expression data into genome-scale metabolic models. Metab Eng 6:285–293

Baynes JW, Dominiczak MH (2004) Medical biochemistry, 2 edn. Mosby

Becker SA, Palsson BO (2008) Context-specific metabolic networks are consistent with experiments. PLoS Comput Biol 4:e1000082

Berg JM, Tymoczko JL, Stryer L (2002) Biochemistry. W H Freeman, New York

Bilu Y, Shlomi T, Barkai N, Ruppin E (2006) Conservation of expression and sequence of metabolic genes is reflected by activity across metabolic states. PLoS Comput Biol 2:e106

Bock K, Gerok W, Matern S (1991) Hepatic Metabolism and Disposition of Endo- and Xenobiotics, Springer

Cairns RA, Harris IS, Mak TW (2011) Regulation of cancer cell metabolism. Nat Rev Cancer 11:85–95

Chatziioannou A, Palaiologos G, Kolisis FN (2003) Metabolic flux analysis as a tool for the elucidation of the metabolism of neurotransmitter glutamate. Metab Eng 5:201–210

Arias IM, Boyer JL, Chisari FV, Fausto N, Schachter D, Shafritz DA (eds) (2001) The Liver Biology and Pathobiology, 4th edn. Lippincott Williams & Wilkins

Daran-Lapujade P, Jansen ML, Daran JM, Gulik W Van, Winde J De, Pronk JT (2004) Role of transcriptional regulation in controlling fluxes in central carbon metabolism of Saccharomyces cerevisiae. A chemostat culture study. J Biol Chem 279:9125–9138

Duarte NC, Becker SA, Jamshidi N, Thiele I, Mo ML, Vo TD, Srivas R, Palsson BO (2007) Global reconstruction of the human metabolic network based on genomic and bibliomic data. Proc Natl Acad Sci U S A 104:1777–1782

Durot M, Bourguignon P-Y, Schachter V (2009) Genome-scale models of bacterial metabolism: reconstruction and applications. FEMS Microbiol Rev 33:164–190

Edwards JS, Palsson BO (2000) The Escherichia coli MG1655 in silico metabolic genotype: its definition, characteristics, and capabilities. Proc Natl Acad Sci U S A 97:5528–5533

Edwards JS, Ibarra RU, Palsson BO (2001) In silico predictions of Escherichia coli metabolic capabilities are consistent with experimental data. Nat Biotechnol 19:125–130

Famili I, Forster J, Nielsen J, Palsson BO (2003) Saccharomyces cerevisiae phenotypes can be predicted by using constraint-based analysis of a genome-scale reconstructed metabolic network. Proc Natl Acad Sci U S A 100:13134–13139

Feist AM, Palsson BO (2008) The growing scope of applications of genome-scale metabolic reconstructions using *Escherichia coli*. Nat Biotech 26:659–667

Folger O, Jerby L, Frezza C, Gottlieb E, Ruppin E, Shlomi T (2011) Predicting selective drug targets in cancer through metabolic networks. Mol Syst Biol 7:501–501

Fong SS, Palsson BO (2004) Metabolic gene-deletion strains of Escherichia coli evolve to computationally predicted growth phenotypes. Nat Genet 36:1056–1058

Frezza C, Zheng L, Folger O, Rajagopalan KN, Mackenzie ED, Jerby L, Micaroni M, Chaneton B, Adam J, Hedley A, Kalna G, Tomlinson IPM, Pollard PJ, Watson DG, Deberardinis RJ, Shlomi T, Ruppin E, Gottlieb E (2011) Haem oxygenase is synthetically lethal with the tumour suppressor fumarate hydratase. Nature 477:225–228

Garfinkel D, Hess B (1964) Metabolic control mechanisms. Vii.A detailed computer model of the glycolytic pathway in ascites cells. J Biol Chem 239:971–983

Gille C, Bolling C, Hoppe A, Bulik S, Hoffmann S, Hubner K, Karlstadt A, Ganeshan R, Konig M, Rother K, Weidlich M, Behre J, Holzhutter H-G (2010) HepatoNet1: a comprehensive metabolic reconstruction of the human hepatocyte for the analysis of liver physiology. Mol Syst Biol 6:411

Goodacre R, Vaidyanathan S, Dunn WB, Harrigan GG, Kell DB (2004) Metabolomics by numbers: acquiring and understanding global metabolite data. Trends Biotechnol 22:245–252

Gropper S, Smith L (2008) Advanced Nutrition and Human Metabolism
Hanahan D, Weinberg RA (2011) Hallmarks of cancer: the next generation. Cell 144:646–674
He F (2005) Human liver proteome project: plan, progress, and perspectives. Mol Cell Proteomics 4:1841–1848
Ibarra RU, Edwards JS, Palsson BO (2002) Escherichia coli K-12 undergoes adaptive evolution to achieve in silico predicted optimal growth. Nature 420:186–189
Jamshidi N, Edwards JS, Fahland T, Church GM, Palsson BO (2001) Dynamic simulation of the human red blood cell metabolic network. Bioinformatics 17:286–287
Jensen PA, Papin JA (2011) Functional integration of a metabolic network model and expression data without arbitrary thresholding. Bioinformatics 27:541–547
Jerby L, Ruppin E (2012) Predicting drug targets and biomarkers of cancer via genome-scale metabolic modeling. Clin Cancer Res 18:5572–5584
Jerby L, Shlomi T, Ruppin E (2010) Computational reconstruction of tissue-specific metabolic models: application to human liver metabolism. Mol Syst Biol 6:401
Jerby L et al (2012) Metabolic associations of reduced proliferation and oxidative stress in advanced breast cancer. Cancer Res 72(22):5712–5720
Joshi A, Palsson BO (1989) Metabolic dynamics in the human red cell. I. A comprehensive kinetic model. J Theor Biol 141:515–528
Kanehisa M, Goto S (2000) KEGG: kyoto encyclopedia of genes and genomes. Nucleic Acids Res 28:27–30
Kauffman KJ, Prakash P, Edwards JS (2003) Advances in flux balance analysis. Curr Opin Biotechnol 14:491–496
Kumar VS, Maranas CD (2009) GrowMatch: an automated method for reconciling in silico/in vivo growth predictions. PLoS Comput Biol 5:e1000308
Lee I-D, Palsson BO (1991) A comprehensive model of human erythrocyte metabolism: extensions to include pH effects. Biomed Biochim Acta 49:771–789
Levine DM, Haynor DR, Castle JC, Stepaniants SB, Pellegrini M, Mao M, Johnson JM (2006) Pathway and gene-set activation measurement from mRNA expression data: the tissue distribution of human pathways. Genome Biol 7:R93
Ma H, Sorokin A, Mazein A, Selkov A, Selkov E, Demin O, Goryanin I (2007) The Edinburgh human metabolic network reconstruction and its functional analysis. Mol Syst Biol 3:135
Mahadevan R, Schilling CH (2003) The effects of alternate optimal solutions in constraint-based genome-scale metabolic models. Metab Eng 5:264–276
Mathews C, Holde KV, Ahern K (2000) Biochemistry. Addison-Welsley Publishing Company
Mckusick VA (2007) Mendelian inheritance in man and its online version, OMIM. Am J Hum Genet 80:588–604
Meijer TW, Kaanders JH, Span Pn, Bussink J (2012) Targeting hypoxia, HIF-1 and tumor glucose metabolism to improve radiotherapy efficacy. Clin Cancer Res 18:5582–5594
Miller DM, Thomas SD, Islam A, Sedoris K (2012) Myc and cancer metabolism. Clin Cancer Res 18:5546–5553
Mulquiney PJ, Kuchel PW (1999) Model of 2,3-bisphosphoglycerate metabolism in the human erythrocyte based on detailed enzyme kinetic equations: computer simulation and metabolic control analysis. Biochem J 34:597–604
Oberhardt MA, Palsson BO, Papin JA (2009) Applications of genome-scale metabolic reconstructions. Mol Syst Biol 5:320
Prendergast GC (2011) Cancer: why tumours eat tryptophan. Nature 478:192–194
Romero P, Wagg J, Green M, Kaiser D, Krummenacker M, Karp P (2004) Computational prediction of human metabolic pathways from the complete human genome. Genome Biol 6:R2
Saier MH Jr, Tran CV, Barabote RD (2006) TCDB: the Transporter Classification Database for membrane transport protein analyses and information. Nucl Acids Res 34:D181–186
Schellenberger J, Palsson B (2009) Use of randomized sampling for analysis of metabolic networks. J Biol Chem 284:5457–5461

Schmid M, Davison TS, Henz SR, Pape UJ, Demar M, Vingron M, Scholkopf B, Weigel D, Lohmann JU (2005) A gene expression map of Arabidopsis thaliana development. Nat Genet 37:501–506

Schuster S, Klamt S, Weckwerth S, Moldenhauer F, Pfeiffer T (2002) Use of network analysis of metabolic systems in bioengineering. Bioprocess Biosyst Eng 24:363–372

Segre D, Vitkup D, Church GM (2002) Analysis of optimality in natural and perturbed metabolic networks. Proc Natl Acad Sci U S A 99:15112–15117

Shlomi T, Berkman O, Ruppin E (2005) Regulatory on/off minimization of metabolic flux changes after genetic perturbations. Proc Natl Acad Sci U S A 102:7695–7700

Shlomi T, Cabili MN, Herrgard MJ, Palsson BO, Ruppin E (2008) Network-based prediction of human tissue-specific metabolism. Nat Biotechnol 26:1003–1010

Shlomi T, Cabili MN, Ruppin E (2009) Predicting metabolic biomarkers of human inborn errors of metabolism. Mol Syst Biol 5:263

Shmueli O, Horn-Saban S, Chalifa-Caspi V, Shmoish M, Ophir R, Benjamin-Rodrig H, Safran M, Domany E, Lancet D (2003) GeneNote: whole genome expression profiles in normal human tissues. C R Biol 326:1067–1072

Sotgia F, Martinez-Outschoorn U, Lisanti M (2011) Mitochondrial oxidative stress drives tumor progression and metastasis: should we use antioxidants as a key component of cancer treatment and prevention? BMC Medi 9:62

Tamada M, Suematsu M, Saya H (2012) Pyruvate kinase M2: multiple faces for conferring benefits on cancer cells. Clin Cancer Res 18:5554–5561

Vander Heiden MG (2011) Targeting cancer metabolism: a therapeutic window opens. Nat Rev Drug Discov 10:671–684

Varma A, Palsson B (1993) Metabolic capabilities of *Escherichia coli*. II. Optimal growth patterns. J Theor Biol 165:503–522

Varma A, Palsson BO (1994) Stoichiometric flux balance models quantitatively predict growth and metabolic by-product secretion in wild-type Escherichia coli W3110. Appl Environ Microbiol 60:3724–3731

Vo TD, Greenberg HJ, Palsson BO (2004) Reconstruction and functional characterization of the human mitochondrial metabolic network based on proteomic and biochemical data. J Biol Chem 279:39532–39540

Ward PS, Thompson CB (2012) Metabolic reprogramming: a cancer hallmark even warburg did not anticipate. Cancer cell 21:297–308

Wiback SJ, Palsson BO (2002) Extreme pathway analysis of human red blood cell metabolism. Biophys J 83:808–818

Wiback SJ, Mahadevan R, Palsson BO (2004) Using metabolic flux data to further constrain the metabolic solution space and predict internal flux patterns: the Escherichia coli spectrum. Biotechnol Bioeng 86:317–331

Wishart DS, Tzur D, Knox C, Eisner R, Guo AC, Young N, Cheng D, Jewell K, Arndt D, Sawhney S, Fung C, Nikolai L, Lewis M, Coutouly M-A, Forsythe I, Tang P, Shrivastava S, Jeroncic K, Stothard P, Amegbey G, Block D, Hau DD, Wagner J, Miniaci J, Clements M, Gebremedhin M, Guo N, Zhang Y, Duggan GE, Macinnis GD, Weljie AM, Dowlatabadi R, Bamforth F, Clive D, Greiner R, Li L, Marrie T, Sykes BD, Vogel HJ, Querengesser L (2007) HMDB: the human metabolome database. Nucl Acids Res 35:D521–526

Yan Q, Sadee W (2000) Human membrane transporter database: a Web-accessible relational database for drug transport studies and pharmacogenomics. AAPS PharmSci 2:E20

Yanai I, Benjamin H, Shmoish M, Chalifa-Caspi V, Shklar M, Ophir R, Bar-Even A, Horn-Saban S, Safran M, Domany E, Lancet D, Shmueli O (2005) Genome-wide midrange transcription profiles reveal expression level relationships in human tissue specification. Bioinformatics 21:650–659

Zur H, Ruppin E, Shlomi T (2010) iMAT: an integrative metabolic analysis tool. Bioinformatics doi:10.1093/bi.oinformatics/btq602

Chapter 15
Modeling of Lipid Membranes and Lipoproteins

Artturi Koivuniemi and Ilpo Vattulainen

Abstract Metabolic states that precede diseases such as the Metabolic Syndrome are associated with abnormal conditions, such as dyslipidemia, reflected in circulation as elevated levels of triglycerides and low levels of high-density lipoproteins. Yet, the underlying molecular phenomena associated with these conditions are not understood. This is quite unfortunate, since understanding the interplay of lipids and proteins in the context of normal and abnormal metabolite profiles would promote development of novel ways to guide dysfunctional metabolic profiles towards normal ones. Here we discuss how molecular simulations can be used to shed light on these issues by modelling the structure, dynamics, and function of biological systems comprised of lipids and proteins. By considering recent simulation studies of lipid membranes, membrane proteins, and lipoproteins we highlight the added value brought out by simulations in unravelling how nano-scale phenomena take place in complex lipid-protein systems. The examples shown here also demonstrate the significant added value of bridging molecular simulations with experiments, and in a biomedical context with clinical studies.

Keywords Coarse-grained modelling · Computer simulation · Lipid membrane · Lipoprotein · Molecular dynamics simulation

15.1 Introduction

Without doubt, the research of complex biological systems is mainly based on experiments. Thanks to methodological development, there is now an exceptionally broad arsenal of different experimental techniques by which one is able to

I. Vattulainen (✉)
Department of Physics, Tampere University of Technology,
PO Box 692, 33101 Tampere, Finland

MEMPHYS—Centre for Biomembrane Physics,
University of Southern Denmark, Odense, Denmark
e-mail: ilpo.vattulainen@tut.fi

A. Koivuniemi
VTT Technical Research Centre of Finland, PO Box 1000, 02044, Espoo, Finland

M. Orešič, A. Vidal-Puig (eds.), *A Systems Biology Approach to Study Metabolic Syndrome*, 299
DOI 10.1007/978-3-319-01008-3_15,

characterize properties of living cellular systems, thereby gaining insight into the principles and mechanisms that govern the function of living matter. However, as any means, also experimental techniques have limitations. One of them is related to the fact that a great deal of processes take place on a molecular level where individual molecules in cells interact with one another, giving rise to a wide variety of phenomena and molecular-scale functions, such as formation of lipid-protein complexes, recognition, and modulation of protein conformation through lipid-protein interactions. As these transient molecular processes take place over scales of the order of nano/microseconds and nanometers, it is not a simple feat to gauge them experimentally. Even the state-of-the-art of super-resolution microscopy techniques applied to soft biological systems is currently limited by a spatial resolution of about 5–20 nm (Donnert et al. 2006). This is one of the main reasons why molecular computer simulations (Vattulainen and Rog 2011; Ayton and Voth 2009) have recently received more and more attention as they can make a difference to considerations of molecular phenomena in living matter.

In principle, there is no limit for the scope of phenomena that can be gauged by computer simulations. There are appropriate techniques for studies of enzyme action, proton diffusion, diffusion of lipids and proteins, ligand-receptor binding, endocytosis, lipid transport managed by lipoproteins, and a large variety of other phenomena in the molecular world of lipids, proteins, and carbohydrates. The fact that there are suitable computational methods to consider even the finest details such as chemical reactions, where one has to deal with electronic properties, and specific atom-scale binding events between ligands and molecular receptors, renders computer simulations the method of choice for elucidating nano-scale phenomena in biological systems. In essence, one can gain a great deal of added value by bridging simulations and modelling to experiments.

In this chapter, we discuss the principles of computer simulations. We focus on atomistic and coarse-grained molecular dynamics simulation techniques as they can provide not only structural but also dynamic information for any complex biological system. Here, the emphasis of discussion is on lipids (Mouritsen 2005). The applications that we present focus on lipid membranes where lipids interact with proteins, and lipoproteins where we also see lipids interacting with proteins. Relevance of molecular simulations to life sciences and clinical biomedicine is also briefly discussed.

15.2 Main Principles of Molecular Simulations and Their Limitations

Molecular simulations are often the method of choice, if detailed insight is needed to understand the principles, which govern the behaviour of molecular systems. Before any simulations can be started, however, one has to decide what is the most appropriate technique to be used. The answer depends on the research topic, which one would like to clarify.

For processes where one has to account for electronic degrees of freedom, the method of choice is definitely quantum mechanics (QM). This is the case for diffusion of protons, action of lipases (enzymes) on lipids, and other chemical processes (Kamerlin et al. 2011). Meanwhile, if the process one would like to understand is not chemical but instead a physical one, then a classical atomistic description often suffices very well. Diffusion of water molecules across lipid membranes, binding of lipases on a lipid substrate, and formation of ordered cholesterol-rich domains in membranes are exemplary processes about this situation (Murtola et al. 2009). The downside of classical atomistic simulations lies in the scales that are feasible to be considered, since currently the state-of-the-art of atomistic simulations is roughly a system of about one million atoms simulated over a millisecond (Shaw et al. 2010). This is of course much more than the picosecond time scale in systems of a few tens of atoms in QM simulations, but the scales of atomistic studies are yet rather modest. To overcome this issue, one can coarse grain the molecular description further by lumping several atoms to a single bead, thus decreasing the computational cost and gaining a speed up that is usually about 10^3–10^6 (Marrink et al. 2004, 2007; Murtola 2009; Ayton and Voth 2009) compared to atomistic simulations. Simulations of coarse-grained (CG) models have become common place recently, allowing considerations of large-scale phenomena such as fusion of liposomes, release of osmotic pressure inside liposomes through gating of mechanosensitive channels, and effects of peoples' living habits on lipoprotein structure (Murtola et al. 2009; Louhivuori et al. 2010; Yetukuri et al. 2010). The scales that can be studied with CG models are also rather extensive, and while they depend on the level of coarse graining in a given model, they usually cover times up to seconds and length scales up to millimetres.

Here we focus on atomistic and CG molecular dynamics (MD) simulations since they are the most commonly used simulation techniques applied to lipid systems. In both cases, there are three corner stones on which the simulation of any system is based. First, the simulation project starts from construction of the initial configuration for a given system, such as an aqueous environment with a high density lipoprotein (HDL) comprised of lipids and proteins. Second, most importantly, one has to derive a description for all intra- and intermolecular interactions. This description known as a force field is the heart of the model since the force field determines the properties of the system. Finally, to find out how HDL evolves in time, its time evolution is determined by solving Newton's equations of motion for particle positions and velocities, as they are integrated a sufficient number of times using interatomic forces computed from the force field.

The force field is the core of MD simulations, thus let us consider it more detail. Other important aspects of MD simulations such as integration of Newton's equations of motion, constraints, thermodynamic ensembles, periodic boundary conditions and treatment of long-range interactions in a principally infinite system are discussed elsewhere in detail (Frenkel and Smit 2001; Gromacs 2010).

Consider interactions described in Fig. 15.1. There are three terms describing bonded interactions in terms of bonding between two particles, bending between three particles bonded to each other, and dihedral (torsional) terms as a four-body interaction, while the remaining two terms reflect non-bonded interactions that here

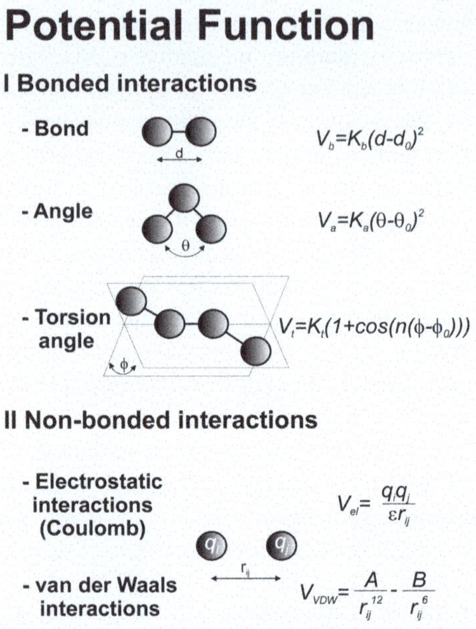

Fig. 15.1 Typical description of potential function terms (*bonding, bending (angle), torsion angle, Coulomb, and van der Waals*) often used in classical molecular dynamics simulations. The parameters associated with these terms are d for bond length, d_0 describing its equilibrium value, and θ being angle in three-body interaction with θ_0 representing its equilibrium value. Further, φ is torsion angle, φ_0 being 0 or 180 degrees, and n represents phase. K_b, K_a, and K_t are force constants of related interactions, q_i and q_j denote partial charges of particles, which are not covalently bonded, and ε is dielectric constant. Finally, r_{ij} is distance between a pair of interacting atoms, and A and B are constants, which depend on the chemical nature of interacting atoms. (Adapted from Vattulainen and Rog 2011)

correspond to electrostatic and Lennard-Jones (van der Waals) interactions. The set of these interactions together with their parameter values constitute the force field.

In principle, the description shown in Fig. 15.1 is an example of a force field, which can be used for atomistic as well as CG models, though the number and the details of the terms as well as their parameter values are often quite different in different force fields. In atomistic simulations, typical force fields used for biomolecular systems are Amber, Charmm, OPLS, and GROMOS (van Gunsteren et al. 2006). The parameters used in these atomistic force fields are physically easy to understand, since they describe, e.g., interatomic distances, or interactions such as Coulombic forces that are intuitively clear to interpret.

In CG simulations, the interpretation of force field parameters is more complicated since due to the CG nature of a molecular description, the parameters also contain an entropic component. Therefore in CG models the force field describes the free energy (sometimes called the effective interaction) rather than the internal energy of the system. Two schemes often used to derive the effective interaction for a CG system are based on matching either forces (Yanting et al. 2009) or structural properties (Murtola et al. 2009) between the coarse-grained and the underlying atomistic systems. Another technique that has also gained considerable attention is to consider the thermodynamic free energy of a CG particle to partition in a water-like or oil-like environment, and to parameterize the coarse-grained model in this spirit. The MARTINI model (Marrink et al. 2004, 2007; Monticelli et al. 2008; Lopez et al. 2009) based on this idea has been very successful in simulations of lipid and protein systems.

While computer simulations using MD have been carried out since the 1950's, there has been continuous improvement in the field. More efficient and accurate algorithms and their implementations have been developed, and a lot of attention has been paid to develop techniques for massively parallel simulations. Several public-domain MD packages have been published, GROMACS (Hess et al. 2008) and NAMD (Phillips et al. 2005) being perhaps the most popular simulation packages used these days in biomolecular sciences. For more discussion on the methodology of molecular simulations (see, e.g., refs. Frenkel and Smit 2001; Gromacs 2010).

15.3 Simulations of Lipid Membranes with Membrane Proteins

Biological membranes are mainly composed of lipids and membrane associated proteins that interact with one another. The fact that proteins are involved in cell functions has been known for a long time, but the idea of lipids being also involved in many functions is more recent. It has become apparent that the role of lipids is important, e.g., in recognition and signaling, and very importantly also in modulating membrane protein function (Lingwood and Simons 2010).

The diversity of lipids in biological membranes is enormous: biological membranes are composed of thousands of chemically distinct lipids, and the lipid composition varies quite substantially between tissues and cell organelles, reflecting their different roles in cellular functions (Yeagle 2005). Despite considerable research, the reasons for this complexity are still far from being understood. What we know is that different lipids affect membrane fluidity and other physical properties quite differently, and properties such as membrane elasticity can in turn affect the function of membrane proteins such as mechanosensitive channels. Meanwhile it has also been realized that lipids can modulate membrane protein and membrane receptor functions in a specific manner, sphingomyelin and cholesterol being just a couple of recently found examples (Contreras et al. 2012; Lingwood and Simons 2010). As there are lipids targeting specific proteins, and considering the large number of different membrane proteins, this is one plausible reason for the diversity of lipids in cells. Another possibility is the need to ensure stable and robust conditions for membrane proteins to carry out their functions even if, e.g., membrane composition and pH changed locally. Without lipid diversity such small local perturbations could lead to the disruption and malfunctioning of the membrane (Simons and Sampaio 2011).

The variety of lipids and their concentration changes are relevant also in the Metabolic Syndrome (MetS). As the composition of membranes changes, it also influences lipid-protein interactions and physical properties of lipid membranes, which in turn modulate the activity of membrane-associated proteins. It is interesting to note that roughly a third of the human genome codes membrane-associated proteins and, still, very little is known about how lipids and proteins play together. This is to a major extent due to the low number of atomic-scale structures that have been determined for integral membrane proteins. This is currently the greatest challenge to foster the understanding of functions associated with biological membranes. As new membrane protein structures are published, what we gain is not only insight of protein structures but also their dynamics in varying pools of lipids interacting with the proteins. These considerations that bridge experiments with molecular simulations can reveal when biological membranes are dysfunctional and lead to diseases, as revealed by pathological symptoms. Combining, e.g., temporal lipidomic analyses with lipid membrane simulations can therefore provide a novel way to better understand mechanistic processes that underlie certain diseases, eventually leading to discoveries of new biomarkers or novel ways to design preventive therapies.

Here we discuss some examples, which show how molecular dynamics simulations can be used to study the properties of lipid membranes with and without membrane proteins.

15.3.1 Simulations of Lipid Membranes

As cell membranes are extremely diverse regarding their lipid composition, the number of different interactions between lipids is staggering and the local domain structure in cell membranes can be very complex. How the structural and dynamical properties of biological membranes arise from the structural details of individual lipids? How the functional groups of different lipids interact in lipid bilayers, and which of these interactions drive the formation of biologically relevant patches or nano-sized domains in cell membranes? These are some of the topical questions that are very hard or even impossible to clarify with the current arsenal of experimental methods. This is one of the reasons for the growing popularity of lipid bilayer simulations since, unlike any other technique, they can provide a detailed view of lipid behavior in complex lipid membranes.

Regarding specific interactions between lipids, one good example where atomistic simulations have provided valuable insight is the interaction between sphingomyelin (SM) and cholesterol molecules (Aittoniemi et al. 2007). They are both important in the formation of lipid rafts known as nano-scale membrane domains, which together with membrane proteins form units with specific functions (Lingwood and Simons 2010). Simulations suggest that the transient hydrogen bonding patterns of SM result in a tilted head group configuration that can shield cholesterol from water molecules (see Fig. 15.2). The tilting of SM head group was shown to increase when cholesterol approaches SM to stand next to it, the driving force being hydrophobic shielding together with the charge pair interaction between them. This mechanism, which in part reflects the so-called umbrella model, complements the previously proposed

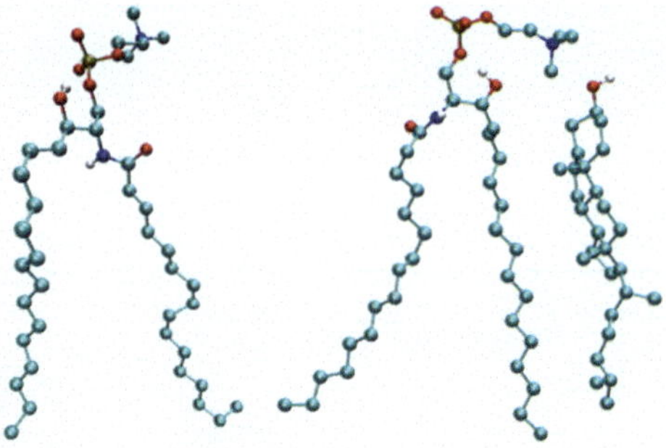

Fig. 15.2 Simulation snapshots representing typical SM orientations (*left*) without, and (*right*) with cholesterol. Note the tilting of the SM head group when cholesterol is next to SM. Other lipids in a membrane and water are not shown for clarity's sake. (Adapted from Aittoniemi et al. 2007)

view that SM and cholesterol like each other due to direct hydrogen bonding only. What likely also plays some role is the fact that cholesterol often prefers saturated lipids instead of unsaturated ones, the interaction in this case being largely entropic.

Glycolipids are the most complex lipid type due to the carbohydrate units in their head groups. In glycolipids, every detail matters. It has been shown through atomistic simulations that single-component lipid bilayers comprised of either glucose- or galactose-based glycerolipids have quite different membrane properties (Rog et al. 2005), fluidity being the primary example since almost all physical properties of membrane are affected by it. Yet, these two sugar head groups are identical except for their stereochemistry. The challenge is to unravel the complex interaction patterns of glycolipids, and atomistic simulations are definitely the method of choice for this quest. In the case of galactosylceramide (GalCer), it was recently shown by Hall et al. that cholesterol molecules form direct hydrogen bonds with the sugar head group of GalCer, resulting again in a tilted head group orientation of GalCer (Hall et al. 2010, 2011). This orientation shields cholesterol molecules from water, suggesting a plausible reason why glycosphingolipids are present in cholesterol-rich lipid rafts. Intriguingly, the same phenomenon has been identified in interactions of cholesterol with GM1, which is one of the glycolipids acting as membrane receptors. A combination of experiments and atom-scale simulations revealed how cholesterol can inhibit the receptor function of GM1 by enforcing GM1 to tilt its complex head group towards the membrane plane, thus rendering it inaccessible for the ligands nearby (Lingwood et al. 2011). In a biological context, the observed cholesterol-induced head group tilting plays a role in erythrocyte blood group presentation and glycolipid receptor function during the activation of sperm fertility (Lingwood et al. 2011).

Simulations of coarse-grained model systems have also improved our understanding of membranes quite considerably. Regarding lipid rafts, CG models have provided insight into the formation of nano-scale domains in membranes composed of saturated and polyunsaturated lipids together with cholesterol (see Fig. 15.3; Risselada and Marrink 2008). Risselada et al. showed how saturated and unsaturated phosphatidylcholines with cholesterol self-assemble spontaneously into coexistence of liquid-ordered and liquid-disordered domains. The findings matched experimental data and emphasized the role of saturated acyl chains in the formation of raft-like domains, since cholesterol prefers saturated over polyunsaturated lipid chains due

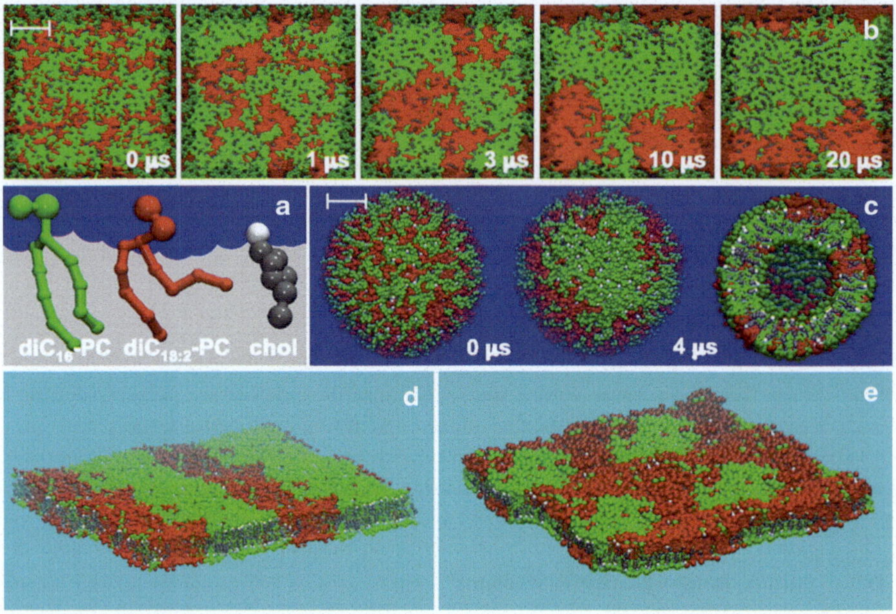

Fig. 15.3 Formation of liquid-ordered domains in ternary lipid mixtures based on CG molecular simulations. **a** Color coding used in describing the different lipids is as follows: *Green* describes saturated lipids, *red* polyunsaturated lipids, and *gray* is used for cholesterol where the hydroxyl group is shown in *white*. **b** Time evolution of phase segregation observed in a planar membrane where the starting structure at time $t = 0$ is a random mixture and the final structure at $t = 20$ µs shows the coexistence of liquid-ordered and liquid-disordered domains. **c** A liposome with a size of about 20 nm showing phase segregation for the same lipid mixture. **d** and **e** Several periodic images (2×2) merged together to show typical patters in terms of stripes and circular domains. The *scale bar* shown in panel *A* corresponds to 5 nm. (Adapted from Risselada and Marrink 2008. Copyright (2008) National Academy of Sciences, U.S.A.)

to stronger van der Waals interactions. Together with an earlier atomistic simulation study (Niemelä et al. 2007), the two pioneering works are the cornerstone of lipid raft simulations, the future aim being to complement these cases with similar considerations in the presence of membrane proteins.

Lipid simulations have also been coupled to clinical studies. Lipidomic analyses of adipose tissue in twin pairs with acquired obesity indicated changes in plasma membrane lipid composition, in particular increased levels of plasmalogen, which is the most abundant ether lipid, and decreasing levels of shorter and more saturated phospholipids (Pietiläinen et al. 2011). In this case, atomistic molecular dynamics simulations provided insightful information (Fig. 15.4) of how plasma membrane thickness and fluidity changed, when the amount of arachidonic acid increased and phosphatidylcholine (PC) head group was replaced by phosphatidylethanolamine (PE). Simulations revealed that the simultaneous increase in levels of polyunsaturated acyl chains and PE head groups does not alter membrane properties like area per lipid and fluidity. Meanwhile, the addition of a vinyl ether bond was observed to

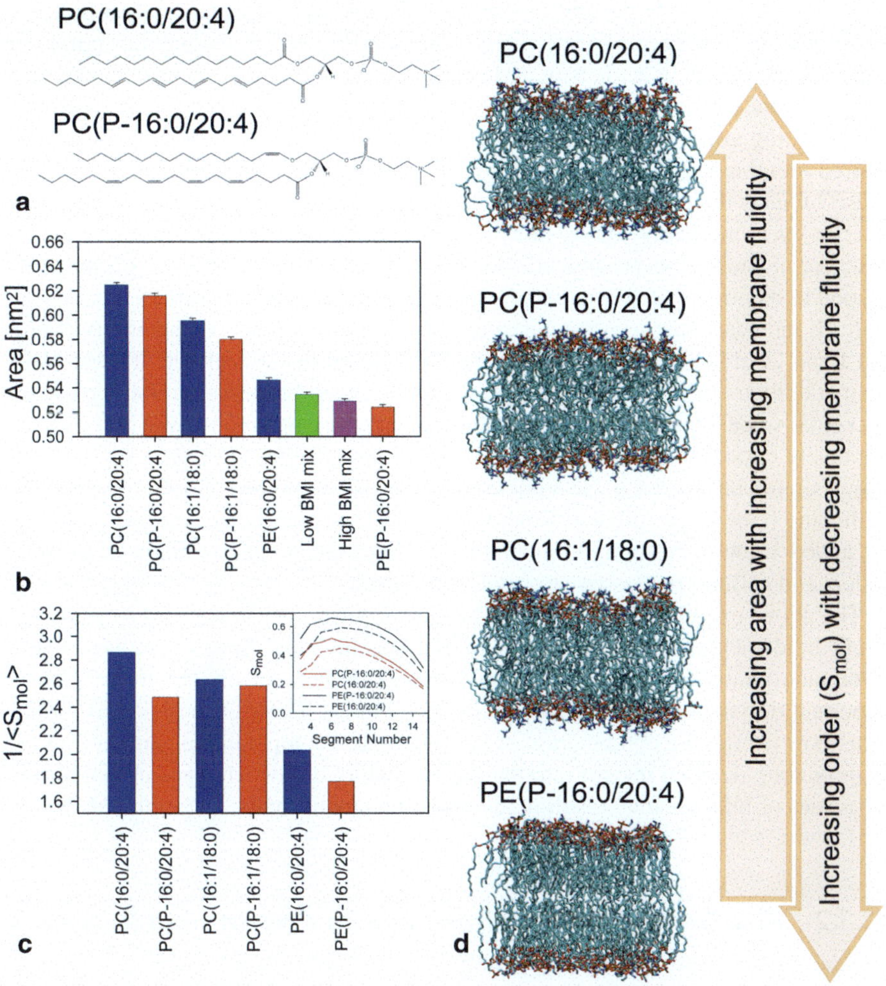

Fig. 15.4 Atomistic molecular dynamics simulations of different lipid membrane systems. Six single-lipid component systems were studied, representing major groups of lipids found to be differentially regulated in acquired obesity (see panel *A*). Two lipid mixes, corresponding to observed concentration changes of most abundant lipids (see panel *B*), were also studied to represent lipid membrane composition in heavy and lean weight-discordant twins. **a** Structures of two PUFA-containing PCs included in the simulations, one plasmalogen and one ester-bonded. **b** Average area per lipid of the eight simulated bilayer systems, indicating the packing or fluidity of the bilayer (smaller area means decreased fluidity). **c** The molecular order parameter (S_{mol}) profiles along selected unsaturated acyl chains. Segment number means the carbon position starting from the ester-vinyl linkage and ending at the methyl group in the end of the chain. **d** Snapshots from the end of four simulations, with the lipids ordered from top to bottom according to decreased fluidity. (Reproduced with permission Pietiläinen et al. 2011)

change the ordering of lipid acyl chains considerably, although the area per lipid was not registered to decrease significantly. The ordering effect of the ether bond was especially profound in acyl chain carbons 2–8. Interestingly, it seems that adipocyte cells are able to adapt to changing fatty acid composition during their expansion by generating vinyl-ether bond plasmalogens with arachidonic acids to maintain the desired membrane properties. Intriguingly, it was further showed that in morbidly obese subjects the proportion of ether lipids containing polyunsaturated fatty acids (PUFAs) was markedly lower, indicating that the mechanism of lipid remodeling observed in healthy obese subjects did not exist. Regarding membrane properties, it is good to mention that adding more double bonds and replacing the PC head group with PE increases the negative curvature of a lipid membrane substantially (Fuller and Rand 2001). Adding ether bonds probably amplifies this effect. If these lipids are markedly increased in planar lipid membranes, membrane frustration is expected to emerge since highly negative spontaneous curvature in a monolayer prefers more negatively curved surfaces. Spontaneous curvature changes have been suggested to play a significant role, e.g., in vesicle fusion events that are important in vesicular trafficking (Chernomordik and Kozlov 2008). Nonetheless, this example highlights the added value of bridging clinical studies, lipidomics, and simulations to better understand cellular processes on a molecular level.

The above examples show how molecular simulations can be used to study different lipid membrane properties including specific lipid-lipid interactions, lipid membrane phase behavior and lateral pressure profile, all of which can play an important role in the functions of cellular membranes. These examples also demonstrate how fruitful the combination of lipidomics and computational simulations can be, providing detailed insight into the understanding of how changes in biophysical properties of bilayers may result in abnormal metabolic states like in the MetS.

15.3.2 Membrane Lipids Interacting with Proteins and Receptors

One of the greatest challenges at the moment is to understand how lipids modulate or even control membrane protein and receptor function. In basic terms one can consider this problem from three perspectives:

- How lipids interact with and modulate membrane proteins and receptors via specific interactions?
- How lipids modulate membrane protein aggregation and complex formation in a manner, where specific lipid-lipid and lipid-protein interactions are important, but generic physical effects such as hydrophobic mismatch also play an important role?
- How the function or conformational state of membrane proteins is affected by membrane-mediated interactions via the lateral pressure profile?

Here we briefly consider recent computational studies as examples that highlight each of these cases.

Specific interactions Can membrane protein and receptor activity be modulated by direct/specific interactions? This idea has been considered quite plausible, since crystallized membrane protein structures often contain lipids as an integral part of the complex (Palsdottir and Hunte 2004; Adamian et al. 2009). Recent studies have strongly supported this view. A beautiful example is the work by Contreras et al. (Contreras et al. 2012), who considered a sphingolipid binding protein (the transmembrane domain of the COPI machinery protein p24) interacting with sphingomyelin (SM). In experiments they found a marked specificity of p24 for sphingomyelin, and SM 18 in particular. Atomistic simulations reported in the same study confirmed the specificity of SM 18: the lifetime of the SM-p24 complex was the largest with fatty acid C18, the lifetimes with fatty acids of different length being five times shorter. The binding mechanism observed in simulations suggests that it is not only the head group which interacts strongly with the protein, but the role of hydrocarbon chains is also important, as they both interact favorably with the protein's SM binding pocket.

More generally speaking, direct lipid interactions can also be important for the proper folding of proteins, and lipid molecules may be an important part of the active site in enzymes. The importance of specific lipid-protein interactions in enzyme function has been realized in X-ray structures of many proteins that contain lipid-binding cavities, for example phosphatidylethanolamine in the photosynthetic reaction centre of *T. tepidum* (Nogi et al. 2000). As for membrane receptors, another very recent example is the study by Lingwood et al., who found cholesterol to modulate membrane receptor conformation and function by specific interactions with the receptor's head group (Lingwood et al. 2011). In the absence of cholesterol, the receptor was visible to its ligand, while in the presence of cholesterol the receptor head group tilted downwards towards cholesterol, rendering itself inaccessible.

Hydrophobic mismatch Integral membrane proteins have a transmembrane domain which is mainly hydrophobic, avoiding contact with water and preferring oil-like environment provided by the lipid hydrocarbon chains. The hydrophobic thickness of transmembrane domains depends on the protein in question, and it is not obvious that it would match the hydrophobic thickness of a membrane which embraces the protein. When the hydrophobic thicknesses of the protein and the membrane do not match, one talks about hydrophobic mismatch.

Positive mismatch refers to proteins whose hydrophobic thickness is thicker than that of the membrane, while negative mismatch corresponds to the opposite case. Obviously, in both cases the mismatch can be expected to affect protein-protein interactions mediated by the membrane and also affect the orientation of a single protein. The topic is broad and extensive discussions of it can be found elsewhere (Jensen et al. 2004; Marsh 2008). Let us here provide just a couple of recent highlights of lipid-protein simulations that have shed light on the consequences of hydrophobic mismatch.

Kaiser et al. combined experiments with atomistic simulations to consider lipid reorganization around integral membrane peptides whose hydrophobic mismatch was varied systematically from negative to positive (Kaiser et al. 2011). The results were quite exciting. It was observed that the organization of lipids around a peptide depends on mismatch, and the strongest effects on protein-induced membrane deformation were observed with negative mismatch. The role of cholesterol was found to be especially important as it controlled the organization of lipids in the peptides' vicinity, in practice controlling elastic stress which emerges from the mismatch. The results indicated how the cholesterol-induced adaptation at the peptide-membrane interface translate into a sorting potential and promote selective lateral segregation of peptides and lipids according to their hydrophobic length.

Ramadurai et al. considered consequences of hydrophobic mismatch from a different perspective (Ramadurai et al. 2010). Their main interest was to better understand how the mismatch affects protein diffusion in the membrane plane. They found that hydrophobic mismatch has a pronounced effect on the lateral diffusion of membrane peptides, and this effect was considerably stronger compared to the effect of lipid head composition. Moving on, when attention is paid to protein-rich conditions, the role of protein-protein interactions becomes a matter of great interest. Parton et al. elucidated this aspect and observed that when the degree of hydrophobic mismatch was varied, the simulations displayed levels of protein aggregation ranging from negligible to extensive (Parton et al. 2011). Protein aggregation was shown to be influenced by, e.g., orientation of the protein, which depends on mismatch.

While a lot has been done, not much is really understood about the hydrophobic mismatch in realistic conditions, corresponding to a protein-rich (crowded) environment in membranes with a realistic many-component lipid composition. Most of the exciting topics still remain to be explored.

Lateral pressure profile and 3-dimensional pressure field inside a membrane Some membrane protein classes such as mechanosensitive channels are known to be sensitive to membrane elasticity and lateral stress along the membrane plane (Lundbak et al. 2010; Ollila et al. 2009). In this spirit, lateral pressure profiles of cell membranes have been suggested to play a role in the activity of membrane proteins (Cantor 1999; Ollila and Vattulainen 2010).

The lateral pressure profile of a lipid membrane describes how pressure inside a membrane is distributed (see Fig. 15.5). The depth-dependent distribution of lateral stress has been shown in several atomistic simulations (Ollila and Vattulainen 2010), and while it is an extremely difficult quantity to gauge by experiments, some measurements have provided evidence that the pressure profile has the same indicated by simulations. Many studies have shown that there are extremely abrupt and strong variations in the stress inside a membrane, a typical gradient being a change in pressure of 1,000 bar over a distance of 0.2 nm. This corresponds to a gradient of 5×10^{12} bar/m. It is quite easy to understand that if the pressure (force) exerted on a membrane protein changes so drastically, it may have an effect on protein conformation and activation too. By molecular dynamics simulations, it has been shown that polyunsaturation of phospholipid membranes changes the pressure profile (see Fig. 5; Ollila et al. 2007) and even more pronounced changes occur when sterols are introduced to the membranes (Ollila et al. 2009). In addition to lipid composition, small metabolites solvated into lipid membranes are also able to change the lateral pressure profiles (Terama et al. 2008). This is especially interesting in the case of

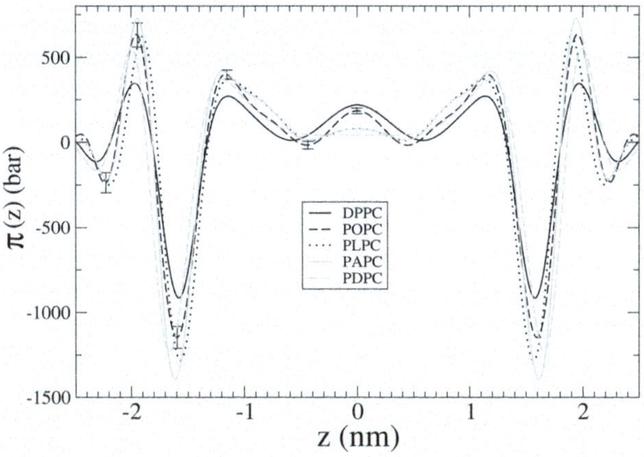

Fig. 15.5 Schematic presentation for lateral pressure profiles of single-component lipid bilayers composed of *DPPC*, *POPC*, *PLPC*, *PAPC*, or *PDPC* lipids having zero, one, two, four, or six double bonds, respectively. The pressure is shown as a function of z, which is the normal coordinate along membrane normal direction, the case $z = 0$ corresponding to the membrane center. The water phase resides for $| z | > 2$ nm. Statistical errors are presented for one leaflet of the POPC system. *Error bars* in the other leaflet and in other systems are similar. (Adapted from Ollila et al. 2007)

anesthetics, where the interaction mechanism of small molecules with biological systems is not understood (Cantor 1997; Terama et al. 2008).

Atomistic simulations have shown that changes in lateral pressure can indeed play a role in membrane protein activation. Changes in molecular composition of a membrane change also the pressure profile of the membrane, which in turn couples to changes in protein conformation (Ollila and Vattulainen 2010). The contribution of the pressure profile to protein activation can be significant, even tens of percent of the total free energy barrier for activation (Ollila and Vattulainen 2010), and this contribution is due to membrane-mediated interactions. Summarizing, it is apparent that in order to understand the role of the lipids in membrane protein function, one cannot neglect the role of generic interactions mediated by membranes.

15.4 Lipoprotein Simulations

Cardiovascular diseases associated with the MetS are the main cause of death in Western countries. It is believed that cardiovascular diseases are partially caused by high concentration of low density lipoprotein (LDL, 'bad' cholesterol) particles in circulation. Somehow LDL particles end up into the arterial intima, which is followed by a complex cascade of events leading to the formation of arterial plaques, the hallmark of atherogenesis. This gives rise to the narrowing of arteries, rupture, clotting, and potential death. Meanwhile, HDL particles (often called "good" cholesterol) are aggregates of lipids and proteins that mainly transport cholesterol and its esters as well as other lipid molecules between tissues in circulation, essentially transporting cholesterol back to the liver for excretion and re-utilization. HDL therefore plays an important role in reverse cholesterol transport, and clinical studies suggest that a low level of HDL in circulation correlates with increased risk of atherosclerosis.

One important question is how the metabolic processes related to shuttling of lipoprotein particles are associated with their molecular composition? Also, are there biophysical properties of lipoprotein particles that could be used to predict individuals' risk of being exposed to cardiovascular diseases?

To date there has been a tremendous effort to elucidate the structure of lipoproteins and to reveal its impact to lipoprotein functionality. This development has been a rocky road due to the lack of experimental methods able to characterize the properties of extremely small and heterogeneous particles. Therefore, to complement experiments, computational approaches have been used to unravel the structure and dynamics of lipoprotein particles, thus giving insight to their structure-function relationship. In this section, we highlight the recent progress in molecular simulations of lipoprotein particles.

15.4.1 Distribution of Lipids in HDL Particles

HDL particles have been studies in quite a few simulations (Koivuniemi and Vattulainen 2012). While the majority of these considerations has assessed the conformational properties of apoA-I in discoidal HDL particles, some attention has also been paid to spherical HDL to study the effect of lipid and protein composition on the distribution of lipids in HDLs (Koivuniemi and Vattulainen 2012). Recently, the idea of how the lipid composition differs between individuals with low and high HDL levels has also been explored (Yetukuri et al. 2010; see Fig. 15.6).

A low level of high density lipoprotein is a significant risk factor for cardiovascular disease. What molecular properties drive the low concentration of HDL? Does lipidomics of HDL have predictive power when combined with biophysical information, such as the distribution of lipids and surface tension of HDL. Yetukuri et al. clarified related questions by studying the lipidomics of individuals with low or high HDL levels (Yetukuri et al. 2010). The average lipidomic profiles of these two groups, determined from clinical studies, were used in molecular dynamics simulations to consider how variations in molecular composition change the structural properties of HDL. Experiments showed that the particle size of HDL was smaller in the low-HDL group compared to the high-HDL group. Simulations supported this finding and also showed that in the low-HDL group the amount of triglycerides (TGs) at the surface of HDL particles increased, suggesting that the lipid distribution in low-HDL individuals could promote the increased catabolism of HDL particles. Interestingly it was also noted that free (unesterified) cholesterol accumulated next to apoA-I proteins. One plausible explanation for this observation is that cholesterol molecules prefer to stand next to apoA-I because of the high curvature of HDL: the highly curved surface would expose hydrophobic cholesterol molecules more to the polar phase than in planar bilayers.

In another study, lipidomics and molecular dynamics simulations were combined to study the effect of drug treatment on lipid and protein composition of HDL particles

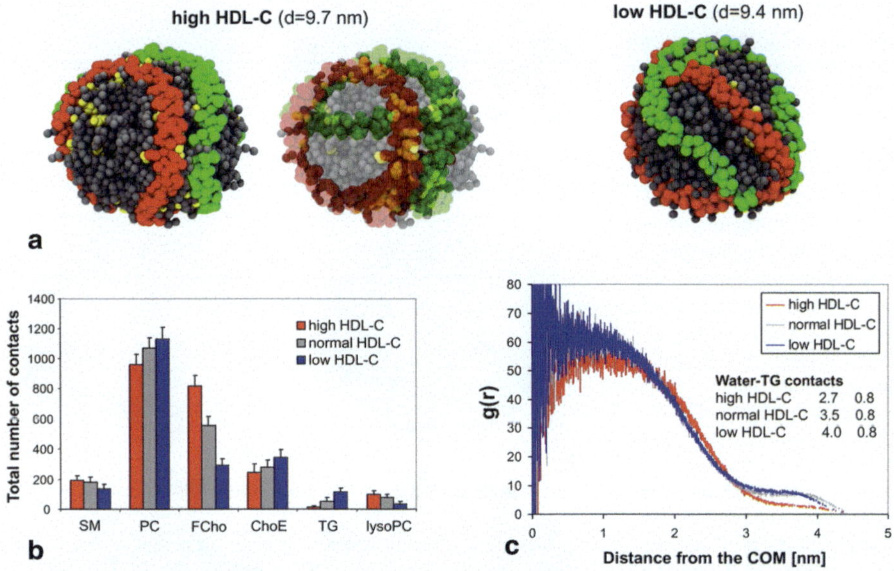

Fig. 15.6 Coarse-grained simulations of HDL particles reconstituted based on lipidomics data. **a** Snapshots from the end of high- and low-HDL simulations (8 CEs). Apo-AIs are colored with *red* and *green*, cholesterol molecules are *yellow*, and all other lipids are *gray*. Water phase was removed from the snapshots for clarity. The snapshot in the middle demonstrates how the cholesterol molecules are localized next to and under apoA-Is in high HDL-C simulation. **b** The number of contacts between apoA-Is and different lipid beads in each simulation (*error bars* indicate ± standard deviation). The number of contacts was not normalized with the number of different lipid constituents. **c** Radial distribution function for TG molecules g(r) with respect to the center of mass (COM) of HDL. When surface lipid concentration decreases, more TGs are able to penetrate the surface monolayer, which is noticed as formation of higher plateaus near 3.5 nm. The number of contacts between TG and water beads (per TG) in different simulations is shown in the inset. (Adapted from Yetukuri et al. 2010)

(Yetukuri et al. 2011). In this case, simulations demonstrated that the amount of triglycerides and cholesteryl esters (CEs) at the surface layer depends on the amount of apolipoproteins (Apos) at the surface. Especially, the novel role of apoA-II in HDL particle remodeling was demonstrated by the finding that the higher amount of apoA-II in HDL decreased the amount of core lipids at the surface of lipoparticles, thus changing core lipid metabolism in circulation.

15.4.2 Structure and Dynamics of LDL Particles

LDL particles are much larger than HDL (diameter varying between 21–25 nm), thus LDL particles are also computationally more demanding to simulate. In addition, apoB-100 that is the protein part of LDL is extremely complex. Its structure has not been solved despite years of substantial research aimed to elucidate the conformations and structural details of apoB-100 around LDL. Consequently the exact mechanism of how LDL particles interact with, e.g., the LDL-receptor is not known. Furthermore,

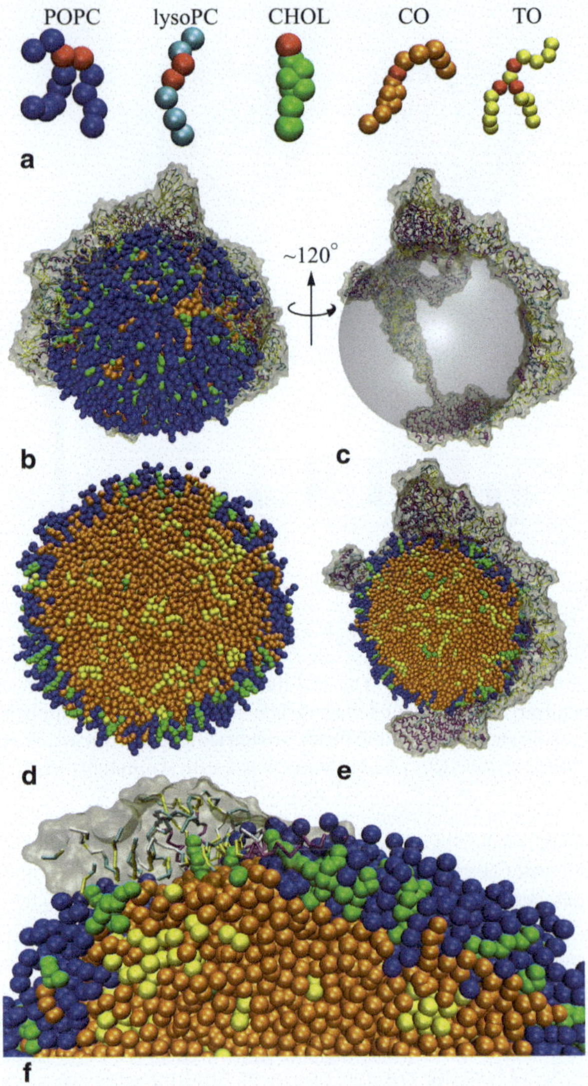

Fig. 15.7 LDL particle structure. **a** Lipids. Red beads show the location of glycerol (*PC, lysoPC*), hydroxyl (*CHOL*), and ester (*CO, TO*) groups. **b** Full LDL from the final structure. Lipids shown in similar colors as in panel A, without red highlights, and PC and lysoPC colored blue. ApoB-100 shown in *tan*, with the lipovitellin domain (first 1000 residues) pointing up. **c** Same as panel B, except that lipids are shown as transparent spheres. **d** LDL simulated without the protein, cut through the middle. Note how the core penetrates to the surface. **e** Same as panel B, except that the lipid part has been cut through the middle. The core penetrates to the lipid surface under the protein. **f** Close-up image of the surface from panel B. (Adapted from Murtola et al. 2011)

an important aspect in lipoprotein metabolism is the activation of apoB-100 during the conversion of VLDL particles to smaller LDL particles.

Starting almost from scratch, Murtola et al. constructed a model for apoB-100 by using bioinformatics tools to define the model structure for different domains of apoB-100, after which it was placed around a pre-equilibrated lipid droplet. This system was then simulated to study the interactions between lipids and apoB-100 (Murtola et al. 2011).

The results (Fig. 15.7) highlighted the importance of lipids in understanding LDL function. Not only the surface lipids but also the core lipids were fund to play an important role in LDL, as interactions between lipids and apoB-100 drive core lipids to the surface. As suggested by Segrest et al., core lipids and especially cholesteryl esters like to gather under the amphiphilic beta-sheet structures. This observation may be biologically significant in the transportation of core lipids by carrier proteins such as the cholesteryl ester transfer protein, which mediates the heteroexchange of cholesteryl esters and triglycerides between apoB-100 containing lipoproteins and HDL. The rather loose packing at the surface and the observed fast diffusion at the LDL surface compared to the core would promote this for lipids close to the surface layer.

Simulation results also showed that the distribution of lipids in LDL cannot be defined in terms of strict and well-defined layers, such as those for core and surface lipids, as usually have been suggested. Instead, the lipid regions are loosely defined. Yet, one can conclude that phospholipids are located at the surface, neutral lipids (triglycerides and cholesteryl esters) are almost entirely in the core, and finally cholesterol molecules reside close to the lipid-water interface between phospholipids and the core region. Regarding the enzymatic processes related to core lipid metabolism, the amount of core lipids at the surface of LDL may play an important role in lipoprotein metabolism. In general, the lipid composition governed by lipid distribution could play a major role in the function of LDL and likely varies between individuals with different diet and living habits. However, it is very difficult to associate this core lipid distribution to certain lipid type, or just to the ratio of core and surface lipids, as there are many factors that are compensating each other. Thus, it is very hard to clarify experimentally how much the core lipid distribution matters in lipoprotein metabolism. One way to clarify this issue would be to construct *in silico* models for different LDL species using lipidomics data from different individuals and link this structural information, e.g., to the different symptoms of MetS.

The above-discussed results provide a solid foundation to design follow-up simulations for elucidation of LDL functions. The models can be used to find, e.g., how enzymatic modifications and lipid oxidation due to free radicals change the molecular interactions in LDL particles, as these processes are known to promote the development of cardiovascular diseases.

15.5 Conclusions

Molecular simulations have become a technique which can provide experiments with substantial added value. First, the constantly developing computer resources and simulation technologies render consideration of realistic systems possible over time and length scales that are comparable to those studied in experiments. Second, the increasing scales imply that simulations can rapidly broaden their profile to new, more challenging phenomena and processes, using a broad arsenal of quantum-mechanical, atomistic, as well as coarse-grained molecular models. The greatest impact emerges from cases where these atomistic or semi-atomistic approaches are used to complement experiments, with an objective to clarify the mechanisms of exciting biomolecular processes, find the principles that govern those, and in general explain why the processes in question take place as they do. What is more, molecular simulations no longer just repeat what has been observed in experiments, but instead they also predict novel phenomena to be tested by experiments. Summarizing, the greatest impact in lipid research will be gained by combining experiments with simulations and theory, focusing their efforts together to unlock the most important questions in life sciences. As this strategy has recently been used more and more in the field, we are confident that also the research of lipid-related syndromes and diseases will benefit from this progress at an increasing pace.

Acknowledgments The Academy of Finland, the European Research Council (Advanced Grant CROWDED-PRO-LIPIDS), and the EU FP7 project ETHERPATHS are thanked for funding.

References

Adamian L, Naveed H, Liang J (2011) Lipid-binding surfaces of membrane proteins: evidence from evolutionary and structural analysis. Biochim Biophys Acta 1808:1092–1102

Aittoniemi J, Niemelä P, Hyvönen MT, Karttunen M, Vattulainen I (2007) Insight into the putative specific interactions between cholesterol, sphingomyelin, and palmitoyl-oleoyl phosphatidylcholine. Biophys J 92:1125–1137

Ayton GS, Voth GA (2009) Systematic multiscale simulation of membrane protein systems. Curr Opin Struct Biol 19:138–144

Cantor RS (1997) The lateral pressure profile in membranes: a physical mechanism of general anesthesia. BioChemistry 36:2339–2344

Cantor RS (1999) Lipid composition and the lateral pressure profile in bilayers. Biophys J 76:2625–2639

Chernomordik LV, Kozlov MM (2008) Mechanics of membrane fusion. Nature Struct Mol Biol 15:675–683

Contreras FX, Ernst AM, Haberkant P, Björlholm P, Lindahl E, Gönen B, Tischer C, Elofsson A, Heijne G von, Thiele C, Pepperkok R, Wieland F, Brügger B (2012) Molecular recognition of a single sphingolipid species by a protein's transmembrane domain. Nature 481:525–529

Donnert G, Keller J, Medda R, Andrei MA, Rizzoli SO, Lührmann R, Jahn R, Eggeling C, Hell SW (2006) Macromolecular-scale resolution in biological fluorescence microscopy. Proc Natl Acad Sci USA 103:11440–11445

Frenkel D, Smit B (2001) Understanding molecular simulation: from algorithms to applications. Academic Press, USA

Fuller N, Rand RP (2001) The influence of lysolipids on the spontaneous curvature and bending elasticity of phospholipid membranes. Biophys J 81:243–254

User Manual GROMACS (2010) Web-site: www.gromacs.org/Documentation/Manual/. Version 4.5.4

Gunsteren WF van, Bakowies D, Baron R, Chandrasekhar I, Christen M, Daura X, Gee P, Geerke DP, Glattli A, Hunenberger PH, Kastenholz MA, Ostenbrink C, Schenk M, Trzesniak D, Vegt NFA van der, Yu HB (2006) Biomolecular modeling: goals, problems, perspectives. Angew Chem Int Ed 45:4064–4092

Hall A, Rog T, Karttunen M, Vattulainen I (2010) Role of glycolipids in lipid rafts: a view through atomistic molecular dynamics simulations with galactosylceramide. J Phys Chem B 114:7797–7807

Hall A, Rog T, Vattulainen I (2011) Effect of galactosylceramide on dynamics of cholesterol-rich lipid membranes. J Phys Chem B 115:14424–14434

Hess B, Kutzner C, Spoel D van der, Lindahl E (2008) GROMACS 4: algorithms for highly efficient, load-balanced, and scalable molecular simulation. J Chem Theory Comput 4:435–447

Jensen MO, Mouritsen OG (2004) Lipids do influence protein function—the hydrophobic matching hypothesis revisited. Biochim Biophys Acta 1666:205–226

Kaiser HJ, Orlowski A, Rog T, Nyholm TKM, Chai W, Feizi T, Lingwood D, Vattulainen I, Simons K (2011) Lateral sorting in model membranes by cholesterol-mediated hydrophobic matching. Proc Natl Acad Sci 108:16628–16633

Kamerlin SCL, Vicatos S, Dryga A, Warshel A (2011) Coarse-grained (multiscale) simulations in studies of biophysical and chemical systems. Ann Rev Phys Chem 62:41–64

Koivuniemi A, Vattulainen I (2012) Revealing structural and dynamical properties of high density lipoproteins through molecular simulations. Soft Matter 8:1262–1267

Lingwood D, Simons K (2010) Lipid rafts as a membrane†organizing principle. Science 327:46–50

Lingwood D, Binnington B, Rog T, Vattulainen I, Grzybek M, Coskul U, Lingwood CA, Simons K (2011) Cholesterol modulates glycolipid conformation and receptor activity. Nature Chem Biol 7:260–262

Lopez CA, Rzepiela A, Vries AH de, Dijkhuizen L, Huenenberger PH, Marrink SJ (2009) The Martini coarse grained force field: extension to carbohydrates. J Chem Theory Comput 5:3195–3210

Louhivuori M, Risselada HJ, Giessen E van der, Marrink SJ (2010) Release of content through mechano-sensitive gates in pressurized liposomes. Proc Natl Acad Sci USA 107:19856–19860

Lundbak JA, Collingwood SA, Ingulfsson HI, Kapoor R, Andersen OS (2010) Lipid bilayer regulation of membrane protein function: gramicidin channels as molecular force probes. J Royal Soc Interf 7:373–395

Marrink SJ, Vries AH de, Mark AE (2004) Coarse grained model for semi-quantitative lipid simulations. J Phys Chem B 108:750–760

Marrink SJ, Risselada HJ, Yefimov S, Tieleman DP, Vries AH de (2007) The MARTINI forcefield: coarse grained model for biomolecular simulations. J Phys Chem B 111:7812–7824

Marsh D (2008) Protein modulation of lipids, and vice-versa, in membranes. Biochim Biophys Acta 1778:1545–1575

Monticelli L, Kandasamy SK, Periole X, Larson RG, Tieleman DP, Marrink SJ (2008) The MARTINI coarse grained force field: extension to proteins. J Chem Theory Comput 4:819–834

Mouritsen OG (2005) Life—as a matter of fat. Springer, Heidelberg

Murtola T, Bunker A, Vattulainen I, Deserno M, Karttunen M (2009) Multiscale modeling of emergent materials: Biological and soft matter. Phys Chem Chem Phys 11:1869–1892

Murtola T, Vuorela TA, Hyvonen MT, Marrink SJ, Karttunen M, Vattulainen I (2011) Low density lipoprotein: structure, dynamics, and interactions of apoB-100 with lipids. Soft Matter 7:8135–8141

Niemelä PS, Ollila S, Hyvönen MT, Karttunen M, Vattulainen I (2007) Assessing the nature of lipid raft membranes. PLoS Comput Biol 3:304–312

Nogi T, Fathir I, Kobayashi M, Nozawa T, Miki K (2000) Crystal structures of photosynthetic reaction center and high-potential iron-sulfur protein from Thermochromatium tepidum: thermostability and electron transfer. Proc Natl Acad Sci USA 97:13561–13566

Ollila OHS, Vattulainen I (2010) Lateral pressure profiles in lipid membranes: dependence on molecular composition. In: molecular simulations and biomembranes: for biophysics to function (Royal Society of Chemistry), pp 26–55

Ollila S, Hyvönen MT, Vattulainen I (2007) Polyunsaturation in lipid membranes: dynamic properties and lateral pressure profiles. J Phys Chem B 111:3139–3150

Ollila OHS, Risselada HJ, Louhivuori M, Lindahl E, Vattulainen I, Marrink SJ (2009) 3D pressure field in lipid membranes and membrane-protein complexes. Phys Rev Lett 102:078101

Palsdottir H, Hunte C (2004) Lipids in membrane protein structures. Biochim Biophys Acta 1666:2–18

Parton DL, Klingelhoefer JW, Sansom MSP (2011) Application of model membrane proteins, modulated by hydrophobic mismatch, membrane curvature, and protein class. Biophys J 101:691–699

Pietiläinen K, Rog T, Seppänen-Laakso T, Virtue S, Gopalacharyulu P, Tang J, Rodriguez-Cuenca S, Maciejewski A, Naukkarinen J, Ruskeepää AL, Niemelä P, Yetukuri L, Tan CY, Velagapudi V, Castillo S, Nygren H, Hyotylainen T, Rissanen A, Kaprio J, Yki-Järvinen H, Vattulainen I, Vidal-Plug A, Oresic M (2011) Association of lipidome remodelling in the adipocyte membrane with acquired obesity in humans. PLoS Biology 9:e1000623

Phillips JC, Braun R, Wang W, Gumbart J, Tajkhorshid E, Villa E, Chipot C, Skeel RD, Kale L, Schulten K (2005) Scalable molecular dynamics with NAMD. J Comput Chem 26:1781–1802

Ramadurai S, Holt A, Schafer LV, Krasnikov VV, Rijkers DTS, Marrink SJ, Killian JA, Poolman B (2010) Influence of hydrophobic mismatch and aminoacid composition on the lateral diffusion of transmembrane peptides. Biophys J 99:1447–1454

Risselada HJ, Marrink SJ (2008) The molecular face of lipid rafts in model membranes. Proc Natl Acad Sci USA 105:17367–17372

Rog T, Vattulainen I, Karttunen M (2005) Modeling glycolipids: take one. Cell Mol Biol Lett 10:625–630

Shaw DE, Maragakis P, Lindorff-Larsen K, Piana S, Dror RO, Eastwood MP, Bank JA, Jumper JM, Salmon JK, Shan Y, Wriggers W (2010) Atomic-level characterization of the structural dynamics of proteins. Science 330:341–346

Simons K, Sampaio JL (2011) Membrane organization and lipid rafts. Cold Spring Harb Perspect Biol 3:a0004697

Terama E, Ollila OHS, Salonen E, Rowat A, Trandum C, Westh P, Patra M, Karttunen M, Vattulainen I (2008) Influence of ethanol on lipid membranes: from lateral pressure profiles to dynamics and partitioning. J Phys Chem B 112:4131–4139

Vattulainen I, Rog T (2011) Lipid simulations: a perspective on lipids in action. Cold Spring Harb Perspect Biol. 3:a004655

Yanting W, Noid WG, Liu P, Voth GA (2009) Effective force coarse-graining. Phys Chem Chem Phys 11:2002–2015

Yeagle PL (2005) Structure of biological membranes. CRC Press, USA

Yetukuri L, Soderlund S, Koivuniemi A, Seppanen-Laakso T, Niemela PS, Hyvonen M, Taskinen MR, Vattulainen I, Jauhiainen M, Oresic M (2010) Composition and lipid spatial distribution of HDL particles in subjects with low and high HDL-cholesterol. J Lipid Res 51:2341–2351

Yetukuri L, Huopaniemi I, Koivuniemi A, Maranghi M, Hiukka A, Nygren H, Kaski S, Taskinen MR, Vattulainen I, Jauhiainen M, Oresic M (2011) High density lipoprotein structural changes and drug response in lipidomic profiles following the long-term fenofibrate therapy in the FIELD substudy. PLoS One 6:e23589

Chapter 16
Computational Statistics Approaches to Study Metabolic Syndrome

Ilkka Huopaniemi and Samuel Kaski

Abstract In this chapter, we review a set of key research problems and methods in analysing 'omics' data, gene expression, proteomics, metabolomics, and lipidomics. We start with the common systems biology approach to study metabolic syndrome, as well as any other disease, namely comparative case-control setting. The setting is usually an over-simplification, since there are other covariates that affect the concentrations of molecules, for instance drug treatments, gender, body mass index (BMI), and time in time-series experiments. Given these covariates, the setting becomes a multi-way experimental design. When multiple data sources are available, such as several 'omics' types, multiple tissues or multiple species, each forms a different data space with different molecules or variables, bringing in the problem of data integration. We start by giving a brief tutorial on the commonly used basic univariate and multivariate statistical approaches applicable if the problem is simplified by stratifying to a case-control design. We then focus on the multi-way setups of the Analysis of Variance (ANOVA) type, and in particular their main difficulty for 'omics' data: the large number of variables compared to the small number of observations. We introduce a recent family of Bayesian methods that is able to deal with multi-way, multi-source data sets and to translate biomarkers between multiple species. The approach is able to handle small sample-size combined with high dimensionality, and it allows a rigorous estimation of uncertainty of the results.

Keywords ANOVA · Data integration · Multi-way experimental design · Omics data · Statistical analysis

---

S. Kaski (✉)
Helsinki Institute for Information Technology (HIIT), Department of Information and Computer Science, Aalto University, Espoo, Finland
e-mail: samuel.kaski@aalto.fi

I. Huopaniemi
The Charles Bronfman Institute for Personalized Medicine,
The Icahn School of Medicine at Mount Sinai, New York, NY, USA
e-mail: ilkka.huopaniemi@gmail.com

Helsinki Institute for Information Technology (HIIT), Department of Information and Computer Science, Aalto University, Espoo, Finland

M. Orešič, A. Vidal-Puig (eds.), *A Systems Biology Approach to Study Metabolic Syndrome*, 319
DOI 10.1007/978-3-319-01008-3_16, © Springer International Publishing Switzerland 2014

16.1 Introduction

The common systems biology approach for studying diseases is a comparative case-control setting: two populations of samples are compared in order to infer what is different between them. In the case of metabolic syndrome, the cases can be for example patients with a diagnosed Type 2 diabetes and controls are healthy individuals. The high-throughput measurement techniques—metabolomics (lipidomics), proteomics, and gene expression—have become a standard tool for such analyses, and they all result in similar concentration-type, continuous-valued, high-dimensional data.

The motivation of such case-control settings is to discover potential biomarkers for disease and to find out physiological changes caused by the disease. If the samples in the case population have a statistically significantly higher (lower) concentration of a certain molecule, this is called an up (down)-regulation and is considered a potential biomarker for disease. Such a biomarker can then end up as a diagnostic marker for the disease after a careful clinical validation. A more advanced example is having patients who later developed into Type 1 diabetes as cases and patients who remained healthy as controls (Orešič et al. 2008), implying that differences between cases and controls are potential early biomarkers for Type 1 diabetes. A third example is studying treatment-response in omics data by having drug-treated patients as cases and placebo-treated patients as controls. In each of the three examples, various standard statistical tools can be used for the statistical testing of the significance of the up or down regulations.

The omics data have made it possible to profile a large number of molecules (variables) in search of the biomarkers, but they have also brought a serious, fundamental statistical difficulty to the core of the data analysis (Benjamini and Hochberg 1995; West 2003): whereas the number of variables (molecules) p is large, the number of samples (patients) n is often small due to economical and ethical reasons or simply because of a small number of patients available. The scientific community is currently in the process of searching for feasible computational approaches to deal with the ‘small n, large p’-conditions, conditions where the assumptions into which a century of work in traditional multivariate statistics have leaned, do not hold.

The ‘small n, large p’ problem permeates all omics data analyses. We will next introduce two other key problems that must be addressed in serious biomedical studies: multi-way experimental designs, stemming from there usually being multiple other covariates in the experimental design in addition to the case-control; and data integration, that is, analysis of measurements of multiple types. A common approach in systems biology has been to stratify the analysis according to the additional covariates and to analyze each data source separately. The different statistical problems that arise in studying a disease based on omics data are summarized in Fig. 16.1.

In Sect. 16.2, we will give a brief tutorial to the commonly used basic univariate and multivariate methods applicable to (stratified) single-source, case-control settings, and when the task is simply to find groups of similarly behaving molecules. The emphasis is on discussing the implications of the main challenge of analyzing omics data: small number of samples (patients) and large number of variables (molecules).

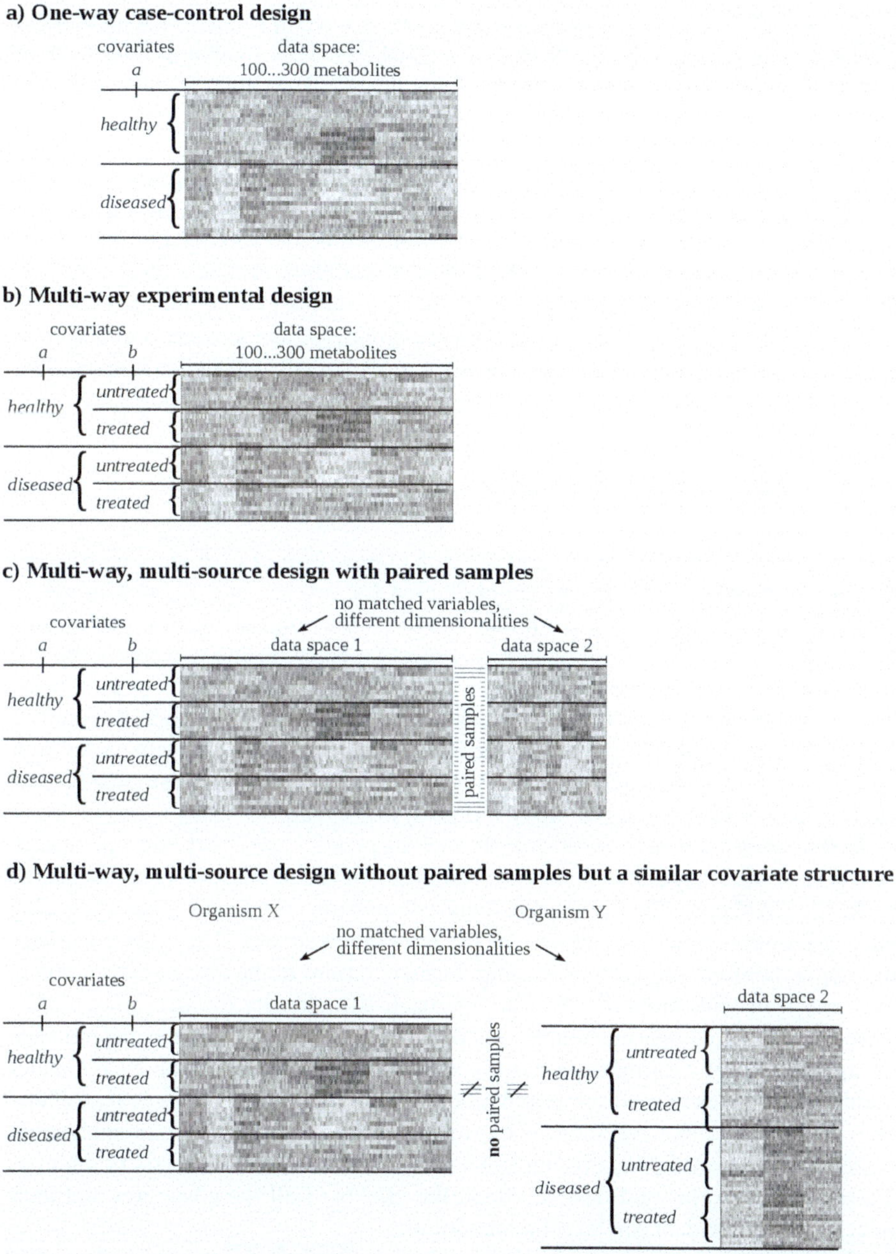

Fig. 16.1 Illustration of the four data analysis tasks in this chapter. **a** Case-control design, but with large dimensionality (variables) compared to number of samples (rows). **b** Multi-way experimental design. **c** Multi-way, multi-source design with paired samples. **d** Multiple data sources without paired samples (translation). The images represent data matrices, where rows are samples and columns are variables. The illustration represents the experimental design of each task, composed of a combination of covariates (disease, treatment) and integration of multiple data sources

In Sect. 16.3, we discuss modeling omics data under a multi-way experimental design as multi-way Analysis of Variance (ANOVA)-type modeling (and General Linear Models); the task here is to model the effects of all the covariates and their interactions. The experimental design may also be a time-series setting and the time-point can also be seen as a covariate.

Finally, in Sect. 16.4 we will discuss a state-of-the-art Bayesian approach for analyzing multi-way, multi-source data. The advantage of the Bayesian approach is that one can build a single unified model for the whole problem involving multiple covariates and multiple data sources. In a Bayesian model, rigorous uncertainty estimates are inherently obtained, which is crucial when the number of samples is small. We will also discuss an extension of the Bayesian approach and a recent non-Bayesian approach for translating biomarkers between model organisms and human studies.

16.1.1 Multi-way Experimental Design

The conceptualization of an experiment as a comparison of case and control populations is in many cases an oversimplification for two reasons: (1) in biomedical experiments there are usually confounding effects that might bias the results unless dealt with properly, and (2) in many experiments it is interesting to actually find the effects of multiple covariates in the data, and additionally the effects of their interactions, which are usually the most interesting effects. A covariate here is simply a variable that annotates a sample. Each sample (patient) is associated with one or multiple covariates, such as disease status or treatment, and the usual task of statistical analysis is to find the effect of the covariate(s) in the data.

In human disease studies, covariates such as gender, age, BMI, drug treatment, or race can have large effects on concentrations of metabolites or other measured molecules. Although the diseased-healthy comparison is here the most interesting question, disease status is only one of the relevant covariates. The other covariates may act as confounding factors, biasing the associations to the disease effect, unless they are dealt with properly. A basic approach is to stratify the data into smaller sets, for instance comparing healthy and diseased males and females in different age groups separately. Stratification has the side effect that it leads to an even smaller number of samples in each sub-population, worsening the ' small n large *p* problem' and, additionally, interpretation of the results of the multiple separate analyses is more tedious. It would be advantageous to formulate the data analysis problem as a multi-way experimental design, where the effects of all the covariates and their interactions in the data can be estimated jointly.

On the other hand, in many experiments the covariates are not seen as confounding but the main research question is to study the effects of many relevant covariates and their interactions. One of the most common biomedical experiment types is study of the effects of various interventions or other factors such as drug, diet, oncogene, knock-out, or the effects of gender and age, as additional risk factors for disease. Treatment groups and other relevant descriptors are therefore a common covariate in experimental designs. For instance, in multi-way experiments on diseased-healthy, drug intervention (such as Huopaniemi et al. 2010a), a direct drug effect may be interpreted as a drug side effect, and the interaction of drug and disease is actually the interesting effect, indicating whether the drug cures the disease. As another example, patients of different genders have been shown to have different omics profiles (Nikkilä et al. 2008; Mostertz et al. 2010), and the interaction effect of disease and gender may be interesting for determining whether different genders have differential disease effects or differential responses to drugs.

In summary, many clinical or biomedical settings can be formulated as multi-way experimental designs, where observations (samples or individuals) are divided into populations of measurements according to multiple covariates.

16.1.2 Measuring and Integrating Multiple Data Sources

There is a growing trend in biomedical research to measure multiple omics data sources from each individual, for the very good reason that the different sources are biologically complementary and the prices of the measurements are becoming tolerable. Other cases where multi-source omics data are measured are measurements from multiple tissues in the same individual, and measurements from both model organisms and men. In the latter case the goal is to translate biomarkers between humans and model organisms, and data sources come from different individuals. In each case, (at least partly) different molecules are measured in the different data sources, which results in an interesting statistical challenge since the data spaces are different.

Multiple data sources from each individual There are two types of applications for multi-tissue experiments: (1) The practical application of predicting the state of a disease tissue from another, more easily collectable tissue, such as blood plasma. The relevant research question then is whether there are shared disease-related effects between blood and disease tissue molecules, and it would be of great diagnostic and scientific interest to be able to map target tissue molecules to blood plasma molecules in the hope that their disease-dependent associations would be preserved in the mapping. (2) More generally, it is informative to search for disease-related changes in multiple tissues, to obtain a more holistic view of the physiological or biological state of the individual.

From the statistics point of view, the task in analyzing both multi-tissue measurements and multiple omics types is integration of data from multiple sources with paired samples and different variable spaces. Pairing here means that multiple sources have been measured from the same patient. Different omics types (mRNA, proteins, metabolites) clearly have different variables and we also assume that different tissues in general have different variables even if the omics type is the same; there might be partially same chemical molecules, but these might have different roles in different tissues.

Although each data source can be (and unfortunately usually is) analyzed with a reductionist approach, that is, separate analysis by standard statistical tools, such analyses miss many of the multivariate effects while producing a huge number of partial results. Interpretation of such results gets tedious when the number of sources and covariates grows. It would be desirable to complement the toolbox with more powerful methods.

Translating biomarkers between humans and model organisms Other organisms are often used in biomedical research as disease models, and drug effects are also often tested on model organisms in pre-clinical phases before proceeding to human clinical studies. A further need for model organisms is that since biopsies are usually difficult or impossible to obtain from humans, multi-tissue studies are mostly limited to model organisms.

A particularly relevant research question is whether disease- and treatment-related findings done in a model organism actually have a correspondence in human clinical studies (Salek et al. 2007; Seo et al. 2007; Sysi-Aho et al. 2011). As different species do not in general have fully the same lipids or proteins, for example, and chemically similar biomolecules may have different roles in different species, a new statistical modeling problem emerges: how to translate findings made in a model organism into human clinical studies, and how to map a disease-related human metabolome to a mouse metabolome (Sysi-Aho et al. 2011), for instance. Such an analysis can be also used to evaluate the validity of a model organism as a model for the disease (Damian et al. 2007). From the statistics point of view, the data analysis problem here is integrating multiple data sources without paired samples.

16.2 Basic Tools Commonly Used in the Exploratory Statistical Analysis of Omics Data

In this chapter, we review briefly tools that can be used when the underlying multi-way (and multi-source) experimental design has been simplified by stratification to a case-control or unsupervised setting in one data source. Most omics data are simply continuous-valued multivariate data, and hence many of the widely known classical tools are applicable after suitable pre-processing. The data typically are relative or absolute concentrations of molecules (usually in log-scale), collected into a data matrix where rows are samples (patients) and columns are variables (molecules). Each sample is annotated by a covariate, such as class label disease or healthy, according to the experimental design. Here we concentrate on exploratory analysis, which aims at hypothesis generation based on the data, for instance at finding potential biomarkers.

Statistical analysis can be done either one variable at time (univariate analysis) or taking into account their relationships (multivariate analysis). Another division of methods is into supervised methods or unsupervised methods. Supervised methods are usable for predicting values of covariates, and applicable to diagnostics or analyzing what is different between the case and control populations. Unsupervised methods do not take the covariate information into account, but they can be used, for example, to discover unknown subgroups of patients or groups of similarly behaving molecules. We first describe the nature of omics data from the point of view of statistics. We then review the most commonly used methods, including univariate statistical tests, unsupervised multivariate methods such as PCA and clustering, and supervised multivariate methods: classification, regression and testing of known groups.

16.2.1 Data and Statistical Challenges

In many respects omics data are just standard continuous-valued data, but what is special is the ubiquitousness of the 'small n, large p'-condition, where n is the number of samples and p the number of variables. The condition causes several problems for data analysis. First, it is difficult to interpret high-dimensional data by visual inspection and hence dimensionality reduction is necessary. Second, univariate testing evokes the multiple testing problem of inflating significance thresholds. Third, most multivariate methods run into technical problems, such as overfitting or singularity of the covariance matrix.

Another problem in omics data are that they are often multicollinear, meaning that there are groups of highly correlated variables (genes, proteins, or metabolites) because of the existence of biochemical networks and pathways regulated similarly. The highly correlated groups of variables result in redundancy of information and, therefore, multicollinearity is a problem for some standard methods. On the other hand, searching for the correlated groups of variables is biologically interesting. Some methods can deal with multicollinear data, and some methods can even take advantage of it.

16.2.2 Univariate Analysis

An univariate statistical test is the basic tool for answering whether two populations, case and control, differ. The Student t-test, one-way Analysis of Variance (ANOVA), and Wilcoxon rank sum test are the most common such tests.

In technical terms, a statistical test tests whether the null hypothesis holds; the simplest null hypothesis is that there is no difference between the means of the one variable in the two populations. The statistical test gives a p-value, which indicates the probability to observe a difference equal to or larger than the actually observed value, under the null distribution where there is no difference and data are drawn from the same population at random. If the p-value is smaller than the a priori defined threshold of 0.05, the null hypothesis is rejected and there is said to be a statistically significant difference.

Univariate statistical tests and p-values are a widely accepted means to measure the statistical significance of the difference of a variable between two populations, and they are the basic approach for detecting potential single-variable biomarkers for disease. A common threshold is $p = 0.05$, but tighter thresholds, such as $p = 0.01$, 0.005 or 0.001 are also used to decrease the chance of detecting **false positives**, erroneously rejected null hypotheses, where there was a difference between the variables by chance. If one has to test only one or a few variables, univariate tests are a well-working approach.

The testing of statistical significance is often accompanied by a study of 'fold change', the actual magnitude of the difference in the variable values between the two populations. Using these two measures together (McCarthy and Smyth 2009) is a more reliable approach for detecting biomarkers, since a statistically significant difference between two populations may not be biologically meaningful if the actual fold change is very small.

Unfortunately, omics data are very high-dimensional (the number of molecules is large), which implies big statistical problems for the univariate approach of testing each variable at time. The reason is the problem of multiple testing: setting a threshold, say $p = 0.05$, for rejecting a null hypothesis means a probability of 5 % of finding a false positive in each single test, and hence testing, say, 10,000 variables from gene expression data results in approximately 500 false positives. This is obviously unacceptable in biomedical data analysis. This problem is commonly alleviated by modifying the p-value to account for the number of statistical tests. False Discovery Rate (FDR) (Benjamini and Hochberg 1995) is the most common multiple testing correction used. Bonferroni correction is an older method, that is too conservative (makes the threshold too low to find markers in practice) when the dimensionality is high.

Another problem is that univariate analysis of high-dimensional data unfortunately only results in long lists of p-values and the similarity between variables (genes, proteins or metabolites functioning together) is not modeled in any way. The need for multivariate modeling, and need for dimensionality reduction stemming from multicollinearity in omics data, are widely acknowledged. The most common measure of similarity between two variables is correlation, usually measured by Pearson correlation.

16.2.3 Clustering

Clustering is a basic statistical tool for partitioning objects to a set of groups of similar objects. One can cluster either samples or variables depending on the data analysis question. Since multicollinearity of omics data is a known fact, clustering the variables (genes, proteins and metabolites) is a standard way of finding similarly behaving clusters of biological molecules and reducing redundancy of information in the data. The most commonly used clustering tools are hierarchical clustering (Ward 1963) and K-means clustering, and Pearson correlations and Euclidean distance are common choices for similarity or distance measures.

Clustering of the patients can be relevant as well. For example, in a population of patients with a metabolic syndrome, such a clustering can be used to detect subtypes of metabolic syndrome. The groups of patients can further be related to clinical variables. An example method for clustering patients from an omics experiment is (Damian et al. 2007).

16.2.4 Unsupervised Component Models: PCA

Principal Component Analysis (PCA) is the most commonly used dimensionality reduction method. It enables detecting similarly behaving groups of variables in the multicollinear omics data, and also enables visualization of the relatedness of samples (patients). PCA is very often used as the first basic analysis, since it does not require using the covariates but it serves as 'the first look at the data'. PCA can also be used in the 'small n, large *p*'-conditions and it greatly reduces the redundant information in multicollinear omics data.

PCA projects the original high-dimensional data onto a few principal components. A principal component is a single score computed for each sample (patient), forming one axis and compressing a large number of correlated variables. It is common to plot a two-dimensional representation of the scores of the two most important principal components to illustrate how different samples (patients) are related. Such a plot could reveal for example that patients with a high blood pressure are closer to patients with a diagnosed type 2 diabetes than to healthy patients with normal blood pressure. Note that PCA is unsupervised, meaning that the covariate information (healthy, diseased) is not used in constructing the principal components, but can be used later to interpret the visualization.

PCA is useful if there is a large number of correlated variables (redundant information), such as in omics data. PCA forms a loading matrix that relates variables to the principal components and the loading matrix can be directly used to assess which variables are correlated.

There are two problems in using PCA. First, since PCA is unsupervised, the association of data to covariates is not taken into account when forming the components, and hence there is no guarantee that it will find biologically relevant variation in the data. For example, if PCA cannot separate diabetic and healthy patients it does not necessarily mean that there is no difference between them. Second, when grouping variables standard PCA will result in each component having a non-zero loading for all of the original variables. This makes interpretation of the components somewhat difficult when there is a large number of variables. This problem has been a motivation for developing sparse approaches, where most loadings are set to zero, and only the most significant loadings are retained.

16.2.5 Supervised Models

Supervised modeling means predicting the value of one variable given other variables. The variable to be predicted can here be one of the covariates; classification then refers to modeling the association of (omics) data to discrete covariates (class labels), such as (healthy, disease) or (treated, untreated). Regression differs from classification in that the variable to be predicted is continuous-valued; below we will focus on classification and refer the reader to statistics textbooks for specifics of regression.

Supervised classification methods can be (and are often) used for multivariate modeling of the association of variables to a discrete covariate. The idea is to train a classifier model that maximally separates (at the simplest) two populations of data points (classes) and can predict the class label of a new sample with an unknown class label.

Classification accuracy is a measure of how well the classifier can predict the class labels of new test samples, and it is a widely accepted measure of the association of data and the covariate. The accuracy is measured by dividing the samples in the data set with known class labels into a training set and test set. The training set is used to build the classifier, which is then used to predict the class labels of the test samples, and the proportion of successes determines the classification accuracy. Many classifiers are interpretable in the sense that it is straightforward to report which variables influence the discrimination between the classes the most.

Some classifiers do dimensionality reduction by forming supervised latent factors when learning the classifier from data. The idea is the same as in PCA scores, but the latent factors have been constructed taking the association with the covariate into account. This latent factor-space is used for the classification task, but the components can also be studied for interpretations. Partial Least Squares Discriminant Analysis (PLS-DA) (Wold et al. 1984) and its advanced version Orthogonal Projection in Latent Structures OPLS (Trygg and Wold 2002) are among the most commonly used classifiers for omics data, since they can form such latent factors in 'small n, large *p*'—conditions and can deal with multicollinear data. Linear discriminant Analysis (LDA) is a similar classical method, but it is based on inverting a data covariance matrix which becomes singular in the 'small n, large *p*'-conditions. Hence a regularization scheme, such as in (Tai and Pan 2007), is needed to use LDA on omics data.

Variable selection is another approach for dimensionality reduction for classifiers, and it also enhances interpretability by highlighting the most significant separating variables. Variable selection is a good approach to search for a small set of significant variables. If the separation between the classes is due to a correlated group of variables (redundant information), variable selection may not work well in the sense that the selected variables may not be stable. Any of the redundant variables is almost equally good explanation, and different variables may get selected in different trials.

Some of the most popular classifiers for omics data using variable selection are Decision Trees and their extension Random Forest (Breiman 2001), as well as LASSO (Tibshirani 1996). Support Vector Machines (SVM) (Vapnik 1995) and K-Nearest Neighbors (KNN) (Tan et al. 2006), on the other hand, are examples of well-performing classifiers where interpretation may be harder.

Using classifiers for biomarker discovery Interpretable classifiers have become a popular method for biomarker discovery, since the standard tool of univariate statistical testing is not a satisfactory approach for high-dimensional data, and there are no widely used multivariate statistical tests for high-dimensional data. Classifiers enable multivariate modeling of the variables contributing to the differences between case and control classes. The justification for using them for biomarker discovery is that variables that have predictive value for a class label (case, control), can be assumed to be biologically relevant. There are, however, several problems in this approach, of which the reader should be aware of.

First, it is important to realize that in classification, the task is to find a sufficient distinction between different classes, which can be achieved by finding only a few, most strongly discriminative variables. When classification is combined with variable selection or assumptions of sparsity, the goal is explicitly to find a minimal set of variables. Then all the relevant variables are not necessarily found.

Second, overfitting is a serious problem for supervised multivariate methods, especially when the number of samples is small and dimensionality high. Overfitting means that a supervised method searches for a maximal separation between the classes in the learning data set, but when there is a large number of variables and small number of samples, the model can fit to whatever variables it can find, which can be simply noise. Serious care to guard against overfitting has to be taken. Resampling methods, such as cross-validation, bootstrapping (Efron 1979), or permutation testing, are a commonly used tool to estimate the uncertainty of the results and therefore guarding against overfitting. Resampling means that one divides the data into training data and test data multiple ($\sim$ 1,000–10,000) times. This is done randomly in bootstrapping, and deterministically in cross-validation. In permutation testing, the classifier is compared to multiple models that are obtained by permuting the class labels randomly; the random classifiers form a null distribution from which the real classifier needs to distinguish itself. The idea underlying these uncertainty estimation methods is to accept only such discriminative variables that are consistently found after these validation steps, which helps considerably against overfitting. Also, classification accuracy needs to be evaluated by resampling to get a reliable estimate.

When using classifiers that form a supervised latent factor space, such as PLS-DA and OPLS, one has to very cautious (Westerhuis et al. 2008) if considering to use the latent factors for further analysis, such as visualization or doing further statistical testing on them. If the same data which were used for estimating the latent factor space are visualized or represented in the resulting latent factor space, the class separability will seem over-optimistic; testing classifiers on learning data will give too optimistic results, drastically so in 'small n, large p'—conditions.

16.2.6 Testing Known Groups

Another commonly used multivariate approach which could perhaps be considered supervised, is testing the joint differential expression of *a priori* known sets of variables. This means simply that one defines known groups of, say, genes that are known to belong to the same pathway, and tests whether the genes in this group are differentially expressed. The standard methods are Gene Set Enrichment Analysis (GSEA) (Subramanian et al. 2005) and Gene Set Analysis (Efron and Tibshirani 2007), and the concept has also been extended for metabolomics data (Kankainen et al. 2011). In an advanced extension (Wang et al. 2008), multi-way experimental designs were taken into account.

The disadvantages of testing known sets are that the a priori defined sets may not capture the interesting biological effects; a correlated cluster may also be only a subset from a gene set (Wu et al. 2009), or composed of subsets from several gene sets. An additional, exploratory drawback is that by testing known sets, no new and potentially interesting sets can be found.

16.2.7 Summary

As a summary, whereas unsupervised methods are at a risk of not retaining biologically relevant covariate-related variation when forming latent factors, supervised methods model the most strongly separating direction in the data and are at a risk of overfitting—especially for small sample-sizes which are typical to omics data.

Another important question that we do not discuss further is model complexity selection. When using any component model, supervised or unsupervised, the choice of how many components are chosen is crucial, yet this is in fact a very difficult, fundamental question and subject to considerable research effort. To point some keywords, searching a cutoff for the number of components according to a percentage of variance explained, is the classical approach for the model complexity selection of unsupervised methods. In classifiers, classification accuracy is often chosen as such a measure: when adding components does not improve classification accuracy, the optimal number of components has been found. More advanced measures include predictive likelihood, Akaike information criterion (AIC), and Bayesian Information Criterion (BIC) and their more recent variants.

16.3 Modeling Data Under a Multi-way Experimental Design

Most of the current omics studies have inherently a multi-way experimental design where, in addition to the case-control comparison, there are additional covariates. Examples include drug treatments, age, gender, BMI, and time in a time-series setting. These additional covariates can be either confounding factors, or it can be interesting to explore the effects of all the covariates and their interactions in the data.

In this chapter, we discuss ANOVA-type analysis, which is the standard approach for analyzing data from multi-way experimental designs. We begin by discussing the univariate multi-way ANOVA, and its multivariate generalization MANOVA. We then explain how these are special cases of the framework of General Linear Models, which is followed by a discussion on Linear Mixed Models. We then discuss the problem of ANOVA and MANOVA in the case of 'small n, large *p*', and proposed solutions that combine a PCA dimensionality reduction and further (M)ANOVA modeling.

16.3.1 ANOVA and MANOVA

The univariate Analysis of Variance (ANOVA) (Fisher 1918) was the first and is currently the most widely known method for the task of modeling the effects of multiple covariates and their interactions in populations of measurements. Multi-way ANOVA is an extension of the one-way ANOVA and Student t-test. The central assumption of multi-way modeling is an additive generative model for explaining the observed covariate-related variation in the data. In the case of two covariates, the equation is

$$\mathbf{x}_j|_{(a,b)} = \mu + \alpha_{\mathbf{a}} + \beta_{\mathbf{b}} + (\alpha\beta)_{ab} + \varepsilon \tag{16.1}$$

Here $\mathbf{x}_j$ is the observed data point for sample j, a and b ($a = 0 \ldots A$ and $b = 0 \ldots B$) are the two independent covariates, such as disease and treatment, and the main effects α_a and β_b and the interaction effect $(\alpha\beta)_{ab}$ model the variation from the baseline level (grand mean) μ. Despite the confusing notation, the interaction effect $(\alpha\beta)_{ab}$ is just another term: the interaction effect of disease and treatment, for example. The ε is a noise term.

Whereas a one-way ANOVA or t-test would only solve the statistical significance of one term, α_a, the result of two-way ANOVA is the statistical significance (p-value) of all the terms α_a, β_b, and the interaction effect $(\alpha\beta)_{ab}$ simultaneously. Using a t-test in the case of a multi-way design results in comparing all the pairwise combinations, which results in tedious interpretation of the results.

If α_a or β_b can only have dichotomous values, such as healthy/diseased or male/female, they are called binary covariates. A covariate can also have multiple levels, such as multiple time points or multiple drug treatments.

There can be an arbitrary number of covariates and therefore main effects, but when the number of covariates grows, the number of interactions grows exponentially. Interpreting the results then becomes tedious and massively more data are needed to estimate the effects reliably.

Multivariate ANOVA (MANOVA) is a multivariate generalization of ANOVA. The MANOVA model is the same as the ANOVA model in Eq. (16.1), except that all the terms are vectors, whereas in the univariate ANOVA-model they are scalars. A Bayesian formulation of ANOVA was presented in (Gelman et al. 2003).

16.3.2 General Linear Model

ANOVA and MANOVA are special cases of the General Linear Model (GLM), which is a general term for the generative model

$$\mathbf{X} = \mathbf{DB} + \text{noise} \tag{16.2}$$

Where $\mathbf{X}$ is the data matrix, $\mathbf{D}$ is a design matrix of known covariates, and $\mathbf{B}$ are the regression coefficients. The statistical significance of the elements of $\mathbf{B}$ can be used as a measure of which covariates in **D** explain the data in **X**. GLM is applicable to both univariate and multivariate data and for discrete and continuous-valued covariates. ANOVA is the special case of GLM in the case of discrete covariates and univariate data; in MANOVA the data are multivariate. In the case of a mixture of discrete and continuous covariates, the corresponding methods are called Analysis of Covariance (ANCOVA) and Multivariate Analysis of Covariance (MANCOVA). The Student t-test, F-test and linear regression are also special cases of GLM. Generalized Linear Model is a generalization of GLM that additionally includes non-linear relationships between data and covariates, such as the logistic function in logistic regression. A century of work in classical statistics has been dedicated to studying the properties of GLM in its various forms. The work has mostly assumed univariate or low-dimensional data and large number of samples, whereas in omics data the opposite case holds: high dimensionality and small number of samples.

16.3.3 Linear Mixed Models and Time-series Modeling

Whereas the standard (M)ANOVA-type linear model consists of standard fixed effects, such as diseases, treatments and time-points, Linear Mixed Models (LMM) extend (M)ANOVA-type analysis such that it can also include random effects. The fixed effects refer to effects such as treatments, that have been chosen to the experimental design and are relevant regarding the chosen research question. Random effects are uninteresting effects that 'model out' uninteresting but real effects in the data, such as repetitions on an individual (Wolfinger et al. 2001), gene-specific effects in a clustering model (Celeux et al. 2005; Ng et al. 2006), or confounding factors due to a population structure (Listgarten et al. 2010). The equation for linear mixed models is often written as

$$\mathbf{X} = \mathbf{DB} + \mathbf{VU} + \text{noise} \tag{16.3}$$

which extends the fixed effects model with the inclusion of random effects **V**. The **U** are the regression coefficients from random effects to the data. Whereas the fixed effect design matrix **D** indeed consists of fixed zeros and ones that relate known covariates to the samples, the random effects **V** are assumed to have been sampled from a Gaussian distribution, and they have to be learned as well.

LMM's have been a popular choice for modeling time-series omics data. In ANOVA-type or LMM-modeling of time-series data, the time-point is usually the fixed effect. A particularly interesting, multivariate application of LMM:s, is clustering genes in time-series omics data (Celeux et al. 2005; Ng et al. 2006). In this approach, cluster-specific time-course is the interesting fixed effect, and the gene-specific variation from the time-course, gene belonging to the cluster, is a random effect. This is also a model-based clustering approach, where the cluster-specific time-course becomes the 'model'. The random effects are used here to model out uninteresting gene-specific time-effects, which helps in forming the clusters.

Beyond the possibilities of ANOVA-type modeling, time-series modeling of omics data is a big field of research of its own (Bar-Joseph 2004), where the central research question is how to optimally take the time-series nature of data into account and predict future time-points.

As an example, one recent approach (Sysi-Aho et al. 2011) tackled the fact that individuals develop to different age-related metabolic states at an individual pace, and age may not be a relevant descriptor *per se* to find common metabolic development states. Hidden Markov Models were used to align time-series omics data into unobserved latent metabolic states. Such an abstraction of underlying metabolic states can help to deal with additional technical constraints of (1) missing time points that are common in clinical follow-up studies, and (2) humans and mice have very different life spans, yet the time-series results from multiple species have to be compared or integrated.

16.3.4 Problems of ANOVA and MANOVA

The general challenges of univariate statistical tests for high-dimensional data also hold for multi-way ANOVA. The problem of multiple testing can be alleviated with multiple testing corrections, such as FDR, and there are also more advanced test statistics developed for multi-way experimental designs (Smyth 2004; McCarthy and Smyth 2009). Access to the information of correlated clusters can be sought by clustering the variables before the analysis, or by grouping the variables according to the p-values obtained from the univariate statistical tests.

MANOVA is the multivariate generalization of ANOVA and defines a formal multivariate statistical test for testing the effects of covariates on populations of measurements, taking correlations between the variables into account. MANOVA was originally designed for low-dimensional data and in $n > p$ conditions. Unfortunately, in 'small n, large p' conditions, ubiquitous in omics data, the covariance (correlation) matrix becomes singular and MANOVA cannot be used. In addition, MANOVA tests the difference between two populations in terms of all the variables simultaneously, and therefore only gives the statistical significance of the overall effect. This result is not sufficiently informative for high-dimensional data, and the highly relevant information of which variables were up/down-regulated has to be deduced by other methods, such as univariate tests. Furthermore, as ANOVA and MANOVA only give the statistical significance of the effect, the direction, up or down, has to be deduced by other means.

The univariate ANOVA and multivariate MANOVA are two extremes for how to solve the multi-way modeling problem, both of them facing both technical and interpretational problems due to the 'small n, large p'- conditions. Few efforts have been presented for multivariate multi-way modeling, essentially solving the general linear model in 'small n, large p' conditions. It is obvious that a compromise between testing a single variable (ANOVA) and all the variables simultaneously (MANOVA) needs to be sought, and the common modeling choice is a dimensionality reduction into low-dimensional latent factor (principal component) space, where the statistical testing is done.

16.3.5 Multivariate Many-step Approaches

To our knowledge, no methods for multivariate multi-way modeling of high-dimensional data with a single model exist in classical statistical literature. One working solution is a many-step approach, PCA-based dimensionality reduction is done first, and the effects of covariates are then studied with statistical tests in the reduced-dimensional principal component space. Examples are ANOVA on the PCA scores (Bratchell 1989), 50-50 MANOVA (Langsrud 2002) where a MANOVA test is done on the PCA scores, and ANOVA-Simultaneous Component Analysis (ASCA) (Smilde et al. 2005).

16.3.6 Multi-class Classification

A multi-way experimental design can in principle be naively considered as a multi-class classification problem, where each combination of covariates, such as 'diseased treated', is considered as an individual class. Even standard classifiers, such as PLS-DA and LDA, can be used in this sense. An example of converting a data analysis problem with a multi-way, multi-source design into a series of multi-class PLS-DA classifiers, each data source at a time, was presented in (Webb-Robertson et al. 2009). In this approach, one naturally loses the information that samples with 'treatment 1 in early time-point' and samples with 'treatment 1 in late time-point' are related due to both having 'treatment 1'. Multi-class classification cannot be used easily to estimate the statistical significance of the effects of multiple covariates and their interactions.

16.4 Bayesian Multi-way, Multi-source Modeling

We have recently developed Bayesian models for omics data with a multi-way, multi-source design. We start this Chapter by introducing briefly the advantages of using Bayesian models and present our approach for multivariate multi-way modeling in the single-source case. We then continue with a short overview of existing approaches to integrating multiple data sources, followed by our Bayesian approach for multi-way, multi-source data that directly tackles the question of how to take the multi-way experimental design into account. Finally, we present an extension of the Bayesian model to the case of translating biomarkers between multiple species.

16.4.1 Bayesian Modeling

The Bayesian modeling paradigm is an increasingly popular method of choice in statistical analysis. The main advantage of using full Bayesian inference is that a joint probability distribution for all the model parameters is defined, which can be used directly as a rigorous uncertainty estimate of the results in form of confidence intervals. Using Bayesian statistics is especially advantageous when the number of available data samples is small, since a principled uncertainty estimation is crucial in such conditions.

Another advantage of Bayesian modeling is that it allows a principled way of combining several modeling contributions, models or components, into a unified hierarchical model. The uncertainty estimation for the whole model is done jointly, and uncertainty estimates propagate appropriately across the model. This is a main improvement over many-step approaches such as PCA + MANOVA. In our case, we need to combine proper dimensionality reduction, multi-way ANOVA-type modeling, and multi-source integration.

16.4.2 Bayesian Multivariate Multi-way Modeling

Our approach to Bayesian multivariate multi-way modeling, introduced in (Huopaniemi et al. 2009; Huopaniemi 2012), is the following: We take advantage of the known fact that there are groups of correlated variables in omics data, and use clusteredness of data as a main assumption which leads to implicit dimensionality reduction. This enables to overcome both the 'small n, large *p*' problem and multicollinearity. We essentially state that effects of covariates should be modeled on groups of variables, instead of single variables. We then use a Factor Analysis (FA)-type approach (similar to PCA) where each latent factor models one cluster of variables. The effects of multiple covariates and their interactions are modeled on the latent factors, each factor representing a group of correlated variables. This is done according to the following generative model:

$$\mathbf{x}_{j\ (a,b)}^{lat} \sim \mathrm{N}\left(\boldsymbol{\alpha}_a + \boldsymbol{\beta}_b + (\boldsymbol{\alpha\beta})_{ab}, \mathbf{I}\right)$$

$$\mathbf{x}_j = \mathrm{N}(\boldsymbol{\mu} + \mathbf{V}\mathbf{x}_{\mathbf{j}}^{\mathbf{lat}}, \Lambda) \tag{16.4}$$

Here, $\mathbf{x}_{\mathbf{j}}^{\mathbf{lat}}$ is a low-dimensional latent factor for individual *j*, $\mathbf{V}$ is a projection matrix that projects the latent-space variation into the actual observed data space, and Λ is diagonal noise.

This is intuitively analogous to doing an ANOVA or MANOVA analysis on the PCA scores; the added benefit is that the full Bayesian inference allows us to propagate uncertainties between the model parts, resulting in uncertainty estimates that take all sources of uncertainty into account. The technical details of the model are discussed in (Huopaniemi et al. 2009). In Fig. 16.2, an example of results is presented, obtained from a lipidomic data set with a two-way setting: healthy/diabetic, male/female. This method was also applied on another multi-way lipidomic data set in (Yetukuri et al. 2011), where the effects of Fenofibrate and placebo where studied on patients with Type 2 diabetes from the FIELD study.

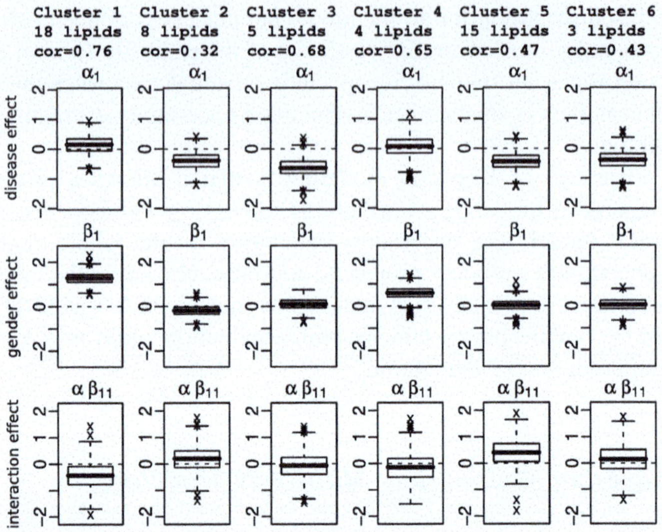

Fig. 16.2 An example of the results from applying the Bayesian multi-way model on a lipidomic dataset with a diseased/healthy, male/female setting. Cluster 3 was down-regulated as a result of disease, and for lipids in clusters 1 and 4, females had a higher concentration. No significant gender-dependent disease effects were found. For each cluster of correlated variables, the statistical significance of the effects of disease, gender, and their interaction is estimated by the posterior distribution being above (below) zero with 95 % confidence indicates an up (down)-regulation. (Reprinted with kind permission from Springer: Huopaniemi et al. 2009 Fig. 4)

Another, related Bayesian approach that uses sparse components in the context of multi-way designs is Bayesian sparse latent factor regression models (West 2003; Carvalho et al. 2008). The work includes a Bayesian approach for the multiple testing problem of univariate multi-way ANOVA-models, together with means of how to supervise latent factors to be predictive of multiple classification or regression problems.

16.4.3 Basic Approaches for Multi-source Data

The standard approach for dealing with omics data coming from multiple tissues or multiple omics types is to analyze each data source separately. As a step towards proper data integration, a few supervised (one-way) and unsupervised approaches have been developed. The unsupervised approaches, such as sparse Canonical Correlation Analysis (CCA) (Parkhomenko et al. 2007) and O2-PLS (Trygg and Wold 2003; Rantalainen et al. 2006) attempt to answer what is shared between the data sources. Another approach (Sysi-Aho et al. 2007; Le Cao et al. 2009; Monni and Tadesse 2009) is regression from one data source to one variable in the other data source at time; as an example, one can do regression by PLS from all the protein concentrations to predict the concentration of each lipid, one at a time. The disadvantage of unsupervised approaches is that they do not take covariates into account in any way. The basic supervised (one-way) approach is to teach a classifier in each data source separately and combine the results, and such integration has been shown to improve classification accuracy over a single data source (Webb-Robertson et al. 2009; Le Cao et al. 2010). In the case of a multi-way design, one has to either stratify the analysis to a (one-way) case-control setting, or predict each covariate at time.

16.4.4 The Bayesian Multi-way, Multi-source Model

We have recently introduced a data integration model for multi-source omics data that tackles directly multi-way experimental designs (Huopaniemi et al. 2010a). The model extends our Bayesian single-source model, having the same dimensionality reduction component, based on clusters of correlated variables. The multiple data sources can be integrated by a CCA-type shared latent variable, which allows a decomposition of the covariate-related variation into shared- and source-specific effects. For example, in Huopaniemi et al. (2010a) we applied the method on a lipidomic lung cancer/healthy, treated/untreated design, where measurements were taken from multiple tissues. We found a shared cancer effect and a shared treatment effect between plasma and lung tissues. We therefore concluded that there was a disease- and treatment-related connection between the tissues.

16.4.5 Translating Biomarkers Between Species

We have also studied the task of translating biomarkers between human studies and model organisms from experiments that have a multi-way experimental design. We have, in particular, concentrated on the case where different species have different variables, and in other words there is no a priori known matching between human and mouse lipidomes, for example.

Previous works on cross-species modeling of omics data (Seo et al. 2007; Lu et al. 2009; Le and Bar-Joseph 2010) have focused on gene expression, where matching between the genes in multiple species is mostly known on the basis of similarity of the DNA sequence (orthologous genes). In the case of lipidomics, for example, the task is to actually find such a matching between the molecules of multiple species from the omics data in a disease context. We have approached this problem with a CCA-type model (Sysi-Aho et al. 2011; Tripathi et al. 2011) and as a Bayesian multi-way, multi-source modeling approach (Huopaniemi et al. 2010b; Suvitaival et al. 2011). The latter approach was developed for the case of a similar multi-way experimental design in both species, such as a longitudinal follow-up study (Sysi-Aho et al. 2011) of patients who later developed into type 1 diabetes, and patients who remained healthy. Again, the definition of integrating the data sources, in other words translation, is a similar response to multiple covariates and their interactions in the two data sources (species) that can be modeled by shared latent effects. The advantages of this approach are that there is a clear interpretation of what defines a translation, and a joint uncertainty estimation of the whole model.

16.5 Conclusions

We have discussed the statistical analysis of omics data, which is often a main difficulty in the systems biology approach to study metabolic syndrome, as well as any other disease. In such a setting, one usually faces the question of having high-dimensional, small sample-size data with a multi-way experimental setting, in which there are confounding factors and other interesting covariates, such as interventions, in addition to the diseased-healthy comparison. The data may additionally come from different data sources, such as multiple omics types, multiple tissues or multiple species, each source having different variables. Due to methodological difficulties, the data analysis is unfortunately often done with a reductionist approach: stratifying the analysis and analyzing each data source separately.

There is a great need of novel statistical tools for proper data integration and dealing with multi-way experimental designs. Developing such methods is currently an active research topic in the machine learning and statistics communities, in addition to the question of how to deal with the problem of high-dimensionality and small sample-size.

We have recently introduced a family of Bayesian methods for modeling multi-way, multi-source omics data. The advantages of the Bayesian approach is that one can build a unified hierarchical model that can deal with the data integration, multi-way modeling and dimensionality reduction jointly. A rigorous uncertainty estimation for the whole model is directly obtained, which is crucial with small sample-sizes.

References

Bar-Joseph Z (2004) Analyzing time series gene expression data. Bioinformatics 20:2493–2503

Benjamini Y, Hochberg Y (1995) Controlling the false discovery rate: a practical and powerful approach to multiple testing. J R Stat Soc Series B Methodol 57:289–300

Bratchell N (1989) Multivariate response surface modeling by principal component analysis. J Chemom 3:579–588

Breiman L (2001) Random forests. Mach Learn 45(1):5–32

Carvalho C, Chang J, Lucas J, Nevins J, Wang Q, West M (2008) High-dimensional sparse factor modeling: applications in gene expression genomics. J Am Stat Assoc 103:1438–1456

Celeux G, Martin O, Lavergne C (2005) Mixture of linear mixed models for clustering gene expression profiles from repeated microarray experiments. Stat Model 5:243–267

Damian D, Orešič M, Verheij E, Meulman J, Friedman J, Adourian A, Morel N, Smilde A, Greef J van der (2007) Applications of a new subspace clustering algorithm (COSA) in medical systems biology. Metabolomics 3:69–77

Efron B (1979) Bootstrap methods: another look at the jackknife. Ann Stat 7:1–26

Efron B, Tibshirani R (2007) On testing the significance of sets of genes. Ann Appl Stat 1:107–129

Fisher R (1918) The correlation between relatives on the supposition of mendelian inheritance. Royal Society of Edinburgh from Transactions of the Society Vol 52, pp 399–433

Gelman A, Carlin JB, Stern HS, Rubin DB (2003) Bayesian Data Analysis, 2nd edn. Chapman & Hall/CRC, Boca Raton

Huopaniemi I (2012) Multivariate Multi-Way Modelling of Multiple High-Dimensional Data Sources. PhD thesis. Aalto University School of Science, Espoo, Finland

Huopaniemi I, Suvitaival T, Nikkilä J, Orešič M, Kaski S (2009) Two-way analysis of high-dimensional collinear data. Data Min Knowl Discov 19:261–276

Huopaniemi I, Suvitaival T, Nikkilä J, Orešič M, Kaski S (2010a) Multivariate multi-way analysis of multi-source data. Bioinformatics 26:i391–398

Huopaniemi I, Suvitaival T, Orešič M, Kaski S (2010b) Graphical multi-way models. In: Balcázar J, Bonchi F, Gionis A, Sebag M (eds) Machine learning and knowledge discovery in databases. Proceedings of European Conference, ECML PKDD 2010, Barcelona, Spain, September 20–24, 2010, vol. 1, Springer, Berlin, pp 538–553

Kankainen M, Gopalacharyulu P, Holm L, Orešič M (2011) MPEA-metabolite pathway enrichment analysis. Bioinformatics 27:1878–1879

Langsrud O (2002) 50-50 multivariate analysis of variance for collinear responses. J R Stat Soc Series D-the Stat 51:305–317

Le Cao K-A, Martin P, Robert-Granie C, Besse P (2009) Sparse canonical methods for biological data integration: application to a crossplatform study. BMC Bioinformatics 10:34

Le Cao K-A, Meugnier E, McLachlan GJ (2010) Integrative mixture of experts to combine clinical factors and gene markers. Bioinformatics 26:1192–1198

Le H-S, Bar-Joseph Z (2010) Cross species expression analysis using a Dirichlet process mixture model with latent matchings. In: Lafferty J et al. (eds) Advances in Neural Information Processing Systems 23, MIT Press, Cambridge, pp 1270–1278

Listgarten J, Kadie C, Schadt E, Heckerman D (2010) Correction for hidden confounders in the genetic analysis of gene expression. Proceedings of the National Academy of Sciences

Lu Y, Huggins P, Bar-Joseph Z (2009) Cross species analysis of microarray expression data. Bioinformatics 25:1476–1483

McCarthy DJ, Smyth GK (2009) Testing significance relative to a fold-change threshold is a treat. Bioinformatics 25:765–771

Monni S, Tadesse M (2009) A stochastic partitioning method to associate high-dimensional responses and covariates. Bayesian Anal 4:413–436

Mostertz W, Stevenson M, Acharya C, Chan I, Walters K, Lamlertthon W, Barry W, Crawford J, Nevins J, Potti A (2010) Age and sex-specific genomic profiles in nonsmall cell lung cancer. J Am Med Assoc 303:535–543

Ng SK, McLachlan GJ, Wang K, Ben-Tovim Jones L, Ng S-W (2006) A mixture model with random-effects components for clustering correlated gene-expression profiles. Bioinformatics 22:1745–1752

Nikkilä J, Sysi-Aho M, Ermolov A, Seppänen-Laakso T, Simell O, Kaski S, Orešič M (2008) Gender dependent progression of systemic metabolic states in early childhood. Mol Syst Biol 4:197

Orešič M, Simell S, Sysi-Aho M, Nanto-Salonen K, Seppänen-Laakso T, Parikka V, Katajamaa M, Hekkala A, Mattila I, Keskinen P, Yetukuri L, Reinikainen A, Lähde J, Suortti T, Hakalax J, Simell T, Hyöty H, Veijola R, Ilonen J, Lahesmaa R, Knip M, Simell O (2008) Dysregulation of lipid and amino acid metabolism precedes islet autoimmunity in children who later progress to type 1 diabetes. J Exp Med 205:2975–2984

Parkhomenko E, Tritchler D, Beyene J (2007) Genome-wide sparse canonical correlation of gene expression with genotypes. BMC Proceedings, vol 1, p S119

Rantalainen M, Cloarec O, Beckonert O, Wilson ID, Jackson D, Tonge R, Rowlinson R, Rayner S, Nickson J, Wilkinson RW, Mills JD, Trygg J, Nicholson JK, Holmes E (2006) Statistically integrated metabonomic-proteomic studies on a human prostate cancer xenograft model in mice. J Proteome Res 5:2642–2655

Salek RM, Maguire ML, Bentley E, Rubtsov DV, Hough T, Cheeseman M, Nunez D, Sweatman BC, Haselden JN, Cox RD, Connor SC, Griffin JL (2007) A metabolomic comparison of urinary changes in type 2 diabetes in mouse, rat, and human. Physiol Genom 29:99–108

Seo DM, Goldschmidt-Clermont PJ, West M (2007) Of mice and men: sparse statistical modelling in cardiovascular genomics. Ann Appl Stat 1:152–178

Smilde AK, Jansen JJ, Hoefsloot HCJ, Lamers R-JAN, Greef J van der, Timmerman ME (2005) ANOVA-simultaneous component analysis (ASCA): a new tool for analyzing designed metabolomics data. Bioinformatics 21:3043–3048

Smyth G (2004) Linear models and empirical Bayes methods for assessing differential expression in microarray experiments. Stat Appl Genet Mol Biol 3: No 1, Article 3

Subramanian A, Tamayo P, Mootha VK, Mukherjee S, Ebert BL, Gillette MA, Paulovich A, Pomeroy SL, Golub TR, Lander ES, Mesirov JP (2005) Gene set enrichment analysis: a knowledge based approach for interpreting genome-wide expression profiles. Proc Natl Acad Sci Unit States Am 102:15545–15550

Suvitaival T, Huopaniemi I, Orešič M, Kaski S (2011) Cross-species translation of multi-way biomarkers. In: Honkela T, Duch W, Girolami M, Kaski S (eds) Proceedings of the 21st International Conference on Artificial Neural Networks (ICANN), Part I, vol 6791 of Lecture Notes in Computer Science, Springer, pp 209–216

Sysi-Aho M, Vehtari A, Velagapudi V, Westerbacka J, Yetukuri L, Bergholm R, Taskinen M-R, Yki-Järvinen H, Orešič M (2007) Exploring the lipoprotein composition using bayesian regression on serum lipidomic profiles. Bioinformatics 23:i519–528

Sysi-Aho M, Ermolov A, Gopalacharyulu PV, Tripathi A, Seppänen- Laakso T, Maukonen J, Mattila I, Ruohonen ST, Vähätalo L, Yetukuri L, Härkönen T, Lindfors E, Nikkilä J, Ilonen J, Simell O, Saarela M, Knip M, Kaski S, Savontaus E, Orešič M (2011) Metabolic regulation in progression to autoimmune diabetes. PLoS Comput Biol 7:e1002257

Tai F, Pan W (2007) Incorporating prior knowledge of gene functional groups into regularized discriminant analysis of microarray data. Bioinformatics 23:3170–3177

Tan P-N, Steinbach M, Kumar V (2006) Introduction to data mining. Pearson Addison-Wesley

Tibshirani R (1996) Regression shrinkage and selection via the lasso. J R Stat Soc Series B 58:267–288

Tripathi A, Klami A, Orešič M, Kaski S (2011) Matching samples of multiple views. Data Min Knowl Discov 23:300–321

Trygg J, Wold S (2002) Orthogonal projections to latent structures (o-pls). J Chemom 16:119–128

Trygg J, Wold S (2003) O2-pls, a two-block (xy) latent variable regression (lvr) method with an integral osc filter. J Chemom 17:53–64

Vapnik V (1995) The nature of statistical learning theory. Springer

Wang L, Zhang B, Wolfinger RD, Chen X (2008) An integrated approach for the analysis of biological pathways using mixed models. PLoS Genet 4:e1000115

Ward J (1963) Hierarchical grouping to optimize an objective function. J Am Stat Assoc 58:236–244

Webb-Robertson B-JM, Mccue LA, Beagley N, Mcdermott JE, Wunschel DS, Varnum SM, Hu JZ, Isern NG, Buchko GW, Mcateer K, Pounds JG, Skerrett SJ, Liggitt D, Frevert CW (2009) A Bayesian integration model of high-throughput proteomics and metabolomics data for improved early detection of microbial infections. Pac Symp Biocomput 2009:451–463

West M (2003) Bayesian factor regression models in the large p, small n paradigm. Bayesian Stat 7:723–732

Westerhuis J, Hoefsloot H, Smit S, Vis D, Smilde A, Velzen E van, Duijnhoven J van, Dorsten F van (2008) Assessment of PLSDA cross validation. Metabolomics 4:81–89. doi:10.1007/s11306-007-0099-6

Wold S, Ruhe A, Wold H, Dunn WJ (1984) The collinearity problem in linear regression. The partial least squares (pls) approach to generalized inverses. SIAM J Sci Stat Comput 5:735–743

Wolfinger R, Gibson G, Wolfinger E, Bennett L, Hamadeh H, Bushel P, Afshari C, Paules R (2001) Assessing gene significance from cDNA microarray expression data via mixed models. J Comput Biol 8(6):625–637

Wu MC, Zhang L, Wang Z, Christiani DC, Lin X (2009) Sparse linear discriminant analysis for simultaneous testing for the significance of a gene set/pathway and gene selection. Bioinformatics 25:1145–1151

Yetukuri L, Huopaniemi I, Koivuniemi A, Maranghi M, Hiukka A, Nygren H, Kaski S, Taskinen M-R, Vattulainen I, Jauhiainen M, Orešič M (2011) High density lipoprotein structural changes and drug response in lipidomic profiles following the long-term fenofibrate therapy in the FIELD substudy. PLoS One 6:e23589

Chapter 17
Towards Modeling of Metabolic Syndrome: Tissue Crosstalk in Lipid Spillover

Sergey Smirnov, Eugeny Metelkin, Nail Gizzatkulov, Oleg Stepanov and Oleg Demin

Abstract In this chapter we propose a framework for development of the mathematical model of Metabolic Syndrome. The cornerstone of the model is "adipose tissue expandability hypothesis", which treats long-term increase in FAs concentration in blood and their subsequent accumulation in muscle and liver as a main cause of decrease in insulin sensitivity and, consequently, MetS development. The key biological facts underlying the construction of MetS model have been collected and analyzed and the main requirements of the model have been formulated. To take these facts into account and meet the requirements formulated, we have developed a novel algorithm enabling us to combine kinetic models, which describe changes in metabolite (lipoproteins) concentrations due to biochemical transformations in blood and tissues, with the mathematical model of the molecular movement in blood flow.

Keywords Lipid metabolism · Lipid spillover · Kinetic modeling · Very low density lipoprotein

17.1 Introduction

Metabolic syndrome (MetS) develops through a long and complex process, which is associated with several organs and tissues. Several experimental data values characterizing various aspects of MetS have been simulated and collected. However, mathematical models describing the development of MetS, and for example, its progression to type 2 diabetes (T2D), have not yet been developed. To a large extent, this is due to the inherent complexity of the processes and regulatory mechanisms related to MetS. To correctly describe the development of MetS, a mathematical model needs to account for the crosstalk between different organs and tissues at various levels such as physiological and biochemical levels.

S. Smirnov (✉) · E. Metelkin · N. Gizzatkulov · O. Stepanov · O. Demin
Institute for Systems Biology SPb, Leninskie Gory 1/75, Moscow, Russia, 119992
e-mail: smirnov@insysbio.ru

M. Orešič, A. Vidal-Puig (eds.), *A Systems Biology Approach to Study Metabolic Syndrome*, 341
DOI 10.1007/978-3-319-01008-3_17, © Springer International Publishing Switzerland 2014

One of the main factors in MetS development is a decrease in tissue sensitivity to insulin (Albu et al. 2010; Felber and Golay 2002; Golay et al. 1984; He et al. 2010; Lin et al. 2009; Muscelli et al. 2008; Pastucha et al. 2010; Pigeon et al. 2009). We use this fact as a basis for our MetS model. However, to construct such a model we first need to identify an experimentally confirmed mechanism that would allow us to establish the relationship between obesity and a decrease in insulin sensitivity (Czech et al. 2010; Katz et al. 2000; Prentki and Madiraju 2011; Stannard and Johnson 2004; Wu et al. 2010).

The aim herein is to outline a framework for further development of the mathematical model describing the occurrence of MetS and its progression to T2D within the conceptual framework of the adipose tissue expandability hypothesis (AEH) (Chap. 1.1). For this purpose, first, we developed a model to describe lipid spillover, which is the non-homogeneous concentration of fatty acids (FA) and lipoproteins in the circulatory system. To account for the space-time distribution of FAs and lipoproteins, we developed a new approach that allows us to combine models describing intra- and extra- cellular processes of biomolecules and lipoprotein particle transformation, where the models describe the transport of blood, molecules, and lipoproteins in the circulatory system.

17.2 Modeling Framework: Description of the Biological System and Model Requirements

Based on AEH, the development of MetS can be associated with obesity and it consists of the following events.

In the first stage of the development of obesity, we assume that the expandability of the adipose tissue is high. This means that excess FAs can be stored in adipose tissue. The term "excess FA" refers to all FAs that originate in consumed food or are produced from glucose in the liver but have not been utilized in energy production. As obesity develops (i.e., patient weight increases), the expandability of adipose tissue decreases. If the energy supply resulting from food intake continues to be higher than the energy consumption required to sustain the organism, then the rate of influx of FAs starts to exceed the rate of FA storage in adipose tissue. This causes the concentration of FAs in the blood to increase and FAs to accumulate in certain tissues such as those of the muscle and liver. Accumulation of FAs in these tissues results in a decrease in their sensitivity to insulin. Muscle cells can utilize both glucose and FAs to produce the energy required to maintain intracellular processes and performance. This means that an increase in FA concentration may result in a decrease in glucose consumption, and as a consequence, a decline in the insulin-dependent transport of glucose in muscles.

Thus, to develop a model that satisfactorily describes MetS on the basis of AEH, we need to account for the following aspects of the disease (Armoni et al. 2007; Bajaj et al. 2002; Everett-Grueter et al. 2006; Hirasawa et al. 2005; Kelley and Mandarino 2000; Kim et al. 2007; Lewis et al. 2002; Schnell et al. 2007; Staehr et al. 2003):

(i) transformation of glucose in the organism
(ii) lipid spillover
(iii) energy balance in the organism
(iv) tissue crosstalk underlying the interconnections between (i), (ii), and (iii).

Transformation of glucose in an organism. Glucose entering the organism through food can either be utilized to produce energy or it can be stored as glycogen in liver and muscle tissues and as FAs in adipose tissue. If glucose uptake exceeds glucose utilization for energy production over a long period, then all excess glucose is transformed into FAs, which are then stored in adipose tissue. The alternative method of glucose storage (as glycogen) cannot compete with the process of glucose transformation into FAs since the glycogen storage capacity of liver and muscle tissues is extremely low. This means that the ability to accumulate excess glucose in the form of glycogen is quickly weakened.

If the rate of glucose uptake exceeds the sum of the rates of (i) glucose consumption to produce energy and (ii) glucose transformation into FAs, then the glucose level in blood increases until it is balanced with the glucose clearance in urine. If the dominance of glucose intake over total glucose consumption occurs over a long period, then the glucose level in blood remains considerably high all the time. This is the main symptom of MetS (or T2D). In conclusion, we can state that MetS (or T2D) is a long-lasting state of an organism when the glucose intake rate permanently exceeds the total rate of its consumption to produce energy, resulting in the storage of glucose as FAs in adipose tissue.

Lipid spillover. Chap. 2.6 describes in detail the processes involved in FA transformation, transport, and homeostasis. Based on this description, we conclude that the fate of FAs entering the organism is highly similar to that of glucose. FAs can either be utilized to produce energy or be stored in adipose tissue. If the rate of FA uptake exceeds the sum of the rates of (i) FA consumption to produce energy and (ii) FA storage in adipose tissue, then the FA level in blood increases, i.e., excess FAs are observed. In accordance with AEH, excessive amounts of FAs in the organism are responsible for the development and progression of MetS.

In the version of the model presented below, we neglect many aspects of lipid spillover that are described in Chap. 2.6. For example, we do not distinguish between VLDL1 and VLDL2 and we do not take into account that there are other lipoproteins involved in lipid spillover, such as IDL and HDL.

Requirements for development of MetS model. Based on the description of the biological system presented above, we can prepare a list of requirements for the development of the MetS model, as follows:

- The model should reproduce the transport of glucose and FAs between different organs and tissues through biological fluids (tissue crosstalk).
- The model should reproduce glucose and FA dynamics for different diet and food intake regimens.

- The model should reproduce the dynamics of the glucose and FA exchange that is dependent on the energy balance (difference between the energy from consumed food and that utilized to meet energy needs) in the organism.
- The model should describe changes in the glucose and FA blood concentration levels resulting from the passage of blood through a selected organ/tissue (or part of the organ/tissue).

17.3 Model Development

As we have indicated in previous sections (see also Chap. 2.6), the main items considered in the MetS development model are the glucose and the FAs attached to various carriers. This means that the model development can be subdivided into three stages:

- development of the lipid spillover model
- development of the glucose homeostasis model
- merging of the lipid spillover and glucose homeostasis models

In this section, we describe the development of the lipid spillover model. Development of the glucose homeostasis model is not presented in this chapter. In accordance with the description in Chap. 2.6, there are several organs/tissues actively involved in lipid spillover. These are the liver, gastrointestinal tract, blood, muscles, and adipose tissue. Underlying lipid spillover is the intensive crosstalk between organs and tissues. The term "tissue crosstalk" refers to the mechanism by which one organ/tissue can influence the operation of another organ/tissue via excretion and transport of active components that can be absorbed and metabolized by the target organ/tissue. Tissue crosstalk is necessary for the cooperation between tissues, which results in homeostasis of glucose and FAs in an organism. The following types of crosstalk are involved in lipid spillover: *liver-adipose*, *liver-muscle*, and *adipose-muscle*.

To be able to describe tissue crosstalk and meet all the requirements discussed above, the MetS model should consider the complete organism as a set of compartments that exchange biological fluids (e.g., blood and lymph) and consist of biological entities (e.g., lipoproteins, molecules, and cells).

To build the compartmental structure, we define a "route" as a set of compartments that are sequentially located. Each route corresponds to a set of blood vessels located in an organ or part of an organ and describes the blood flow through the organ without branching or merging into the bloodstream.

A route is characterized by the following basic features:

- blood flow velocity along the route (i.e., the volume of the blood passing through the route per unit time)
- total volume of blood in a route (i.e., the sum of blood volumes in all compartments of the route at any given time)
- diameter of the blood vessels along the route

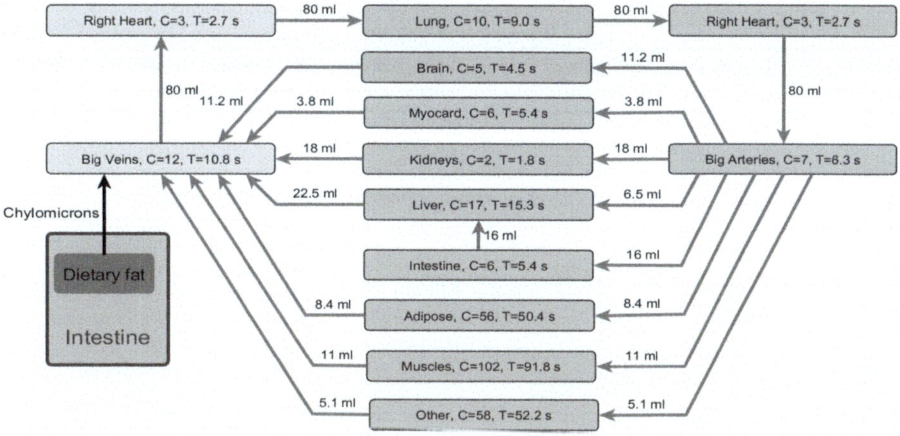

Fig. 17.1 Schematic of routes taken in lipid spillover model (physiological module). The circulatory system routes in the lipid spillover model are represented in this schematic. The number above each arrow indicates the volume of blood traveling through the routes per second (i.e., per cardiac beat). The following information is indicated in the rectangles: (i) name of the route, (ii) number of compartments included in the route (C), and (iii) residence time of the blood portion in the route (T)

On the basis of these characteristics, we can calculate the residence time of the blood in the compartments of a route. The residence time is equal to the ratio of the total volume of blood in the route and the blood flow velocity along the route.

The number of compartments in a route should be selected such that it is possible to reproduce both the anatomical location of various organs/tissues in an organism and their linkage via the circulatory system. Further, there should be a sufficient number of compartments to enable a detailed description of the space-time distribution of the biological entities (lipoproteins, molecules, or cells) along the route. Moreover, the number of compartments should be kept as low as possible to reduce the complexity of the model. To meet these requirements, we selected a compartment volume such that the blood residence time of the compartment was equal to one second. Therefore, the number of compartments in a route is equal to the blood residence time of the route, expressed in seconds. However, since the residence time and total blood volume differ for each route, the number of compartments and their volumes also differ.

Splitting the circulatory system into a set of routes is based on a specific biological task, and therefore, it changes for each model. The route schematic for the lipid spillover model is given in Fig. 17.1. A detailed description of the routes is presented in Table 17.1.

To model biochemical reactions that occur in either biological fluids (blood or lymph) or various organs/tissues interconnected via the circulatory system, we developed the step-by-step algorithm (SBSA). Within the framework of this algorithm, movement of blood along the route is combined with changes in concentrations of transported biological entities (molecules and lipoproteins), which is a result of

Table 17.1 Routes and compartment structure of model

Route	Model #	Number of compartments	Compartment volume (ml)	Transient time (sec)	Total blood volume in route (ml)
Right heart	No model[a]	3	80	2.7	240
Lung	No model[a]	10	80	9	800
Left heart	No model[a]	3	80	2.7	240
Big arteries	No model[a]	7	80	6.3	560
Brain	No model[a]	5	11.2	4.5	56
Myocard	Model 3	6	3.8	5.4	22.8
Kidneys	No model[a]	2	18	1.8	36
Liver	Model 4	17	22.5	15.3	382.5
Intestine	No model[a]	6	16	5.4	96
Adipose	Model 2	56	8.4	50.4	470.4
Muscles	Model 3	102	11	91.8	1122
Other	No model[a]	58	5.1	52.2	295.8
Big veins	Model 1	12	80	10.8	960

[a] "No model" means that only transfer of molecules with blood flow without any chemical transformation occur in these organs. Therefore, any kinetic model was not defined for these organs

chemical reactions and interactions at the walls of the blood vessels, in the following manner:

- Blood movement along a route is divided into discrete steps. The duration of each step corresponds to the blood residence time in a compartment.
- For the duration of a single step, blood is considered an immovable fluid. Changes in concentrations of biological entities in the selected compartment result from biochemical reactions and interactions at cells of only the vessel wall.
- At the end of a single step, the content of the *i-th* compartment is transported infinitely fast to the *(i+ 1)-th* compartment. This is continued in subsequent steps.

Physical implementation of SBSA In this subsection, we focus on a system that can be represented by a network of interconnected compartments with fluid flowing through them. In this system, there are molecules (or particles), which can (1) react with each other in accordance with stoichiometry and reaction rates, (2) interact with immovable parts of the system (for example, the compartment walls), and (3) be transferred from one compartment to another with the fluid.

As a model of fluid flow, we consider an incompressible fluid that fills the complete volume of the compartments and has a laminar flow. Each point in the fluid flows at the same rate. The equation describing changes in metabolite (particle) concentrations in compartment s can be written as

$$\frac{\partial \overline{C}_s}{\partial t} + r_s \frac{\partial \overline{C}_s}{\partial x} = \widehat{N}_s \overline{v}_s + \overline{u}_s \tag{17.1}$$

Here, $\overline{C}_s$ is the vector of metabolite concentrations, r_s is the flow rate through the compartment s (if there is no flow in a compartment, then r = 0), $\widehat{N}_s$ is the stoichiometric matrix, $\overline{v}_s$ is the vector of reaction rates, $\overline{u}_s$ is the vector of the influxes of molecules/particles from outside (for example, drug administration), t is time, and x is the variable describing the transfer along the compartment(s) resulted from the flowing fluid. The boundary condition describing the junction between compartments can be written as

$$r_s \frac{\partial \overline{C}_s}{\partial x}(0,t) = \sum\nolimits_j r_j \frac{\partial \overline{C}_j}{\partial x}(x_{\mathrm{max,j}}, t) \tag{17.2}$$

where index j corresponds to all compartments that supply the flowing fluid to compartment s. This boundary condition imposes the law of conservation of molecules/particles at the border between each compartment. The law of conservation of flowing fluid is given by

$$r_s = \sum_j r_j$$

Thus, the mathematical model (described by Eqs. 1 and 2) represents a set of advection equations interconnected with boundary conditions describing the continuity of the flowing fluid and the topology of the linked compartments.

To solve the system numerically, we have (1) assumed that the characteristic time for chemical reactions is much less than that of fluid transfer and (2) used an upwind differencing scheme for the construction of finite difference equations on the basis of differential equations (1 and 2).

We subdivide each compartment into smaller parts, allowing us to neglect the concentration gradient of volume $\Delta \mathrm{x} = \mathrm{V}_f$ where V_f is the fixed volume of the sub-compartment. Using the selected difference scheme, we obtain

$$\frac{\partial \overline{C}_s}{\partial x} \approx \frac{\overline{C}_f - \overline{C}_{f-1}}{V_f} \tag{17.3}$$

Where f is the index of the current sub-compartment of compartment s. Taking into account Eq. 1 and the boundary condition (Eq. 2), we arrive at

$$\frac{d\overline{C}_f}{dt} = -r_f \frac{\overline{C}_f}{V_f} + \sum\nolimits_i r_i \frac{\overline{C}_i}{V_i} + \widehat{N}_f \overline{v}_f + \overline{u}_f \tag{17.4}$$

where $\overline{C}_f$ is the function/variable, which depends on time only and index $\boldsymbol{i}$ corresponds to all compartments that supply the flowing fluid to compartment $\boldsymbol{f}$.

The value of $\overline{C}_f$ for each iteration of k is calculated in two stages: (1) we solve the system of equations without taking into account the contribution of fluid transfer and (2) we introduce corrections, which describe the contribution of the flowing fluid to changes in the concentrations of molecules/particles as well as the contributions from external sources.

The first stage describes the system without any influxes from other compartments:

$$\frac{d\overline{C'_f}}{dt} = \widehat{N}_f \overline{v}_f + \overline{u}_f \tag{17.5}$$

$$\overline{C'_f}(0) = \overline{C}_{f,k'} \tag{17.6}$$

These equations describe the chemical transformation of molecules/particles in the system but do not account for the contribution of their transfer with flowing fluid from one compartment to another. Concentrations of molecules/particles calculated in the previous iterations are taken as initial conditions. These equations give us the following solution at any small time period (ΔT):

$$\overline{C'_{f,k+1}} = \overline{C'_f(\Delta T)}$$

The second stage is to correct the solution derived in the first stage for the contribution of the flowing fluid:

$$\frac{\overline{C}_{f,k+1} - \overline{C'_{f,k+1}}}{\Delta T} = -r_f \frac{\overline{C'_{f,k+1}}}{V_f} + \sum_i r_i \frac{\overline{C}'_{k+1i}}{V_i} \tag{17.7}$$

Hence,

$$\overline{C}_{f,k+1} = \overline{C}'_{f,k+1} - r_f \frac{\overline{C'_{f,k+1}}}{V_f} \Delta T + \sum_i r_i \frac{\overline{C}'_{i,k+1}}{V_f} \Delta T \tag{17.8}$$

Mathematical implementation of SBSA As a mathematical object, the complete system can be represented as a series of compartments interconnected by edges. The compartments are split into routes that correspond to different organs (see Table 17.1). The edges are described by numeric characteristics that describe the movement between compartments in terms of a table of interconnections. The biological processes in each compartment are described by a kinetic model. Different compartments can be associated with the same or a different kinetic model.

The movement of blood and biological entities located in a specific portion of the blood is described in a pulse manner with a period equal to ΔT. It is assumed that the movement of the blood between neighboring compartments happens instantaneously at the end of the time period. This allows us to separate the solution of the system into two independent stages. The first stage includes the solution of the kinetic model in each compartment without taking into account the contributions to metabolite concentrations resulting from the movement of the blood. The second stage calculates corrections to metabolite concentrations resulting from the movement of the blood.

Here, we describe the calculation in detail for the first step (t_0, $t_0 + \Delta T$). At the beginning of the step (i.e., when $t = t_0$), we know all the metabolite concentrations in all compartments of the system ${C_{f,0}}'^{i}$, where ${C_{f,0}}'^{i}$ is the concentration of the *i-th* metabolite in compartment *f* at $t = t_0$. These data are supplied from the previous step in the calculation. Then, we show that within the framework of SBSA, it is possible to calculate the metabolite concentrations ${C_{f,1}}'^{i}$ at $t_1 = t_0 + \Delta T$. In the first stage of the calculations, we solve all kinetic models numerically in all compartments during the time interval [t_0, $t_0 + \Delta T$]. As a result, we have intermediate values of the metabolite concentrations ${C_{f,1}}'^{i}$ at $t = t_1$. In the second stage of the calculations, these intermediate concentrations ${C_{f,1}}'^{i}$ are corrected to account for the movement of the blood and biological entities between compartments (see Eq. 8). From these calculations, we now have the required values of the metabolite concentrations ${C_{f,1}}'^{i}$.

Table 17.2 Description of variables of model 1

Variable	Description	Initial Value (at time point zero)	Units
Timer	Auxiliary variable represents time in special function that reproduce the dynamics of dietary fat in small intestine	0	sec
F1	Auxiliary variable, that describes the increase of dietary fat in small intestine	0	Dimentionless
F2	Auxiliary variable, that describes the decrease of dietary fat in small intestine	1	Dimentionless
Dietary_Fat	Dose of dietary fat, expressed as concentration in intestine lumen at time point zero	100	μM
FA_ent	Concentration of free fatty acids in cytosole of enterocytes	0	μM
ChM	Concentration of Chylomicrons in blood	0	μM
VLDL	Concentration of VLDLs in blood	0	μM
VLDL_rem	Concentration of VLDL remnants in blood	0	μM
ChM_rem	Concentration of Chylomicron remnants in blood	0	μM
FA_flow	Concentration of free fatty acids in blood	0	μM

Kinetic models describing biochemical transformations of metabolites/particles The SBSA model consists of (i) basic (physiological) and (ii) kinetic modules. The physiological module is responsible for the description of the movement of biological fluids and the transfer of biological entities between various organs/tissues. The mathematical apparatus of the module is given by Eqs. (7) and (8). The kinetic module is responsible for the description of biochemical processes in each organ/tissue. The mathematical apparatus of the module is given by Eqs. (5) and (6). As explained in the "*Mathematical implementation of SBSA*" subsection, all biological processes in each compartment are described by a kinetic model (see Table 17.1). Different compartments can be associated with the same or different kinetic models. Therefore, the number of different kinetic models included in the kinetic module is determined by the number of different organs/tissues included in the construction of the SBSA-based model. In this section, we describe the different kinetic models associated with the SBSA model of lipid spillover. These are

Model 1: Formation of chylomicrons (ChMs) and their influx in *vena cava* from the lymphoid system through the *thoracic duct*.
Model 2: Hydrolysis of TGs of ChMs and VLDLs in adipose tissue and the release/uptake of fatty acids from/in the tissue.
Model 3: Hydrolysis of TGs of ChMs and VLDLs in muscle and the release/uptake of fatty acid from/in the tissue.
Model 4: Uptake of free FAs and FAs released from ChMs and VLDL remnants and VLDL formation in the liver.

Below, we describe the kinetic models in further detail.

Model 1 describes ChMs formation from the fat of food in enterocytes and the release of the ChMs into the blood via the lymphatic system (thoracic duct). *Model 1* includes 10 variables (see detailed description in Table 17.2), whose dynamics are

Table 17.3 Description of rate laws of model 1

Reaction	Rate Law	Comments
R1: = timer;	V(1) = k_time;	Auxiliary reaction for agreement between the times of dietary fat dynamics and the internal time of model
R2: = f1;	If *timer* > *f1_lim* then *V*(2) = k_f1 else *V*(2) = 0	V(2) describes the increase of auxiliary variable f1
R3: = f2;	If *timer* >;*f2_lim* then *V*(3) = k_f2 else V(3) = 0	V(3) describes the increase of auxiliary variable f2
R4:Dietary_Fat = FA_ent;	V(4) = k_df_fa*Dietary_Fat*f2*f1^2/ (Km_f1 + f1^2);	Fatty acids absorption, the part of formula in bold reproduces the dietary fat dynamics in small intestine
R5:FA_ent = ;	V(5) = k_fa_cm*FA_ent;	Fatty acids inclusion in chylomicrons
R6: = ChM;	V(6) = k_fa_cm*FA_ent/N_fa_cm;	Chylomicron production
R7:VLDL = ;	V(7) = k_vldl_deg*VLDL;	VLDLs degradation in blood
R8:VLDL_rem = ;	V(8) = k_vldlr_deg*VLDL_rem;	VLDL remnants degradation in blood
R9:ChM = ;	V(9) = k_cm_deg*ChM;	Chylomicrons degradation in blood
R10:ChM_rem = ;	V(10) = k_cmr_deg*ChM_rem;	Chylomicron remnants degradation in blood
R11:FA_flow = ;	V[11] = k_fafl_deg*FA_flow;	Free Fatty Acids degradation in blood

described by the following ODE system:

$$
\begin{aligned}
d\text{timer}/dt &= \text{V}[1] \\
d\text{f1}/dt &= \text{V}[2] \\
d\text{f2}/dt &= \text{V}[3] \\
d\text{Dietary\_Fat}/dt &= -V[4] \\
d\text{FA\_ent}/dt &= \text{V}[4] - \text{V}[5] \\
d\text{ChM}/dt &= \text{V}[6] - \text{V}[9] \\
d\text{VLDL}/dt &= -\text{V}[7] \\
d\text{VLDL\_rem}/dt &= -\text{V}[8] \\
d\text{ChM\_rem}/dt &= -\text{V}[10] \\
d\text{FA\_flow}/dt &= -\text{V}[11],
\end{aligned}
\tag{17.9}
$$

The right-hand sides of the system described by Eq. (9) are given by the rate laws listed in Table 17.3. The values of the rate law parameters are listed in Table 17.4.

Model 2 describes FA exchange in capillaries of adipose tissue. *Model 2* includes 10 variables (see detailed description in Table 17.5), whose dynamics are described by the following ODE system:

$$
\begin{aligned}
d\text{FA\_flow}/dt &= -\text{V}[1] \\
d\text{FA\_exch}/dt &= \text{V}[1] - \text{V}[2] + \text{V}[3] + \text{V}[7] + \text{V}[9] \\
d\text{FA\_stor}/dt &= \text{V}[2] - \text{V}[3]
\end{aligned}
$$

Table 17.4 Description of parameters of model 1

Parameter	Description	Value	Units
k_time	Auxiliary coefficient of agreement between the times of dietary fat dynamics and the internal time of model	1	Dimensionless
k_f1	Auxiliary coefficient of rate of increase of dietary fat in small intestine after fat load	1	Dimensionless
f1_lim	Time after fat load, when the f1 variable stops to increase	60	sec
k_f2	Auxiliary coefficient of rate of decrease of dietary fat in small intestine after fat load	1	Dimensionless
f2_lim	Time after fat load, when the f2 variable stops to increase	7200	sec
k_df_fa	Rate constant of fatty acids absorptions from small intestine to enterocytes	1	1/sec
Km_f1	Concentration of fatty acids in intestine lumen, that causes half-maximal rate of absorption (EC50)	1	mM
N_fa_cm	Number of fatty acids molecules in one chylomicrons	10000	Molecules/particle
k_fa_cm	Rate constant of fatty acids inclusion in chylonicrons (chylomicrons production)	1	1/sec
k_vldl_deg	Rate constant of VLDLs degradation in blood	0	1/sec
k_vldlr_deg	Rate constant of VLDL remnants degradation in blood	0	1/sec
k_cm_deg	Rate constant of Chylomicrons degradation in blood	0	1/sec
k_cmr_deg	Rate constant of Chylomicron remnants degradation in blood	0	1/sec
k_fafl_deg	Rate constant of Free Fatty Acids degradation in blood	0	1/sec

Table 17.5 Description of variables of model 2

Variable	Description	Initial Value (at time point zero)	Units
FA_flow	Concentration of free fatty acids in blood	0	μM
FA_exch	Concentration of free fatty acids in proteoglycan matrix	0	μM
FA_stor	Concentration of stored fatty acids in adipocytes	0	μM
ChM	Concentration of Chylomicrons in blood	0	μM
BS	Concentration of binding sites on endothelium for VLDLs and Chylomicrons	5	μM
VLDL	Concentration of VLDLs in blood	0	μM
ChM_rem	Concentration of Chylomicron remnants in blood	0	μM
VLDL_rem	Concentration of VLDL remnants in blood	0	μM

$$
\begin{aligned}
d\mathrm{ChM}/dt &= -\mathrm{V}[4] \\
d\mathrm{BS}/dt &= -\mathrm{V}[4] - \mathrm{V}[5] + \mathrm{V}[6] + \mathrm{V}[8] \\
d\mathrm{ChM\_BS}/dt &= \mathrm{V}[4] - \mathrm{V}[6] \\
d\mathrm{VLDL}/dt &= -\mathrm{V}[5] \\
d\mathrm{VLDL\_BS}/dt &= \mathrm{V}[5] - \mathrm{V}[8] \\
d\mathrm{ChM\_rem}/dt &= \mathrm{V}[6] \\
d\mathrm{VLDL\_rem}/dt &= \mathrm{V}[8],
\end{aligned}
\tag{17.10}
$$

Table 17.6 Description of rate laws of model 2

Reaction	Rate law	Comments
R1:FA_flow = FA_exch;	V(1) = k_fa_tr_fl_ex*FA_flow-k_fa_tr_ex_fl*FA_exch;	Free fatty acids exchange between blood plasma and proteoglycan matrix
R2:FA_exch = FA_stor;	V(2) = k_fa_uptake*FA_exch;	Free fatty acids uptake by adipoytes
R3:FA_stor = FA_exch;	If *FA_stor > FA_stor_lim* then *V*(3) = *k_fa_release* else *V*(3) = *k_fa_release*FA_stor/ FA_stor_lim*	Rate of fatty acids mobilisation from adipocytes is not depend on concentration of stored fatty acids (first order reaction) if this concentration is under lower limit (FA_stor_lim). Otherwise, this rate is limited with lack of stored fatty acids
R4:ChM + BS = ChM_BS;	V(4) = k_cm_bs*ChM*BS;	Chylomicrons binding to their binding sites on endothelium
R5:VLDL + BS = VLDL_BS;	V(5) = k_vldl_bs*VLDL*BS;	VLDLs binding to their binding sites on endothelium
R6:ChM_BS = BS + ChM_rem;	V(6) = k_cm_hyd*ChM_BS*LPL;	Triglyceride hydrolysis by lipoprotein lipase (LPL) and Chylomicron remnants formation
R7: = FA_exch;	V(7) = k_cm_hyd*ChM_BS*LPL*(N_fa_cm-N_fa_cmr);	Triglyceride hydrolysis by lipoprotein lipase (LPL) and fatty acids release from Chylomicrons
R8:VLDL_BS = BS + VLDL_rem;	V(8) = k_vldl_hyd*VLDL_BS*LPL;	Triglyceride hydrolysis by lipoprotein lipase (LPL) and VLDL remnants formation
R9: = FA_exch;	V(9) = k_vldl_hyd*VLDL_BS*LPL*(N_fa_vldl-N_fa_vldlr);	Triglyceride hydrolysis by lipoprotein lipase (LPL) and fatty acids release from VLDL

The right-hand sides of the system described by Eq. (10) are given by rate laws listed in Table 17.6. The values of the rate law parameters are listed in Table 17.7.

Model 3 describes FA exchange in capillaries of muscles. *Model 3* includes 10 variables (see detailed description in Table 17.8) whose dynamics are described by the following ODE system:

$$
\begin{aligned}
d\mathrm{FA\_flow}/dt &= -\mathrm{V}[1] \\
d\mathrm{FA\_exch}/dt &= \mathrm{V}[1] - \mathrm{V}[2] + \mathrm{V}[3] + \mathrm{V}[7] + \mathrm{V}[9] \\
d\mathrm{FA\_stor}/dt &= \mathrm{V}[2] - \mathrm{V}[3] \\
d\mathrm{ChM}/dt &= -\mathrm{V}[4] \\
d\mathrm{BS}/dt &= -\mathrm{V}[4] - \mathrm{V}[5] + \mathrm{V}[6] + \mathrm{V}[8] \\
d\mathrm{ChM\_BS}/dt &= \mathrm{V}[4] - \mathrm{V}[6] \\
d\mathrm{VLDL}/dt &= -\mathrm{V}[5] \\
d\mathrm{VLDL\_BS}/dt &= \mathrm{V}[5] - \mathrm{V}[8] \\
d\mathrm{ChM\_rem}/dt &= \mathrm{V}[6] \\
d\mathrm{VLDL\_rem}/dt &= \mathrm{V}[8],
\end{aligned}
\tag{17.11}
$$

Table 17.7 Description of parameters of model 2

Parameters	Description	Value	Units
k_fa_tr_fl_ex	Rate constant of free fatty acids transport from blood plasma to proteoglycan matrix	1	1/sec
k_fa_tr_ex_fl	Rate constant of free fatty acids transport from proteoglycan matrix to blood plasma	1	1/sec
k_fa_uptake	Rate constant of free fatty acids absorption from proteoglycan matrix by adipocytes	1	1/sec
k_fa_release	Rate constant of free fatty acids release from adipocytes to proteoglycan matrix	0	1/sec
FA_stor_lim	Minimal concentration of stored free fatty acids in adipocytes that not limited fatty acids release	1	μM
k_cm_bs	Rate constant of Chylomicrons binding to their binding sites on endothelium	1	1/sec
k_vldl_bs	Rate constant of Chylomicrons binding to their binding sites on endothelium	1	1/sec
k_cm_hyd	Rate constant of triglyceride hydrolysis by lipoprotein lipase (LPL) and fatty acids release from Chylomicrons	1	1/sec
k_vldl_hyd	Rate constant of triglyceride hydrolysis by lipoprotein lipase (LPL) and fatty acids release from Chylomicrons	1	1/sec
LPL	Lipoprotein lipase (LPL) concentration	10	μM
N_fa_cm	Number of fatty acids molecules in one chylomicron	10000	Molecules/particle
N_fa_cmr	Number of fatty acids molecules in one chylomicron remnant	100	Molecules/particle
N_fa_vldl	Number of fatty acids molecules in one VLDL	5000	Molecules/particle
N_fa_vldlr	Number of fatty acids molecules in one VLDL remnant	100	Molecules/particle

Table 17.8 Description of variables of model 3

Variable	Description	Initial Value (at time point zero)	Units
FA_flow	Concentration of free fatty acids in blood	0	μM
FA_exch	Concentration of free fatty acids in proteoglycan matrix	0	μM
FA_stor	Concentration of stored fatty acids in muscle cells	0	μM
ChM	Concentration of Chylomicrons in blood	0	μM
BS	Concentration of binding sites on endothelium for VLDLs and Chylomicrons	5	μM
VLDL	Concentration of VLDLs in blood	0	μM
ChM_rem	Concentration of Chylomicron remnants in blood	0	μM
VLDL_rem	Concentration of VLDL remnants in blood	0	μM

The right-hand sides of the system described by Eq. (11) are given by rate laws listed in Table 17.9. Values of the rate law parameters are listed in Table 17.10.

Table 17.9 Description of rate laws of model 3

Reaction	Rate law	Comments
R1:FA_flow = FA_exch;	V(1) = k_fa_tr_fl_ex*FA_flow-k_fa_tr_ex_fl*FA_exch;	Free fatty acids exchange between blood plasma and proteoglycan matrix
R2:FA_exch = FA_stor;	V(2) = k_fa_uptake*FA_exch;	Free fatty acids uptake by adipoytes
R3:FA_stor = ;	V(3) = k_energy_exp*FA_stor	Free fatty acids usage for energy production
R4:ChM + BS = ChM_BS;	V(4) = k_cm_bs*ChM*BS;	Chylomicrons binding to their binding sites on endothelium
R5:VLDL + BS = VLDL_BS;	V(5) = k_vldl_bs*VLDL*BS;	VLDLs binding to their binding sites on endothelium
R6:ChM_BS = BS + ChM_rem;	V(6) = k_cm_hyd*ChM_BS* LPL;	Triglyceride hydrolysis by lipoprotein lipase (LPL) and Chylomicron remnants formation
R7: = FA_exch;	V(7) = k_cm_hyd*ChM_BS* LPL*(N_fa_cm-N_fa_cmr);	Triglyceride hydrolysis by lipoprotein lipase (LPL) and fatty acids release from Chylomicrons
R8:VLDL_BS = BS + VLDL_rem;	V(8) = k_vldl_hyd*VLDL_BS* LPL;	Triglyceride hydrolysis by lipoprotein lipase (LPL) and VLDL remnants formation
R9: = FA_exch;	V(9) = k_vldl_hyd*VLDL_BS* LPL*(N_fa_vldl-N_fa_vldlr);	Triglyceride hydrolysis by lipoprotein lipase (LPL) and fatty acids release from VLDL

Table 17.10 Description of parameters of model 3

Parameters	Description	Value	Units
k_fa_tr_fl_ex	Rate constant of free fatty acids transport from blood plasma to proteoglycan matrix	1	1/sec
k_fa_tr_ex_fl	Rate constant of free fatty acids transport from proteoglycan matrix to blood plasma	1	1/sec
k_fa_uptake	Rate constant of free fatty acids absorption from proteoglycan matrix by adipocytes	1	1/sec
k_energy_exp	Rate constant of free fatty acids usage for energy production	0	1/sec
k_cm_bs	Rate constant of Chylomicrons binding to their binding sites on endothelium	1	1/sec
k_vldl_bs	Rate constant of Chylomicrons binding to their binding sites on endothelium	1	1/sec
k_cm_hyd	Rate constant of triglyceride hydrolysis by lipoprotein lipase (LPL) and fatty acids release from Chylomicrons	1	1/sec
k_vldl_hyd	Rate constant of triglyceride hydrolysis by lipoprotein lipase (LPL) and fatty acids release from Chylomicrons	1	1/sec
LPL	Lipoprotein lipase (LPL) concentration	10	μM
N_fa_cm	Number of fatty acids molecules in one chylomicron	10000	Molecules/particle
N_fa_cmr	Number of fatty acids molecules in one chylomicron remnant	100	Molecules/particle
N_fa_vldl	Number of fatty acids molecules in one VLDL	5000	Molecules/particle
N_fa_vldlr	Number of fatty acids molecules in one VLDL remnant	100	Molecules/particle

Table 17.11 Description of variables of model 4

Variables	Description	Initial Value (at time point zero)	Units
ChM_rem	Concentration of Chylomicron remnants in blood, expressed as concentration of included fatty acids	0	μM
FA_liv	Concentration of free fatty acids in hepatocytes	0	μM
VLDL_rem	Concentration of VLDL remnants in blood, expressed as concentration of included fatty acids	0	μM
FA_flow	Concentration of free fatty acids in blood	0	μM
VLDL	Concentration of VLDLs in blood, expressed as concentration of included fatty acids	0	μM
ChM	Concentration of Chylomicrons in blood, expressed as concentration of included fatty acids	0	μM

Table 17.12 Description of rate laws of model 4

Reaction	Rate laws	Comments
R1:ChM_rem = ;	V(1) = k_cmr_upt*ChM_rem;	Chylomicron remnants uptake by hepatocytes
R2: = FA_liv;	V(2) = k_cmr_upt*ChM_rem* N_fa_cmr;	Uptake of free fatty acids from Chylomicron remnants
R3:VLDL_rem = ;	V(3) = k_vldlr_upt*VLDL_rem;	VLDL remnants uptake by hepatocytes
R4: = FA_liv;	V(4) = k_vldlr_upt*VLDL_rem* N_fa_vldlr;	Uptake of free fatty acids from VLDL remnants
R5:FA_flow = FA_liv;	V(5) = k_fa_upt*FA_flow;	Free fatty acids uptake by hepatocytes
R6:FA_liv = ;	V(6) = k_fa_energy*FA_liv;	Free fatty acids usage for energy production (beta-oxidation)
R7:FA_liv = ;	V(7) = k_fa_vldl_syn*FA_liv;	Fatty acids inclusion in VLDLs
R8: = VLDL;	V(8) = k_fa_vldl_syn*FA_liv/ N_fa_vldl;	VLDL production
R9:ChM = ;	V(9) = k_cm_deg*ChM;	Chylomicrons degradation in blood

Model 4 describes FA exchange in liver tissue. *Model 4* includes 6 variables (see detailed description in Table 17.11) whose dynamics are described by the following ODE system:

$$\begin{aligned} d\text{ChM\_rem}/dt &= -\text{V}[1] \\ d\text{FA\_liv}/dt &= \text{V}[2] + \text{V}[4] + \text{V}[5] - \text{V}[6] - \text{V}[7] \\ d\text{VLDL\_rem}/dt &= -\text{V}[2] \\ d\text{FA\_flow}/dt &= -\text{V}[5] \\ d\text{VLDL}/dt &= \text{V}[7] \\ d\text{ChM}/dt &= -\text{V}[9], \end{aligned} \tag{17.12}$$

The right-hand sides of the system described by Eq. (12) are given by rate laws listed in Table 17.12. Values of rate law parameters are listed in Table 17.13.

Each of the models of the kinetic module has been associated with compartments of the physiological module (see Table 17.1).

Table 17.13 Description of parameters of model 4

Parameters	Description	Value	Units
k_cmr_upt	Rate constant of Chylomicron remnants uptake by hepatocytes	1	1/sec
N_fa_cmr	Number of fatty acids molecules in one chylomicron remnant	100	Molecules/particle
k_vldlr_upt	Rate constant of VLDL remnants uptake by hepatocytes	1	1/sec
N_fa_vldlr	Number of fatty acids molecules in one VLDL remnant	100	Molecules/particle
k_fa_upt	Rate constant of free fatty acids uptake by hepatocytes	1	1/sec
k_fa_energy	Rate constant of free fatty acids usage for energy production (beta-oxidation)	0.1	1/sec
k_fa_vldl_syn	Rate constant of VLDL production	1	1/sec
N_fa_vldl	Number of fatty acids molecules in one VLDL	5000	Molecules/particle
k_cm_deg	Rate constant of Chylomicrons degradation in blood	0	1/sec

17.4 Outcome of the Model

Comparison of SBSA with CAT/ACAT algorithm In this section, we compare two different algorithms used to describe the movement of biological entities with blood flow through blood vessels. These are the CAT/ACAT algorithm (Huang et al. 2009) and the SBSA presented in this chapter.

The CAT/ACAT algorithm was initially developed to describe the absorption and transit of non-degradable and highly soluble drugs in the gastrointestinal tract (Agoram et al. 2001; Grass 1997). The process of a drug passing through the small intestine is considered as a flow through a series of segments. Each segment can be described as a single compartment with linear transfer kinetics. Each compartment can have different volumes and flow-rates, but they all have the same transit rate constant. CAT/ACAT (or similar) algorithms have been implemented in various software packages (simulations-plus.com 2012; system-biology.com 2012).

SBSA was developed to describe fast processes that occur in the bloodstream (or other flowing biological fluids). The term "fast process" refers to processes that result in a change in the concentration of some metabolite (or any other biological entity) located in blood with a characteristic time less than the turnover time of the circulatory system. An example of a fast process is the exchange of oxygen and carbon dioxide between erythrocytes and tissues in the bloodstream. The blood in arterioles is oxygen-rich and the blood in venules is saturated with carbon dioxide. This means that the O_2/CO_2 ratio decreases significantly from the beginning of the capillaries to their ends. This spatial distribution of the O_2/CO_2 ratio is provided by fast O_2/CO_2 exchange in capillaries. SBSA was developed for the purpose of describing the spatially non-homogeneous distribution of biological entities.

To compare the SBSA and CAT/ACAT algorithm we selected a model describing the movement of a portion of blood through a tissue. Blood transports non-homogeneously distributed biological entities. We assumed that within the framework of the model, during the movement, the biological entity could interact with the tissue. Chylomicrons (ChM) are considered to be biological entities transported within the bloodstream. Adipose tissue was selected as a tissue through which ChMs are transported. In accordance with description in sub-section "Lipid spillover" of the section "Modeling framework – description of biological system and model requirements," ChMs circulating in blood are able to attach to endothelial cells that pave the capillaries of adipose tissue. To account for the process in our model, we assumed that the rate of attachment of ChMs to the cells is described in accordance with the irreversible mass action law with a corresponding binding rate constant (k_cm_bs):

$$v\_sorption = k\_cm\_bs*[ChM]*[BS],$$

where BS is the concentration of binding sites for ChMs or VLDLs located on the surface of the endothelium wall.

The value of the rate constant (k_cm_bs = 0.008 s^{-1}) was selected so as to provide a 20 % attachment of ChMs to the endothelial cells paving the capillaries of adipose tissue during a single passage through the part of adipose tissue considered in the model. This is in line with experimental results of measurements of arterio-venous differences in ChM concentration across subcutaneous adipose tissue and skeletal muscle *in vivo* (Evans et al. 1999).

The initial distribution of ChMs (i.e., dependence of ChM concentration on spatial coordinate or compartment number) was selected in the following way. Dependence of ChM concentration is bell-shaped for the first 16 compartments of the model and equal to zero for all other compartments.

On the basis of the model, we determine the answers to the following questions:

- Does the shape of the spatial distribution change over time?
- How does the ChM concentration change in blood? Is this change fully determined by the value of *k_cm_bs*?

We answered these questions by applying the SBSA and CAT/ACAT algorithm to the model described above. This means that the movement of the portion of blood through the vessel and the distribution of the concentration of ChMs at each moment of time was modeled and simulated in accordance with these algorithms. To simulate the dependence of ChM concentration on time and compartment number (spatial coordinate) within the framework of the SBSA, we applied the model described in the "Model development" section. To compare the results obtained from the framework of the SBSA-based model with those obtained from the CAT/ACAT algorithm, we simulated the passage of ChMs through the "adipose" route in the SBSA-based model. In accordance with the data given in Table 17.1 and Fig. 17.1, this route includes 56 compartments.

To simulate the dependence of ChM concentration on time and compartment number within the framework of the CAT/ACAT algorithm, we applied the model described below:

$$d\mathbf{ChM_1}/d\mathbf{t} = -k\_transport^*[ChM_1],$$

$$d\mathbf{ChM_i}/d\mathbf{t} = k\_transport^*\left[ChM_{i-1}\right] - k\_transport^*\left[ChM_i\right], \mathrm{i} = 2, \ldots 12$$

$$d\mathbf{ChM}i/d\mathbf{t} = k\_transport^*\left[\mathrm{ChM_{i-1}}\right] - k\_transport^*\left[\mathrm{ChM_i}\right] - k\_cm\_bs^*\left[\mathrm{ChM_i}\right], \mathrm{i} = 13, \ldots 44,$$

$$d\mathbf{ChM_i}/d\mathbf{t} = k\_transport^*\left[\mathrm{ChM_{i-1}}\right] - k\_transport^*\left[\mathrm{ChM_i}\right], \mathrm{i} = 45, \ldots, 55$$

$$d\mathbf{ChM_{56}}/d\mathbf{t} = k\_transport^*[\mathrm{ChM_{55}}],$$

where *k_transport* is the rate constant of ChM transport from the upstream compartment to the downstream one. It is worth noting that values of the parameters responsible for the transport of ChMs between compartments, within the framework of the CAT/ACAT model, were selected so as to provide an exact match of the time of passage of ChMs through the adipose tissue in the CAT/ACAT model with the time of blood flow passage through adipose tissue within the framework of the SBSA model (which is equal to 56 s):

$$k\_transport = 1\mathrm{s}^{-1}$$

Simulation results obtained within the framework of the SBSA and CAT/ACAT algorithm are shown in Figs. 17.2 and 17.3. The following differences between simulations were identified:

(1) Despite the decrease in the maximal value of ChM concentration with the compartment number, the shape of the time profile of ChM concentration calculated in various compartments did not change in the SBSA-based model (Fig. 17.2a). On the contrary, we observed a broadening of the time profile in the CAT/ACAT algorithm-based model (Fig. 17.2b). In fact, ChMs were distributed among 40 compartments at $t = 30$ s, although they were initially distributed among only 17 compartments (Fig. 17.3).
(2) The value of ChM concentration at the peak of the time dynamics decreased by 20 % with an increase in compartment number in the SBSA-based model (Fig. 17.2a). In the CAT/ACAT algorithm-based mode (Fig. 17.2b), we observed a more significant decrease (more than two times the initial level of ChMs) in the concentration.

In accordance with clause (2), we can state that the decrease in ChM concentration resulting from the single passage of the selected portion of blood through adipose tissue, calculated within the framework of the CAT/ACAT algorithm, is significantly more than that calculated on the basis of the SBSA. To understand the nature of the significant differences in simulation results produced by the CAT/ACAT and SBSA-based models, we simulated the spatial-temporal distribution of ChM concentration under the condition of absence of ability of ChMs to attach to the endothelial cells, i.e., we assumed the value of $k\_cm\_bs = 0$. Evidently, under these conditions, the concentration of ChMs during the movement with blood flow cannot decrease since there is no process for ChM consumption in the model. The results shown in Fig. 17.4a

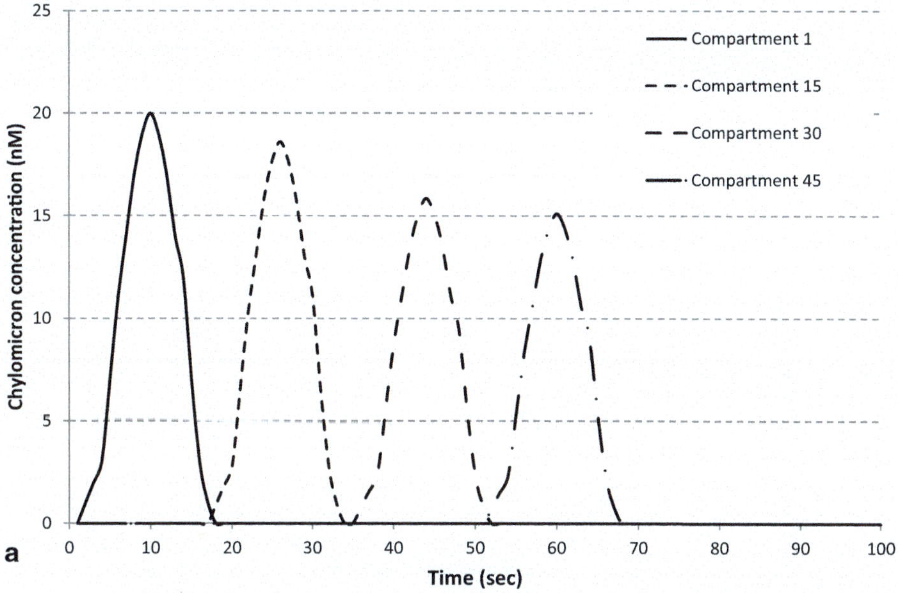

a

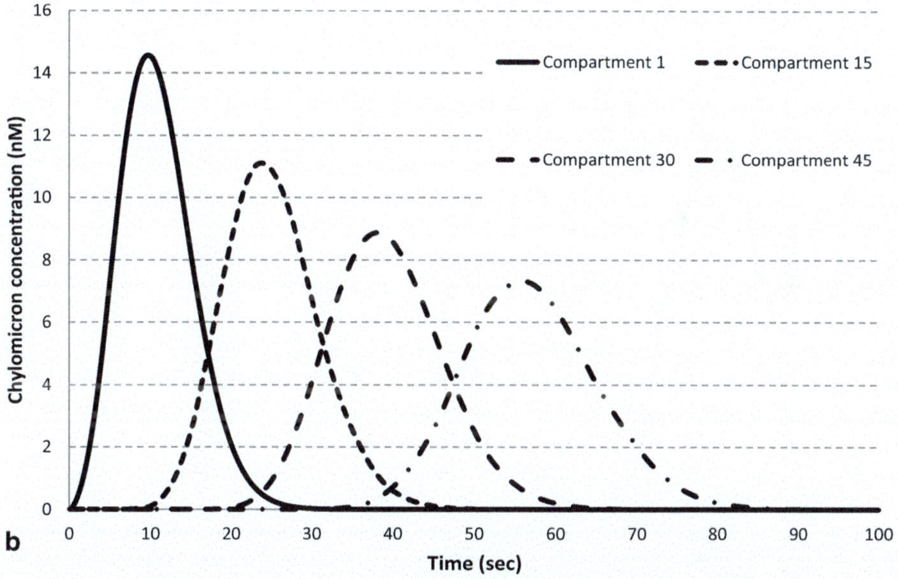

b

Fig. 17.2 Simulations of time courses of ChM concentration for various compartments of the "Adipose" route. Simulations were performed to compare models developed within the framework of the SBSA (**a**) with those of the CAT/ACAT algorithm (**b**). All initial values and parameters were identical for both versions of the model. In both the models, the "Adipose" route was considered. The route consisted of 56 compartments. At the $t = 0$, a bell-shaped distribution of ChMs was observed in 12 upstream compartments located immediately before the start of the "Adipose" route. In both CAT/ACAT and SBSA-based models, the initial concentration of ChMs in any of the 56 compartments of the "Adipose" route was zero. Time courses for ChM concentration were calculated in the 1st, 10th, 20th, 30th, 40th, and 50th compartments of the "Adipose" route. The kinetic parameter responsible for binding ChMs to the vessel wall cells was equal to 0.08 s^{-1}

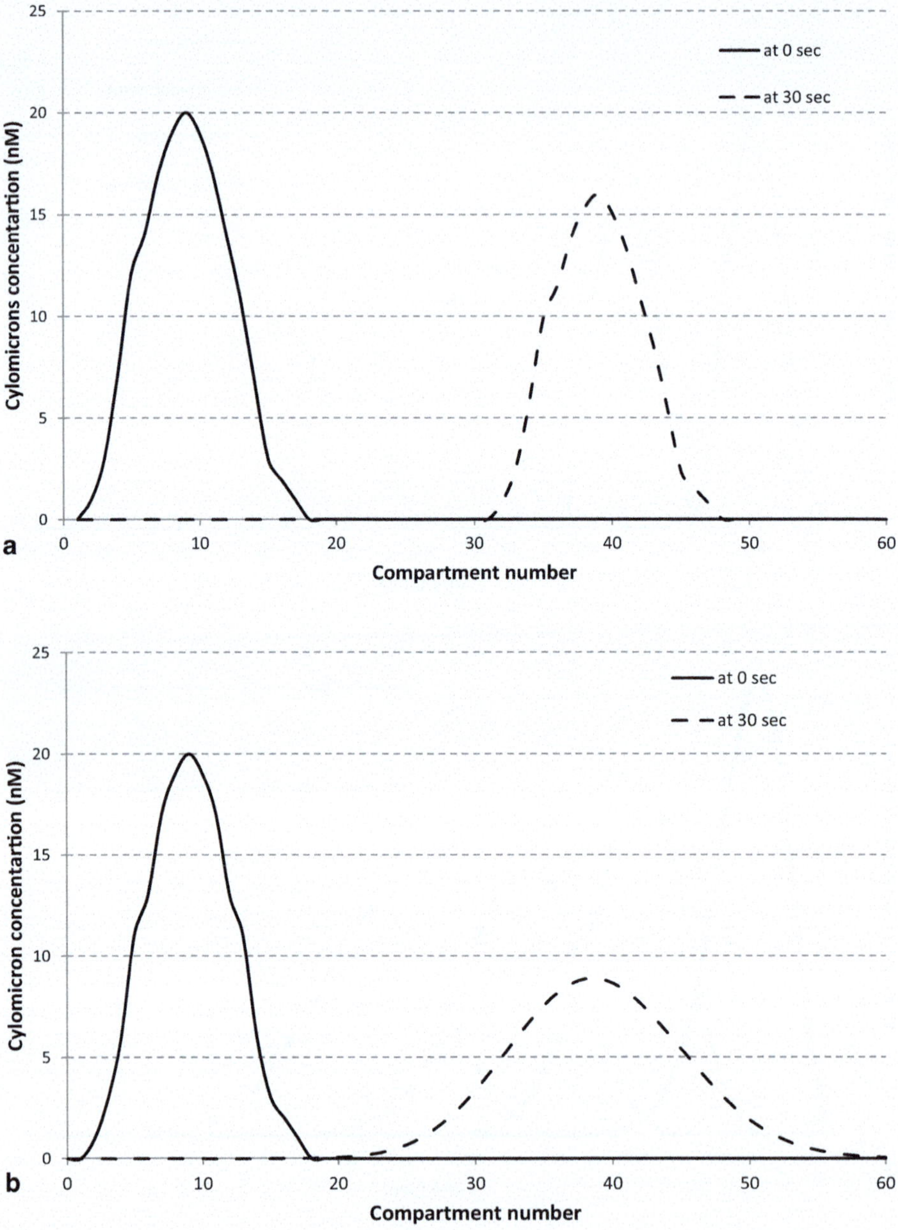

Fig. 17.3 Simulations of spatial distribution of ChM concentration along the "Adipose" route. Simulations were performed to compare models developed within the framework of the SBSA (**a**) with those of the CAT/ACAT algorithm (**b**). All initial values and parameters were identical for both versions of the model. In both models, the "Adipose" route was considered. The route consisted of 56 compartments. At $t = 0$, a bell-shaped distribution of ChMs was observed in the first 17 compartments of the "Adipose" route. The maximum concentration of ChMs was equal to 20 nM. The peak of the initial distribution corresponded to the 9th compartment. The spatial distribution of ChM concentration along the "Adipose" route was simulated for an initial time $t = 0$ (*solid curve*) and $t = 30$ s (*dashed curve*). We assumed that the kinetic parameter responsible for the binding of ChMs to the vessel wall cells was equal to 0.08 s^{-1}

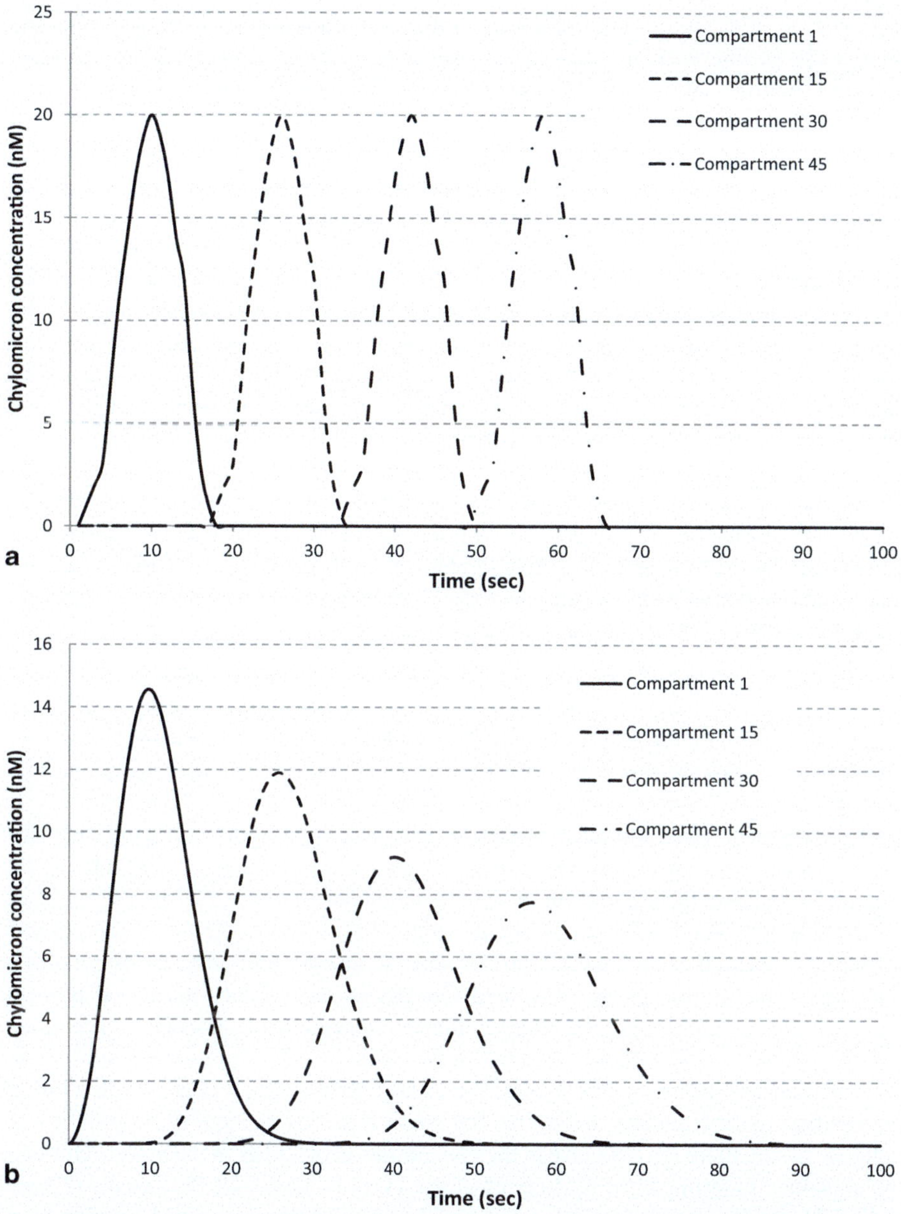

Fig. 17.4 Time course simulations of ChM concentration for various compartments of the "Adipose" route. Simulations were performed to compare models developed within the framework of the SBSA (**a**) with those of the CAT/ACAT algorithm (**b**). All initial values and parameters were identical for both versions of the model. In both the models, the "Adipose" route was considered. The route consisted of 56 compartments. At $t = 0$, a bell-shaped distribution of ChMs was observed in the 12 upstream compartments located immediately before the start of the "Adipose" route. In both CAT/ACAT and SBSA-based models, the initial concentration of ChMs in any of the 56 compartments of the "Adipose" route was zero. Time courses for ChM concentration were calculated in the 1st, 10th, 20th, 30th, 40th, and 50th compartments of the "Adipose" route. The kinetic parameter responsible for binding ChMs to the vessel wall cells was 0

demonstrate that simulations obtained within the framework of the SBSA-based model are in line with the conclusion. In fact, there is no decrease in the peak value of the time series of ChM concentration calculated for various compartments and the shape of the time profile does not change with an increase in compartment number. However, the simulations obtained within the framework of CAT/ACAT algorithm-based model (Fig. 17.4b) demonstrate both a significant decrease in the peak value of the time series of ChM concentration with an increase in compartment number and a broadening of the time profile with an increase in compartment number. By comparing Fig. 17.2b and 17.4b, we can conclude that the time series of ChM concentrations calculated in various compartments do not depend significantly on the ability of ChMs to attach to the endothelial cells, and this has been taken into account in the CAT/ACAT model. Even under the condition of $k\_cm\_bs = 0$, a significant decrease in the peak value of ChM concentrations and a broadening of the time profile is observed with an increase in compartment number.

This means that the CAT/ACAT algorithm cannot be used to describe the dynamics of the movement of any biological entity with blood flow. In contrast, the SBSA-based model enables us to correctly simulate both the movement of a biological entity with blood flow and the dynamic changes in its concentration resulting from chemical reactions or interactions with cells of the vessel wall. Hence, we propose using the SBSA to model the dynamics of lipid spillover.

Prediction of spatial distribution of FAs and lipoproteins in adipose and liver tissues In this section, we illustrate how the SBSA-based model of lipid spillover can describe the time dynamics of free FAs, ChMs, and ChMRs in the selected portion of blood passing through various blood vessels located in adipose tissue and liver tissue.

In our SBSA-based model, adipose tissue is described by 56 compartments and the residence time of a portion of blood is estimated as 50.4 s (see Table 17.1). Figure 17.5 shows the simulation of time dynamics of free FAs, ChMs, and ChMRs in the selected blood portion passing through adipose tissue. At $t = 0$, the portion of blood is located at the entrance to the adipose arteries and contains only ChMs, i.e., the concentration of free FAs and ChMRs is zero. Then, the blood portion sequentially passes through the adipose arteries (from 0 to 11 s), adipose capillaries (from 12 to 45 s), and adipose veins (from 46 to 50.4 s). The simulation results show that the concentrations of free FAs, ChMs, and ChMRs do not change when blood passes through the arteries and veins, but when it passes through the capillaries of adipose tissue, there is a decrease in the concentration of ChMs in the blood. This can be explained in terms of the hydrolysis of the TGs of ChMs by LPL (see also Chap. 2.6). LPL catalyzes the production of free FAs and ChMRs from ChMs, and in fact, it was seen that the concentrations of free FAs and ChMRs in the blood increased as they passed through the capillaries of adipose tissue. It was also observed that as the passage time for the portion of blood increased through the capillaries of adipose tissue, the changes in the concentrations of free FAs, ChMRs, and ChMs became more pronounced.

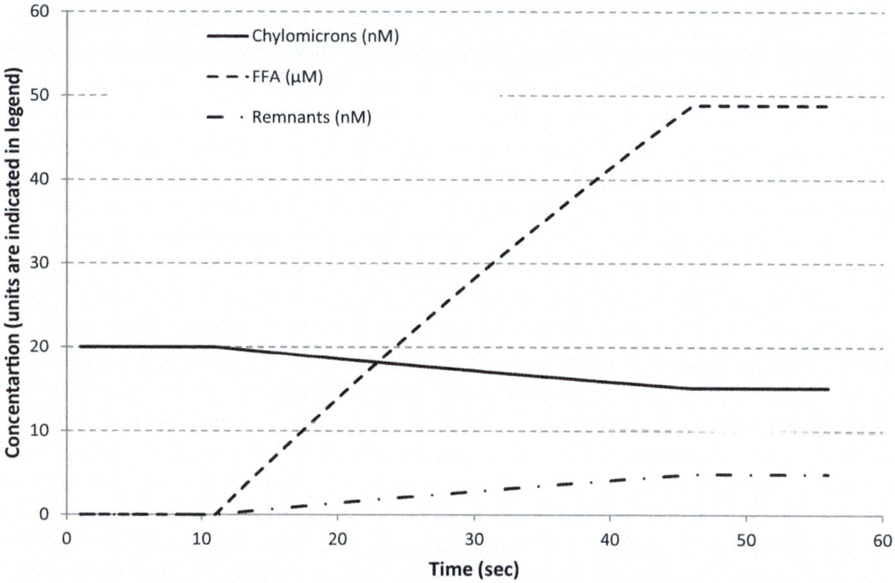

Fig. 17.5 Spillover of fatty acids from chylomicrons to FFA fraction in adipose tissue. These simulations represent the dynamics of the concentration of ChMs, ChMRs, and free FAs in the blood portion flowing along the "Adipose" route. We selected the portion of the blood located in the first compartment of the "Adipose" route at $t = 0$. Initial concentrations of ChMs, ChMRs, and free FAs in the blood portion are equal to 20 nM, 0.0 nM, and 0.0 μM, respectively. Concentrations of all other molecules/particles were assumed to be equal to zero. With each cardiac beat (i.e., each second), the portion of blood is transported to the next downstream compartment. Graphs represent changes in ChM, ChMR, and free FA concentrations in the blood portion flowing along the "Adipose" route

In a similar manner, we simulated the time dynamics of free FAs, ChMRs, and VLDLs in the selected blood portion passing through liver tissue (see Fig. 17.6). Liver tissue can be described as 17 compartments, and the residence time of a portion of blood is estimated as 15.3 s (see Table 17.1). At $t = 0$, the portion of blood is located at the entrance to the liver arteries and contains non-zero concentrations of free FAs, ChMRs, and VLDLs. Then, the blood portion sequentially passes through the liver arteries (from 0 to 3 s), liver capillaries (from 4 to 13 s), and liver veins (from 14 to 15.3 s). The simulation results show that the concentrations of free FAs, ChMRs, and VLDLs do not change when blood passes through the arteries and veins, but passage through the capillaries of liver enriches the blood with VLDLs and decreases concentrations of free FAs and ChMRs. This can be explained in terms of the absorption of free FAs and ChMRs by hepatocytes (see also Chap. 2.6). Hepatocytes use the absorbed FAs and ChMRs to assemble VLDLs with subsequent release to the blood stream. It was seen that the concentration of VLDLs increased as the blood passed through the capillaries of liver tissue. It was also observed that as the passage time for the portion of the blood increased through the capillaries of liver tissue, the changes in the concentrations of free FAs, ChMRs, and VLDLs became more pronounced.

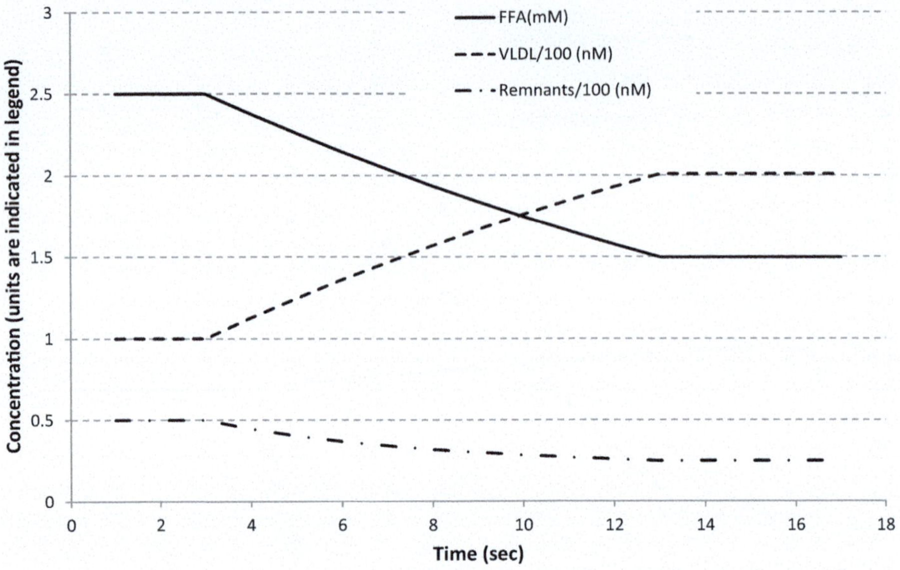

Fig. 17.6 Spillover of fatty acids from FFA fraction and chylomicron remnants to VLDL in liver. These simulations represent the dynamics of the concentrations of VLDLs, VLDLRs, and free FAs in the blood portion flowing along the "Liver" route. We selected the portion of blood located in the first compartment of the "Liver" route at $t = 0$. Initial concentrations of the VLDLs, VLDLRs, and free FAs in the blood portion are equal to 100 nM, 50 nM, and 2.5 mM, respectively. Concentrations of all other molecules/particles were assumed to be equal to zero. With each cardiac beat (i.e., each second), the portion of blood is transported to the next downstream compartment. Graphs represent changes in VLDL, VLDLR, and free FA concentrations in the blood portion flowing along the "Liver" route

17.5 Conclusions

A new algorithm was developed, which enabled us to combine kinetic models describing changes in metabolite (lipoprotein) concentrations that occurred due to biochemical transformations in blood and tissues, with the mathematical model of the molecular movement in blood flow. The algorithm may be applied to integrate a variety of experimental data and to understand the regulatory mechanisms underlying MetS development and progression.

Acknowledgements This work was supported by the FP7 ETHERPATHS project (Grant Number 222639).

References

Agoram B, Woltosz WS, Bolger MB (2001) Predicting the impact of physiological and biochemical processes on oral drug bioavailability. Adv Drug Deliv Rev 50(1):41–67

Albu JB, Heilbronn LK, Kelley DE, Smith SR, Azuma K, Berk ES, Pi-Sunyer FX, Ravussin E (2010) Look AHEAD Adipose research group. Metabolic changes following a 1-year diet and exercise intervention in patients with type 2 diabetes. Diabetes 59(3):627–633

Armoni M, Harel C, Karnieli E (2007) Transcriptional regulation of the GLUT4 gene: from PPAR-gamma and FOXO1 to FFA and inflammation. Trends Endocrinol Metab 18(3):100–107

Bajaj M, Pratipanawatr T, Berria R, Pratipanawatr W, Kashyap S, Cusi K, Mandarino L, DeFronzo RA (2002) Free fatty acids reduce splanchnic and peripheral glucose uptake in patients with type 2 diabetes. Diabetes 51(10):3043–3048

Czech A, Tatoń J, Piatkiewicz P (2010) Cellular glucose transport disturbances in the pathogenesis and therapy of type 2 diabetes mellitus. Endokrynol Pol 61(3):292–302

Evans K, Clark ML, Frayn KN (1999) Effects of an oral and intravenous fat load on adipose tissue and forearm lipid metabolism. Am J Physiol 276(2 Pt 1):241–248

Everett-Grueter C, Edgerton DS, Donahue EP, Vaughan S, Chu CA, Sindelar DK, Cherrington AD (2006) The effect of an acute elevation of NEFA concentrations on glucagon-stimulated hepatic glucose output. Am J Physiol Endocrinol Metab 291(3):449–459

Felber JP, Golay A (2002) Pathways from obesity to diabetes. Int J Obes Relat Metab Disord 26(2):39–45

Golay A, Felber JP, Meyer HU, Curchod B, Maeder E, Jéquier E (1984) Study on lipid metabolism in obesity diabetes. Metabolism 33(2):111–116

Grass G (1997) Simulation-models to predict oral-drug absorption from in-vitro data. Advanced Drug Delivery Review 23(1–3):199–219

He M, Su H, Gao W, Johansson SM, Liu Q, Wu X, Liao J, Young AA, Bartfai T, Wang MW (2010) Reversal of obesity and insulin resistance by a non-peptidic glucagon-like peptide-1 receptor agonist in diet-induced obese mice. PLoS One 5(12):e14205

Hirasawa A, Tsumaya K, Awaji T, Katsuma S, Adachi T, Yamada M, Sugimoto Y, Miyazaki S, Tsujimoto G (2005) Free fatty acids regulate gut incretin glucagon-like peptide-1 secretion through GPR120. Nat Med 11(1):90–94

Huang W, Lee SL, Yu LX (2009) Mechanistic approaches to predicting oral drug absorption. AAPS J 11(2):217–224

Katz A, Nambi SS, Mather K, Baron AD, Follmann DA, Sullivan G, Quon MJ (2000) Quantitative insulin sensitivity check index: a simple, accurate method for assessing insulin sensitivity in humans. J Clin Endocrinol Metab 85(7):2402–2410

Kelley DE, Mandarino LJ (2000) Fuel selection in human skeletal muscle in insulin resistance: a reexamination. Diabetes 49(5):677–683

Kim JY, Wall E van de, Laplante M, Azzara A, Trujillo ME, Hofmann SM, Schraw T, Durand JL, Li H, Li G, Jelicks LA, Mehler MF, Hui DY, Deshaies Y, Shulman GI, Schwartz GJ, Scherer PE (2007) Obesity-associated improvements in metabolic profile through expansion of adipose tissue. J Clin Invest 117(9):2621–37

Lewis GF, Carpentier A, Adeli K, Giacca A (2002) Disordered fat storage and mobilization in the pathogenesis of insulin resistance and type 2 diabetes. Endocr Rev 23(2):201–229

Lin E, Davis SS, Srinivasan J, Sweeney JF, Ziegler TR, Phillips L, Gletsu-Miller N (2009) Dual mechanism for type-2 diabetes resolution after Roux-en-Y gastric bypass. Am Surg 75(6):498–502. (discussion 502–503)

Muscelli E, Mari A, Casolaro A, Camastra S, Seghieri G, Gastaldelli A, Holst JJ, Ferrannini E (2008) Separate impact of obesity and glucose tolerance on the incretin effect in normal subjects and type 2 diabetic patients. Diabetes 57(5):1340–1348

Pastucha D, Talafa V, Malincikova J, Cihalik C, Hyjanek J, Horakova D, Janout V (2010) Obesity, hypertension and insulin resistance in childhood–a pilot study. Biomed Pap Med Fac Univ Palacky, Olomouc. Czech Repub 154(1):77–81

Pigeon E, Riou ME, St-Onge J, Couillard E, Tremblay A, Marette A, Weisnagel SJ, Joanisse DR (2009) Validation of a simple index (SIisOGTT) of insulin sensitivity in a population of sedentary men. Diabetes Metab 35(5):398–403

Prentki M, Madiraju SR (2011) Glycerolipid/free fatty acid cycle and islet β-cell function in health, obesity and diabetes. Mol Cell Endocrinol 353(1–2):88–100

Schnell S, Schaefer M, Schöfl C (2007) Free fatty acids increase cytosolic free calcium and stimulate insulin secretion from beta-cells through activation of GPR40. Mol Cell Endocrinol 263(1–2):173–180

Simulations Plus, Inc. (2012) Simulations-plus. http://www.simulations-plus.com. Accessed 10 Aug 2012

Staehr P, Hother-Nielsen O, Landau BR, Chandramouli V, Holst JJ, Beck-Nielsen H (2003) Effects of free fatty acids per se on glucose production, gluconeogenesis, and glycogenolysis. Diabetes 52(2):260–267

Stannard SR, Johnson NA (2004) Insulin resistance and elevated triglyceride in muscle: more important for survival than "thrifty" genes? J Physiol 554(3):595–607

Systems Biology (2012) Systems Biology. http://www.systems-biology.com. Accessed 10 Aug 2012

Wu CZ, Pei D, Su CC, Hsiao FC, Chu YM, Lee LH, Wang K, Hsieh AT, Lin JD, Hsia TL (2010) Comparison of oral glucose insulin sensitivity with other insulin sensitivity surrogates from oral glucose tolerance tests in Chinese. Ann Acad Med Singapore 39(1):4–8

Index

M. Orešič, A. Vidal-Puig (eds.), *A Systems Biology Approach to Study Metabolic Syndrome*, DOI 10.1007/978-3-319-01008-3,

M

P

The manufacturer's authorised representative in the EU is Springer
Nature Customer Service Centre GmbH, Europaplatz 3, 69115 Heidelberg,
Germany. If you have any concerns regarding our products, please
contact ProductSafety@springernature.com

Printed and bound by CPI Group (UK) Ltd, Croydon, CR0 4YY
30/09/2026
02225233-0010